Endothelial Cell Biology

in Health and Disease

Endothelial Cell Biology

in Health and Disease

Edited by
Nicolae Simionescu
and
Maya Simionescu

Institute of Cellular Biology and Pathology
Bucharest, Romania

Plenum Press · New York and London

Library of Congress Cataloging in Publication Data

Endothelial cell biology in health and disease / edited by Nicolae Simionescu and
  Maya Simionescu.
      p.   cm.
  Includes bibliographies and index.
  ISBN-13: 978-1-4612-8254-9   e-ISBN-13: 978-1-4613-0937-6
  DOI: 10.1007/978-1-4613-0937-6
  1. Endothelium—Physiology. 2. Endothelium—Pathophysiology. I. Simionescu, N.
(Nicolae). II. Simionescu, Maya. [DNLM: 1. Endothelium—cytology. 2. Endothelium
—physiology. 3. Endothelium—physiopathology. QS 532.5.E7 E563]
QP88.45.E52   1988
611'.0187—dc19
DNLM/DLC   88-15935
for Library of Congress   CIP

© 1988 Plenum Press, New York
Softcover reprint of the hardcover 1st edition 1988
A Division of Plenum Publishing Corporation
233 Spring Street, New York, N.Y. 10013

Contributors

Michael P. Bevilacqua • Vascular Research Division, Department of Pathology, Brigham and Women's Hospital and Harvard Medical School, Boston, Massachusetts 02115

Ramzi S. Cotran • Department of Pathology, Brigham and Women's Hospital and Harvard Medical School, Boston, Massachusetts 02115

Pamela Cowin • Division of Membrane Biology and Biochemistry, Institute of Cell and Tumor Biology, German Cancer Research Center, D-6900 Heidelberg, Federal Republic of Germany. *Present address:* Department of Cell Biology, New York University, School of Medicine, New York, New York 10016

Scott A. Curriden • Scripps Clinic and Research Foundation, La Jolla, California 92037

Charles T. Esmon • Thrombosis/Hematology Research Program, Oklahoma Medical Research Foundation, Oklahoma City, Oklahoma 73104

Alfred P. Fishman • Cardiovascular–Pulmonary Division, University of Pennsylvania, Philadelphia, Pennsylvania 19104

Werner W. Franke • Division of Membrane Biology and Biochemistry, Institute of Cell and Tumor Biology, German Cancer Research Center, D-6900 Heidelberg, Federal Republic of Germany

Michael A. Gimbrone, Jr. • Vascular Research Division, Department of Pathology, Brigham and Women's Hospital and Harvard Medical School, Boston, Massachusetts 02115

Christine Grund • Division of Membrane Biology and Biochemistry, Institute of Cell and Tumor Biology, German Cancer Research Center, D-6900 Heidelberg, Federal Republic of Germany

Dean A. Handley • Department of Physiology, Columbia University, New York, New York 10027

John H. Hansen-Flaschen • Cardiovascular–Pulmonary Division, University of Pennsylvania, Philadelphia, Pennsylvania 19104

Ronald L. Heimark • Department of Pathology, University of Washington, Seattle, Washington 98195

I. Joris • Department of Pathology, University of Massachusetts Medical School, Worcester, Massachusetts 01655

Hans-Peter Kapprell • Division of Membrane Biology and Biochemistry, Institute of Cell and Tumor Biology, German Cancer Research Center, D-6900 Heidelberg, Federal Republic of Germany

Morris J. Karnovsky • Department of Pathology, Harvard Medical School, Boston, Massachusetts 02115

Nicholas A. Kefalides • Connective Tissue Research Institute and Department of Medicine, University of Pennsylvania and University City Science Center, Philadelphia, Pennsylvania 19104

Nika V. Ketis • Department of Pathology, Harvard Medical School, Boston, Massachusetts 02115

M. C. Kowala • Department of Pathology, University of Massachusetts Medical School, Worcester, Massachusetts 01655

Caecilia Kuhn • Division of Membrane Biology and Biochemistry, Institute of Cell and Tumor Biology, German Cancer Research Center, D-6900 Heidelberg, Federal Republic of Germany

David J. Loskutoff • Scripps Clinic and Research Foundation, La Jolla, California 92037

Joseph A. Madri • Department of Pathology, Yale University School of Medicine, New Haven, Connecticut 06510

G. Majno • Department of Pathology, University of Massachusetts Medical School, Worcester, Massachusetts 01655

James A. Marcum • Department of Pathology, Beth Israel Hospital and Harvard Medical School, Boston; and Department of Biology, Massachusetts Institute of Technology, Cambridge, Massachusetts 02139

Peter P. Nawroth • Department of Physiology and Cellular Biophysics, College of Physicians and Surgeons, Columbia University, New York, New York 10032

J. J. Nunnari • Department of Pathology, University of Massachusetts Medical School, Worcester, Massachusetts 01655

G. E. Palade • Department of Cell Biology, Yale University School of Medicine, New Haven, Connecticut 06510

Jordan S. Pober • Department of Pathology, Brigham and Women's Hospital and Harvard Medical School, Boston, Massachusetts 02115

Thomas J. Podor • Scripps Clinic and Research Foundation, La Jolla, California 92037

Bruce M. Pratt • Department of Pathology, Yale University School of Medicine, New Haven, Connecticut 06510

Eugene M. Renkin • Department of Human Physiology, School of Medicine, University of California, Davis, California 95616

Robert D. Rosenberg • Department of Medicine, Beth Israel Hospital and Harvard Medical School, Boston; and Department of Biology, Massachusetts Institute of Technology, Cambridge, Massachusetts 02139

Russell Ross • Department of Pathology, University of Washington, Seattle, Washington 98195

Stephen M. Schwartz • Department of Pathology, University of Washington, Seattle, Washington 98195

Maya Simionescu • Institute of Cellular Biology and Pathology, 79691 Bucharest, Romania

Nicolae Simionescu • Institute of Cellular Biology and Pathology, 79691 Bucharest, Romania

David M. Stern • Department of Physiology and Cellular Biophysics, College of Physicians and Surgeons, Columbia University, New York, New York 10032

Roger C. Wagner • School of Life and Health Sciences, University of Delaware, Newark, Delaware 19716

Judith Yannariello-Brown • Department of Pathology, Yale University School of Medicine, New Haven, Connecticut 06510

T. Zand • Department of Pathology, University of Massachusetts Medical School, Worcester, Massachusetts 01655

Preface

Although blood capillaries were first observed through a flea-lens microscope by Malpighi in 1661, 200 more years elapsed before the cellular nature of the vessel wall was conclusively demonstrated. Beginning with the middle of the 19th century, our knowledge of the histological organization of blood vessels has steadily increased. However, the endothelium, which for a long time was considered to be just an inert barrier lining, had been barely explored until three decades ago. Since then, there has been an upsurge of interest in the fine structure and function of endothelial cells. Intense *in vivo* and *in vitro* investigations have revealed that the endothelial cell is a key element in a wide variety of normal activities and diseases. A large number of investigators and laboratories have been attracted to endothelial cell research, thus supporting the expansion of the continuously growing and diversifying field of endotheliology. The number of articles published annually on this subject has increased from a few score at the beginning of the 1970s to more than a thousand in recent years, and an increasing number of journals, books, societies, and symposia focused primarily on the vascular endothelium have marked the last decade.

It is currently widely recognized that the ubiquitous endothelial cell is morphologically and physiologically highly differentiated along the vascular tree, as a biological adaptation to local needs. The endothelial cell, which is very active metabolically, is endowed with sensory and synthetic machinery through which it contributes crucially to the regulation of the entire cardiovascular system and to systemic homeostasis as well. Endothelial dysfunctions may be at the root of not only some vascular abnormalities, such as atherosclerosis and thrombosis, but also certain systemic diseases.

The modern exploration of endothelial cell structure and function is a largely interdisciplinary exercise. As in many other similar fields, there are still problems arising from the different nomenclatures of those investigating the endothelium from various perspectives, including physicists, mathematicians, engineers, physiologists, morphologists, biochemists, and cell biologists. On some issues, more rapid progress will only be possible to the extent that each side learns the other's language. Endothelial cell research is a good example of a field that can only be tackled by interdisciplinary collaboration and communication or by working in mixed research teams, for the grueling process of trying to train people capable of spanning all the disciplines involved seems unrealistic.

That is why this book has two goals. One is to act as a stimulus for those readers with an interest in endothelial cells to join the venture. The other is to promote broader communication among experts in different disciplines in the combined effort that is obviously necessary to confront a problem as complex and important as the roles of the endothelial cell in health and disease.

The chapters are not reviews in the usual sense but rather research-oriented accounts by some of those who have contributed measurably to our knowledge of endothelial cell biology.

Truly comprehensive coverage of the subjects remains an unattainable aim, and the volume thus contains—for selected topics—mere snapshots of our incomplete but continuously expanding understanding of the endothelial cell.

Nicolae Simionescu and Maya Simionescu

Bucharest

Contents

III. Endothelial Cell Growth and Differentiation

6. *Endothelial Morphogenesis*

   Ronald L. Heimark and Stephen M. Schwartz

10. The Biochemistry and Physiology of Anticoagulantly Active Heparin-like Molecules

James A. Marcum and Robert D. Rosenberg

11. The Fibrinolytic System of Cultured Endothelial Cells

Scott A. Curriden, Thomas J. Podor, and David J. Loskutoff

VI. Endothelial Cell Procoagulant Activity

12. Vascular Endothelium: Functional Modulation at the Blood Interface

Michael A. Gimbrone, Jr., and Michael P. Bevilacqua

VII. *Endothelial Cell Response to Stress Factors*

VIII. Endothelial Cell in Atherogenesis

17. Endothelial Injury and Atherosclerosis

Russell Ross

18. Prelesional Changes of Arterial Endothelium in Hyperlipoproteinemic Atherogenesis

Nicolae Simionescu

19. Response of Blood Vessel Cells to Viral Infection

Nicholas A. Kefalides

*Endothelial
Cell Biology*

Structure–Function Correlations in Endothelial Cells

The Microvascular Endothelium Revisited

G. E. Palade

I. WELL-ESTABLISHED KNOWLEDGE

The starting premise of a review written nine years ago (Palade *et al.*[56]) was that the vascular endothelium is a simple squamous, highly attenuated epithelium differentiated to permit large-scale exchanges of micro- and macromolecules between the blood plasma and the interstitial fluid of higher, multicellular organisms. In general terms, this differentiation is expressed by (1) a remarkable attenuation of the constituent cells (to < 0.3 μm for extensive parts of their cell bodies), (2) junctional complexes simplified to occluding and adhering zonules, and (3) a large population of vesicles, ~ 70 nm in diameter, referred to as caveolae or plasmalemmal vesicles associated, as the name implies, with the plasmalemma on both the luminal and abluminal front of the cells. At that time it was already well established that additional, structural elements regularly appear as expression of further differentiation in specific microvascular beds. The fenestrae, for instance, were known to be circular openings ~ 60 nm in diameter, most of them apertured, present within the attenuated cytoplasm of the endothelium of capillary and post-capillary venules in the microvascular beds of different mucosae and of exocrine and endocrine glands. Still another characteristic feature was the presence of large (≥ 100 nm) fenestrae without diaphragms in a restricted but physiologically important capillary bed, that of renal glomerular capillaries, and finally the existence of larger, aperture-free, relatively irregular discontinuities or lacunae, in the endothelium lining the sinusoids of the liver and bone marrow.[78]

The endothelium of all vessels was known to be supported by an external, tightly applied basement membrane which, in the case of capillaries and post-capillary venules, was split into distinct leaflets to accommodate, within the space thus created, special cells called pericytes. Sinusoids were known to have no (or only a rudimentary) basement membrane in most organisms.

Work done over the last nine years has repeatedly confirmed and reinforced these basic findings. It has also added new and significant details to our under-

G. E. Palade • Department of Cell Biology, Yale University School of Medicine, New Haven, Connecticut 06510.

standing of the vascular endothelium and, in the process, has generated a few controversies of rather radical nature which remain to be resolved hopefully in the near future.

II. NEW INFORMATION

A. Plasmalemma Proper and Plasmalemmal Vesicles

Information on the chemistry of the plasmalemma of the type already obtained for other cell populations by cell fractionation procedures has not been available for many years for any kind of endothelial cell. Moreover, the same applies for all other subcellular components of endothelial cells. Only recently has some success been achieved in isolating plasmalemmal fractions from bovine aortic endothelium[40] (see below). Difficulties still to be overcome concern logistics (i.e., small amounts of starting material) and, to some extent, unsatisfactory homogenization procedures.

As a substitute, the surface chemistry of the plasmalemma has been investigated using rather general probes (lectins, cationic or cationized proteins) and testing for binding *in situ* (by perfusion) before and after treatment (again by perfusion) with potentially relevant enzymes.[77,85] This approach has revealed the presence of anionic binding sites at high density over the entire luminal plasmalemma and their absence from the membrane of plasmalemmal vesicles, infundibula leading to such vesicles, and stomatal diaphragms, when present. The probe used in such studies was cationized ferritin (CF) with pI values ranging from 6.8 to 8.4, but other cationic probes gave similar results. Pretreatment with Pronase or papain removed all binding sites which, therefore, appear to be contributed primarily by anionic proteins, not glycolipids or gangliosides. Treatment with neuraminidase and glycosaminoglycan-degrading enzymes was less informative.[77]

In general, CF labeling was quasi-continuous—although patchy on some endothelia—and the decoration usually involved two superimposed layers of ferritin molecules which was taken to indicate that acidic glycoproteins protrude for at least 20 nm above the lipid bilayer of the plasmalemma. The distribution of anionic sites was found to be the same on both (luminal and abluminal) domains of the plasmalemma.[74]

Lectins [tagged with horseradish peroxidase (HRP)] showed a rather different distribution of their glycoprotein "receptors": the plasmalemma proper, as well as the plasmalemmal vesicles and their diaphragms were labeled by probes recognizing N-acetylglucosamine, sialic acid as well as galactose residues.[73]

The surface of the endothelial cells appears, therefore, as a mosaic of differentiated microdomains with striking differences in anionic site concentration between plasmalemma proper and plasmalemmal vesicles. It was this difference, especially the absence of detectable anionic sites on the membrane of the plasmalemmal vesicles, that led to the assumption that plasma proteins (practically all anionic) must be transported across the endothelium by plasmalemmal vesicles.[77]

Even at the rather superficial level of analysis represented by CF binding, regional differences in the vasculature became apparent: anionic sites are present at higher density on the luminal aspect of the endothelial plasmalemma in capillaries and arterioles than in venules, and on the latter their surface density decreases with the age of the animal (mouse).[76] CF of decreasing pI values has increased access to the plasmalemmal vesicles of the continuous endothelium of muscle (diaphragm) capillaries and venules,[76] but it has no access to those of the fenestrated endothelium of mucosal and glandular capillaries.[85] Moreover, it decorates heavily the introit (neck) of the plasmalemmal vesicles in the aortic endothelium.[1] All this goes to say that the surface chemistry of the endothelium is regionally modulated and that, depending on their position in the vasculature, similar endothelial structures, like plasmalemmal vesicles and their infundibula, have dissimilar chemistry supposedly reflecting differences in their functions.

The existence of differentiated microdomains in a mosaic membrane provided with a continuous lipid bilayer, fluid at the temperature of the immediate ambient, raises interesting problems: Randomization is prevented which means that the differentiated chemistry of the microdomains must be maintained by protein interactions either in the membrane itself or between the membrane and stabilizing infrastructures located on the cytoplasmic side of the plasmalemma. These infrastructures are probably similar but not necessarily as extensive as that found in erythrocytes.[45] When cultured endothelial cells are split open and the cytoplasmic aspect of their plasmalemma is examined by high-resolution scanning electron microscopy (using thin metal films of Cr or Ta), a fibrillar network is visualized under the plasmalemma proper and a characteristic surface structure is detected on plasmalemmal vesicles, which appears as a series of meridian ridges disposed between two centers (or poles) located in a plane parallel to, and immediately under, the plasmalemma.[57] The chemical nature of this infrastructure (which is clearly distinct from that of coated pits and coated vesicles) is unknown. The fibrillar meshwork under the plasmalemma proper probably consists in part of spectrin (or fodrin) and actin. Spectrin has been localized by immunocytochemistry in endothelial cells,[34] and spectrin, vimentin, and actin were recently identified by immune overlays in plasmalemmal fractions obtained from bovine aortic endothelium.[40] These fractions contain, in addition, "analogues" of ankyrin and 4.1 protein. Chemical similarity of the corresponding infrastructures is not surprising, since erythrocytes and endothelia subject each other to comparable stresses in the circulation (especially in capillaries). Both membranes are in need of stabilization and the stabilizing procedures used for one are expected to appear, perhaps with some modulations, in the other. A number of integral membrane glycoproteins of high molecular weight were also detected in the plasmalemmal fractions mentioned above.

B. Diaphragmed Fenestrae

When the surface chemistry approach already mentioned was applied to the fenestrated endothelium of the microvasculature of the pancreas and intestinal mucosa, it was discovered that the luminal surface of the fenestral diaphragms binds CF avidly: these diaphragms are the first structural elements to be labeled by perfusion and the last to lose the label in intact animals.[85] Pretreatment with heparinase[77] and heparitinase[75] as well as Pronase[77] prevented labeling by CF. Perfusion with tagged lectins revealed little or no binding. Taken together, these tests indicate that the anionic sites of the fenestral diaphragms are contributed mostly by heparan sulfate proteoglycans. This new finding showed that differentiated microdomains are in fact more diversified in fenestrated endothelia than in their continuous, nonfenestrated counterparts. It also suggested that the efflux of anionic plasma proteins is limited or prevented at the level of the fenestrae, which would explain why such capillaries are not as leaky as expected given the size and the surface density of their fenestrae. Cationic proteins, below certain dimensions, are expected to pass through the fenestrae, and earlier results indicated that this is the case with HRP.[15] Small cationic polypeptides, e.g., protamine and polylysine, were recently shown to increase markedly (up to fivefold) lymph flow and protein efflux in the intestine presumably by neutralizing the negative charge on the fenestral diaphragms.[32] Interestingly enough, the abluminal aspect of these diaphragms cannot be decorated with cationic probes,[74] which means that these structures are highly asymmetric and definitely polarized.

Incidentally, the discovery of the high concentration of heparan sulfate proteoglycans on the luminal aspect of the fenestral diaphragms puts in doubt some earlier speculations[56] according to which endothelial fenestrae derive from transendothelial channels by collapse to minimal dimensions and elimination of all but one diaphragm (see Section II.C). The luminal stomatal diaphragms of transendothelial channels have no detectable anionic sites and the charge present on the abluminal diaphragm remains unknown. The old assumption may still be valid provided future work proves that the abluminal diaphragm bears heparan sulfate proteoglycans.

It was already known that the fenestral diaphragms have radial sixfold or eightfold[47] symmetry—a knob in the center with spokes radiating from it and anchoring into a polygonal rim. More recently, wedge-shaped pores have been visualized in between the spokes by metal rotary shadowing[3] and by scanning electron microscopy.[58] But the actual dimensions of these pores and their degree of patency remain to be checked by adequate probes.

As in the case of plasmalemmal vesicles, regional modulations have already been recorded for fenestral diaphragms. To begin with, some fenestrae may have no apertures[15] and others have diaphragms with heterogeneous concentrations of anionic sites.[2] Those found in the choriocapillaries of the eye appear to have anionic sites provided by heparin rather than the heparan sulfate proteoglycans[60]; in addition, they have receptors for wheat germ agglutinin[59] and are impermeable to hemeproteins larger than 3 nm in diameter.[61]

C. Transendothelial Channels

The existence of transendothelial channels is well documented in the literature. What is still debated is not their occurrence, but their surface density over the endothelium and, in relation to this parameter, the extent of their actual participation in blood plasma–interstitial fluid exchanges. Structurally speaking, they are patent openings or pores, uninterrupted by hydrophobic barriers; their patency is presumed to be limited only by strictures along the channel pathway or by the porosity of their stomatal diaphragms.

At present, it appears quite clearly that the morphology of the transendothelial channels is not the same in all types of endothelia. In the continuous endothelium of capillaries and venules, the channels are formed by single vesicles or short chains of (two to four) vesicles that open simultaneously on both fronts of the endothelium.[82] The openings of these channels do not appear to have diaphragms. Hence, size-limiting barriers are probably provided by their generally narrow openings (20 to 30 nm) or by remnant strictures at the level of vesicle–vesicle fusion. Transendothelial channels in fenestrated endothelia generally consist of a single vesicle with diaphragmed openings on both sides.[15] Structurally these diaphragms are comparable to those of the fenestrae and to the stomatal diaphragms of plasmalemmal vesicles: all have comparable radial symmetry. However, the luminal diaphragms of the channels do not have anionic binding sites—and in this respect they are similar to stomatal rather than fenestral diaphragms. The albuminal diaphragm of the channel is not accessible to the probes usually used; hence, the nature of local charges is unknown.

Recent work has shown that channels, like fenestrae, occur in clusters located most of the time in the highly attenuated parts of the endothelial cells.[50,58] The functional meaning of this clustering as well as that of the fenestrae remains unknown.

D. Coated Pits and Coated Vesicles

It was already well established that endothelial cells have a relatively small population of coated pits and coated vesicles and that their surface density is considerably higher in the endothelia of the hepatic sinusoids[29,97] than in those of other microvascular beds. Information more recently obtained indicates that, in fenestrated endothelia, coated pits have a higher surface density of anionic sites, removable by proteinases but not by the glycosaminoglycan-degrading enzymes so far tested.[74] The nature of these sites and the type of membrane proteins that provide them remain unknown.

The coated pits of the hepatic sinusoidal endothelia appear to be different: they bind both cationic and anionic probes within the same unit,[29] which again brings forward, as a leitmotif, the occurrence of regional modulations affecting structurally similar elements of the endothelium in different vascular beds.

E. Endothelial Pockets

These structures were discovered by using in parallel scanning and transmission electron microscopy.[51] They are relatively large pockets 0.5 to 1 μm in size, and have roughly the shape of "empty" half domes or half cylinders applied on the luminal surface of the endothelium. Their luminal top is a thin (0.1 μm) flap of fenestrated cytoplasm, and their abluminal bottom consists of an equally thin and fenestrated cytoplasmic layer. The cavity or pocket thus delimited appears to be closed toward both the lumen and the pericapillary spaces. The diaphragms of the luminal openings are free of fixed anionic sites, just like the diaphragms of the transendothelial channels. The abluminal diaphragms are generally not accessible to probes. When they are (presumably due to accidental ruptures), they appear to be negatively charged.[51] It is somehow surprising that new structures can still be found after so many years of electron microscopy. This finding, like others before it, adds for the moment another entry to the list of endothelial structures for which a defined function is lacking.

The systematic investigation of a number of microvascular beds provided with fenestrated endothelia has revealed another type of modulation detected this time at the structural level. The fraction of highly attenuated endothelium, the surface density of fenestrae and channels, and the ratio of fenestrae to channels vary characteristically from one microvascular bed to another with the renal peritubular capillaries and the pancreatic capillaries at the high end and the low end of the series, respectively.[50]

Although the inquiry is still limited in both extent and depth, it is already evident that extensive modulations occur at the structural and surface chemistry level within the broad categories of continuous and fenestrated microvascular endothelia. These modulations most probably affect the relative chemistry of the local blood plasma–interstitial fluid exchanges.

III. CONTROVERSIES

A. Plasmalemmal Vesicles, Volume Density

Earlier studies had established the surface density of open plasmalemmal vesicles on both sides of the endothelium as well as the fractional volume they occupy within endothelial cells.[8] Moreover, in the microvasculature of a striated muscle, the murine diaphragm, some of these parameters were studied at the level of different segments of the microvasculature, i.e., arterioles, capillaries, and venules.[83] More recently a careful, technically excellent morphometric study has covered these and other parameters in the microvascular bed of the amphibian mesentery,[11] an object chosen because it is amenable to permeability studies carried on in individually perfused capillaries.[20] In all these cases, the starting preparations were chemically fixed with aldehydes (glutaraldehyde) followed by OsO_4. More recent and less systematic studies on a diversity of preparations ranging from cultured endothelia[48,64] to the microvasculature of the swimming bladder of the eel[91] claim, however, that the volume density of these vesicles is substantially smaller in fresh, quick-frozen specimens (subsequently fixed by freeze-substitution), the reduction in number of vesicles ranging up to ~ 70%. In addition, in certain specimens a population of considerably smaller vesicles (diameter ~ 30 nm) was detected under such conditions.[91] Chemical fixation is known to be a slow process.[6] In the twilight zone, between the beginning and the completion of the process, it is conceivable that changes in the distribution of certain structures and their membranes may occur, especially if the affected structures are part of a dynamic system, as the plasmalemmal vesicles are supposed to be. This matter definitely requires further attention because of the limited scope of the reports so far published and because the results obtained by similar procedures vary from no change[27] to drastic change.[48,64]

B. Plasmalemmal Vesicles, Intracellular Distribution

It was originally postulated that the plasmalemmal vesicles of endothelial cells function in the transport of water and solutes between the blood plasma and the interstitial fluid.[55] The postulate was supported by evidence obtained in experiments in which the tracers used (e.g., colloidal gold particles and ferritin)[8] were directly and individually detected by electron microscopy. Upon introduction into the systemic circulation of rats, these tracers appeared in the pericapillary spaces and while in transit through the endothelium they were restricted to plasmalemmal vesicles and did not appear in other structures such as intercellular junctions. The rate of transport was, however, slow, which was taken as an indication that plasmalemmal vesicles could not play a significant role in transendothelial exchanges. Moreover, at that time the pore theory of capillary permeability was in general favor among physiologists and that theory postulated the existence of small (diameter $\sim$ 12 nm) and large (diameter $\sim$ 50 nm) patent pores across the capillary walls through which molecules could move either by diffusion or in response to pressure differentials.[42] In addition, most physiologists assumed that these pores were located at the level of intercellular junctions, a suggestion originally advanced by Zweifach and Chambers.[99] There was no easy way to connect vesicles to the postulated pore systems.

The work with tracers was rapidly expanded primarily as a result of the introduction of "mass tracers," i.e., proteins with peroxidatic activity which could be detected via a convenient histochemical procedure worked out by Graham and Karnovsky.[31] The prototype of the new tracers was HRP,[38] but other peroxidases of different sizes and properties (e.g., myoglobin,[81] hemepeptides[82]) were used in the hope of finding better matches between probes and small pores. The new tracers were indirectly detected by a peroxidatic reaction that had the inherent advantage of amplifying the signal, i.e., the electron-opaque product of the reaction used for detection by electron microscopy. But this advantage was counteracted by the low resolving power of the histochemical procedures. Without membrane boundaries, e.g., in the pericapillary spaces, reactants and reaction products could diffuse beyond the range of dimensions of the structures of interest. Moreover, the signal was amplified throughout the diffusion area. All these mass tracers were transported much more rapidly than ferritin (in less than 40 sec) across the microvascular endothelium. They labeled the plasmalemmal vesicles and the intercellular junctions without the time resolution needed to establish convincingly which of these structures represented the primary exit pathway and which were labeled by "back-diffusion" from the pericapillary spaces. The evidence in favor of one efflux pathway, i.e., plasmalemmal vesicles,[81,82] was not strong enough to convince the proponents of the other pathway, i.e., intercellular junctions.[39,98] Hence, neither one nor the other interpretation of these experimental findings was generally accepted.

Ferritin continued to be used as a probe, primarily by investigators of amphibian mesenteric capillaries. Their evidence suggested that this tracer is transported across the endothelium by plasmalemmal vesicles that do not function as individual shuttles (as originally postulated[55]) but work in relays of two or more vesicles, the concentration of the initial tracer load being reduced at each successive vesicle fusion.[16,17]

It was at this juncture that the first careful three-dimensional reconstructions from serial thin sections of the capillary endothelium were published by Frokjaer-Jensen, Bundgaard, and their collaborators. The results showed that in amphibian[25] as well as in mammalian[13,26] capillary endothelia, the vast majority of the vesicles are organized in branched chains or dendritic structures open at one front or the other of the endothelium. Only a small fraction ($\sim$ 2%) were found free in the cytoplasmic matrix. Given this distribution, the authors concluded that the vesicles cannot function in transendothelial transport because a "shuttle system" would require the presence of a larger fraction than 2% of the vesicular population in transit through the cytoplasm. Accordingly, they postulated that the vesicles must have some other (perhaps secretory) function and cannot participate in exchanges[9,12] which were again ascribed to pores located at the level of the intercellular junctions.

The endothelial samples examined in all these studies were of necessity of small volume and, therefore, not necessarily representative. Moreover, they came from chemically fixed specimens. But the evidence was clearly of high enough quality to establish convincingly the pattern of vesicular distribution mentioned above.* There was nothing wrong with the structural analysis of the results; in fact, it approached technical perfection. Yet, the conclusions drawn by the authors were open to question. They ignored all results previously obtained with various tracers, especially particulate tracers, like ferritin. More importantly, they were based on the assumption that the vesicular distribution found in fixed specimens applies without any doubt to the distribution *in vivo*. As already mentioned, chemical fixation is a slow process. Even when effected by perfusion it takes from 9 to 18 sec to arrest the transport of HRP across the endothelium.[6] During the progressive slowdown of all cellular activities, the distribution and the interactions of subcellular components may change. Vesicles may come to rest, for instance, in dendritic structures which may represent their state at minimum free energy. It is, therefore, more reliable to follow a tracer of physiological significance and appropriate dimensions, and to identify the pathways it follows during its efflux, irrespective of modifications possibly introduced by preparation procedures in the structures that constitute the relevant pathway. This approach substitutes a functional analysis to a strictly structural analysis of endothelial specimens.

---

* A slightly larger fraction of free vesicles, 6–7% was found in studies using tannic acid as tracer.[54] Incidentally, less extensive vesicular fusion was detected earlier in a three-dimensional reconstruction of capillary endothelium in mammalian muscle.[7]

As already mentioned, the discovery of differentiated microdomains in microvascular endothelia led to the assumption that plasmalemmal vesicles are involved in the transcellular transport of anionic plasma proteins among which the most likely candidate is albumin. This assumption was recently put to test by Ghitescu et al.[30] in perfusion experiments using albumin–colloidal gold complexes as tracers. The results revealed a striking association of these probes to the membrane of plasmalemmal vesicles, in contrast to their limited binding to the plasmalemma proper. Since the binding of albumin–gold complexes was found to be saturable and competable (by free albumin), the authors postulated that their findings revealed the existence of a receptor-mediated transcytosis of albumin.[30] Transcytosis was demonstrated in lung and myocardium capillaries but the rate was slow: it took 5 min or more to detect albumin–gold complexes in the pericapillary spaces. Moreover, a detectable fraction of the tracer was routed to endosomes presumably for eventual degradation in lysosomes.

When albumin transport was reinvestigated by Milici et al.[52] in the microvasculature of the murine myocardium using monomeric heterologous albumin (bovine) and detection by antialbumin followed by a gold-tagged secondary antibody, the transport of albumin was found to be massive, and rapid (< 15 sec), with no detectable diversion to endosomes or lysosomes. The tracer was found to be restricted to plasmalemmal vesicles while in transit through the endothelium. Its efflux did not appear to involve intercellular junctions. These results were obtained by using immunocytochemical procedures applied to the surface of thin frozen sections or sections of specimens embedded in a hydrophilic (LR white) resin. In the albumin–gold experiments, the probe behaves as a multifunctional ligand and this probably explains the slow rate of transcytosis and the partial diversion of the probes to endosomes. The size of these complexes (14.5 nm) is large enough[52] to make them exclusive probes for the postulated large-pore system of capillary physiologists. Monomeric albumin is, however, an adequate probe for both postulated (small and large) pore systems, but given the large difference in surface density between the two types of pores, it is expected to reveal primarily the structural equivalents of the small-pore system.

The experiments mentioned above promise to shift again the attention of investigators from intercellular junctions to plasmalemmal vesicles but the results are too recent for final assessment. It remains to be seen to what extent they will be confirmed by work done in other laboratories and to what extent they will influence the thinking of capillary physiologists. At present, the most likely solution of the pore versus vesicle dilemma is to assume: (1) that vesicles function as shuttles as well as transendothelial channel generators, (2) that these channels are transient and fragile structures (only a small fraction survive specimen preparation), and (3) that local structural details, e.g., size of strictures and porosity of diaphragms, when present, determine their behavior as either small or large pores. Again, it is too early to assume that we have finally arrived at the conclusion of a long and vexing controversy, but there is hope that, with more adequate tracers and better resolution in our detection procedures than available in the past, we are approaching a generally acceptable solution of an old and difficult problem. Recent data from Renkin's laboratory indicate that a large fraction of transported albumin cannot be accounted for by the sum of small- and large-pore fluxes across the endothelium.[62,63] This large fraction probably represents vesicular transport.

C. Intercellular Junctions

It has been known for some time that the intercellular junctions of the vascular endothelium are less elaborate than those of other epithelia.[56] Desmosomes (maculae adherentes) are generally missing; adhering zonules are well developed but their adequate demonstration depends on preparation procedures*; and occluding zonules (tight junctions) appear limited to a few points of membrane contact or apparent fusion in sections, and a few strands (some of them free-ending) in replicas of freeze-cleaved preparations.[72,98] Segmental variations of the junctions have been identified and among them the most striking is found at the level of the postcapillary venular endothelium in which the strands of the occluding zonules are replaced by discontinuous ridges (or crises) in freeze-cleaved preparations.[72] Many (but not all) of these venular junctions appear open and permeated by mass tracers (see below) in sectioned specimens. It is at the level of the postcapillary venular endothelium that inflammatory reactions begin in response to a variety of mediators (histamine, serotonin,[44] bradykinin, leukotrienes[5]) by focal separation of the endothelial cells[44] most probably facilitated by the apparent weakness of the junctions.

Communicating junctions (gap junctions) are not found in capillary and postcapillary venular endothelia, but are present in the endothelium of muscular venules, veins, arterioles, and arteries.[72] In the latter two instances, they are well developed and characteristically intercalated among the strands of the occluding zonules. In general, it appears that the development of the junctions reflects the intensity of the physical forces to which the endothelium is locally subjected. Since, by the nature of their construction, communicating junctions are also cell-to-cell adhering elements, it may be assumed that their characteristic association with occluding zonules reinforces intercellular adhesion and stabilizes areas of intercellular communication in arterial endothelia.

The permeability of the intercellular junctions of the microvascular endothelium, especially capillary endothelium, is still a controverted issue. As already discussed, the high hydraulic conductivity of blood capillaries and their differential permeability to hydrophilic macromolecules have led to the postulate that their walls are fitted with water-filled channels or pores of two different dimensions and surface densities. Since at the time it seemed unlikely that such channels could cut across the body of endothelial cells, it was assumed that the pores are accommodated at the level of intercellular junctions most probably as focal gaps between adjacent cells. If cylindrical, the expected diameters would be ~ 12 and

---

* The fibrillar cytoplasmic densities that accompany these junctional elements are better preserved or better visualized in preparations postfixed in OsO_4 in acetate–veronal buffer or in specimens embedded in hydrophilic resins (LR white).[52]

~ 50 nm for small and large pores, respectively; if slit-shaped, they would be gaps of approximately two-thirds those dimensions.

Morphological studies could not settle the issue since the reported findings ranged from practically all tight junctions closed[7] to many junctions opened to a gap of ~ 4 nm.[39,98] A recent three-dimensional reconstruction study from serial sections of myocardial capillaries has established the presence of such focal gaps ~ 4 nm wide.[10] When tracers first began to be used, those qualifying as large-pore probes (ferritin,[8] dextrans[80]) were not found in junctions but were present in plasmalemmal vesicles. When probes for both small and large pores, such as peroxidatic mass tracers, were introduced, they were detected cytochemically in both plasmalemmal vesicles and junctions. For reasons already discussed, the earliest exit pathway could not be convincingly established to everyone's satisfaction and the issue remained open. The discovery of transendothelial channels had little impact on pore location, primarily because of their apparently low surface density. The three-dimensional reconstruction studies focused attention on junctions because most vesicles were found fused in dendritic structures[9] and because gaps were detected along intercellular junctions.[10] It deserves to be stressed, however, that (1) the width of these gaps was less than expected even for small pores and (2) measurements of intermembrane gaps cannot be taken at face value in electron microscopy.[56] The results of recent experiments with monomeric albumin[52] call again the attention of investigators to plasmalemmal vesicles and transendothelial channels as likely candidates for exit sites for small-pore probes. As with plasmalemmal vesicles, this is another case in which the results of a functional analysis may prove to be more reliable than those of a strictly structural investigation.

At present it appears that the sites of albumin exit are not at the level of the intercellular junctions of the capillary endothelium and the same seems to apply, within the limits of detection of cytochemical procedures, for other macromolecules tested down to hemepeptides, i.e., down to probes of ~ 2 nm diameter. Water and hydrophilic solutes smaller than 2 nm most probably move through junctions as they do in other "leaky" epithelia.[28] The question that remains to be generally settled, by direct rather than indirect evidence, is the permeability of the junctions for macromolecules of diameters larger than 2 nm and smaller then 10 nm. Without such evidence, it would be highly unlikely that the junctions are the site of the small pores postulated by the pore theory of capillary permeability.

In contradistinction to the situation found in capillaries, the junctions of the endothelium of postcapillary venules are structurally partially open and functionally permeable to HRP and hemoglobin.[83] In this case also a discrepancy is found between the results of structural and functional analyses. All junctions appear "rudimentary" (and potentially leaky) especially in replicas of freeze-cleaved preparations, but the tracers used permeate only a fraction of the junctions which decreases as the size of the probe increases. The open endothelial junctions of the postcapillary venules can be the cause of some confusion since venules are not always easy to identify in sectioned specimens. Only in the diaphragm muscle can they be reliably identified primarily on account of their differential distribution.[84]

IV. NEW DEVELOPMENTS

A. Fiber Matrix

It has been known for some time that plasma proteins, especially albumin, regulate capillary permeability: in the presence of very low concentrations (< 1% of normal) of albumin, the hydraulic conductivity of perfused capillaries increases markedly.[22] The increase is effectively reversed by concentrations of albumin considerably lower than normal.

The effect has been attributed to the interactions of albumin with the surface of the endothelium, more precisely with the ectodomains of plasmalemmal proteins. These interactions presumably create a fibrillar matrix that acts as an added permselective barrier to the endothelium.[21] This "fiber matrix" hypothesis does not specify the location of the barrier, but clearly relates it with the postulated pores of the endothelium, presumably with intercellular junctions. Relevant experiments carried out by Schneeberger and Hamelin[68] suggest that a low concentration of plasma proteins (obtained by extensive replacement of the blood plasma with fluorocarbon emulsions), the access of ferritin to plasmalemmal vesicles and its transport to the pericapillary spaces are increased. Since no concurrent changes are detected in the organization of the intercellular junctions, these findings suggest that the fiber matrix operates at the introit of plasmalemmal vesicles.

In the experiments with monomeric albumin already mentioned,[52] albumin was found to bind with low affinity to the entire surface of the endothelium. It was easily removed by prefixation flashes with buffered saline from the luminal surface as well as from plasmalemmal vesicles open to the lumen. In comparison with the luminal plasmalemma, the amount of albumin detected in the junctions was negligible. These results, like the findings of Schneeberger and Hamelin,[68] suggest that the fiber matrix effect may be related to plasmalemmal vesicles rather than intracellular junctions. At present, cultured endothelial cells derived from the capillaries of the epididymal fat pad of the rat[92] are used to define the binding parameters of albumin to the surface of endothelial cells. The binding appears to be saturable, competable, and specific since other cells, i.e., fibroblasts, bind albumin in considerably lower amounts.[67] The intent is to identify the endothelial membrane proteins that bind albumin and to find out if some of these proteins function as high-affinity "receptors" for albumin transcytosis as postulated in Ghitescu et al.[31]

B. Other Transport Systems

Low-density lipoproteins (LDL) were shown to be transported *in situ* across the endothelium of rat arteries and pulmonary microvasculature[53,79,87] by coated pits, coated vesicles, and plasmalemmal vesicles, but the receptors, if present, were not identified. Hashida et al.[33] used a two-chamber system and showed that cultured porcine aortic endothelia transport LDL in a temperature- and energy-dependent process and Shasby et al.[70] obtained evidence that albumin is transported across cultured endothelium derived from porcine pulmonary arteries. Transport depends on metabolic energy and, intriguingly enough, it is faster toward the lumen than toward the subendothelial spaces. The structural elements that effect transport were not identified in either study.[33,70] Assuming that plasmalemmal vesicles are involved, the findings imply that vesicular transcytosis is, as expected, an energy-dependent process.

Receptor-mediated transport of insulin and transferrin has been postulated for brain capillaries,[24,37,41] but the carriers have not been identified. As in the case of LDL, they could be coated vesicles and/or plasmalemmal vesicles. Transferrin binding has been localized preferentially to coated pits/vesicles in the endothelium of the sinusoids of the bone marrow.[86]

Williams et al.[96] and Wagner et al.[93] have used rather extensively a model system for the study of endothelial transport activities. It is based on endothelia or microvascular segments isolated from the epididymal fat pad of the rat and assays the uptake as well as the subsequent discharge of a variety of proteins by and from these cells.[93,96] The same probe is used for uptake and discharge, but the fluorescent tag on the probe is changed from one step to the other. The results indicate that isolated endothelia are capable of both endocytosis and exocytosis of a variety of proteins from and to the incubation medium. Together the two operations may represent transcytosis, but since the protocol involves a single chamber, the overall nature of the process remains to be established. The system is ingenious and versatile but some of the results obtained are puzzling: albumin, for instance, is not taken up by these endothelial cells but chemically glycosylated albumin is actively endocytosed.[95] As already mentioned, in situ experiments indicate that albumin is taken up and transported across the endothelium.[30,52]

Although the information so far obtained is still limited and fragmentary, it suggests quite strongly that the vascular endothelium is provided with specific transendothelial transport systems for a variety of macromolecules of physiological significance, such as metabolite carriers (albumin, LDL), metal carriers (transferrin), and hormones (insulin). Questions that remain to be addressed in the future concern the mechanisms that control association and dissociation of the ligands to and from their receptors, and the possibility that endothelial cells may have two sets of receptors: one for their own use and the other for transcytosis. They may be able to manage with a single set of receptors, but then they should be able to use different carriers and different routes for the two activities. The same still incomplete evidence also suggests that transcytosis of macromolecules and by implication vesicular transport is an energy-dependent process. These are all interesting and attractive prospects that deserve to be further investigated.

C. Polarization

Like all other epithelia so far investigated, the vascular endothelium appears to be polarized in the sense that the chemistry of the luminal domain of the plasmalemma is different from that of its abluminal counterpart. But the information in hand is still limited and difficult to rationalize into a functionally meaningful pattern.

The Na, K-ATPase of brain capillaries is preferentially located in the abluminal domain,[4] whereas other phosphatases appear to be present on the luminal domain.[46] Different antigens have been localized by immunocytochemistry to different domains: the 140k glycoprotein podocalyxin is found on the luminal plasmalemma[35]; another glycoprotein, podoendin, is detected on the same front and in plasmalemmal vesicles,[36] and the same applies to still another unidentified antigen.[66] In humans but not in other species, only the glycoproteins of the luminal plasmalemma bind fucose-specific lectins.[49]

The main difficulty in rationalizing the polarization of the vascular endothelium stems from the fact that the media in the two compartments it separates (i.e., the blood plasma and the interstitial fluid) equilibrate rapidly and are qualitatively of similar chemistry. The polarization may reflect transport mechanisms of still unknown nature (vectorial?), differential cell–cell interactions, the necessity of response to local conditions (as in inflammation), or vectorial discharge of secretory products (see next section).

D. Biosynthetic Activities

Endothelial cells synthesize extracellular matrix components like collagen IV, fibronectin, and proteoglycans, used to construct their associated basement membrane.[65] These components are produced and processed as in any other secretory cell by the endoplasmic reticulum–which appears highly developed in the endothelia of growing capillaries. Matrix components are probably discharged abluminally. Coagulation reactants like factor VIII, the von Willebrand factor, are apparently synthesized continuously and stored in highly polymerized form[89,90] in Weibel–Palade bodies,[94] the local equivalent of secretion granules in other cell types. The discharge in this case is probably on the luminal front of the cells. More recent entries to the list of secretory products include fibroblast and endothelial cell growth factors,[69,88] PDFG-like growth factors,[18,19] granulocyte–macrophage colony-stimulating factor,[71] plasminogen activator, and factors that control the latter's activity.[43] The kinetics of synthesis, the type of secretion (constitutive or regulated?), and the compartment of preferential discharge (if applicable) remain unknown. Still to be elucidated in adequate detail is the participation of the vascular endothelium in the overall process of blood coagulation (e.g., control of thrombin activity[23]).

V. CONCLUDING REMARKS

The main function of the vascular endothelium is the mediation and control of transendothelial exchanges between the blood plasma and the interstitial fluid. It is undoubtedly a "vital" function judging by the dependency of all tissues in advanced multicellular organisms on the uninterrupted maintenance of these exchanges. It is also an evolutionarily important function: without it, evolution would have stopped at some level before vertebrates.

The dominant character of this function is clearly expressed in the differentiation of the endothelium at the level of the entire cell as well as at the level of its specific structures and differentiated microdomains. Each of the latter appears to be involved in some specific aspect of transendothelial exchanges. Within the framework of this general function, there are multiple and significant local modulations which affect the structure and the chemistry of a variety of differentiated microdomains. The two aspects, structure and chemistry, are not necessarily tightly coupled; certain structures, plasmalemmal vesicles for instance, may differ in chemistry from one location to another in the vasculature, notwithstanding their apparent identity at the structural level. Transport appears to be their common function, but rates, directions, and especially chemistry of transported molecules may vary.

Structures involved in transendothelial exchanges, e.g., plasmalemmal vesicles and transendothelial channels, are highly dynamic entities that appear to be significantly affected by chemical fixation procedures. Accordingly, their functional analysis is expected to give more reliable results than strictly structural studies.

In addition to its principal function of mediator and controller of transendothelial exchanges, the vascular endothelium is involved in many other subsidiary, though important, activities which include hemostasis, defense reactions (inflammatory response), and angiogenesis during development in the embryo and the infant, and during repair or pathological events in the adult. It is possible that the vascular endothelium participates also in the presentation of antigens in immune reactions.[14] Given the importance and multiplicity of endothelial functions, it is somehow surprising to find out how incomplete or unsatisfactory is the information available in this field. Research on the vascular endothelium is lagging far behind many other fields in cellular and molecular biology. The area is clearly underdeveloped but obviously important biomedically and biologically, which means that within its confines there is a lot to uncover, understand, and put to good use, hopefully in the near future.

ACKNOWLEDGMENTS. The expert secretarial assistance of Marybeth Hicks is gratefully acknowledged. Laboratory and library research for this paper has been supported by NIH Grant HL-17080 and a gift from RJR Nabisco, Inc.

REFERENCES

1. Baldwin, A. L., and Chien, S., 1983, Endothelial transport of anionized and cationized ferritin in the rabbit aorta, *Fed. Proc.* **42:**5924.
2. Bankston, P. W., and Milici, A. J., 1983, A survey of the binding of polycationic ferritin in several fenestrated capillary beds: Indication of heterogeneity in the luminal glycocalyx of fenestral diaphragms, *Microvasc. Res.* **26:**36.

3. Bearer, E. L., and Orci, L., 1985, Endothelial fenestral diaphragms: A quick-freeze, deep-etch study, *J. Cell Biol.* **100**:418.
4. Betz, A. L., Firth, J. A., and Goldstein, G. W., 1980, Polarity of blood brain barrier: Distribution of enzymes between the luminal and abluminal membranes of the brain capillary endothelial cell, *Brain Res.* **192**:17.
5. Bjork, J., Hedquist, P., and Arfors, K. E., 1982, Increase in vascular permeability induced by leukotriene B_4 and the role of polymorphonuclear leukocytes, *Inflammation* **6**:189.
6. Boyles, J., L'Hernault, N., Laks, H., and Palade, G. E., 1981, Evidence for vesicular shuttle in heart capillaries, *J. Cell Biol.* **91**:418a.
7. Bruns, R. R., and Palade, G. E., 1968a, Studies on blood capillaries. I. General organization of blood capillaries in muscle, *J. Cell Biol.* **37**:244.
8. Bruns, R. R., and Palade, G. E., 1968b, Studies on blood capillaries. II. Transport of ferritin molecules across the wall of muscle capillaries, *J. Cell Biol.* **37**:277.
9. Bundgaard, M., 1980, Transport pathways in capillaries—in search of pores, *Annu. Rev. Physiol.* **42**:325.
10. Bundgaard, M., 1984, The three-dimensional organization of tight junctions in a capillary endothelium revealed by serial-section electron microscopy, *J. Ultrastruct. Res.* **88**:1.
11. Bundgaard, M., and Frokjaer-Jensen, J., 1982, Functional aspects of the ultrastructure of terminal blood vessels: A quantitative study of consecutive segments of the frog mesentery microvasculature, *Microvasc. Res.* **23**:1.
12. Bundgaard, M., Frokjaer-Jensen, J., and Crone, C., 1979, Endothelial plasmalemmal vesicles as elements in a system of branching invaginations from the cell surface, *Proc. Natl. Acad. Sci. USA* **76**:6439.
13. Bundgaard, M., Hageman, P., and Crone, C., 1983, The three-dimensional organization of plasmalemmal vesicular profiles in the endothelium of rat heart capillaries, *Microvasc. Res.* **25**:358.
14. Burger, D. R., Vetto, R., Hamblin, A., and Dumonde, D. C., 1982, T-lymphocyte activation by antigen presented by HLA-DR compatible endothelial cells, in: *Pathobiology of the Endothelial Cells* (H. L. Nossel and H. J. Vogel, eds.), Academic Press, New York, p. 387.
15. Clementi, F., and Palade, G. E., 1969, Intestinal capillaries. I. Permeability to peroxidase and ferritin, *J. Cell Biol.* **41**:33.
16. Clough, G., 1982, The steady-state transport of cationized ferritin by endothelial cell vesicles, *J. Physiol.* (*London*) **328**:389.
17. Clough, G., and Michel, C. C., 1981, The role of vesicles in the transport of ferritin through frog endothelium, *J. Physiol.* (*London*) **315**:127.
18. Collins, T., Ginsburg, D., Boss, J. M., Orkin, S., and Pober, J. S., 1985, Cultured human endothelial cells express platelet-derived growth factor B chain: cDNA cloning and structural analysis, *Nature* **316**:748.
19. Collins, T., Pober, J. S., Gimbrone, M. A., Jr., Hammacher, A., Betsholtz, C., Westermark, B., and Heldin, C. H., 1987, Cultured human endothelial cells express platelet-derived growth factor A chain, *Am. J. Pathol.* **126**:7.
20. Curry, F. E., Mason, J. C., and Michel, C. C., 1976, Osmotic reflection coefficients of capillary walls to low molecular weight solutes measured in single perfused capillaries of the frog mesentery, *J. Physiol.* (*London*) **261**:319.
21. Curry, F. E., and Michel, C. C., 1980, A fiber matrix model of capillary permeability, *Microvasc. Res.* **20**:96.
22. Danielli, J. F., 1940, Capillary permeability and oedema in the perfused frog, *J. Physiol.* (*London*) **98**:109.
23. Esmon, T. C., Esmon, N. L., Sangstad, J., and Owen, W. G., 1982, Activation of protein C by a complex between thrombin and endothelial cell surface protein, in: *Pathobiology of the Endothelial Cell* (H. L. Nossel and H. J. Vogel, eds.), Academic Press, New York, p. 121.
24. Frank, H. J. L., and Pardridge, W. M., 1983, Insulin binding to brain microvessels, *Adv. Metab. Disord.* **10**:291.
25. Frokjaer-Jensen, J., 1980, Three-dimensional organization of plasmalemmal vesicles in endothelial cells: An analysis by serial sectioning of frog mesenteric capillaries, *J. Ultrastruct. Res.* **73**:9.

26. Frokjaer-Jensen, J., 1984, The plasmalemmal vesicular system in striated muscle capillaries and in pericytes, *Tissue Cell* **16**:31.
27. Frokjaer-Jensen, J., and Reese, T. S., 1986, Three-dimensional organization of the plasmalemmal vesicular system in directly frozen, freeze-substituted frog mesenteric capillaries, *Physiol. J.* **371**:84p.
28. Fromter, E., and Diamond, J., 1972, Route of passive ion permeation in epithelia, *Nature New Biol.* **235**:9.
29. Ghitescu, L., and Fixman, A., 1984, Surface charge distribution on the endothelial cells of liver sinusoids, *J. Cell Biol.* **99**:639.
30. Ghitescu, L., Fixman, A., Simionescu, M., and Simionescu, N., 1986, Specific binding sites for albumin restricted to plasmalemmal vesicles of continuous capillary endothelium: Receptor mediated transcytosis, *J. Cell Biol.* **102**:1304.
31. Graham, R. C., and Karnovsky, M. J., 1966, The early stages of adsorption of injected horseradish peroxidase in the proximal tubules of the mouse kidney: Ultrastructural cytochemistry by a new technique, *J. Histochem. Cytochem.* **14**:291.
32. Granger, D. N., Kvietys, P. R., Perry, M. A., and Taylor, A. E., 1986, Charge selectivity of rat intestinal capillaries: Influence of polycations, *Gastroenterology* **91**:1443.
33. Hashida, R., Anamizu, C., Kimura, J., Ohkuma, S., Yoshida, Y., and Takano, T., 1986, Transcellular transport of lipoprotein through arterial endothelial cells in monolayer culture, *Cell Struct. Funct.* **11**:31.
34. Heltianu, C., Bogdan, I., Constantinescu, E., and Simionescu, M., 1986, Endothelial cells express a spectrin-like cytoskeletal protein, *Circ. Res.* **58**:605.
35. Horvat, R., Havorka, A., Dekan, G., and Kerjaschki, D., 1986, Endothelial cell membranes contain podocalyxin—the major sialoglycoprotein of visceral glomerular epithelial cells, *J. Cell Biol.* **102**:484.
36. Huang, T. W., and Langlois, J. C., 1985, Podoendin, a new cell surface protein of the podocytes and endothelium, *J. Exp. Med.* **162**:245.
37. Jeffries, W. A., Brandon, M. R., Hunt, J. V., Williams, A. F., Gatter, K. C., and Mason, D. Y., 1984, Transferrin receptor on endothelium of brain capillaries, *Nature* **312**:162.
38. Karnovsky, M. J., 1967, The ultrastructural basis of capillary permeability studied with peroxidase as a tracer, *J. Cell Biol.* **35**:213.
39. Karnovsky, M. J., 1979, Morphology of capillaries with special reference to muscle capillaries, in: *Capillary Permeability, Alfred Benzon Symposium II* (C. Crone and N. A. Lassen, eds.), Munskgaard, Copenhagen and Academic Press, New York, pp. 341–350.
40. Ketis, N. V., Hoover, R. L., and Karnovsky, M. J., 1986, Isolation of bovine aortic endothelial cell membranes: Identification of membrane-associated cytoskeletal proteins. *J. Cell. Physiol.* **128**:162.
41. King, G. E., and Johnson, S. M., 1985, Receptor-mediated transport of insulin across endothelial cells, *Science* **227**:1583.
42. Landis, E. M., and Pappenheimer, J. R., 1963, Exchanges of substances through the capillary walls, in: *Handbook of Physiology*, Vol. II (W. F. Hamilton and P. Dow, eds.), American Physiological Society, Washington, D.C., pp. 1035–1073.
43. Loskutoff, D. J., Levin, E., and Mussoni, L., 1982, Fibrinolytic components of cultured endothelial cells, in: *Pathobiology of the Endothelial Cell* (H. L. Nossel and H. J. Vogel, eds.), Academic Press, New York, p. 167.
44. Majno, G., and Palade, G. E., 1961, Studies on inflammation. I. The effect of histamine and serotonin on vascular permeability; an electron microscopic study, *J. Biophys. Biochem. Cytol.* **11**:571.
45. Marchesi, V. T., 1985, Stabilizing infrastructure of cell membranes, *Annu. Rev. Cell Biol.* **1**:531.
46. Marchesi, V. T., and Barrnett, R. J., 1963, The demonstration of enzymatic activity in pinocytic vesicles of blood capillaries with the electron microscope, *J. Cell Biol.* **17**:547.
47. Maul, G. G. Y., 1971, Structure and formation of pores in fenestrated capillaries, *J. Ultrastruct. Res.* **36**:768.
48. McGuire, P. G., and Twietmeyer, T. A., 1983, Morphology of rapidly frozen aortic endothelial cells: Glutaraldehyde fixation increases the number of caveolae, *Circ. Res.* **53**:424.

49. Miettinen, M., Holthofer, H., Lehto, V. P., and Virtanen, A. I., 1983, *Ulex europeus* I lectin as a marker for tumors derived from endothelial cells, *Am. J. Clin. Pathol.* **79**:83.

50. Milici, A. J., L'Hernault, N., and Palade, G. E., 1985, Surface densities of diaphragmed fenestrae and transendothelial channels in different murine capillary beds, *Circ. Res.* **56**:709.

51. Milici, A. J., Peters, K. R., and Palade, G. E., 1986a, The endothelial pocket: A new structure in fenestrated endothelia, *Cell Tissue Res.* **244**:493.

52. Milici, A. J., Watrous, N., and Palade, G. E., 1986b, Immunogold localization of exogenous albumin in murine myocardial capillaries, *J. Cell Biol.* **103**:194a.

53. Nistor, A., and Simionescu, M., 1986, Uptake of low density lipoproteins by the hamster lung: Interactions with capillary endothelium, *Am. Rev. Respir. Dis.* **134**:1266.

54. Noguchi, Y., Yamamoto, T., and Shibata, Y., 1986, Distribution of endothelial vesicles in the microvasculature of skeletal muscle and brain cortex of the rat, as demonstrated by tannic acid tracer analysis, *Cell Tissue Res.* **246**:487.

55. Palade, G. E., 1960, Transport in quanta across the endothelium of blood capillaries, *Anat. Rec.* **136**:254.

56. Palade, G. E., Simionescu, M., and Simionescu, N., 1979, Structural aspects of the permeability of the microvascular endothelium, *Acta Physiol. Scand.* **463**:11.

57. Peters, K. R., Carley, W. W., and Palade, G. E., 1985, Endothelial plasmalemma vesicles have a characteristic striped bipolar surface structure, *J. Cell Biol.* **101**:2233.

58. Peters, K. R., and Milici, A. J., 1983, High resolution SEM of the luminal surface of a fenestrated capillary endothelium, *J. Cell Biol.* **97**:336a.

59. Pino, R. M., 1986, The cell surface of a restrictive fenestrated endothelium. I. Distribution of lectin-receptor monosaccharides on the choriocapillaries, *Cell Tissue Res.* **243**:145.

60. Pino, R. M., 1986, The cell surface of a restrictive fenestrated endothelium. II. Dynamics of cationic ferritin binding and the identification of heparin and heparan sulfate domains on the choriocapillaries, *Cell Tissue Res.* **243**:157.

61. Pino, R. M., and Essner, E., 1981, Permeability of rat choriocapillaries to hemeproteins: Restriction of tracers by a fenestrated endothelium, *J. Histochem. Cytochem.* **29**:281.

62. Renkin, E. M., 1987, Transport pathways and processes, in: *Endothelial Cell Biology* (N. Simionescu and M. Simionescu, eds.), Plenum Press, New York, pp.

63. Renkin, E. M., Gustafson-Sgro, M., and Silbey, L., 1987, Coupling of albumin flux to volume flow in skin and skeletal muscle of anesthetized rats, *Fed. Proc.* **46**:190.

64. Robinson, J. M., Hoover, R. L., and Karnovsky, M., 1984, Vesicle (caveolae) number is reduced in cultured cells prepared for electron microscopy by rapid-freezing, *J. Cell Biol.* **99**:287a.

65. Sage, H., Pritzl, P., and Bornstein, P., 1981, Secretory phenotypes of endothelial cells in culture: Comparison of aortic, venous, capillary and corneal endothelium, *Arteriosclerosis* **1**:427.

66. Schlingemann, R. O., Dingjan, G. M., Emeis, T. T., Blok, J., Warnaur, S. O., and Ruiter, D. J., 1985, Monoclonal antibody PAL-E specific for endothelium, *Lab. Invest.* **52**:71.

67. Schnitzer, J., Carley, W. W., and Palade, G. E., 1987, Albumin binding proteins of rat microvessel endothelial cells, *J. Cell Biol.* **105**:326a.

68. Schneeberger, E., and Hamelin, M., 1984, Interaction of serum proteins with lung endothelial glycocalyx: Its effect on endothelial permeability, *Am. J. Physiol.* **247**:H206.

69. Schweigerer, L., Neufeld, G., Friedman, J., Abraham, J. A., Fiddes, J. C., and Gospodarowicz, D., 1987, Capillary endothelial cells express basic fibroblast growth factor that promotes their own growth, *Nature* **352**:257.

70. Shasby, D. M., and Shasby, S. S., 1985, Active transendothelial transport of albumin: Interstitium to lumen, *Circ. Res.* **57**:903.

71. Sieff, C. A., Tsai, S., and Faller, D. V., 1987, Interleukin I induces cultured human endothelial cell production of granulocyte–macrophage colony-stimulating factor, *J. Clin. Invest.* **79**:48.

72. Simionescu, M., Simionescu, N., and Palade, G. E., 1975, Segmental differentiation of cell junctions in the vascular endothelium: The microvasculature, *J. Cell Biol.* **67**:863.

73. Simionescu, M., Simionescu, N., and Palade, G. E., 1982a, Differentiated microdomains on the luminal surface of capillary endothelium: Distribution of lectin receptors, *J. Cell Biol.* **94**:406.

74. Simionescu, M., Simionescu, N., and Palade, G. E., 1982b, Preferential distribution of anionic

sites on the basement membrane and the abluminal aspect of the endothelium in fenestrated capillaries, *J. Cell Biol.* **95**:425.

75. Simionescu, M., Simionescu, N., and Palade, G. E., 1984, Partial chemical characterization of the anionic sites in the basal lamina of fenestrated capillaries, *Microvasc. Res.* **28**:352.
76. Simionescu, M., Simionescu, N., Santoro, F., and Palade, G. E., 1985, Differentiated microdomains on the luminal plasmalemma of murine muscle capillaries: Segmental variations in young and old animals, *J. Cell Biol.* **100**:1396.
77. Simionescu, M., Simionescu, N., Silbert, J. E., and Palade, G. E., 1981, Differentiated microdomains on the luminal surface of the capillary endothelium. II. Partial characterization of their anionic sites, *J. Cell Biol.* **90**:614.
78. Simionescu, N., and Simionescu, M., 1983, The cardiovascular system, in: *Histology, Cell and Tissue Biology* (L. Weiss, ed.), Elsevier Biomedical, New York, p. 371.
79. Simionescu, N., and Simionescu, M., 1985, Interactions of endogenous lipoproteins with capillary endothelium in spontaneously hyperlipidemic rats, *Microvasc. Res.* **30**:314.
80. Simionescu, N., Simionescu, M., and Palade, G. E., 1972, Permeability of intestinal capillaries: Pathways followed by dextrans and glycogens, *J. Cell Biol.* **53**:365.
81. Simionescu, N., Simionescu, M., and Palade, G. E., 1973, Permeability of muscle capillaries to exogenous myoglobin, *J. Cell Biol.* **57**:424.
82. Simionescu, N., Simionescu, M., and Palade, G. E., 1975, Permeability of muscle capillaries to small hemepeptides: Evidence for the existence of patent transendothelial channels, *J. Cell Biol.* **64**:586.
83. Simionescu, N., Simionescu, M., and Palade, G. E., 1978, Open junctions in the endothelium of the postcapillary venules of the diaphragm, *J. Cell Biol.* **79**:27.
84. Simionescu, N., Simionescu, M., and Palade, G. E., 1978, Structural basis of permeability in sequential segments of the microvasculature of the diaphragm. I. Bipolar microvascular fields; II. Pathways followed by microperoxidase across the endothelium, *Microvasc. Res.* **15**:1, **15**:17.
85. Simionescu, N., Simionescu, M., and Palade, G. E., 1981, Differentiated microdomains on the luminal surface of the capillary endothelium. I. Preferential distribution of anionic sites, *J. Cell Biol.* **90**:605.
86. Soda, R., and Tavassoli, M., 1984, Transendothelial transport (transcytosis) of iron–transferrin complex in bone and marrow, *J. Ultrastruct. Res.* **88**:18.
87. Vasile, E., Simionescu, M., and Simionescu, N., 1983, Visualization of the binding endocytosis and transcytosis of low density lipoproteins in the arterial endothelium in situ, *J. Cell Biol.* **96**:1677.
88. Vlodavsky, I., Folkman, J., Sullivan, R., Friedman, R., Ishai-Michaeli, R., Sasse, J., and Klagsbrun, M., 1987, Endothelial cell-derived basic fibroblast growth factor: Synthesis and deposition into subendothelial extracellular matrix, *Proc. Natl. Acad. Sci. USA* **84**:2292.
89. Wagner, D. D., and Marder, V. J., 1983, Biosynthesis of von Willebrand protein by human endothelial cells: Identification of a large precursor polypeptide chain, *J. Biol. Chem.* **258**:2065.
90. Wagner, D. D., and Marder, V. J., 1984, Biosynthesis of von Willebrand protein by human endothelial cells: Processing steps and their intracellular localization, *J. Cell Biol.* **99**:2123.
91. Wagner, R. C., and Andrews, S. B., 1985, Ultrastructure of the vesicular system in rapidly frozen capillary endothelium of the rete mirabile, *J. Ultrastruct. Res.* **90**:172.
92. Wagner, R. C., and Matthews, M. A., 1975, The isolation and culture of capillary endothelium from epididymal fat, *Microvasc. Res.* **10**:286.
93. Wagner, R. C., Robinson, C. S., Cross, P. J., and Devenny, J. J., 1983, Endocytosis and exocytosis of transferrin by isolated capillary endothelium, *Microvasc. Res.* **25**:387.
94. Weibel, E. R., and Palade, G. E., 1964, New cytoplasmic components in arterial endothelia, *J. Cell Biol.* **23**:101.
95. Williams, S. K., Devenny, J. J., and Bitensky, M. W., 1981, Micropinocytic ingestion of glycosylated albumin in isolated capillary endothelium: Possible role in the pathogenesis of diabetic microangiopathy, *Proc. Natl. Acad. Sci. USA* **78**:2293.
96. Williams, S. K., Greener, D. A., and Solenski, N. J., 1984, Endocytosis and exocytosis of protein in capillary endothelium, *J. Cell. Physiol.* **120**:157.
97. Wisse, E., 1972, An ultrastructural characterization of the endothelial cell of the rat liver sinusoid

under normal and various experimental conditions, as a contribution to the distinction between endothelial and Kupffer cells, *J. Ultrastruct. Res.* **38**:528.

98. Wissig, S. L., 1979, Identification of the small pore in muscle capillaries, *Acta Physiol. Scand. Suppl.* **463**:33.

99. Zweifach, B. W., 1972, Capillary filtration and mechanisms of edema formation, *Pfluegers Arch.* **336**:S81.

Ultrastructural Studies of Capillary Endothelium

Compartmental Tracing, High-Voltage Electron Microscopy, and Cryofixation

Roger C. Wagner

I. INTRODUCTION

The vast majority of the body's tissues are irrigated by continuous capillaries which exhibit neither fenestrae nor interendothelial discontinuities. The endothelium comprising their walls regulates exchange of metabolites and gases in a manner which is critical for the maintenance of tissue homeostasis. The permeability properties of capillaries must result from both the characteristics of the endothelium as a tissue barrier and the dynamic activities of the individual endothelial cells. And yet there is no comprehensive theory which adequately reconciles the permeability behavior of continuous capillaries with ultrastructural correlates in the capillary wall.

It is apparent that materials traverse the endothelium by both convective and dissipative mechanisms. The transport of solutes in response to Starling forces implies the presence of open channels or pores in the capillary wall that restrict movement of solutes in proportion to their molecular size. The prime candidate for such pores are the interendothelial clefts. These vary considerably, however, in their degree of openness in capillaries from different tissues,[48] and along the length of individual capillaries.[6] Their conductivity to peroxidative tracers has been reported to be either nil[35] or limited to the smallest of these tracers.[23,46,47] Larger tracers to which capillaries are permeable such as ferritin[9] cannot use these clefts as routes across the capillary wall. Transendothelial channels of fused vesicles would constitute intraendothelial thoroughfares capable of conducting all but the largest blood solutes.[34] Their incidence of detection is very low, however, implying that they are very rare or that their temporal existence is fleeting.

Roger C. Wagner • School of Life and Health Sciences, University of Delaware, Newark, Delaware 19716.

At normal and low blood pressures, transcapillary transport of solutes is largely dissipative and a significant proportion of this occurs through nonhydraulically conductive pathways.[29] Such pathways are likely to be discontinuous (nonpatent) since transport through them occurs independently of transendothelial pressure gradients in the capillary wall. It is possible that endothelial vesicles in a constant state of flux could convey solutes across the capillary wall in a nonhydraulically conductive fashion by a process termed transcytosis.[36] The evidence that such a mechanism functions in living capillaries is compelling. Electron-dense solutes too large to penetrate between endothelial cells progressively label vesicles across the endothelium with increasing times before fixation and deposit them in the interstitium. The exact mechanism and kinetics of transcytosis, however, remains obscure and its quantitative contribution to overall capillary permeability is not known.

Exceptions to strict permeability–molecular size relationships have also prompted a reexamination of pore size as the sole discriminating mechanism regulating capillary transport. A fiber matrix[10] of unspecified composition but exhibiting exclusion phenomena and specific affinities for solutes has been theorized to exist within pathways to account for differential transport anomalies. Adsorptive[41] and receptor-mediated[17] transcytosis have also been proposed as mechanisms of selective endothelial transport. Molecular determinants such as glucosylation[45] and cationization[11] have been shown to impart a selective transport preference to solutes so endowed. Selective mechanisms involving molecular interactions with endothelial vesicle membranes or extracellular fiber systems are not complete models for a continuously functioning capillary transport system, however. Such interactions must be reversible on the opposite side of the capillary wall if the transport system(s) are to be recharged and in a condition to accept new solute.

Any successful theory of capillary permeability must embody both physiological parameters characterizing the behavior of living capillaries and the underlying cellular-structural mechanisms responsible for that behavior. If this is to be accomplished, ultrastructural studies must provide a means to reveal endothelial structure in three dimensions and also provide for a more accurate representation of the functional state of capillaries in living tissues.

II. THREE-DIMENSIONAL ANALYSIS OF THE CAPILLARY WALL

Electron microscopy is a bountiful source of information in the study of cellular structure–function relationships. Conventional electron microscopy of sectioned material, however, is limited to providing two-dimensional images from which three-dimensional organization must be extrapolated. This has especially hindered the search for ultrastructural correlates of permeability in the walls of capillaries: Interendothelial junctions which appear occlusive in a particular plane of section may open to provide a patent pathway in another plane. Vesicles which appear to be free and discrete cytoplasmic entities may connect with other vesicles or the endothelial plasma membrane in another plane of section. In order to over-come these limitations, innovative techniques which provide three-dimensional information about the capillary have recently been developed. The data derived using these techniques have revealed organization which is not apparent in single sections but which is not necessarily consistent among the various approaches employed. Functional interpretations arising from these studies also vary sub-stantially and are inherently limited by the lack of dynamic data in static electron images.

A. Reconstruction of Ultrathin Serial Sections

Reconstruction of a three-dimensional structure from consecutive sections through the capillary wall can reveal endothelial organization in depth. If sections of conventional thickness (50–100 nm) are taken, however, random cuts may omit certain connections between endothelial vesicles (50–70 nm diameter) which would then be perceived as having no association between them. It is therefore necessary to obtain ultrathin (~ 15 nm) sections in order to reconstruct an image which accurately represents intervesicle associations. The same requirements are necessary to ascertain the coherency of interendothelial junctions along the length of a capillary.

This technique has been pioneered by M. Bundgaard and J. Frokjaer-Jensen of the Panum Institute in Copenhagen. They have obtained series of ultrathin sections from capillary walls in various tissue types and have shown that a sub-stantial error is inherent in assuming that vesicles which appear unconnected in a single section are actually free entities.[4,5,12–14] In fact, their data indicate that essentially all but a vanishing number of endothelial vesicles are conjoined to each other and ultimately with the plasma membrane but never communicate across the capillary wall to form a continuous channel. They have termed this organi-zation the "plasmalemmal vesicular system" which describes sessile, racemose clusters of fused membranous compartments which open onto either blood or interstitial compartments but not both simultaneously. This interpretation would imply that vesicles cannot provide hydraulically conductive channels nor can they be involved in discontinuous transport involving fission and fusion events between them. Similar ultrathin serial section reconstruction of junctional areas[6] has re-vealed that occlusive regions between adjacent endothelial cells are actually intermittent[33] along the length of the capillary providing for open but tortuous channels between the blood and interstitium. The sum of these observations sug-gests that the only portal across the wall of continuous capillaries is interendoth-elial and diminishes active participation of the endothelial cell and its vesicular system in transcapillary transport.

B. Compartmental Labeling with Tannic Acid

Three-dimensional information can also be derived from two-dimensional electron images if contiguous membrane compartments are labeled prior to preparation for electron microscopy. Compartments open to the cell exterior can be labeled by agents which adhere to the membrane surface and are detectable in sectioned material. Ruthenium red,[24] horseradish peroxidase,[18] lanthanum,[30] and tannic acid[16] have been used effectively in this manner to delineate irregular cell surfaces and compartments open to them.

Tannic acid is a mixture of glucosides with several orthophenol radicals which are negatively charged at neutral pH.[16] It combines with basic proteins at the cell surface and subsequently chelates heavy metal ion stains in sections, resulting in increased electron density at sites of tannic acid deposition. Tannic acid thus functions as a stain mordant and is not intrinsically electron dense. This mordanting effect is manifested in an enhanced electron density of the outer leaflet of the membrane bilayer (see Fig. 5a,c). The molecular components of tannic acid have a mean molecular weight of about 1700 and thus its penetrability will mimic that of the smaller blood solutes but preclude its diffusion through an unbroken membrane. In fact, exclusion of tannic acid from the cytoplasmic compartment can verify the intactness of the entire plasma membrane of a fixed cell. If the plasma membrane has been breached osmotically or during excision of tissue, tannic acid penetrates throughout the cytoplasmic compartment and heavily mordants all intracellular membrane systems and cytoplasmic proteins.

Tannic acid is applied either to prefixed tissue or concurrently with glutaraldehyde. In capillary endothelium *in situ*[33,40] or capillaries freshly isolated from tissue,[38] this results in mordanting of the exposed endothelial cell surfaces and spaces open to the extracellular compartment. Many compound and simple vesicular structures which would otherwise be assumed free in the cytoplasm are labeled with tannic acid, indicating their continuity with the exterior out of the plane of section. However, a substantial number of vesicles also appear to exclude tannic acid (Figs. 2 and 3). Unlabeled vesicles are present even if tannic acid is applied to both luminal and abluminal surfaces of the capillaries are sufficiently thin (a bilayer is evident throughout the circumference of a vesicular profile), an all-or-none effect is evident with regard to vesicle labeling (Fig. 3). Thus, omission of tannic acid from unlabeled vesicles is total and differential labeling is unlikely to result from a progressive restriction of tannic acid diffusion throughout the vesicular system.

It is difficult to explain these observations if all vesicles communicate directly with the surface of the endothelial cells. It is possible that tannic acid is halted by some barrier other than membrane which is not revealed by thin serial section reconstructions. However, if such a barrier exists, it would likely restrict movement of blood solutes as well which in effect would render the vesicular system a discontinuous entity. It has been shown that the larger tracer ferritin is progressively diluted in vesicles with increasing distance from the luminal surface of the capillary.[19] However, this phenomenon occurs in living capillaries in which any dynamic activity of the vesicular system would still be operative. In any event, it is difficult to reconcile the apparent interconnectedness of the plasmalemmal vesicular system as revealed by serial section reconstruction with the behavior of either intravital tracers or compartmental labels within it.

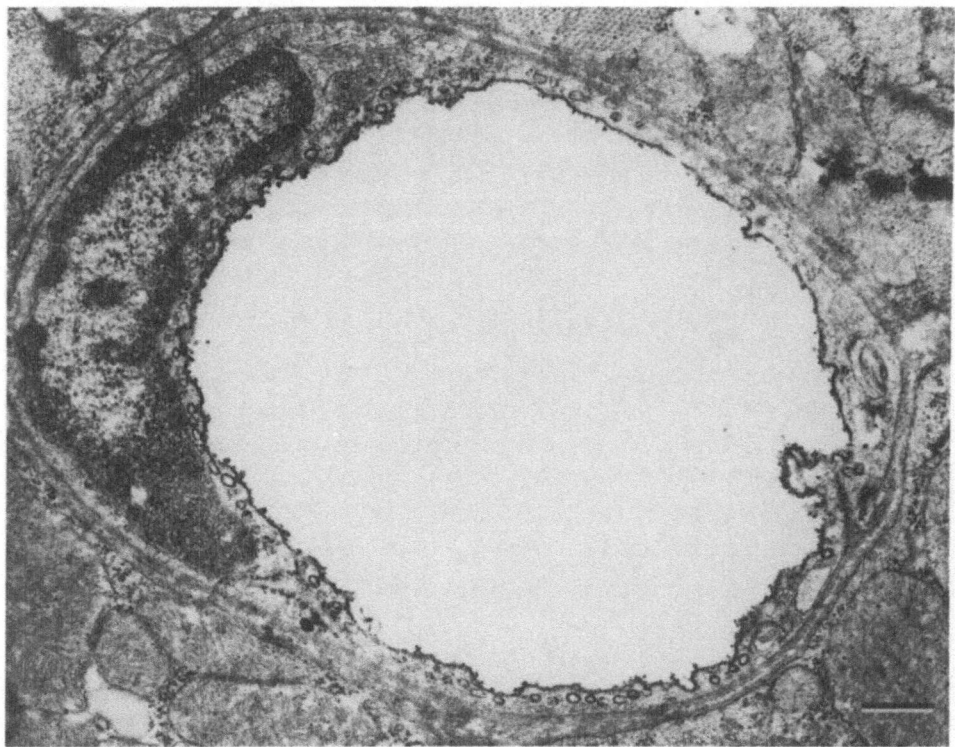

Figure 1. Cross section through a mouse diaphragm muscle capillary which has been perfused-fixed and then perfused with tannic acid. Tannic acid fails to penetrate the capillary wall and is not detectable on the abluminal membrane or in the interstitium. Bar = 0.89 μm.

C. Occlusion of Tannic Acid by Interendothelial Junctions

If tannic acid is perfused through previously fixed capillaries of the diaphragm muscle (Fig. 1) or the arterial segments of the rete mirabile (Fig. 2), it fails to penetrate the capillary wall and no mordanting effect is detectable on the abluminal membrane or in the interstitium. If hydraulically conductive pores are faithfully preserved by fixatives, they would be expected to conduct a solute as small as tannic acid across the capillary wall. Penetration of tannic acid even some distance away from the eventual plane of section should result in mordanting of the abluminal membrane, the basement membrane, and interstitial connective tissue.

The arterial capillary segments of the rete mirabile exhibit interendothelial junctions characteristic of zonula occludens. The outer leaflets of the membrane bilayer converge to a single electron-dense line (Fig. 4). If these occlusive seals are intermittent along the length of the capillary, the channels they provide through the capillary wall should conduct tracers as small as tannic acid. It is possible that binding of tannic acid to membranes in severely constricted regions such as those present between adjacent endothelial cells is cumulative and clogs normally

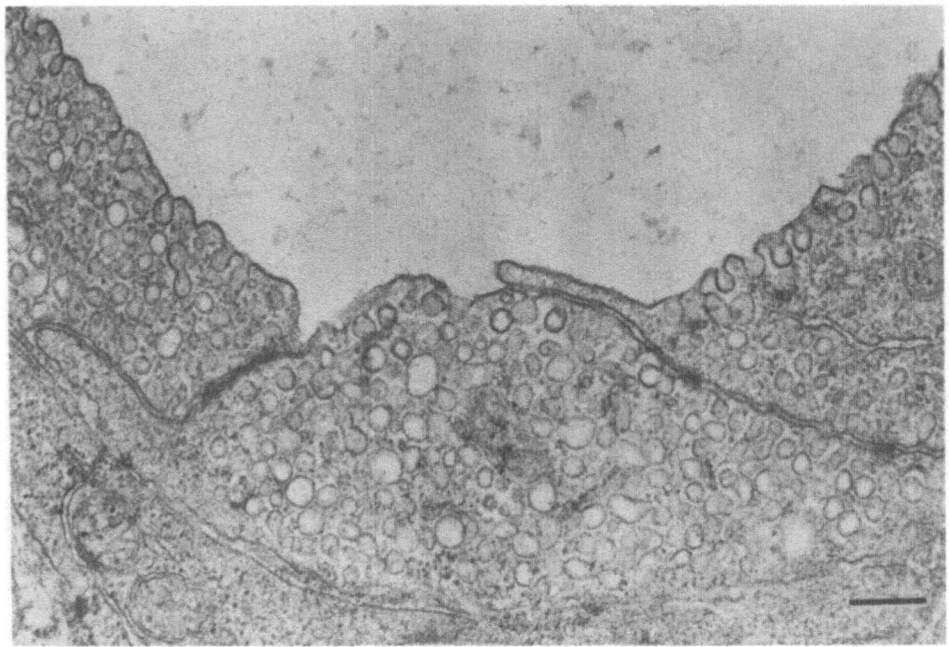

Figure 2. Luminal surface of an arterial capillary segment of the rete mirabile of the eel which has been perfused-fixed and then perfused with tannic acid. Luminal caveolae and profiles of vesicles having access to tannic acid outside of the plane of section exhibit mordanting but not abluminal membranes. Numerous vesicular profiles are devoid of tannic acid and it appears to penetrate only part way down the interendothelial clefts. Bar = 0.30 μm.

open pathways. However, interruption of tannic acid mordanting occurs precisely at the point of contact between the membranes (Fig. 5). It is unlikely that this is coincidental. Rather, it is a strong indication that the tight junctions are the agents of occlusion.

If tannic acid is applied to both luminal and abluminal compartments, it labels both plasma membranes, vesicles attached to them (caveolae), and vesicles to which it had access outside of the plane of section (Fig. 3). It also heavily labels interstitial connective tissue, which attests to its mobility within the extracapillary regions and high affinity for connective tissue components. Many cytoplasmic vesicles exclude tannic acid and it penetrates interendothelial clefts from both sides but is abruptly excluded from the region of tightly juxtaposed endothelial membranes (Fig. 3).

Serial section reconstruction of junctional regions between capillary endothelial cells indicates that the zonula occludens are organized as irregular networks of lines of contact between adjacent membranes.[6] However, tortuous pathways circumventing these lines of contact apparently provide for an open channel connecting blood and interstitial compartments. Indeed, intravital peroxidative tracers have been reported to be able to penetrate these channels[22,47] and this

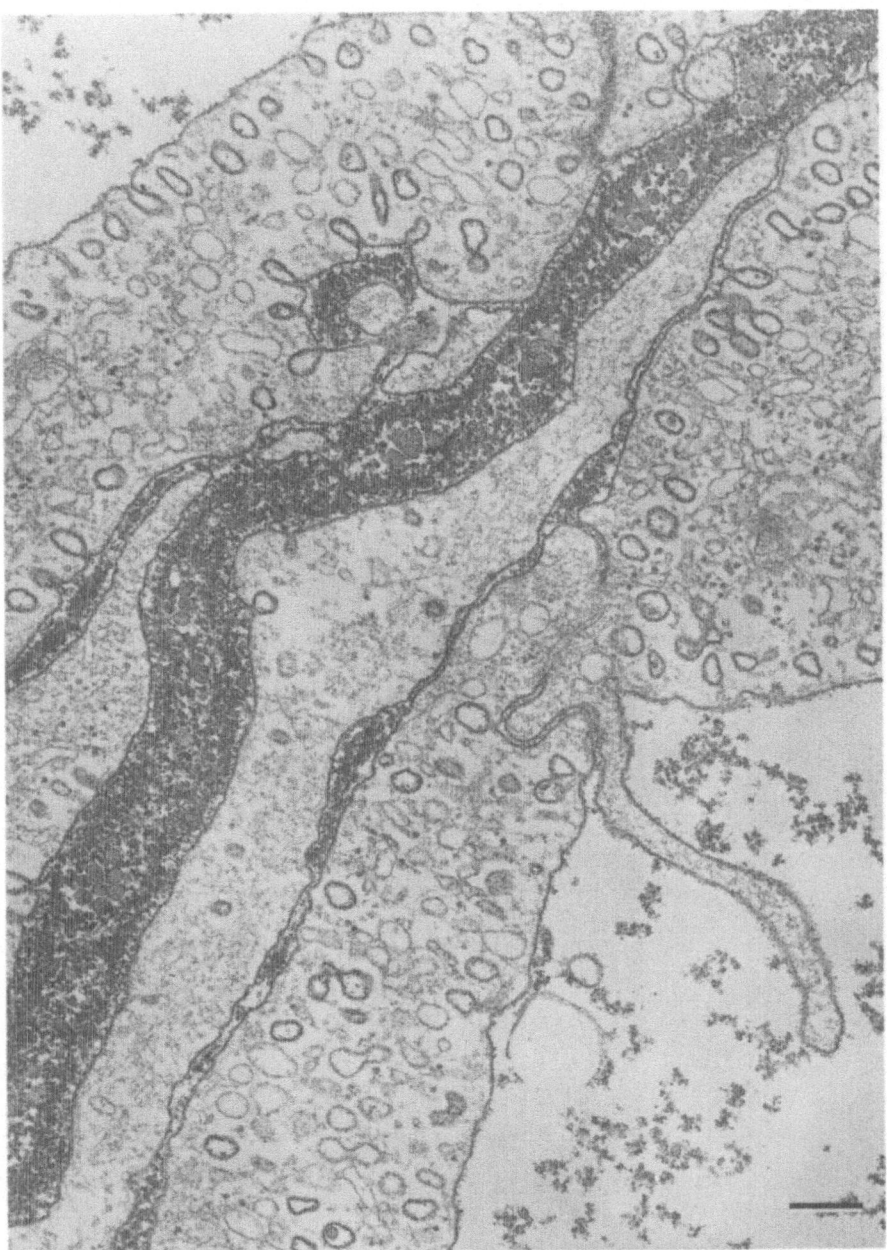

Figure 3. Section through two adjacent arterial capillary segments of the rete which have been perfused with tannic acid and in which tannic acid has also been allowed to penetrate the interstitial spaces. Vesicular profiles exhibit an all-or-none effect with regard to tannic acid labeling. Tannic acid penetrates interendothelial clefts from both sides but is abruptly excluded from a region which is occlusive in three dimensions. Bar = 0.22 μm.

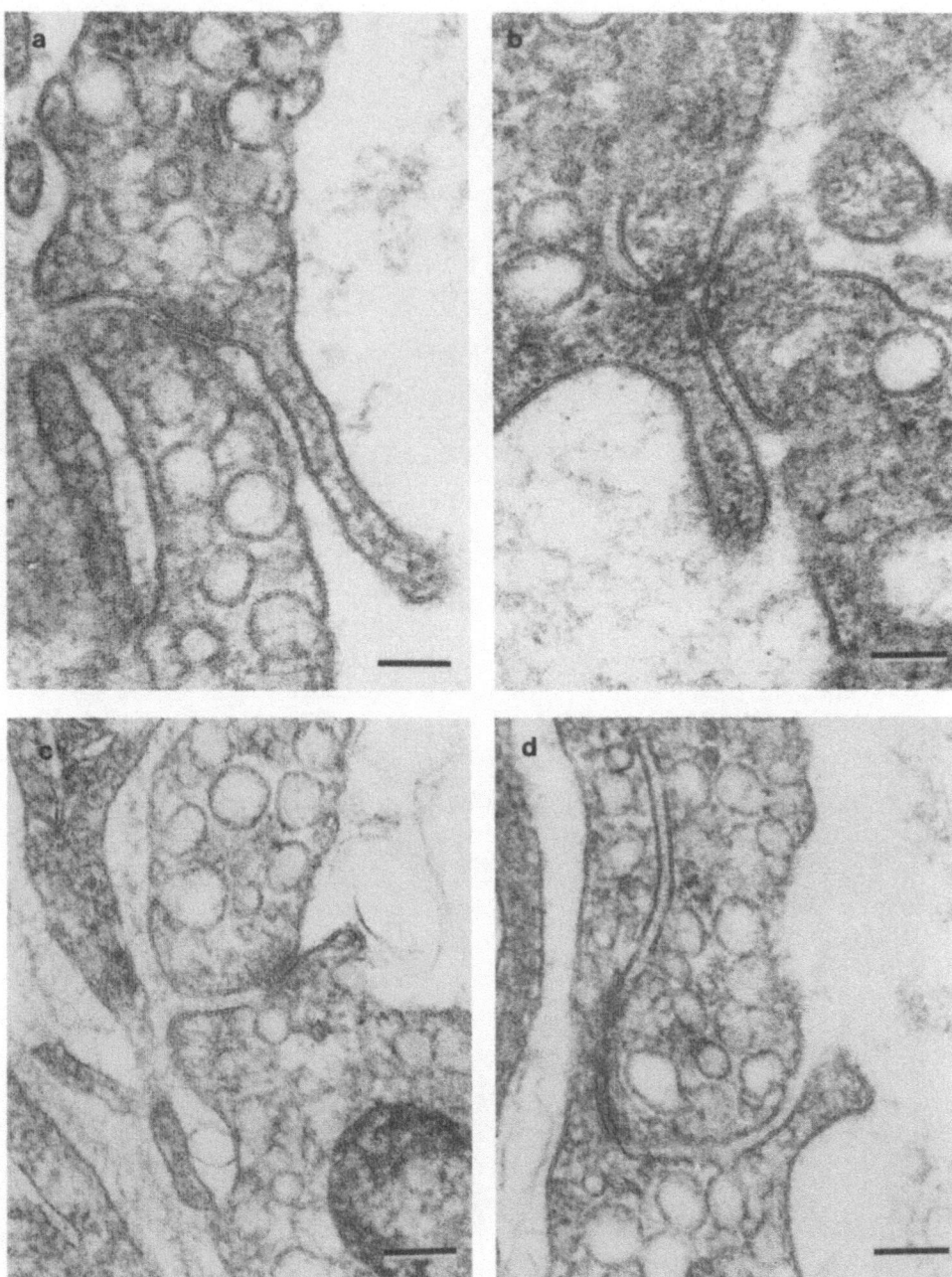

Figure 4. Junctional regions of arterial capillary segments of the rete mirabile of the eel. Occlusive seals are formed in regions where the outer leaflets of the adjacent membrane bilayers converge to a single electron-dense line. Bars = (a) 0.12, (b, c) 0.13, (d) 0.15 μm.

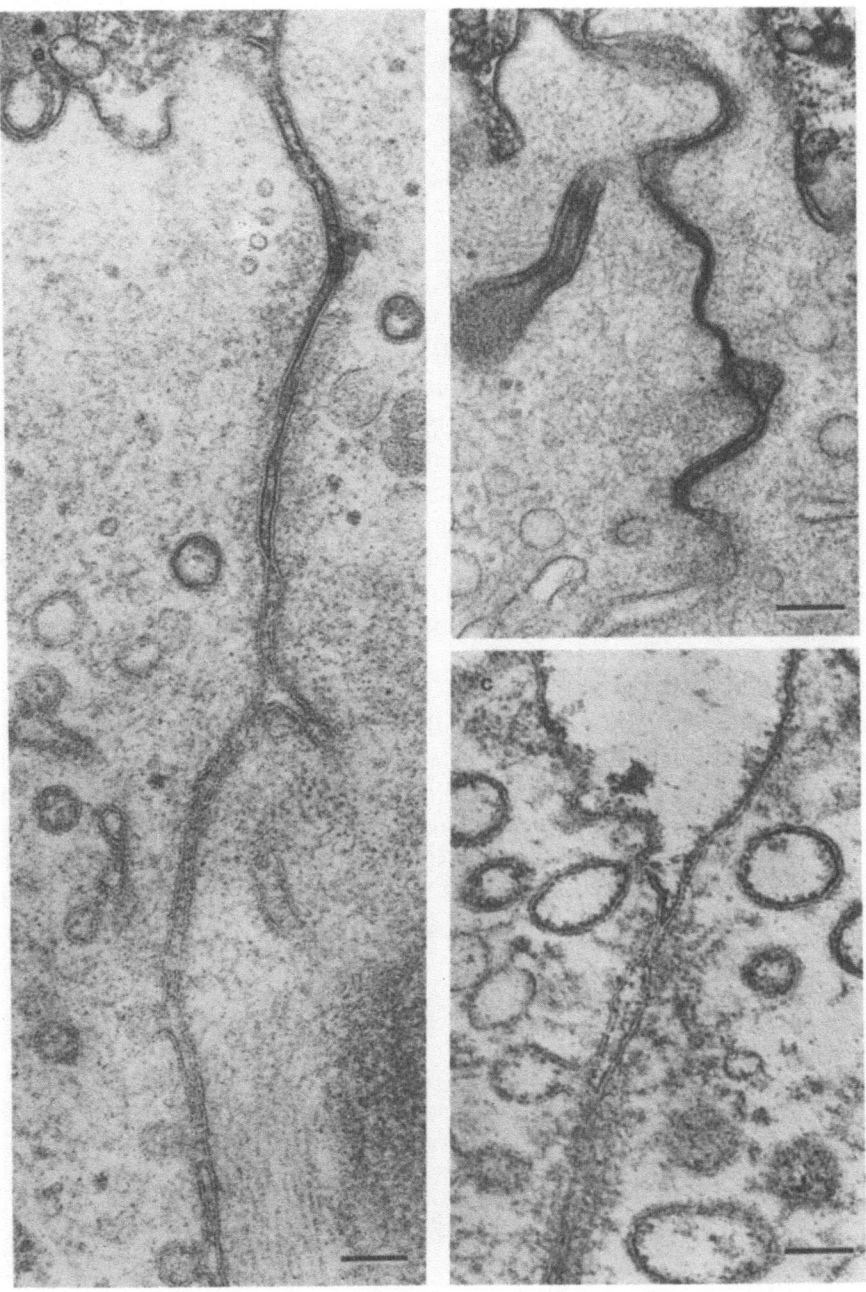

Figure 5. When tannic acid is perfused through fixed rete capillary segments, it penetrates into severely constricted interendothelial regions but is interrupted precisely at the point of contact between adjacent membranes. It must be assumed that this reflects a three-dimensional phenomenon of occlusion representing a considerable length of the capillary wall. Bars = (a) 0.09, (b) 0.12, (c) 0.06 μm.

occurs prior to tissue fixation. Since tannic acid must be applied concurrent with or subsequent to aldehyde fixation of capillaries, fixation may act to close previously open channels. However, the open pathways revealed by serial section reconstruction in fixed capillary walls would seem to mitigate against this possibility.

D. High-Voltage Stereo Electron Microscopy of Thick Sections

An alternative method of obtaining a three-dimensional image of the capillary wall is by stereoscopic viewing of thick sections. Proper transillumination of sections of appropriate thickness (500–1000 nm—green interference color) requires high electron accelerating voltages. These are obtainable with either megavolt electron microscopes (HVEMs) or more conventional microscopes capable of more than 100 kV. Stereopair micrographs taken at 30,000–50,000 × and tilt angle differentials of 4–8° with a goniometer stage (Figs. 7 and 8) can be viewed with appropriate stereoimaging or stereoprojection devices. Three-dimensional perception of stereopairs involves the neurological phenomenon of cortical fusion and thus cannot be conveyed in two-dimensional space. However, this approach obviates some of the rigors and sampling limitations of ultrathin serial section reconstruction by permitting visualization of vesicle associations in a single section several vesicle diameters thick.

The usefulness of stereomicroscopy in three-dimensional analysis of the vesicular system depends upon obtaining the proper relationship between section thickness and the relative volume density of vesicles within the section. Sections less than one vesicle thick (Fig. 6a) allow visualization of vesicle associations only within a limited plane. If the section is too thick or the vesicle density too high (Fig. 6c), overlapping of vesicular structures and membranes obscures their interrelationships and many connections between vesicular compartments remain

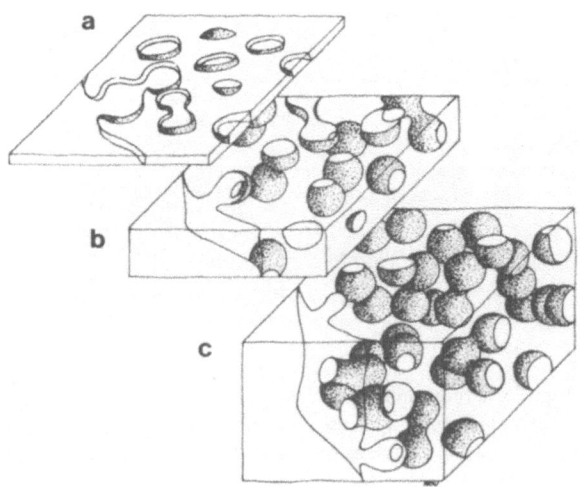

Figure 6. The effectiveness of perceiving vesicle interrelationships by stereomicroscopy depends on relative thickness and volume density of vesicles within the section. Sections less than one vesicle thick (a) allow visualization of vesicle associations only within a limited plane. Excessive vesicle density or section thickness (c) results in occulting of nearly all associations. Several, but not all, vesicle associations can be seen, however, if the relationship between section thickness and vesicle density is ideal (b).

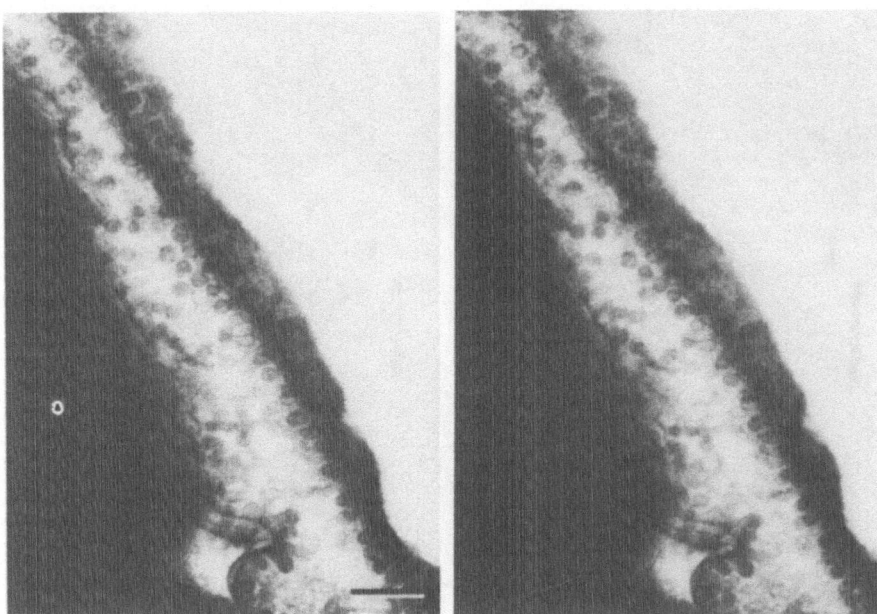

Figure 7. High-voltage stereopair of a thick section through the wall of a rat diaphragm capillary in which tannic acid has been applied to both luminal and abluminal compartments. The cytoplasmic compartment as well as several vesicular structures appear to exclude tannic acid. Some single vesicles are oriented so as to reveal no apparent connections with each other or the surface membrane. Bar = 0.32 μm.

hidden. For similar reasons, stereomicroscopy is ineffectual in determining the tightness of junctions since surfaces in close apposition meander through a thick section and overlapping regions are far too electron dense to show any detail. Under ideal conditions of section thickness and vesicle volume density (Fig. 6b), however, many vesicular forms and their associations can be visualized. But even under these conditions, it is not possible to classify all vesicular structures as simple, compound, free, or attached, which precludes any meaningful morphometric analysis.

High-voltage stereo electron microscopy has been performed on capillaries of the rat diaphragm muscle[42] which were perfused with Karnovsky's fixative and tannic acid and in which tannic acid was allowed to penetrate the interstitial spaces. Tannic acid is particularly prominent in association with erythrocytes and residual protein in the capillary lumen, interstitial collagen, and basement membrane (Figs. 7 and 8). The basement membrane appears as a dense feltlike mat which is somewhat translucent in that underlying attached vesicles can be seen (Fig. 7). The cytoplasmic compartment as well as several vesicular structures appear to exclude tannic acid. The contrast differential between tannic acid-labeled and unlabeled compartments is not as sharp as in thin sections. This is probably due to the fact that an enhanced amount of intrinsic contrast is imparted

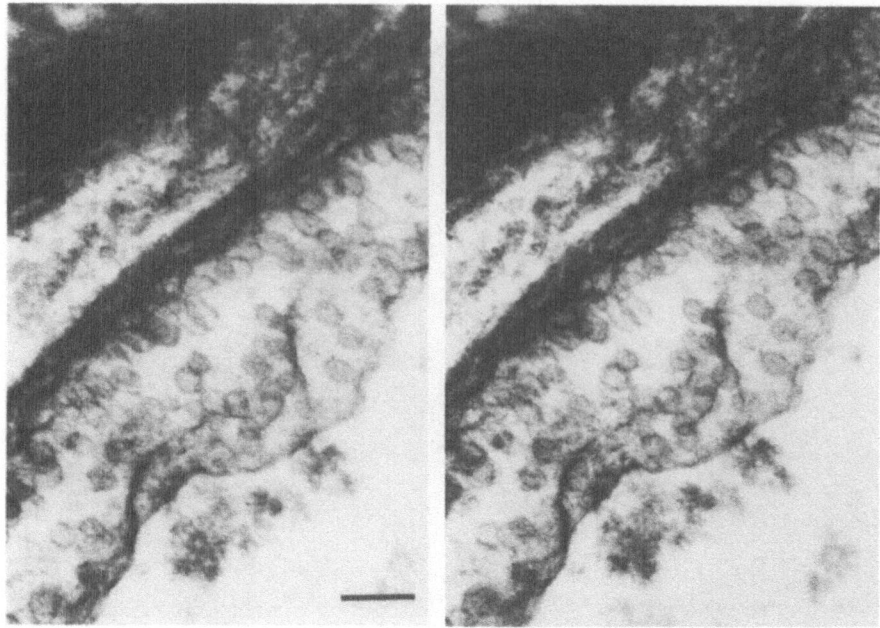

Figure 8. A higher-magnification stereopair of a diaphragm muscle capillary exhibiting a variety of simple and compound vesicular structures. These range from squat spherical caveolae to oblate forms with elongated necks. Teardrop vesicles appear to be emanating from attached caveolae. Bar = 0.20 μm.

by overlapping membranous structures and that membranous compartments are viewed throughout their depth.

A variety of simple and compound vesicular structures are evident and a few are so oriented as to reveal no apparent connections with other vesicles or the surface membrane (Fig. 7). That these vesicles lie totally within the section can be verified by reference to structures which lie above and below the vesicles (Fig. 7). The majority of vesicles attached to both membrane surfaces are simple caveolae and are much more numerous on the albuminal membrane. These vary in form, however, from squat spherical caveolae to oblate forms with elongated necks. A number of racemose clusters of apparently fused vesicles which communicate with the surface are also evident. In some cases single vesicles appear attached to the distal end of a caveolus by a teardrop configuration and in others two spherical vesicles are closely juxtaposed without any apparent continuity between them. These forms are suggestive of dynamic interactions in progress at the time of fixation. Transendothelial channels of fused vesicles were not detectable in the samples observed although their patency would be difficult to verify in these stereoimages.

The total volume of a capillary wall sampled in a thick section is comparable to that reconstructed from a series of ultrathin sections. Serial section reconstruction is quantitatively superior to stereoviewing of thick sections in that all vesicle associations within a sample are eventually revealed. However, although they are not of common occurrence, unambiguous detection of free vesicular entities can be readily made in thick sections.

It has been assumed that if vesicular compartments are in a state of flux in living tissue, some totally discrete compartments should be present in fixed tissue. It is apparent that such free entities are extremely rare in reconstructed images but more readily observed in thick sections. It would be overly presumptuous to assert that the proportion of attached versus free vesicles in fixed tissues implies anything definite about their dynamic activity or the lack thereof in living tissues. Fusion events and accompanying exchange of vesicular contents may be extremely rapid and low detectability of free vesicular entities may be as much a consequence of their fleeting existence in time as their scarcity in space.

III. FIXATION OF CAPILLARY ENDOTHELIUM FOR ELECTRON MICROSCOPY

The morphological features of the capillary wall just described rest upon the assumption that aldehyde-fixed endothelium has a sufficiently close resemblance to living tissue so that structure–function relationships can be made. Likewise the use of intravital tracers in determining the routes of transport across the capillary wall and kinetics of vesicular activity are based on the assumption of rapid immobilization of vesicular events by chemical fixation. Thus, current concepts of the function of endothelial cell vesicles[39] and interendothelial pathways in capillary transport have been largely derived from ultrastructural analysis of aldehyde-fixed tissues. The rapidity and fidelity with which chemical fixation preserves the structure of capillaries, however, has not been adequately investigated.

A. Chemical Fixation with Aldehydes

The fixation properties of glutaraldehyde stem from its bifunctional interaction with proteins by forming inter- and intramolecular links between amino acids yielding rigid heteropolymers. Presumably this immobilizes proteins and protein-containing structures such as membranes and prevents their displacement, denaturation, and extraction during subsequent treatment for electron microscopy. The speed and effectiveness with which this occurs will be a function of the time required for the fixative to penetrate all regions of a tissue and the plasma membrane of cells. Monofunctional formaldehyde is smaller than glutaraldehyde and penetrates membranes faster, but its interaction with proteins is less permanent and can be reversed by subsequent washing. Fixation times can be shortened by perfusing vessels with fixatives and using a combination of glutaraldehyde and formaldehyde. It has been shown[20] that glutaraldehyde is likely to penetrate individual cells in 0.01 sec. However, Brownian motion and the more directed saltatory motion of cellular organelles persist for 30–60 sec after application of glutaraldehyde to cells in tissue culture.[3] A significant amount of enzyme activity is also retained in glutaraldehyde-fixed cells.[32] Immobilization of cellular organelles by chemical fixatives will thus probably vary considerably depending on their nature and proximity regardless of uniform penetration of fixative.

There are at least two ways in which chemical fixation may lead to erroneous interpretation of capillary endothelial cell form and function: (1) by causing events and interactions between membranes which might not otherwise occur and (2) by permitting normal movements and interactions between membranes long after they have been presumed stopped. It has been shown[21] that membrane lipids which interact poorly with glutaraldehyde may remain mobile for periods sufficient to allow fission or fusion of adjacent membranes long after other cell components have been stabilized. Clearly, a more rapid and noninvasive means of immobilizing ultrastructural components of endothelial cells is needed in order to confirm or dispal uncertainties associated with chemical fixation.

B. Rapid Freezing and Freeze Substitution

Cryofixation is an alternative means of preserving endothelial cell structure and can serve as a suitable control for any spurious effects of glutaraldehyde fixation. Rapid freezing with liquid helium-cooled devices arrests all molecular motion in less than 1 msec,[28] which means that assumptions concerning time and rapidity of fixation can be very precise. Rapid freezing can also be performed without the aid of cryoprotectants, which eliminates chemical intervention as a variable.

Several recent studies have employed a variety of tissues, methods of freezing, and ultrastructural analyses to assess differences between endothelial vesicular systems in aldehyde and fresh-frozen endothelium (Table 1). A definite trend is evident: The numerical density of vesicular structures in endothelial cells is significantly increased in glutaraldehyde-fixed tissues as compared to fresh-frozen samples.[4,7,25,26,31] It is apparent that endothelial membrane systems are considerably more labile to the effects of glutaraldehyde than was previously assumed. Perturbation of membrane systems and their interactions by chemical fixatives in other cell types has also been demonstrated by utilizing rapid freezing as a control.[1,8,27]

One of the most rapid and effective means of achieving cryofixation in the absence of cryophylaxis is with a metal-mirror, rapid-freezing device.[19,28,37] In this technique, freshly excised tissue affixed to an inverted stage is dropped onto a polished copper block which has been cooled by liquid helium (< 20 K). Tissue thus frozen can either be fractured and replicated or freeze-substituted and embedded in plastic for thin section analysis. Procedures for freeze-substitution[2] accomplish the replacement of water ice in the tissue with organic solvents (acetone) which are liquid at low temperatures and provide for eventual embedment in plastic resin. Included in the acetone are the fixatives acrolein, osmium tetroxide, and glutaraldehyde which are applied in separate incubations as the tissue is passively warmed to room temperature over several days. These fixatives stabilize structure originally immobilized by freezing so that changes do not occur as the tissue is slowly warmed. In a typical experiment, separate tissue samples are

Table 1. Chemical versus Cryofixation of Endothelial Vesicles[a]

Tissue source, method of fixation, and reference	Vesicles/μm^3 cytoplasm	Caveolae[b]/μm^2 luminal surface	Caveolae[b]/μm^2 abluminal surface
Lung capillaries[25]			
Perfused-fixed with aldehyde	420	60(70)[s,a]	55(80)[s,a]
Frozen by immersion in ethylene glycol	250	55(75)[s,a]	55(65)[s,a]
Mouse diaphragm capillaries[7]			
Immersed-fixed with aldehyde, frozen in acetone–nitrogen slush (-80°C)	1760	136[s,b]	
Unfixed and frozen as above	999	54[s,b]	
Aortic endothelium[26]			
Perfused-fixed with aldehyde, frozen by immersion in liquid nitrogen (-197°C)		68[f,b]	
Perfused-fixed with aldehyde, rapidly frozen against liquid nitrogen-cooled block (-197°C)		62[f,b]	
Unfixed, rapidly frozen as above		20[f,b]	
Rete mirabile capillaries[43]			
Immersed-fixed with aldehyde, rapidly frozen against liquid helium-cooled block (>20 K)	639	60[s],57[f]	94[s],92[f]
Unfixed and frozen as above	217	20[s],35[f]	28[s],44[f]
Cultured human venous endothelium[49]			
Immersed-fixed with aldehyde, rapidly frozen against liquid helium-cooled block (>20 K)	36	8[s]	
Unfixed and frozen as above	7	8[s]	
Rat cremaster muscle[31]			
Immersed-fixed with aldehyde, rapidly frozen against liquid helium-cooled block (>20 K)	455		
Unfixed and frozen as above	158		

[a] a, density of caveolae on both thick and thin side of capillary wall; b, specification as to luminal or abluminal surface not available; s, data from sectioned material; f, data from freeze-fractured material.

[b] Vesicle profiles which are fused with the surface membrane or cross fractured through regions of continuity between vesicles and the surface membrane (pits).

treated with buffered glutaraldehyde, or buffer alone and both are then frozen and freeze-substituted identically. The only variable is the presence or absence of glutaraldehyde in the original incubation. Dehydrated and chemically stabilized samples from freeze-substitution are then embedded in plastic by conventional means.

Without cryoprotectants, only a zone of about 15–20 μm from the freezing surface is sufficiently free of ice crystal damage so that suitable ultrastructural analysis can be performed. This is an inherent limitation of this approach which relates to the rapidity of heat removal and the speed of ice crystal growth. Nevertheless, excellent preservation is attainable in very thin tissue samples, cell monolayers, or tissues in which prior trimming positions blood vessel endothelium close to the freezing surface and within the zone of good tissue preservation. Since this zone is so limited, a high density of blood vessels within the tissue minimizes extensive searching of the sample for well-preserved capillary profiles.

C. The Endothelial Vesicular System: Chemically versus Cryofixed Capillaries

The rete mirabile (red bodies) of the eel swim bladder are ideal for this purpose (Fig. 9) since they consist of venous and arterial capillaries in a close-packed array. A high density of capillary endothelium can be positioned adjacent to the freezing surface and within the zone of maximum freezing quality by slicing the rete parallel to the long axis of the vessels prior to freezing. This close-packed arrangement of capillaries also permits inspection of numerous capillary endothelial cells within a single section (Fig. 10) which enhances confidence that structural parameters observed are widespread throughout the tissue. In retia that are rapidly frozen in this configuration, a gradient of preservation quality is evident.

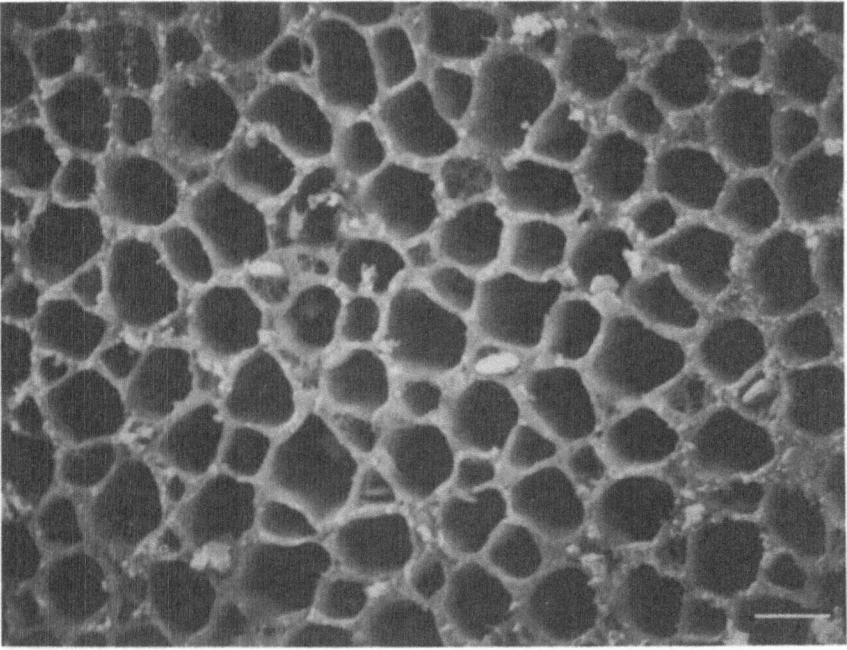

Figure 9. Scanning electron micrograph of a critical point dried rete mirabile which has been cut to reveal capillaries in cross section. Both large venous and smaller arterial capillary segments are present in a close-packed array which is ideal for positioning a high density of capillary endothelium within a limited zone of high-quality freezing. Bar = 15.6 μm.

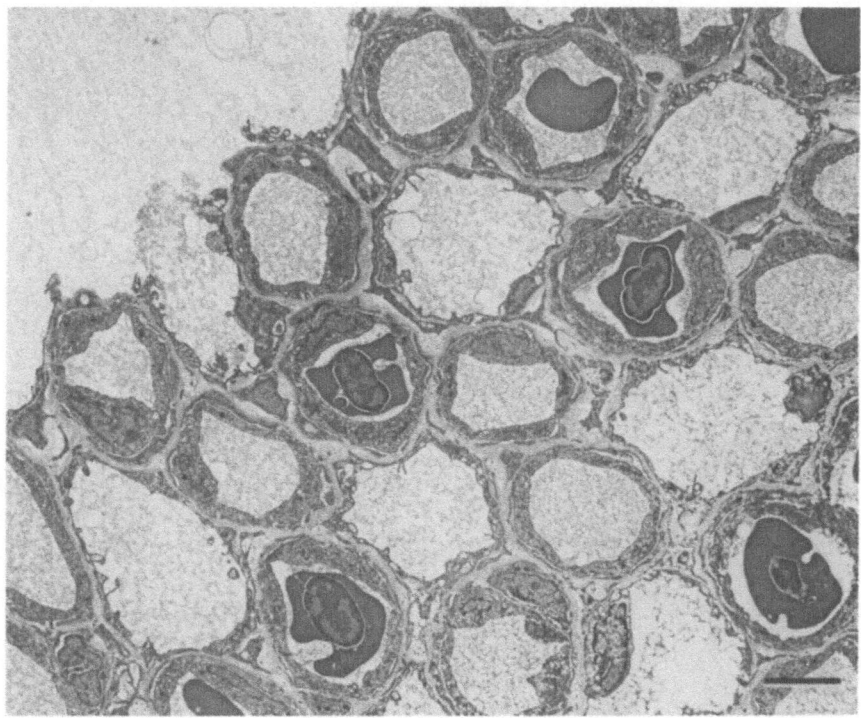

Figure 10. Low-magnification survey view of a rete mirabile which was sliced parallel to the long axis of the capillaries prior to rapid freezing. The sliced surface (upper left) constitutes the area of first contact with the freezing block. Good tissue preservation is evident two to three capillary diameters from the freezing surface but a gradual increase in ice crystal damage is evident deeper into the tissue (lower right). Bar = 2.90 μm.

Minimal freezing effects are observable at least two to three capillary diameters (15–20 μm) from the freezing surface but gradual increases in ice crystal damage are evident deeper into the tissue (Fig. 10).

Analysis of arterial capillary segments frozen in this manner[44] indicates that glutaraldehyde-fixed capillaries have a significantly larger population of vesicular profiles (Fig. 11) than those which had been fresh frozen (Fig. 12). Morphometric analysis substantiates this impression since prefixed capillaries have nearly three times the numerical density of endothelial vesicles as fresh-frozen capillaries (Table 2). A significant increase in the number of attached vesicles (caveolae) is also evident in glutaraldehyde-fixed capillaries and the magnitude of change parallels that of the total vesicle population (Table 2). These observations are consistent with several other studies using different endothelial cell systems and a variety of other chemical and cryofixation techniques (Table 1). It is clear that these large differences are real and due only to exposure to aldehyde fixatives and not to tissue source or methods of freezing.

The time constraints involved would seem to preclude membrane synthesis

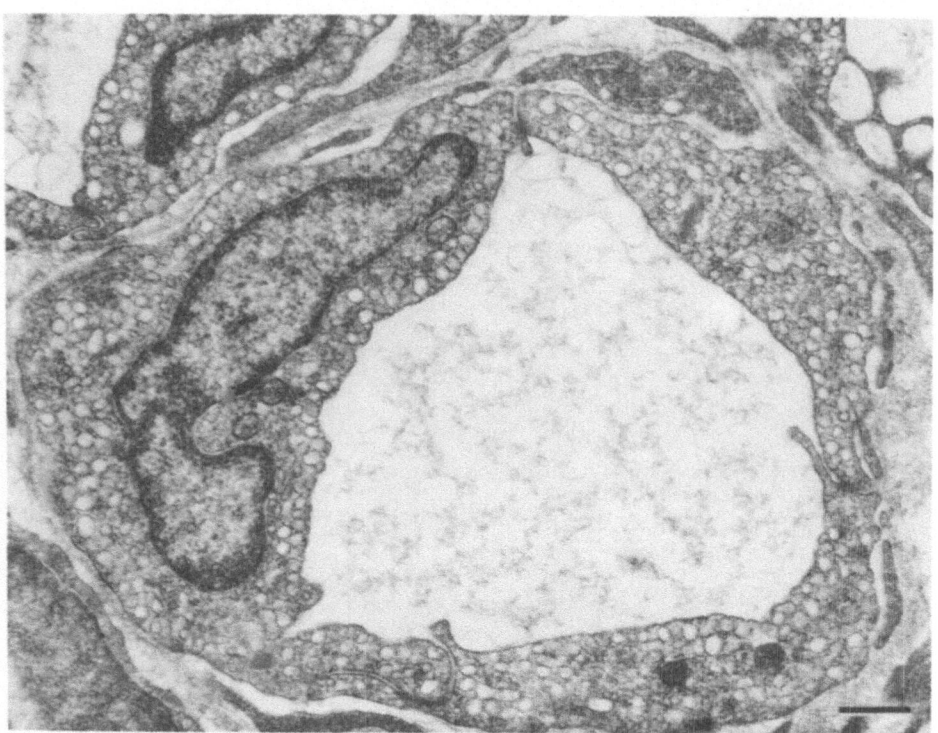

Figure 11. Cross section of an arterial capillary segment which had been fixed with glutaraldehyde prior to rapid freezing. Numerous vesicular profiles crowd the cytoplasm and caveolae line both luminal and abluminal surfaces. Bar = 0.67 μm.

Table 2.   Morphometry of Prefixed and Fresh-Frozen Rete Capillaries[a]

	Prefixed capillaries	Fresh-frozen capillaries
Vesicles/μm³ cytoplasm	639 (33.9)	217 (30.0)
Caveolae/μm² luminal membrane[b]	63ₛ (2.4)	20ₛ (2.7)
	57f (3.9)	35f (3.6)
Caveolae/μm² abluminal membrane[b]	94ₛ (4.0)	28ₛ (3.4)
	92f (3.7)	44f (4.5)
Cytoplasmic area (μm²/cap. profile)	8.0 (1.3)	8.9 (1.3)
Abluminal perimeter (μm)	18.0 (1.5)	19.6 (1.3)
Luminal perimeter (μm)	14.2 (1.4)	13.9 (2.4)
Junctional lengths (μm)	0.6 (0.2)	0.7 (0.1)
Perimeter/area ratio (surface area/volume)	4.0 (0.5)	3.8 (0.4)

[a] Numerical densities were calculated assuming an average section thickness of 600 Å. Numerical counts are given with standard error of the mean in parentheses. s, data from sectioned specimens; f, data from freeze fracture replicas.
[b] See Table 1.

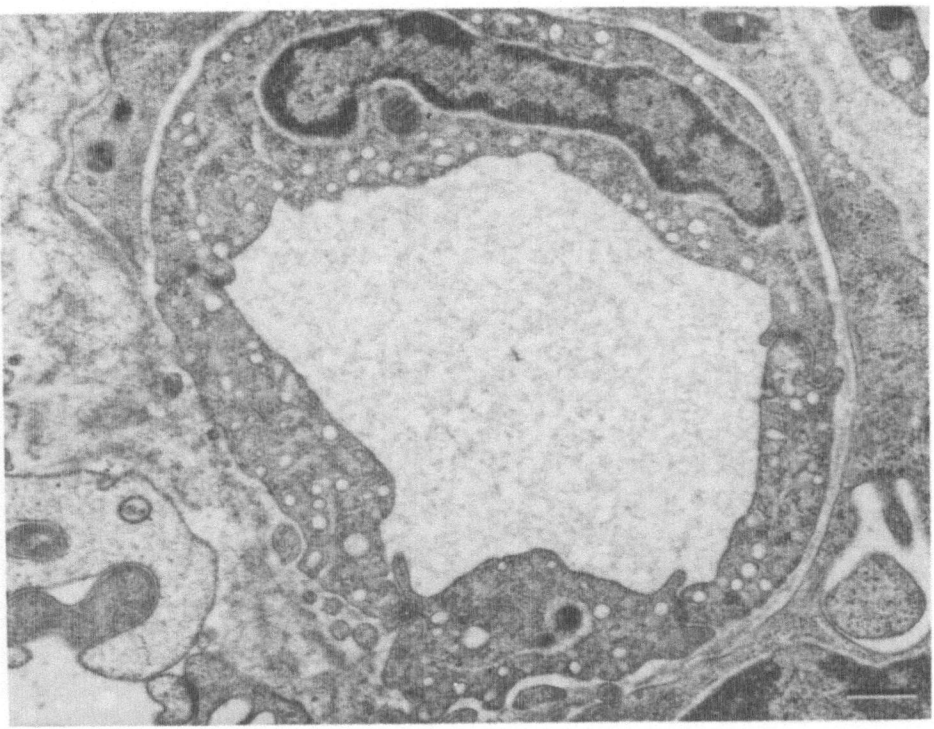

Figure 12. A similar profile of a capillary segment which was fresh frozen without prior exposure to glutaraldehyde. The density of vesicular profiles is markedly reduced and several smaller membranous profiles are also evident. Bar = 0.50 μm.

as a source of additional membrane. This would suggest that the additional vesicles present in glutaraldehyde-fixed endothelium are comprised of membrane derived from some preexisting pool. When other parameters of prefixed and fresh-frozen capillaries are compared, the cytoplasmic volume, surface membrane, and junctional lengths do not vary significantly between the two methods of fixation (Table 2). It would appear that the additional membrane must derive from an intracellular source. Membranous profiles (20–30 nm diameter) which are actually tubular in three dimensions (Fig. 13) and cisternal membrane systems are present in large numbers in fresh-frozen retia (Fig. 14) but exceedingly rare in those which have been prefixed. It is possible that accretion of subdivision of these membrane systems is induced by glutaraldehyde resulting in additional vesicular entities of the standard size (50–70 nm diameter). These tubules and cisternae which may uniquely characterize rete capillaries are possibly unstable entities which are modified in form by chemical fixation.

Ultrathin serial section analysis of fresh-frozen retia indicates that, however large the difference in numbers of vesicular structures, their extent of interconnectedness is unaffected.[14] The artifactual consequences of glutaraldehyde fixation therefore appear to be more quantitative than qualitative. It is possible that

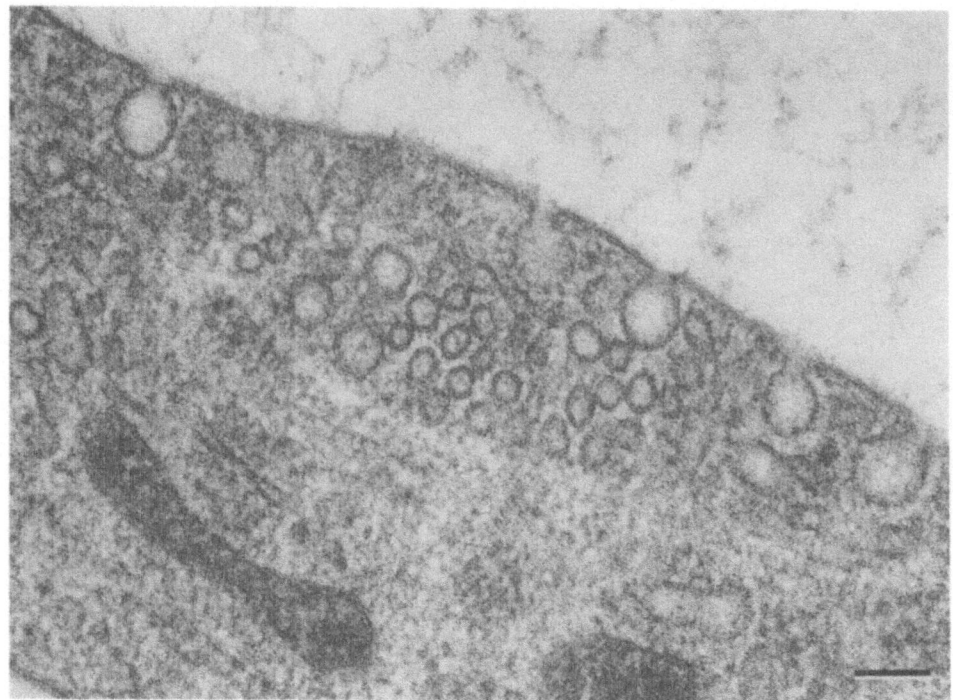

Figure 13. High-magnification micrograph of a fresh-frozen rete capillary wall exhibiting numerous membranous profiles (30–40 nm diameter) which are actually tubular and arranged parallel to each other. They exhibit various intimate associations with each other and caveolae at the cell surface. Bar = 0.11 μm.

an equilibrium between vesicles and other intraendothelial membrane systems is sustained in living capillary endothelial cells. This equilibrium is perturbed by the intervention of glutaraldehyde resulting in a redistribution of membrane into the standard endothelial vesicle configuration. If this is so, then the relative number of vesicles induced by glutaraldehyde will be in proportion to the amount of intraendothelial membrane from which they can be generated. Indeed, in thin-walled capillaries of the frog mesentery, no difference in volume density of vesicles has been detected between prefixed and fresh-frozen samples.[15]

It is evident that glutaraldehyde markedly alters the ultrastructural features of the capillary wall, either by disturbing an equilibrium between intraendothelial membrane systems or by active intervention in membrane interactions. The vesicular system as a real element of the capillary wall is not an artifact since it is perfectly preserved by rapid freezing. However, if vesicle dynamics and interactions persist for some time after chemical fixation, the kinetics of vesicle labeling with electron-dense tracers may involve longer times than those calculated. Generation of additional vesicles intracellularly may also require reassessment of that proportion of them which sequester electron-dense tracers and participate in transendothelial transport.

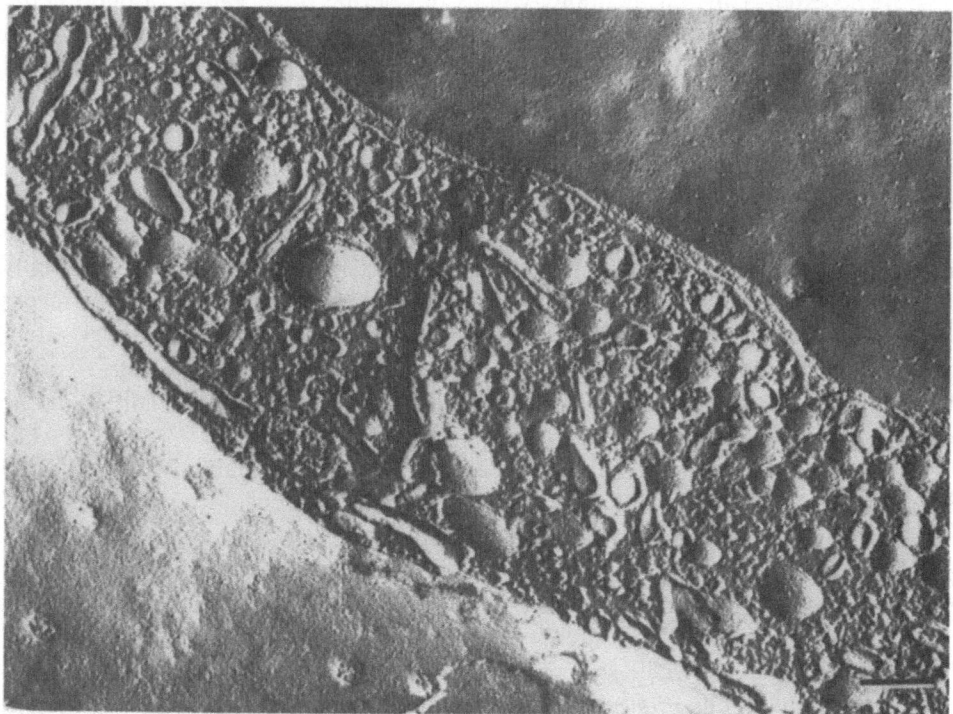

Figure 14. A replica of a cross fracture through the wall of a fresh-frozen capillary exhibiting several tubular and cisternal membrane systems in the cytoplasm. Bar = 0.22 μm.

Unfortunately, cryofixation and subsequent freeze-substitution seriously limit the variety of tracer experiments which can be performed to test these possibilities. The frozen condition of the tissue and subsequent treatment with non-aqueous solvents preclude the use of tracers whose staining properties depend on their peroxidative activities or other reactions in aqueous media. Also, tannic acid cannot be used as a compartmental tracer since it must be applied in conjunction with glutaraldehyde. However, tracers which are intrinsically electron dense such as ferritin or colloidal gold or those which stain intensely with heavy metal stains such as dextrans and glycogens can be used as intravital tracers which can be visualized in freeze-substituted material.

IV. STRUCTURE–FUNCTION RELATIONSHIPS IN THE CAPILLARY WALL

It would appear as though an uncertainty principle is in effect with regard to experimental approaches to permeability–structure relationships in the capillary wall. It is possible to measure the dynamics of solute movement across capillaries but in order to study the underlying structural correlates, the kinetics of the system must be arrested. Current technology does not permit observation of endothelial cell activity at a resolution necessary to discern activity of the vesicular system in living cells. Uncertainties involved in conventional preparation of vascular systems for electron microscopy cast into doubt not only the precise time of immobilization of structures but also the fidelity with which living structures are preserved. It is also evident that extrapolation of three-dimensional organization of the capillary wall from two-dimensional electron images can lead to serious errors of interpretation. Another source of experimental uncertainty is that tissue must be excised prior to fixation or perfusion-fixed *in situ*, both of which not only interrupt blood flow but alter osmotic and hydrostatic pressure gradients across the capillary wall. Almost nothing is known regarding the effects of oxygen deprivation and cellular osmotic stress on the structure of endothelial cells. This is primarily due to the necessity of interrupting blood flow and chemically invading endothelial cells prior to electron microscopy.

Experimental methods and analytical tools are now available which will minimize many of these uncertainties and provide a more accurate representation of the functional state of the capillary endothelium. Holistic experimental approaches must be embraced in which intravital measurements and observations can be coupled with ultrastructural analysis of the capillary wall. In certain thin tissues such as mesentery or cremaster muscle, it should be possible to rapidly freeze capillary systems *in situ* without interruption of the circulation. In this manner, instantaneous immobilization of living capillaries will provide for a true representation of ultrastructural correlates of permeability with a minimum of experimental intervention. Three-dimensional serial section reconstruction, stereomicroscopy of thick sections, or appropriate tracers will eliminate ambiguities associated with observation of sectioned material. These approaches involve a high degree of skill and effort but will pay high dividends in the search for structural counterparts of capillary permeability.

V. CONCLUDING REMARKS

There is no comprehensive theory which adequately reconciles the permeability behavior of continuous capillaries with underlying cellular-structural correlates in the capillary wall. This may be due, in part, to inherent limitations of two-dimensional electron microscopy and artifactual consequences of chemical fixation. Serial section reconstruction and compartmental labeling with tannic acid have revealed interconnections between endothelial vesicles with each other and the cell surface which are not apparent in single sections. However, both the exclusion of tannic acid from some vesicles and high-voltage stereomicroscopy of thick sections indicate the existence of vesicles in fixed tissue which do not have access to extracellular substances. Tannic acid also does not penetrate between capillary endothelial cells and appears to be occluded by tight junctions. Comparison of chemically fixed and rapidly frozen capillaries indicates that glutaraldehyde induces a substantial increase in the number of endothelial vesicles.

It is hypothesized that chemical intervention of glutaraldehyde results in a redistribution of membrane in which additional vesicles are generated from preexisting intraendothelial membrane systems. Assumptions concerning the dynamic activity of the vesicular system or the lack thereof, however, cannot be derived from these studies.

ACKNOWLEDGMENTS. Supported by USPHS Grants HL-16666 and HL-33936.

REFERENCES

1. Bretscher, M. S., and Whytock, S., 1977, Membrane-associated vesicles in fibroblasts, *J. Ultrastruct. Res.* **61:**215–217.
2. Bridgeman, P. C., and Reese, T. S., 1981, The structure of cytoplasm in directly-frozen cultured cells. I. Filamentous meshworks in the cytoplasmic ground substance, *J. Cell Biol.* **99:**1655–1668.
3. Buckley, I. K., 1973, Studies in fixation for electron microscopy using cultured cells, *Lab. Invest.* **29:**398–410.
4. Bundgaard, M., Frokjaer-Jensen, J., and Crone, C., 1979, Endothelial plasmalemmal vesicular profiles as elements in a system of branching invaginations from the cell surface, *Proc. Natl. Acad. Sci. USA* **76:**6439–6442.
5. Bundgaard, M., Hagman, P., and Crone, C., 1983, The three dimensional organization of plasmalemmal vesicular profiles in the endothelium of rat heart capillaries, *Microvasc. Res.* **25:**358–368.
6. Bundgaard, M., 1984, The three dimensional organization of tight junctions in a capillary endothelium revealed by serial section electron microscopy, *J. Ultrastruct. Res.* **88:**1–17.
7. Casley-Smith, J. R., 1985, Vesicular form and fusion—as revealed by freeze-immobilization and stereoscopy of semi-thin sections, *Prog. Appl. Microcirc. Res.* **9:**6–20.
8. Chandler, J. E., and Heuser, J. E., 1980, Arrest of membrane fusion events in mast cells by quick freezing, *J. Cell Biol.* **86:**666–674.
9. Clough, G., and Michel, C. C., 1981, The role of vesicles in the transport of ferritin through frog endothelium, *J. Physiol. (London)* **315:**127–142.
10. Curry, F. E., and Michel, C. C., 1980, A fiber matrix model of capillary permeability, *Microvasc. Res.* **20:**96–99.
11. Devenny, J. J., and Wagner, R. C., 1985, Transport of immunoglobulin G by endothelial vesicles in isolated capillaries, *Microcirculation, Endothelium and Lymphatics* **2:**15–26.
12. Frokjaer-Jensen, J., 1980, Three-dimensional organization of plasmalemmal vesicles in endothelial cells: An analysis by serial sectioning of frog mesenteric capillaries, *J. Ultrastruct. Res.* **73:**9–20.
13. Frokjaer-Jensen, J., 1984, The plasmalemmal vesicular system in striated muscle capillaries and in pericytes, *Tissue Cell* **16:**31–42.
14. Frokjaer-Jensen, J., Wagner, R. C., Andrews, S. B., Hagman, P., and Reese, T. S., 1988, Three-dimensional organization of the plasmalemmal vesicular system in directly-frozen capillaries of the rete mirabile of the eel, *Cell and Tiss. Res.* (in press).
15. Frokjaer-Jensen, J., and Reese, T. S., 1985, The plasmalemmal vesicular system in mesothelial cells of frog mesentery; comparison of chemical fixation and direct freezing, *Microvasc. Res.* **29:**221a.
16. Futaesaku, Y., Mizuhira, V., and Nakamura, H., 1972, The new fixation method using tannic acid for electron microscopy and some observations of biological specimens, *Acta Histochem. Cytochem.* **4:**155–156.
17. Ghitescu, L., Fixman, A., Simionescu, M., and Simionescu, N., 1986, Specific binding sites for albumin restricted to plasmalemmal vesicles: Receptor-mediated transcytosis, *J. Cell Biol.* **102:**1304–1311.
18. Graham, R. C., and Karnovsky, M. J., 1966, The early stages of adsorption of injected horseradish

peroxidase in the proximal tubules of the mouse kidney: Ultrastructural cytochemistry by a new technique, *J. Histochem. Cytochem.* **14**:291–302.

19. Heuser, J. E., Reese, T. S., Dennis, M. J., Jan, Y., Jan, L., and Evans, L., 1979, Synaptic vesicle exocytosis captured by quick freezing and correlated with quantal transmitter release, *J. Cell Biol.* **81**:275–300.

20. Hopwood, D., 1967, Some aspects of fixation with glutaraldehyde, *Am. J. Anat.* **106**:82–92.

21. Johnson, T. J. A., and Rash, J. E., 1981, Glutaraldehyde fixation chemistry: Substituted pyridine polymers result from glutaraldehyde–amine reactions, *J. Cell Biol.* **91**:271a.

22. Karnovsky, M. J., 1967, The ultrastructural basis of capillary permeability studied with peroxidase as a tracer, *J. Cell Biol.* **35**:213–236.

23. Karnovsky, M. J., 1968, The ultrastructural basis of transcapillary exchange, *J. Gen. Physiol.* **52**:64–95.

24. Luft, J. H., 1971, Ruthenium red and violet. I. Chemistry, purification, methods of use for electron microscopy and mechanism of action, *Anat. Rec.* **171**:346–368.

25. Mazzone, R. W., and Kornblau, S. M., 1981, Pinocytic vesicles in the endothelium of rapidly-frozen rabbit lung, *Microvasc. Res.* **21**:193–211.

26. McGuire, P. G., and Tweitmeyer, T. A., 1983, Morphology of rapidly-frozen aortic endothelial cells: Glutaraldehyde fixation increases the number of caveolae, *Circ. Res.* **53**:424–430.

27. Ornberg, R. L., and Reese, T. S., 1981, Beginning of exocytosis captured by rapid-freezing of *Limulus* amebocytes, *J. Cell Biol.* **90**:40–54.

28. Plattner, H., and Bachman, L., 1982, Cryofixation: A tool in biological ultrastructure research, *Int. Rev. Cytol.* **79**:237–304.

29. Renkin, E. M., 1978, Transport pathways through the capillary endothelium, *Microvasc. Res.* **15**:123–135.

30. Revel, J. P., and Karnovsky, M. J., 1967, Hexagonal array of subunits in intercellular junctions of the mouse heart and liver, *J. Cell Biol.* **33**:C7–C12.

31. Robinson, J. M., Hoover, R. L., and Karnovsky, M. J., 1984, Vesicle (caveolae) number is reduced in cultured endothelial cells prepared for electron microscopy by rapid freezing, *J. Cell Biol.* **99**:287a.

32. Sabatini, D. D., Bensch, K., and Barrnett, R. J., 1963, Cytochemistry and electron microscopy: The preservation of cellular structures and enzyme activity by aldehyde fixation. *J. Cell Biol.* **17**:19–32.

33. Simionescu, N., and Simionescu, M., 1976, Galloyl glucoses of low molecular weight as mordant in electron microscopy, *J. Cell Biol.* **70**:608–621.

34. Simionescu, N., Simionescu, M., and Palade, G. E., 1975, Permeability of muscle capillaries to small heme peptides: Evidence for the existence of patent transendothelial channels, *J. Cell Biol.* **64**:586–607.

35. Simionescu, N., Simionescu, M., and Palade, G. E., 1978, Structural basis of permeability in sequential segments of the microvasculature in the diaphragm. II. Pathways followed by microperoxidase across the endothelium, *Microvasc. Res.* **15**:17–36.

36. Simionescu, N., 1983, Cellular aspects of transcapillary exchange, *Physiol. Rev.* **63**:1536–1579.

37. Van Harreveld, A., and Crowell, J., 1964, Electron microscopy after rapid freezing on a metal surface and substitution fixation, *Anat. Rec.* **149**:381–386.

38. Wagner, R. C., 1976, The effect of tannic acid on electron images of capillary endothelial cell membranes, *J. Ultrastruct. Res.* **57**:132–139.

39. Wagner, R. C., and Casley-Smith, J. R., 1981, Endothelial vesicles, *Microvasc. Res.* **21**:267–298.

40. Wagner, R. C., and Robinson, C. S., 1982, Tannic acid tracer analysis of permeability pathways in the capillaries of the rete mirabile: Demonstration of the discreetness of endothelial vesicles, *J. Ultrastruct. Res.* **81**:37–46.

41. Wagner, R. C., Robinson, C. S., Cross, P. J., and Devenny, J. J., 1983, Endocytosis and exocytosis of transferrin by isolated capillary endothelium, *Microvasc. Res.* **25**:387–396.

42. Wagner, R. C., and Robinson, C. S., 1984, High voltage electron microscopy of capillary endothelial vesicles, *Microvasc. Res.* **28**:197–205.

43. Wagner. R. C., 1985, Application of high voltage electron microscopy to visualize the three-

dimensional structure of the vesicular system in thick sections, *Prog. Appl. Microcirc. Res.* **9**:1–5.

44. Wagner, R. C., and Andrews, S. B., 1985, Ultrastructure of the vesicular system in rapidly-frozen capillary endothelium of the rete mirable, *J. Ultrastruct. Res.* **90**:172–182.

45. Williams, S. K., Devenny, J. J., and Bitensky, M. W., 1981, Micropinocytic ingestion of glycosylated albumin by isolated microvessels: Possible role in diabetic microangiography, *Proc. Natl. Acad. Sci. USA* **78**:2393–2397.

46. Wissig, S. L., and Williams, M. C., 1978, Permeability of muscle capillaries to microperoxidase, *J. Cell Biol.* **76**:341–359.

47. Wissig, S. L., 1979, Identification of the small pore in muscle capillaries, *Acta Physiol. Scand.* **463**:33–44.

48. Wissig, S. L., and Charonis, A. S., 1984, Capillary ultrastructure, in: *Edema* (N. C. Staub and A. E. Taylor, eds.), Raven Press, New York, pp. 117–142.

49. Wood, M. R., Wagner, R. C., Andrews, S. B., Greener, D. A., and Williams, S. K., 1987, Rapidly-frozen human endothelial cells in culture: Comparison of vesicles in prefixed vs. fresh frozen samples, *Microcirculation, Endothelium and Lymphatics* **3**:323–358.

Transport Functions of Endothelial Cells

Transport Pathways and Processes

Eugene M. Renkin

I. INTRODUCTION

The goal of this chapter is to chart procedures for identifying endothelial transport pathways and mechanisms for water and solutes. An excellent description of the discovery of the microcirculation, of the identification of the capillaries as the main sites of interchange of fluid and solutes, and of the development of the concept of capillary permeability has been written by E. M. Landis.[25] In the first half of this century, it was recognized that capillaries in most organs of the body allowed relatively free exchange of water and small solutes (crystalloids), but restricted the escape of plasma proteins and other macromolecules (colloids). In 1896, Ernest Starling[53] explained the retention of vascular volume as the result of a balance of the osmotic pressure of plasma colloids and the hydrostatic pressure of capillary blood. Imbalance of hydrostatic and colloid osmotic forces results in fluid movement into or out of the bloodstream. This process is called ultrafiltration. The fluid moved (ultrafiltrate) contains water and small solutes in equilibrium with plasma, but only traces of plasma proteins. The direction and rate of fluid movement are proportional to sign and magnitude of the difference between hydrostatic and effective osmotic forces (positive = outward).[24,35] Exchange of water and solutes smaller than plasma proteins was attributed to passive diffusion.[22,23] Differences in capillary permeability from organ to organ were recognized: the capillaries of the brain were known to be nearly impermeable to ions and small molecules, while the sinusoids of the liver and spleen were known to be permeable to plasma proteins. It was also recognized that capillaries in most organs allowed small ("trace") amounts of plasma proteins to escape into the interstitial fluid, and that the lost protein was returned to the bloodstream by the lymphatics.[16,29]

Knowledge of endothelial structure before 1950 was limited by the resolution of light microscopy: in capillaries of most organs, a layer of thin cells on a basement membrane; in brain capillaries, thicker cells; in liver and spleen sinusoids, large openings between cells, with an incomplete basement membrane. Compar-

Eugene M. Renkin • Department of Human Physiology, School of Medicine, University of California, Davis, California 95616.

isons of capillary permeability with that of epithelial membranes drew attention to the possible role of intercellular junctions as pathways for substances which could not penetrate the cells themselves. Chambers and Zweifach[5] suggested that the junctions were filled with a gellike "intercellular cement" which acted as a molecular sieve.

II. THE "PORE" THEORY OF CAPILLARY PERMEABILITY

It was from this background that a simple, passive, geometrically defined model of capillary transport was developed by Pappenheimer and his collaborators.[26,33,34,43,47] Their model was based on experimental measurements of capillary ultrafiltration and diffusion, and analysis by then-current physicochemical theories of fluid movement and molecular diffusion. The physiological evidence on which the original pore model was based may be summarized as follows:

1. The osmotic effects of lipid-insoluble solutes are graded with respect to the *size* of the transported particles (molecules or ions). Small solutes which penetrate capillary walls exert a transient osmotic influence on fluid movement which increases in magnitude and duration as molecular size increases, approaching full osmotic pressure and lasting effect for serum albumin and larger molecules.
2. Transcapillary diffusion rates of such lipid-insoluble molecules are similarly graded with respect to molecular size. The time constants of decay of osmotic effect and transcapillary diffusion rate are the same.
3. Small lipid-soluble molecules have smaller shorter-lasting osmotic effects, and diffuse faster than lipid-insoluble molecules of similar size.[41,42]

Because of the close parallelism of transient osmotic effects and diffusion rates for lipid-insoluble solutes, Pappenheimer and his colleagues concluded that the pathways used for ultrafiltration and diffusion were the same. Because these effects were dependent on molecular size, they concluded that these pathways acted as a molecular sieve, offering progressively increasing resistance to permeation of larger molecules, reaching a cutoff at the size of serum albumin. They identified this pathway with the endothelial cell junctions. They attributed the faster diffusion of small lipid-soluble molecules and their weaker osmotic effects to the ability of these substances to penetrate the endothelial cell lipid membranes, as well as through the junctions.

Pappenheimer *et al.*[34] modeled the junctional pathways as uniform cylindrical pores through the intercellular cement of the junctional regions, or as long slits between the cells. Applying hydrodynamic theory (Poiseuille's law for circular pores or its equivalent for parallel-walled slits) and diffusion theory (Fick's law) to their measurements of fluid and solute transport rates in isolated, blood-perfused hindlimbs of cats, they were able to calculate an effective pore diameter or slit width for the capillary membrane, and to estimate the total cross-sectional area of the hypothetical pores or slits. Their original figures were 60 Å for pore diameter, 37 Å for slit width. These values appeared consistent with impermeability to serum albumin, effective molecular diameter 72 Å. The open area of pores or slits was estimated to be less than 0.2% of total capillary surface, consistent with a junctional location.

The methods used by Pappenheimer *et al.* were not sensitive enough to measure the very slow transport rates of plasma proteins or other large molecules through capillary walls in the perfused hindleg. Subsequent development of methods for evaluating solute permeabilities from measurements of lymph flow and lymph/plasma concentration ratios showed that a model with only one pore size or slit width (or a narrow distribution of either) could not account for both small- and large-molecule permeabilities.[17,19,29] A second population of "large pores," much larger in size but smaller in number, was required to account for the transport of molecules too large to go through the "small pores." Differences in permeability to macromolecules in different organs were attributed to differing proportions of large and small pores.[2,56] Estimated diameters of large pores range from 360 Å to over 1000 Å. Equivalent dimensions of "wide slits" could also be calculated, but this has rarely been done, because the very small number of large pores—only a few per capillary—suggests widely separated locations, rather than a continuous cleft.

Elaboration of capillary transport theory and the development of multipore models has progressed rapidly since the 1950s. Physiological measurements of capillary transport have been carried out for a large diversity of solutes in a wide variety of organs and tissues, even in single microvessels. These new developments have augmented the strength and applicability of the fundamental idea that the predominant transport pathway in most capillaries is extracellular, and that the predominant transport mechanisms involve passive interactions between extracellular structures and transported ions and molecules. Receptor-mediated transport of glucose by capillaries of the brain was demonstrated by Crone,[6] and evidence suggesting receptor-mediated transport of amino acids and other brain metabolites has been reported.[37] Recent observations of Pardridge *et al.*[38] on isolated brain capillaries suggest receptor-mediated transport of insulin. However, no instances of carrier-mediated transport outside the central nervous system have been confirmed to date. In general, newer evaluations (and reevaluations of older data) have tended to increase small-pore diameters or slit widths to values somewhat higher than those originally proposed (e.g., to 90 and 56 Å respectively for the cat's hindlimb) and to reduce the total cross-sectional areas of the pores or slits (to <0.02% for the cat's hindlimb). Recent experimental studies have shown that interactions of serum albumin (and possibly other plasma proteins) with capillary endothelium are responsible for maintaining normal permeabilities,[28] and that electrical charges on large molecules may exert an influence secondary to molecular size on their permeation.[1,14,15,20,40] Distributions of charged sites in the transport pathways have been postulated. A new extension of the "pore" concept to pores or slits filled with fiber matrices (reminiscent of Chambers and Zweifach's intercellular cement hypothesis) has been proposed as a more suitable vehicle for dealing with chemical interactions and charge effects. Modeling the "small pore"

Table 1.   Hydraulic (L_p) and Electrical (L_e) Conductivities of Capillary Endothelia Compared with Some Epithelial Membranes[a]

	L_p (cm^3 sec^{-1} dyne^{-1})	L_e (ohm^{-1} cm^{-2})
Capillary endothelia		
Fenestrated		
Renal glomerulus (frog, mammal)	1.5×10^{-8}	—
Intestinal mucosa (mammal)	1.3×10^{-9}	—
Nonfenestrated		
Mesentery (frog, mammal), omentum (mammal)	5.0×10^{-10}	5.4×10^{-1}
Heart (mammal)	8.6×10^{-11}	—
Skeletal muscle (frog)	7.5×10^{-11}	3.6×10^{-2}
Skeletal muscle (mammal)	2.5×10^{-11}	—
Tight-junction		
Brain (frog, mammal)	3.0×10^{-13}	5.0×10^{-4}
Transport epithelia		
"Leaky" (low resistance)		
Proximal renal tubule (rat)	2.0×10^{-10}	2.0×10^{-1}
Jejunum (rat)	9.0×10^{-12}	3.3×10^{-2}
"Tight" (high resistance)		
Urinary bladder (turtle)	3.6×10^{-13}	1.5×10^{-3}
Skin (frog)	4.5×10^{-13}	4.0×10^{-4}

[a] Compiled from a variety of sources, notably: House,[21] Renkin,[44] Crone and Levitt,[8] Michel,[31] Oleson and Crone,[32] Curry and Frøkjaer-Jensen.[11]

pathway as a fibrous matrix with appropriate sieving characteristic yields closer agreement between hydraulic conductivity and solute permeabilities than simple pore or slit models.[12] These developments have been reviewed by Renkin and Curry,[46] Crone and Levitt,[8] Michel,[31] Curry,[10] and Taylor and Granger.[55,56] Tables 1 to 4 summarize some current evaluations of capillary permeability and their interpretation in terms of the "pore" model.

III. MORPHOLOGICALLY IDENTIFIED TRANSPORT PATHWAYS

In the late 1950s, more detailed information about the structure of vascular endothelia began to be acquired by electron micrography. These observations confirmed the special characteristics of capillary endothelia with respect to other transport epithelia: thinner cells, simpler junctional structures with fewer interconnecting bonds, consistent with the concept of passive junctional transport. At the same time, cellular structures were discovered which might also contribute to transport: surface caveoli and cytoplasmic vesicles which appeared capable of taking up fluid and solutes from the plasma or interstitial fluid and transferring them to the opposite side either by moving back and forth, as ferries or shuttles, or by transiently fusing to form large channels. Some endothelial cells in visceral organs exhibited special openings ("fenestrae"), some of which might be closed by a thin diaphragm, and others open, exposing the basement membrane. The

Table 2. Capillary Permeabilities (P; cm sec^{-1} × 10^6) and Reflection Coefficients (σ) to Small, Lipid-Insoluble Solutes[a]

Substance	MW	D_{37} (cm^2 sec^{-1} × 10^5)	d_e (Å)	Mammalian skeletal muscle		Mammalian heart muscle		Mammalian GI mucosa		Mammalian brain		Frog mesentery	
				P	σ	P	σ	P	σ	P	σ	P	σ
KCl	—	—	4.4	35	—	60	—	77	—	—	(1.0)	670	—
NaCl	—	1.8	4.6	27	0.06	27	0.07	—	—	4.6	0.44	—	0.07
Urea	60	1.9	5.2	7.4	0.12	7.1	0.09	—	—	0.2	0.89	—	0.07
Sucrose	342	0.72	9.4	5.4	0.20	—	0.19	—	—	—	(1.0)	—	0.12
Raffinose	504	0.58	11.4	1.2	0.61	2.3	0.22	6	—	—	—	—	—
Inulin	5,500	0.22	30	—	—	—	0.39	3.3	—	—	(1.0)	—	—
Serum albumin	69,000	0.093	72	0.01	0.89	0.03	0.80	0.03	0.90	<0.0001	(1.0)	—	0.82

a Data from Renkin,[44] Renkin and Curry,[46] Crone and Levitt,[8] Crone et al.[7]

Table 3. Capillary Permeabilities (P; cm sec^{-1} × 10^8) and Reflection Coefficients (σ) to Large Molecules[a]

Substance	MW × 10^{-3}	D_{37} (cm^2 sec^{-1} × 10^7)	d_e (Å)	Mammalian skeletal muscle		Mammalian heart muscle		Mammalian GI mucosa		Mammalian brain	
				P	σ	P	σ	P	σ	P	σ
Serum albumin	69	9.3	72	4.7	0.89	2.9	—	2.9	0.90	0.008	—
Transferrin	90	7.6	86	6.3	0.89	—	—	—	—	0.013	—
Haptoglobin	100	7.2	92	3.1	0.91	1.4	—	1.4	—	—	—
Immunoglobulin G	160	5.9	112	3.3	0.91	—	—	—	0.95	0.002	—
α$_2$-Macroglobulin	820	3.7	182 }	1.6	0.94	0.7	—	0.7	0.98	0.0004	—
Fibrinogen	340	3.1	216 }			—	—	—	—	—	—

[a] Modified after Renkin.[44]

Table 4. Capillary Permeability: Two-Pore Model Parameters[a,b]

	d_{SP} (Å)	d_{LP} (Å)	K_{SP}	K_{LP}	K_{CM}	$\dfrac{N_{LP}}{N_{SP}}$
Continuous, nonfenestrated endothelium						
Dog paw (skin, skeletal muscle)	94	390	0.82	0.13	0.05	1:2000
Dog lung	160	400	0.80	0.16	0.04	1:200
Continuous, fenestrated endothelium						
Cat small intestine	92	400	0.90	0.05	0.05	1:6400
Dog colon	106	360	0.71	0.17	0.12	1:550
Discontinuous						
Cat liver	180	660	0.20	0.80	—	1:46

[a] Examples slightly modified from Taylor and Granger.[56]
[b] Symbols: d, diameter; N, number; K, relative water conductivity. Subscripts: SP, small pores; LP, large pores; CM, cell membrane (endothelial).

tight-junction structure of brain capillaries was confirmed, as has the presence of large openings between cells in hepatic and splenic sinusoids. More recent studies have shown special localizations of charged and chemically reactive sites on the surfaces of some of these structures. Further details of endothelial morphology and surface chemistry are presented elsewhere in this volume (see also Ref. 52). Figure 1 is a physiologist's sketch of endothelial structures which might be transport pathways. The problem to be addressed in the remainder of this chapter is the relation between measured transport rates and the hypothetical transport pathways (pores, slits, gel matrix) derived from them, and the structures shown in

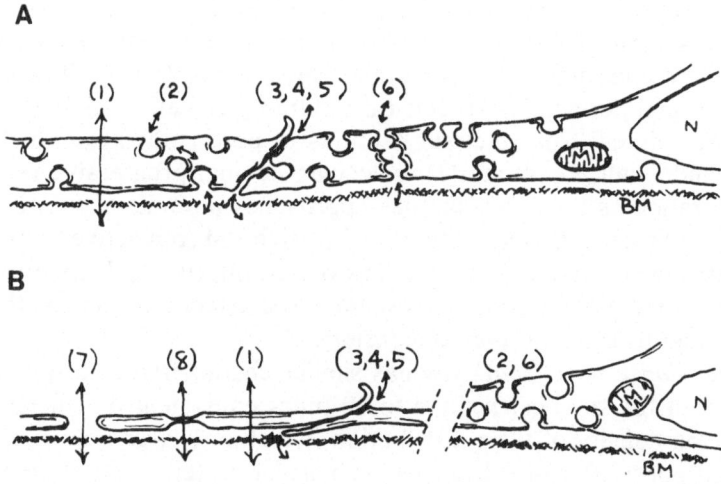

Figure 1. Diagrammatic sketches of presumptive transport pathways in (A) unfenestrated and (B) fenestrated microvascular endothelium. N, cell nucleus; M, mitochondrion; BM, basal lamina. The numbered pathways (1–8) are identified in the text. (From Renkin[44] with permission of the American Heart Association.)

Fig. 1. Discussion will be restricted to capillaries of the continuous (nonfenes-trated) type found in skin, muscles, and most organs of the body, and to fenes-trated capillaries of visceral organs. The numbers in the following paragraphs correspond to those in Fig. 1.

(1) *The cell membrane pathway* consists of three barriers in series: plasma membrane, cytoplasm, plasma membrane. Analogy with permeability of cells in general suggests that water and small, lipid-soluble substances may use this path-way. Permeation of solutes other than water requires *both* water and lipid-solu-bility. High capillary permeability (even of brain capillaries) to respiratory and anesthetic gases is attributed to the accessibility of this pathway.

(2) *The vesicular pathway*, operating in the shuttle mode (moving back and forth between surfaces), would tend to equilibrate luminal and abluminal fluids. If equal numbers of vesicles move in both directions (any other hypothesis raises problems of cell membrane stability), there is no net fluid movement, and if no connection from one side to the other is made, there is no movement of fluid under pressure. Vesicular shuttling is thus similar to Brownian movement, and its transport characteristics are those of diffusion: transport rate being propor-tional to concentration difference and rate of vesicular turnover.[51]

Garlick and Renkin[17] showed that it was possible to account for capillary transport of macromolecules quantitatively either by diffusion through large pores or by vesicular shuttling. Vesicular transport tends to fall off as molecular size becomes comparable to vesicle size, due to steric exclusion. They concluded that vesicular transport could not replace the *small* pore system, because enormously high rates of vesicular turnover (= extremely short transit times) would be re-quired to account for observed small solute transport rates, and because vesicles could not account for net fluid movements across capillary walls.

(3) *The "small pore" pathway* is one of three pathways localized at the in-tercellular junctions. It has been variously modeled as pores, slits, or fibrous matrix, its essential feature being a continuous route of extracellular fluid from one side of the endothelium to the other. Such channels will allow convective flow of water and permeating solutes (ultrafiltrate) as well as diffusion exchange of water and solutes. Characteristic features of pore pathways (which apply to both large and small pores) are (1) molecular sieving, progressive restriction of solute penetration as its molecular size approaches pore diameter, slit width, or interfiber spacings, and (2) interactions of diffusive and convective transport, most notably coupling of solute transport to flow of ultrafiltrate ("solvent drag"). These characteristics make it possible to test for the existence of pore pathways in a membrane, and to evaluate pore dimensions.

(4) *The "large pore" pathway* has similar characteristics to the small pore system, except for size, generally 5 to 10 times equivalent dimensions of small pores, and the density of their distribution, which is 1 large pore to every 200–10,000 small pores. Because transport of plasma proteins and other macromol-ecules across capillary walls is at least partially coupled to ultrafiltration, large pores must be an important pathway for their transport.[3,48,49]

(5) *A cell surface layer pathway through the junctions.* Scow *et al.*[50] have suggested that large lipid-soluble molecules which are not appreciably soluble in

water may be carried through intercellular junctions by lateral diffusion in the lipid surface membrane of the adjacent cells. Such substances (triglycerides, long-chain fatty acids) can be brought to the endothelial cell surface bound to albumin molecules in plasma, and picked up outside the capillary by interstitial fluid albumin. Transport of such substances is much more rapid than of the carrier albumin. Interaction of the albumin–free fatty acid complex with a cell surface receptor may facilitate transport of fatty acids by decreasing the affinity of the carrier for solute. Similar mechanisms may be effective for other protein-bound solutes.[36]

(6) *A nonjunctional channel equivalent to a large pore* may be formed transiently by fusion of two or more vesicles. Except for their transient existence, they should exhibit the same characteristics as junctional large pores. Channels of fused vesicles have been reported, but the frequency of their occurrence is very low.

(7) *Open fenestrae* in visceral capillaries are too large (700–800 Å diameter) to restrict escape of any but the cellular components of the blood. Such openings occur in the renal glomerular capillaries, where the basement membrane or overlying capsular epithelium provides the restrictive barrier. In the case of the renal glomerular capillaries, permeability to plasma proteins and other macromolecules is lower than that of capillaries in other organs.[14]

(8) *Closed fenestrae* in visceral capillaries are closed by a specialized diaphragm which appears to differ structurally from cell surface membranes. In many visceral organs (e.g., intestine), some fenestrae appear to be open, some closed. Permeability of fenestrated capillaries to plasma proteins does not appear to be higher than that of nonfenestrated capillaries; rather, they are characterized by high permeability to water and ions and small molecules (Tables 2 and 3). Fenestrated capillaries also have all the transport pathways of nonfenestrated endothelium.

IV. A PATHOPHYSIOLOGICAL TRANSPORT PATHWAY

In addition to those enumerated above, which are characteristic of normal capillaries, another less restrictive extracellular pathway may be opened transiently in postcapillary venules and capillaries by local tissue injury and by a number of agencies associated with injury.[27] This pathway has the form of large (up to 1 μm wide) intercellular gaps, and is believed to result from endothelial cell contraction and junctional separation. The number of such openings at full activation is comparable to that of normally occurring large pores, but their distribution may be different, and they are much larger. The basement membrane remains intact, and appears still to function as a barrier to escape of large molecules, though a less restrictive one than the intact endothelial membrane.[4,54]

V. RELATION OF ENDOTHELIAL STRUCTURE TO ITS TRANSPORT FUNCTION

Microvascular endothelia are mosaic membranes with multiple transport pathways, heterogeneously distributed. While some substances may be restricted to a single pathway, others may penetrate by several different routes. For such substances, it is not sufficient only to identify the routes of transport. To characterize the system, it is necessary to determine partition of transport among alternate pathways, and to evaluate the forces acting on transport through each available pathway. Much work remains to be done before this goal can be fully achieved. We shall have to increase and refine our knowledge of the morphological characteristics of microvascular endothelia, and of their solute transport rates and the forces which control them. Enough is known, however, to make a start. In the remainder of this chapter, I shall try to evaluate evidence bearing on three problems of current interest:

1. How is bulk flow of water through capillary walls partitioned between cell membrane and pore pathways?
2. How is transport of lipid-insoluble solutes divided between small and large pore pathways?
3. What evidence is there for transport of large molecules through pathways other than pores, and what fraction of total macromolecular transport is carried by such pathways?

A. Partition of Water Flows between Cell Membrane and Pore Pathways

Water is moved in response to differences in capillary hydrostatic and solute osmotic pressures. Only solutes to which all membrane pathways are completely impermeable exert their full osmotic force; others exert a partial force, the magnitude of which is dependent on their permeability relative to water. Physical chemists define a *reflection coefficient* (σ) which represents the fraction of total osmotic force exerted by a specific solute acting on a given membrane. For complete impermeability, $\sigma = 1$; for equal permeability to water, $\sigma = 0$. In capillary endothelia, σ's for small lipid-insoluble solutes such as NaCl and urea have values below 0.1; with increasing molecular size, σ increases steadily until for the plasma proteins, σ's approach 0.9–1.0.

In multiple-pathway membranes, membrane reflection coefficients (σ_m) are weighted averages of the σ's for each pathway, the weighting factors being the relative water conductivities of each pathway, to that of the whole membrane (K):

$$\sigma_m = \sigma_c K_c + \sigma_{sp} K_{sp} + \sigma_{lp} K_{lp} \text{ etc.} \tag{1}$$

where the subscripts c, sp, and lp represent cell membrane, small pore, and large pore pathways. If the relations of σ to molecular size (or other properties) can be established experimentally or theoretically for individual pathways, some of the K's in Eq. (1) may be calculated. For example, we may assume that even for small lipid-insoluble substances, σ_c approaches 1, while σ_{sp} and σ_{lp} approach zero. Thus, the second and third terms on the right drop out, and $K_c = \sigma_m/\sigma_c$. Curry et al.[13] (see also Michel[31]) measured reflection coefficients of several small solutes

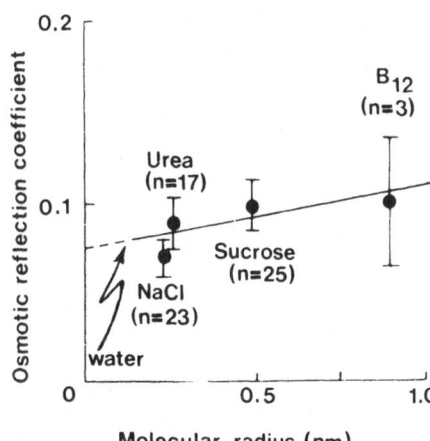

Figure 2. Relationship between the osmotic reflection coefficient (σ) and effective molecular radius for small, hydrophilic solutes in frog mesenteric capillaries. (From Michel[30] with permission.)

in capillaries of frog mesentery. Figure 2 is a plot of σ_m as a function of effective molecular radius. By extrapolating their data to the molecular size of water (arrow) to ensure that σ_{sp} and σ_{lp} were zero, they were able to estimate $K_c = 0.08$, i.e., that approximately 8% of total water flux went through the cell membrane pathway and 92% through the intercellular (pore) pathways. Estimates of K_c for other endothelia generally fall in the range 0.02–0.10, though more carefully obtained data on small solutes are needed to confirm these estimates.

B. Partition of (Lipid-Insoluble) Solute Fluxes between Small and Large Pore Pathways

The principal basis for partition is molecular size, although secondary effects of electrical charge and possibly of other factors (shape, chemical interactions) have been postulated. Plots of solute permeabilities (P) and solvent drag coefficients ($1 - \sigma$) against a suitable measure of molecular size (usually the Einstein–Stokes effective hydrodynamic radius or diameter) show two phases: a steep fall with increasing size in the "small molecule" range ($d_e < 60$ Å), with an extension or "tail" of much smaller slope above this range. Examples for mammalian skeletal muscle capillaries (taken from Tables 2 and 3) are shown in Fig. 3. By extrapolating the tails of the curves to the size of a water molecule ($d_e = 2.2$ Å), and subtracting the ordinates of the extrapolated curve from the measured values, estimates of the partition of solute fluxes and convective flows between small and large pore pathways can be made. The plot of P versus d_e (Fig. 3A) shows the partition for diffusion and/or vesicular turnover.[17] The data points for low-molecular-weight solutes fall on the theoretical curve for restricted diffusion through 80-Å-diameter pores. The macromolecular data points are fitted by alternative models: diffusion through 2000-Å pores and exchange (transcytosis) of 500-Å-internal-diameter vesicles. Both fit equally well; from this analysis alone there is no way to tell them apart, or if both are operative, to estimate the contribution of each to the total transport. For large pores alone, a total area of less than 2%

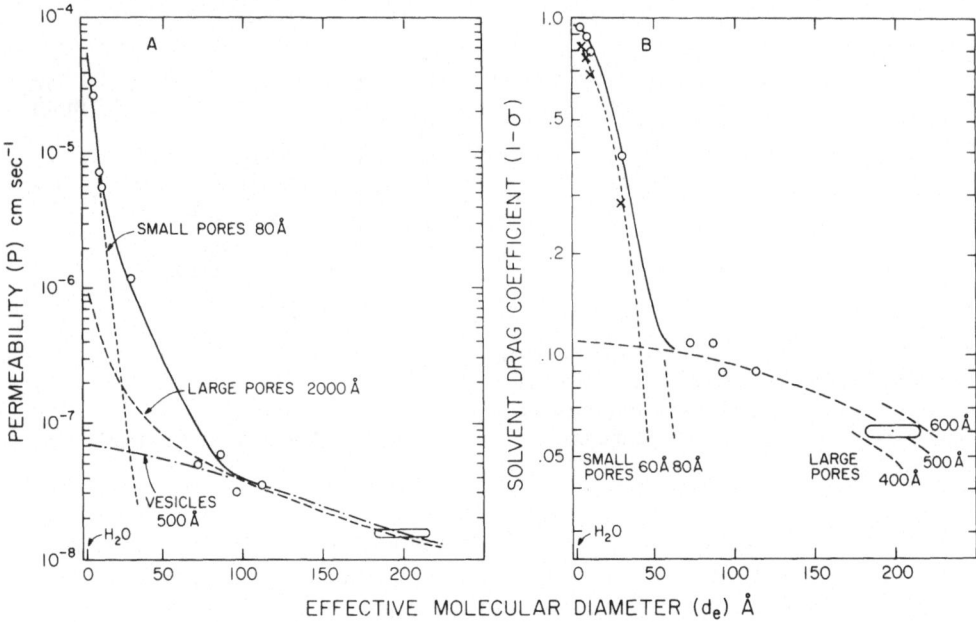

Figure 3. Graphical analyses of solute transport through microvascular endothelium. The open circles are experimental data from Tables 2 and 3. The dashed and dotted lines are theoretical curves, based on models and parameters as indicated. See Curry[10] and Taylor and Granger[56] for details of the models used. (A) *Partition of diffusive fluxes.* Data for small solutes are fitted only to the pore model, those for macromolecules to both pore and vesicle models. (B) *Partition of convective fluxes.* Data for small solutes are fitted only to the pore model, which underestimates small pore size because it neglects diffusive solute transport. Data for macromolecules are fitted to the pore model. The full curve for 500-Å pores is shown. Portions of the curves for 400-Å and 600-Å pores originating from the same ordinate intercept (0.11) are also shown, to illustrate the sensitivity of pore size evaluation. The sum of the ordinate intercepts (at 2 Å, the diameter of a water molecule) is not equal to 1.0 because about 5% of the convective flow goes through the cell membrane pathway.

of the small pore area is predicted from comparison of their ordinate intercepts. For vesicles alone, the ordinate intercept is equal to the volume turnover rate: 7 $\times$ 10^{-8} cm³/sec per cm² endothelial surface in this case. This plot yields reliable values for small pore and vesicular parameters (size and collective area or turnover rate), but overestimates the size and underestimates the number of large pores, due to neglect of convective transport through them.

A better evaluation of the convective role of the large pore pathway is given by a plot of $(1 - \sigma)$, the solvent drag coefficient, against d_e (Fig. 3B). Large pore diameter is estimated to be 500 Å, about one-fourth the value obtained from the *P* versus d_e curve. Small pore size is underestimated as 60 Å, due to neglect of diffusion. Vesicular transport, if present, is not represented because vesicular exchange does not result in a net volume flow. The ordinate intercepts of the theoretical curves for small and large pores represent the relative convective capacities of the two pathways. The large pores carry 11% and the small 84% of total ultrafiltrate flow. The remaining 5% is carried by the cell membrane pathway.

These graphs attempt to deal with convection and diffusion separately in a somewhat arbitrary manner. Under physiological conditions, both convective and diffusive transport occur simultaneously. The third question (below) deals with their interaction, and illustrates a procedure for evaluating their relative contributions under different conditions.

Note: Data for lipid-soluble solutes are not shown in Fig. 3. For small lipid-soluble molecules, at least, P's lie above the curve for lipid-insoluble molecules of comparable size. If it is assumed that their transport through intercellular pathways is the same for the same molecular diameter, the differences between data points and curves should indicate how much of the solute flux goes through the cell membrane pathway.

C. Transport of Large Molecules through Nonpore Pathways: Vesicular Transport?

Uptake of macromolecular and particulate tracers from plasma or interstitial fluid by endothelial cell vesicles has been demonstrated in capillaries of many organs and tissues,[27,52] but it has not been possible to estimate the amount of such transport from morphological studies. Attempts to evaluate solute fluxes due to vesicular exchange have followed two different approaches:

1. Suppression of vesicular exchange by temperature reduction. This procedure is based on the assumption that vesicular turnover has a greater dependence on temperature than diffusion or ultrafiltration. Temperature coefficients for water flow and for exchange of small molecules and ions across microvascular endothelium are close to values observed for passive flow or diffusion *in vitro*, i.e., proportional to the change in viscosity of water.[9,57] Rippe et al.[49] reported that the effect of temperature on the accumulation of radiolabeled albumin by tissues of perfused rat hindquarters was not significantly different from the passive value, and concluded that there could be little vesicular contribution to the transport of this substance.

2. Comparison of observed macromolecular fluxes with predictions for pore model systems. Transport through porous membranes by simultaneous (and interacting) convection and diffusion is described by a nonlinear relation[10,39]:

$$\frac{J_s}{C_p} = J_v(1 - \sigma)\frac{1 - (C_T/C_p)e^{-Pe}}{1 - e^{-Pe}} \tag{2}$$

where J_s/C_p is the clearance of the solute under study, C_p being its concentration in plasma and C_T in capillary filtrate. J_v is the volume flow (filtration rate), σ is the solute reflection coefficient, and e represents the base of natural logarithms. The exponent Pe is a modified Péclet number, equal to $J_v(1 - \sigma)/PS$, where P is the permeability and S the surface area of the endothelial barrier. Solute clearances are measured experimentally over a range of volume flows produced by increasing microvascular pressures stepwise. The characteristics of small and large pores are obtained from the high volume flow data, by the method illustrated in Fig. 3B, and total solute clearance through the pores (large and small) is calculated by means of Eq. (2) (see Ref. 45 for details of the procedure). If observed transport is greater than predicted, the difference can be taken as an estimate of vesicular or other nonpore transport. An obvious defect of this approach is its dependence on precise modeling. The present author[45] compared two sets of published experimental data with the pore models based on them. In both cases, observed transport was greater than the sum of calculated small-pore and large-pore transport, suggesting the presence of another transport pathway. The comparisons are shown in Fig. 4:

A. Dog paw (continuous, nonfenestrated capillaries; data of Renkin *et al.*[48]). The pore parameters (given in the legend) were determined by best theoretical fit to data at high values of volume flow (J_v). Slit and fiber matrix models were also tested, with similar results. At the lowest volume flow, observed transport of serum albumin was about 80% greater than the sum of calculated small pore and large pore fluxes. The absolute magnitude of the difference remained the same at all measured J_v values, though the proportion fell as convective flux through the pores increased. The discrepancy was observed also for larger plasma proteins, though it was proportionally less in magnitude, and tended to decrease at higher volume flows.

B. Cat ileum (continuous, fenestrated capillaries; data of Granger and Taylor[18]). The discrepancies between measured and predicted transport at low volume flows (J_v) were considerably larger than for the dog paw, indicating a larger nonpore component (accounting for two-thirds of the albumin flux at low J_v). As in the dog paw, nonpore clearance decreased with molecular size, consistent with steric exclusion from vesicles of 400–600 Å internal diameter.

VI. CONCLUDING REMARKS

Exchange of water and solutes across microvascular endothelium takes place through several cellular and intercellular pathways (Fig. 1). Although many studies of transport rates of water and a wide variety of solutes have been made and extensive investigations of endothelial morphology have been carried out, correlation of measured transport with observed structures remains obscure. It is generally agreed that water and small lipid-soluble solutes diffuse through the entire endothelial surface and that bulk flow of water and diffusive and convective transport of small lipid-insoluble molecules take place through intercellular channels. However, there is no general agreement on pathways and mechanisms for exchange of large molecules. Physiologists measure transport of plasma proteins or other large molecules and interpret their observations in terms of convection and diffusion through a large pore (or coarse fiber matrix) pathway. Morphologists demonstrate localization or binding of EM-visualizable macromolecules and interpret their findings in terms of vesicular exchange or confluence. Few studies have combined both methodologies to resolve the dilemma.

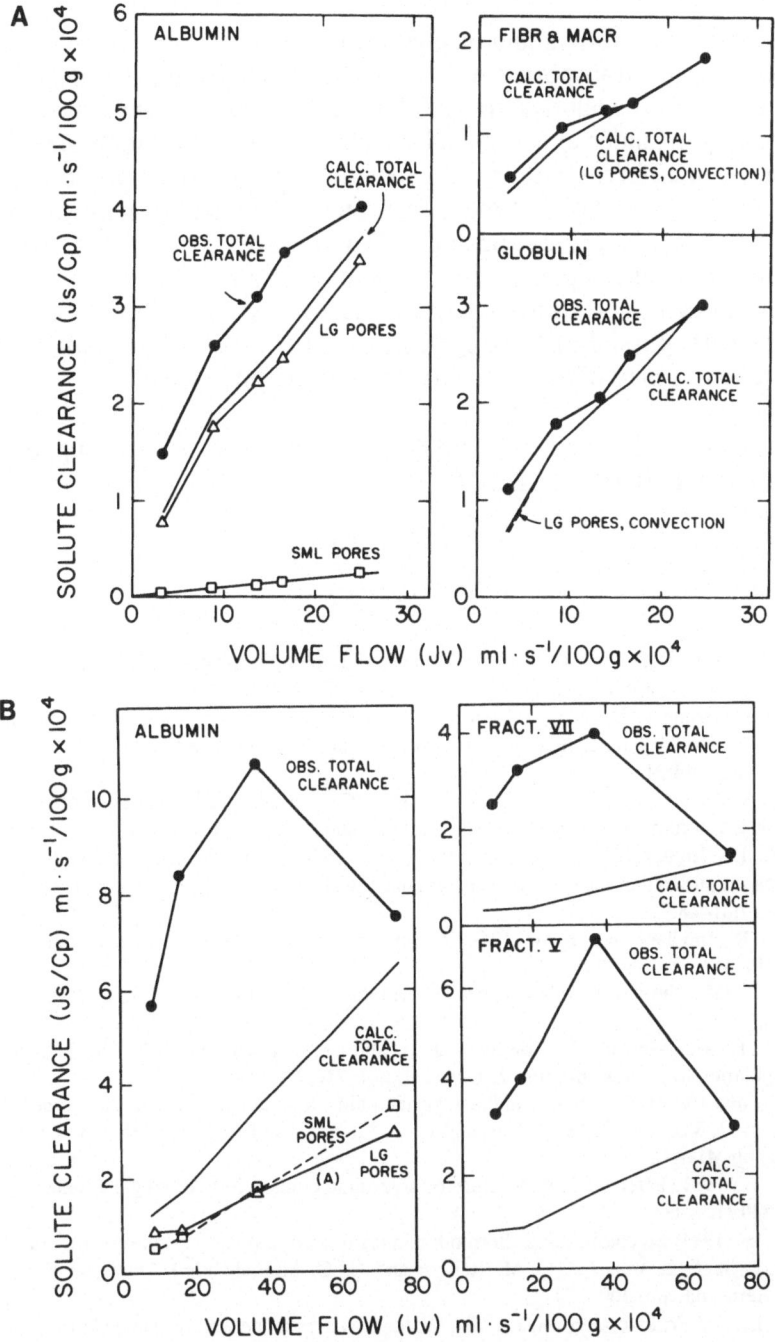

Figure 4. Graphical analysis of endothelial transport of plasma proteins by simultaneous convection and diffusion. See Curry[10] and Renkin[45] for details of procedure and models. Solute clearance = solute flux divided by plasma concentration (J_s/C_p). (A) Dog paw; data of Renkin *et al.*[48] Small pores 80 Å diameter, large pores 500 Å. (B) Cat ileum; data of Granger and Taylor.[18] Small pores 92 Å, large pores 400 Å. The effective diameter of protein fractions V and VII is 90 and 192 Å, respectively. In both preparations, the sum of modeled small and large pore transport is substantially less than observed blood-to-lymph clearances.

To identify the pathways and mechanisms by which a specific substance or class of substances is transported, it is necessary to evaluate the transport characteristics of each presumptive transport pathway, and to compare them with measured transport under experimental conditions chosen to emphasize the distinctions between different pathways. Application of this procedure to water and to large molecules suggests that more than one morphological pathway may be involved in transport of a given substance. Water moves through cell membranes and intercellular channels; plasma proteins and other macromolecular solutes are transported through intercellular channels and by another pathway not coupled to fluid transport, possibly involving vesicles. The relative contributions of the alternate pathways may differ in endothelia of different organs.

ACKNOWLEDGMENT. The author's study of this topic has been supported by USPHS Grants HL-10936 and HL-18010.

REFERENCES

1. Areekul, S., 1969, Reflection coefficients of neutral and sulphate-substituted dextran molecules in isolated perfused rabbit ear, *Acta Soc. Med. Ups.* **74:**129–138.
2. Brace, R. A., Granger, D. N., and Taylor, A. E., 1977, Analysis of lymphatic protein flux data. II. Effects of capillary heteroporosity on estimates of reflection coefficients and PS products, *Microvasc. Res.* **14:**215–226.
3. Brace, R. A., Granger, D. N., and Taylor, A. E., 1978, Analysis of lymphatic protein flux data. III. Use of the nonlinear flux equation to estimate sigma and PS, *Microvasc. Res.* **16:**297–303.
4. Carter, R. D., Joyner, W. L., and Renkin, E. M., 1974, Effect of histamine and some other substances on molecular selectivity of the capillary wall to plasma proteins and dextran, *Microvasc. Res.* **7:**31–48.
5. Chambers, R., and Zweifach, B. W., 1947, Intercellular cement and capillary permeability, *Physiol. Rev.* **27:**436–463.
6. Crone, C., 1965, The permeability of brain capillaries to nonelectrolytes, *Acta Physiol. Scand.* **64:**407–417.
7. Crone, C., Frøkjaer-Jensen, J., Friedman, J. J., and Christensen, O., 1978, The permeability of single capillaries to potassium ions, *J. Gen. Physiol.* **71:**195–220.
8. Crone, C., and Levitt, D., 1984, Capillary permeability to small solutes, in: *Handbook of Physiology*, Sect. 2, Vol. IV (E. M. Renkin and C. C. Michel, eds.) American Physiological Society, Bethesda, pp. 411–466.
9. Curry, F. E., 1981, Effects of temperature on hydraulic conductivity of single capillaries, *Am. J. Physiol.* **240:**H29–H32.
10. Curry, F. E., 1984, Mechanics and thermodynamics of transcapillary exchange, in: *Handbook of Physiology*, Sect. 2, Vol. IV (E. M. Renkin and C. C. Michel, eds.) American Physiological Society, Bethesda, pp. 309–374.
11. Curry, F. E., and Frøkjaer-Jensen, J., 1984, Water flow across the walls of single muscle capillaries in the frog, *Rana pipens*, *J. Physiol.* (*London*) **350:**293–307.
12. Curry, F. E., and Michel, C. C., 1980, A fiber matrix model of capillary permeability, *Microvasc. Res.* **20:**96–99.
13. Curry, F. E., Mason, J. C., and Michel, C. C., 1976, Osmotic reflexion coefficients of capillary walls to low molecular weight hydrophilic solutes measured in single perfused capillaries of the frog mesentery, *J. Physiol.* (*London*) **261:**319–336.
14. Deen, W. M., Bridges, C. R., and Brenner, B. M., 1983, Biophysical basis of glomerular permeability, *J. Membr. Biol.* **71:**1–10.

15. Deen, W. M., Satvat, B., and Jamieson, J. M., 1980, Theoretical model for glomerular filtration of charged solutes, *Am. J. Physiol.* **238**:F126–F139.

16. Drinker, C. K., 1946, Extravascular protein and the lymphatic system, *Ann. N.Y. Acad. Sci.* **46**:807–821.

17. Garlick, D. G., and Renkin, E. M., 1970, Transport of large molecules from plasma to interstitial fluid and lymph in dogs, *Am. J. Physiol.* **219**:1595–1605.

18. Granger, D. N., and Taylor, A. E., 1980, Permeability of intestinal capillaries to exogenous macromolecules, *Am. J. Physiol.* **238**:H457–H464.

19. Grotte, G., 1956, Passage of dextran molecules across the blood–lymph barrier, *Acta Chir. Scand. Suppl.* **211**:1–84.

20. Haraldsson, B., Eckholm, C., and Rippe, B., 1983, Importance of molecular charge for passage of endogenous macromolecules across capillary walls, studied by serum clearance of lactate dehydrogenase isoenzymes, *Acta Physiol. Scand.* **117**:123–130.

21. House, C. R., 1974, *Water Transport in Cells and Tissues*, Arnold, London.

22. Krogh, A., 1929, *The Anatomy and Physiology of Capillaries*, rev. ed., Yale University Press, New Haven, Conn.

23. Kruhøffer, P., 1946, The significance of diffusion and convection for distribution of solutes in interstitial space, *Acta Physiol. Scand.* **11**:37–47.

24. Landis, E. M., 1927, Micro-injection studies of capillary permeability. II. The relation between capillary pressure and the rate at which fluid passes through the walls of single capillaries, *Am. J. Physiol.* **82**:217–238.

25. Landis, E. M., 1964, The capillary circulation, in: *Circulation of the Blood, Men and Ideas* (A. P. Fishman and D. W. Richards, eds.), Oxford University Press, London, pp. 355–406. (Reprinted by the American Physiological Society, 1982.)

26. Landis, E. M., and Pappenheimer, J. R., 1963, Exchange of substances through the capillary walls, in: *Handbook of Physiology*, Sect. 2, Vol. II (W. F. Hamilton and P. Dow, eds.), American Physiological Society, Washington, D.C., pp. 961–1034.

27. Majno, G., 1963, Ultrastructure of the vascular membrane, in: *Handbook of Physiology*, Sect. 2, Vol. II (W. F. Hamilton and P. Dow, eds.), American Physiological Society, Washington, D.C., pp. 2293–2375.

28. Mason, J. C., Curry, F. E., and Michel, C. C., 1977, The effects of proteins upon the filtration coefficients of individually perfused frog mesenteric capillaries, *Microvasc. Res.* **13**:185–202.

29. Mayerson, H. S., 1963, The physiologic importance of lymph, in: *Handbook of Physiology*, Sect. 2, Vol. II (W. F. Hamilton and P. Dow, eds.), American Physiological Society, Washington, D.C., pp. 1035–1073.

30. Michel, C. C., 1981, Flow of water through the capillary wall, in: *Water Transport Across Epithelia: Barriers, Gradients and Mechanisms* (H. H. Ussing, N. Bindslev, N. A. Larsen, and O. Sten-Kundsen, eds.), Munksgaard, Copenhagen, pp. 268–279.

31. Michel, C. C., 1984, Fluid movements through capillary walls, in: *Handbook of Physiology*, Sect. 2, Vol. IV (E. M. Renkin and C. C. Michel, eds.), American Physiological Society, Bethesda, pp. 375–409.

32. Oleson, S.-P., and Crone, C., 1983, Electrical resistance of muscle capillary endothelium, *Biophys. J.* **42**:31–41.

33. Pappenheimer, J. R., 1953, Passage of molecules through capillary walls, *Physiol. Rev.* **33**:387–423.

34. Pappenheimer, J. R., Renkin, E. M., and Borrero, L. M., 1951, Filtration, diffusion and molecular sieving through peripheral capillary membranes; a contribution to the pore theory of capillary permeability, *Am. J. Physiol.* **167**:13–46.

35. Pappenheimer, J. R., and Soto-Rivera, A., 1948, Effective osmotic pressure of the plasma proteins and other quantities associated with the capillary circulation in the hindlimbs of cats and dogs, *Am. J. Physiol.* **152**:471–491.

36. Pardridge, W. M., and Landau, E., 1984, Tracer kinetic model of blood–brain barrier transport of plasma protein-bound ligands, *J. Clin. Invest.* **74**:745–752.

37. Pardridge, W. M., and Oldendorf, W. H., 1977, Transport of metabolic substrates through the blood–brain barrier, *J. Neurochem.* **28**:5–12.

38. Pardridge, W. M., Eisenberg, J., and Yang, J., 1985, Human blood–brain barrier insulin receptor, *J. Neurochem.* **44:**1771–1778.

39. Patlak, C. S., Goldstein, D. A., and Hoffman, J. F., 1963, The flow of solute and solvent across a two-membrane system, *J. Theor. Biol.* **5:**426–442.

40. Perry, M. A., Benoit, J. N., Kvietys, P. R., and Granger, D. N., 1983, Restricted transport of cationic macromolecules across intestinal capillaries, *Am. J. Physiol.* **245:**G568–G572.

41. Renkin, E. M., 1952, Capillary permeability to lipid-soluble molecules, *Am. J. Physiol.* **168:**538–545.

42. Renkin, E. M., 1953, Capillary and cellular permeability to some compounds related to antipyrine, *Am. J. Physiol.* **173:**125–130.

43. Renkin, E. M., 1954, Filtration, diffusion and molecular sieving through porous cellulose membranes, *J. Gen. Physiol.* **38:**225–243.

44. Renkin, E. M., 1977, Multiple pathways of capillary permeability, *Circ. Res.* **41:**735–743.

45. Renkin, E. M., 1985, Capillary transport of macromolecules: Pores and other endothelial pathways, *J. Appl. Physiol.* **58:**315–325.

46. Renkin, E. M., and Curry, F. E., 1978, Transport of water and solutes across capillary endothelium, in: *Membrane Transport in Biology*, Vol. IV (G. Giebisch, D. C. Tosteson, and H. H. Ussing, eds.), Springer-Verlag, Berlin, pp. 1–45.

47. Renkin, E. M., and Pappenheimer, J. R., 1957, Wasserdurchlässigkeit und Permeabilität der Capillarwände, *Ergeb. Physiol. Biol. Chem. Exp. Pharmakol.* **49:**59–126.

48. Renkin, E. M., Watson, P. D., Sloop, C. H., Joyner, W. L., and Curry, F. E., 1977, Transport pathways for fluid and large molecules in microvascular endothelium of the dog's paw, *Microvasc. Res.* **14:**205–214.

49. Rippe, B., Kamiya, A., and Folkow, B., 1979, Transcapillary passage of albumin, effects of tissue cooling and increases in filtration and plasma colloid osmotic pressure, *Acta Physiol. Scand.* **105:**171–187.

50. Scow, R. O., Blanchette-Mackie, E. J., and Smith, L. C., 1976, Role of capillary endothelium in the clearance of chylomicrons; a model for lipid transport from blood by lateral diffusion in cell membranes, *Circ. Res.* **39:**149–162.

51. Shea, S. M., Karnovsky, M. J., and Bossert, W. H., 1969, Vesicle transport across endothelium; simulation of a diffusion model, *J. Theor. Biol.* **24:**30–42.

52. Simionescu, M., and Simionescu, N., 1984, Ultrastructure of the microvascular wall: Functional correlations, in: *Handbook of Physiology*, Sect. 2, Vol. IV, American Physiological Society, Bethesda, pp. 41–101.

53. Starling, E. H., 1896, On the absorption of fluids from the connective tissue spaces, *J. Physiol. (London)* **19:**312–326.

54. Svensjö, E., Arfors, K.-E., and Grega, G. J., 1979, Morphological and physiological correlation of bradykinin-induced macromolecular efflux, *Am. J. Physiol.* **236:**H600–H606.

55. Taylor, A. E., and Granger, D. N., 1983, Equivalent pore modeling: Vesicles and channels, *Fed. Proc.* **42:**2440–2445.

56. Taylor, A. E., and Granger, D. N., 1984, Exchange of macromolecules across the microcirculation, in: *Handbook of Physiology*, Sect. 2, Vol. IV, (E. M. Renkin and C. C. Michel, eds.) American Physiological Society, Bethesda, pp. 467–520.

57. Wolf, M. B., and Watson, P. D., 1985, Effect of temperature on transcapillary water movement in isolated cat hindlimb, *Am. J. Physiol.* **249:**H792–H798.

4

Receptor-Mediated Transcytosis of Plasma Molecules by Vascular Endothelium

Maya Simionescu

I. INTRODUCTION

Vascular endothelium is endowed with a wide spectrum of functions among which of paramount importance is the exchange of substances between blood and tissues. Endothelium possesses all the basic properties of an epithelium highly specialized for transport: it is polarized, metabolically very active, and exhibits a structure and biochemistry subtly differentiated according to the activities of the tissue in which it resides.[8,107] As demonstrated by work carried out especially in the last decade, the participation of endothelial cells in the plasma–interstitial fluid exchanges is a more refined and dynamic phenomenon than initially considered.

In broad terms, the transport of plasma macromolecules across vascular endothelium (transcytosis) is governed by three groups of factors: (1) plasma driving forces, e.g., hydrostatic pressure and oncotic pressure, (2) physicochemical properties of the permeant molecules, i.e., size, shape, charge, chemistry, and concentration (generating gradients for exchange diffusion), and (3) properties and activities of the endothelial cells. Further permeation of molecules through the vessel wall largely depends on the physicochemical state of the basal lamina, extracellular matrix, and interstitial fluid. Within the temporal and topological conditions particular to each tissue, the interplay between these factors and the metabolic requirements of the host cells results in a broad endothelial heterogeneity and a variety of transport mechanisms which are still incompletely understood. Attempts to generalize the data obtained in a given vascular bed to all endothelia are equally unsound and misleading.

In the early 1950s and for an extended period of time afterwards, the capillary wall, which means essentially the endothelium, was considered a passive partition interposed between plasma and interstitial fluid. It was postulated that this partition is provided with two pore systems (small and large) with fixed hydraulic conductivities across which the permeant molecules pass according to their size.[43,85] Grown out of analogy with the physics of artificial membranes, the pore

Maya Simionescu • Institute of Cellular Biology and Pathology, 79691 Bucharest, Romania.

theory has lately become inadequate: it fails to explain the complexity and diversity of endothelial transport functions. Palade's pioneering work on the fine structure of endothelial cells has directed attention to the postulate of plasmalemmal vesicle involvement in the quantal transport of molecules across the endothelium.[82,83] The lack of ultrastructural confirmation of porelike structures in endothelial cells led to the extensive use of tracers of various dimensions injected *in vivo* and detected by electron microscopy. Despite some inherent limitations, this technique brought valuable information on the interactions with, and the pathways taken by, such probes across various endothelia (for reviews see Refs. 84, 104, 112, 113). However, some controversies still persist as to the role of vesicles[14,30] and junctions[13,21] in the transport of macromolecules.

A new twist was added to this issue when both physiologic and electron microscopic experiments revealed that in addition to size, the charge of molecules is an important parameter for their transport across endothelia.[16,22,33,34,94] Concomitantly, it was demonstrated that endothelial cell surface charge and chemistry is not homogeneously distributed but consists of biochemically differentiated microdomains generated by the preferential distribution of anionic sites and some glycoconjugates. These microdomains correspond exactly to the structural elements involved in endothelial transport such as plasmalemmal vesicles, channels, fenestrae, and their associated diaphragms.[100,101,107,108,110,112,116,122] The magnitude and kinetics of charge discrimination in endothelial transport remain to be elucidated in more detail.

Reviving old findings that showed that endothelial permeability increases as the concentration of plasma proteins decreases below a critical level, some recent data were interpreted as indicating the existence on the endothelial luminal front of a ''fiber matrix'' made up essentially of the endothelial glycocalyx and plasma albumin. This feltwork would actually control endothelial porosity and thus its permeability to macromolecules.[33] This hypothesis still remains to be confirmed and characterized in cellular and biochemical terms.

The last decade has seen an ever-increasing awareness in understanding the molecular mechanisms that are at the basis of endothelial functions including transport. Much effort has been invested in defining, from biochemical and cell biological standpoints, the dynamic control of endothelial transport and its multiple role in homeostasis. Numerous current studies have employed physiological tracers (i.e., actual plasma molecules) that have augmented markedly the physiologic and pathologic significance of the findings. As a result, growing evidence indicates that in addition to size and charge, endothelial cells can select plasma molecules according to their chemical properties.[103] In some capillary beds, certain plasma proteins are taken up preferentially by a specific process and are transported by receptor-mediated transcytosis. The new findings on the selective transport, in specific locations, of lipoproteins,[137,139,140] albumin,[36,38,117] transferrin,[123,124,131] insulin,[62-64] and ceruloplasmin[132] have revealed new parameters of vascular permeability: site specificity and chemical specificity in the transport of plasma proteins.

All these new data indicate that endothelial cells can actively regulate vascular permeability to molecules by refined mechanisms which, within the wide endothelial heterogeneity, impart to the transport processes a high complexity.

II. ENDOTHELIAL CELL REGULATION OF VASCULAR PERMEABILITY

There are several ways by which endothelial cells can regulate the vascular permeability to macromolecules. The mechanisms involved are largely reflected in the distinct structural and functional features that characterize endothelial cells in their various locations.

A. Endothelial Phenotypic Heterogeneity

1. Structural Heterogeneity

As a monolayer lining the entire circulatory system, the endothelium has an estimated total area of about 7000 m^2 corresponding to approximately 6×10^{13} endothelial cells that can be constituted in a solid mass of ~1 kg.[154] This ubiquitous simple squamous variety of epithelium, under the influence of local physiologic conditions, has undergone segmental differentiations involving the occurrence and frequency of its basic cellular constituents: plasmalemmal vesicles, channels, fenestrae, and their diaphragms. These dynamic modulations are expressed in a large phenotypic heterogeneity with distinctive features for the endothelium of large vessels versus that of microvessels, and among the latter significant differences between the arteriolar, capillary, and venular endothelium. Nonhomogeneity, connected with the specific functions of the surrounding tissue, is manifested at the level of various capillary endothelia which have been defined as continuous, fenestrated, or discontinuous (sinusoid). Broad modulations also exist within the same type of capillary endothelium, e.g., within the continuous endothelium, the extremes are the brain capillaries (with very few plasmalemmal vesicles) and the heart capillaries (particularly rich in such vesicles). In addition to variations in the numerical densities of plasmalemmal vesicles, channels and fenestrae, heterogeneity can also involve other cellular constituents directly or indirectly involved in transport such as coated pits–coated vesicles, the endocytotic tubular system,[157] or the cytoskeleton.[31]

2. Heterogeneity within a Single Cell

The endothelial cell luminal membrane has a net negative charge that accounts in part for its nonthrombogenic surface.[24] Yet, the surface charge is heterogeneously distributed generating microdomains devoid of strong anionic sites. These biochemically differentiated microdomains are characteristically associated with structures potentially involved in endocytosis (coated pits) and transcytosis (plasmalemmal vesicles, transendothelial channels, and fenestrae).[101,104,107,116,122] Plasmalemmal vesicles devoid of strong anionic sites seem to be devised to facilitate the transport of anionic molecules, which is the case for most plasma proteins. A special case is represented by the alveolar capillary endothelium, the avesicular zone of which has very few anionic sites. It was suggested that this may be associated with the gas exchange through the alveolar–capillary unit.[100] The exact nature and magnitude of the electrostatic interactions at the plasma–endothelial interface and the endothelial–interstitial fluid interface as well as their effects on normal and abnormal permeability remain to be further clarified.

3. Endothelial Cell Polarity

There is increasing evidence that endothelial plasma membrane constituents, especially proteins, exhibit luminal–abluminal polarity. Some proteins are present exclusively on the luminal surface[51,52] whereas other constituents prevail on one endothelial front or another.[76]

Of potential relevance for transendothelial transport is the asymmetric distribution on the fenestral diaphragms of anionic sites which occur in high density on the luminal aspect and are virtually absent on the abluminal surface of these structures.[109,115,122] The influence of the underlying matrix on the expression of endothelial phenotype is discussed by Madri *et al.* in Chapter 8. Endothelial cell polarity should be a crucial factor in endothelial permeability in general and in particular in the capillary endothelia of brain, eye,[92] and endocrine glands, where special metabolites cross the endothelium mostly from the abluminal side to the vascular lumen.

B. Enzymatic and Receptor-Mediated Regulation

Activities at the endothelial cell surface contribute to metabolic changes of some circulating plasma molecules either rendering them permeant through the endothelium (lipoproteins), more active (angiotensin II), or less potent (bradykinin). The latter two effects are due to the endothelium-associated angiotensin-converting enzyme. Chylomicrons are presumed to be trapped by the proteoglycans of the endothelial cell surface. The lipoprotein lipase (LPL), associated with the latter, hydrolyzes triglycerides of chylomicrons and very-low-density lipoproteins (VLDL), and thus is partially responsible for the plasma concentration of the lipolytic products of chylomicrons and VLDL, namely the high-density lipoprotein (HDL) and low-density lipoprotein (LDL). The LPL activity is genetically controlled and disturbances in its synthesis are partially responsible for abnormal concentrations of plasma VLDL and HDL.[106]

The endothelial cell has the ability to regulate vascular permeability by three groups of receptors or binding sites (Fig. 1).

1. *Vasomediator Receptors.* These are represented by endothelial cell membrane receptors for histamine, serotonin, bradykinin, and probably prostaglandins and angiotensin II. The pioneering work of Majno and Palade[71] has revealed that venules are particularly sensitive to histamine, which can induce endothelial contraction with subsequent junctional leakage and plasma extravasation.[59,70,130] This effect is transitory and restricted to small postcapillary venules, some of which

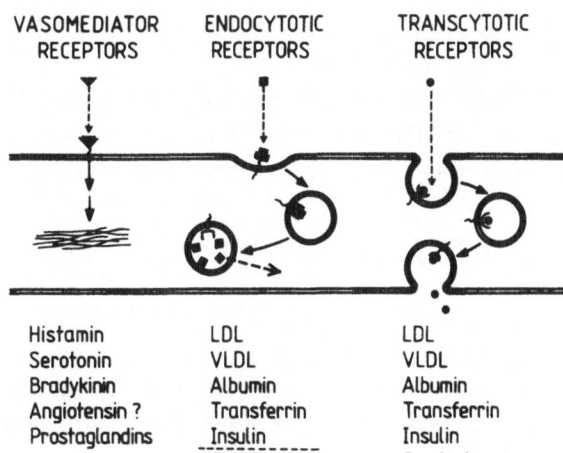

VASOMEDIATOR RECEPTORS	ENDOCYTOTIC RECEPTORS	TRANSCYTOTIC RECEPTORS
Histamin	LDL	LDL
Serotonin	VLDL	VLDL
Bradykinin	Albumin	Albumin
Angiotensin ?	Transferrin	Transferrin
Prostaglandins	Insulin	Insulin
	"Scavenger" r	Ceruloplasmin

Figure 1. Schematic representation of endothelial cell membrane receptors involved in the regulation of vascular permeability.

(~30%) constitutively display junctions open to a gap of 6 nm.[121] Recently, using a histamine–ferritin conjugate, it was demonstrated that venular endothelium has a characteristic high density of histamine (H_2) receptors concentrated in the parajunctional zone commonly associated with cytoskeletal elements.[2,46] In arteries, endothelial histamine receptors may mediate the transmission of the stimulus to the underlying smooth muscle cell. In frog brain venules, serotonin was shown to produce a selective increase in venular permeability to small ions presumably by the stimulation of serotonin receptors.[81] However, the increase in macromolecular transit produced by the inflammation mediators is a transient effect, restricted to small venules only, and is not operant in all microvascular beds (e.g., pulmonary venules).

2. *Endocytotic Receptors.* Under this name we designate the endothelial cell membrane specific binding sites for plasma constituents which are taken up and transported to the lysosomal compartment where they are degraded to smaller components to be used for the metabolic needs of the cell. To this category belong the receptors for LDL, β-VLDL, insulin, insulin growth factors, and transferrin. Chemically modified beta-lipoproteins or plasma proteins are taken up by a scavenger receptor. Though the intracellular routes taken by each of these ligands have not been fully elucidated, generally they reach the endosomal and lysosomal compartments, as in other cell types.

3. *Transcytotic Receptors.* By this name we designate the endothelial receptors that bind and transport plasma molecules across the cell to be delivered to the host tissue. There is increasing evidence emerging from physiological and cell biological studies that in some capillary beds certain plasma macromolecules are taken up specifically and transported through the endothelium by receptor-mediated transcytosis. Endothelial specific transporters have been implied in the transcytosis of LDL,[45,137,139,140] β-VLDL,[138] transferrin,[28,124,131] insulin,[62–64] and other proteins.

It appears that the endothelial cell is endowed with the dual machinery needed to transport the same molecular species both by endocytosis and by transcytosis (e.g., LDL, VLDL, insulin).

There is as yet no clear account as to the influence of other types of receptors (e.g., receptors for procoagulants and anticoagulants, receptors for components of extracellular matrix) on endothelial permeability.

III. MOLECULAR INTERACTIONS AT PLASMA–ENDOTHELIAL INTERFACE

The primary interaction of plasma molecules with the endothelium occurs at its luminal cell surface. As a result of this, permeant molecules are sorted, taken up, gated, and transported either to an intracellular compartment by endocytosis or across the cell by transcytosis.

The blood–endothelial interface is a dynamic composite of the cell coat (glycocalyx) of the endothelial luminal membrane and plasma proteins temporarily associated with this.

A. Endothelial Cell Surface (ECS)

The luminal front of endothelium occupies a special strategic position being continuously exposed to the dynamic internal milieu: it senses, transmits, and participates in the adjustments to deviations in homeostasis. The ECS actually consists of the ectodomains of membrane proteins, glycoproteins, glycolipids, and proteoglycans. It is still uncertain if the latter are true membrane components or are only attached to the glycocalyx. The ECS contains both cationic and anionic sites[34,122] but the prevalence of the latter imparts to the luminal front a net negative charge. At a finer analysis, it was found that neither the anionic sites nor the major oligosaccharide moieties are uniformly disposed on the ECS. Their preferential distribution generates chemically differentiated microdomains characteristically associated with endothelial structures involved in endocytosis (e.g., coated pits very rich in acidic residues), or transcytosis. Among the latter, plasmalemmal vesicles, transendothelial channels, and their diaphragms are devoid of strong anionic sites of low pK_a values, whereas the fenestral diaphragms display a high density of anionic groups on their luminal aspect and lack such residues on their abluminal face.[109]

ECS molecules can either be synthesized by the endothelial cell or are produced by other cells and secondarily attached to the ECS where they perform their activity (e.g., LPL). Components of endothelial glycocalyx may influence the transport as a sieving meshwork, a locally differentiated charge barrier, or by functioning as specific binding sites for certain plasma molecules to be endocytosed or transcytosed.

1. Proteoglycans

Sulfated proteoglycans are among the main contributors to the net negative charge of vascular endothelium. The luminal aspect of fenestral diaphragm is particularly rich in heparan sulfate, which seems to be absent from the abluminal aspect.[109,111] The LPL is attached to ECS proteoglycans where it hydrolyzes chylomicrons and VLDL. Heparin and related compounds can modulate the receptor-mediated uptake of LDL.[139] There is a suggestion that keratan sulfate contributes to the tightness of endothelial intercellular junctions (unpublished observations).

2. Sialoglycoconjugates

These components contribute largely to the net negative charge of plasma membrane, but unlike proteoglycans are absent from fenestral diaphragms. Sialoconjugates detectable with ferritin hydrazide are virtually excluded from the membrane of plasmalemmal vesicles and channels and occur at low density on most coated pits.[77,78] Enzymatic removal of sialic acid from the ECS of rabbit carotid arteries enhanced the uptake of LDL and fibrinogen.[42] It has been shown that ECS sialoconjugates are particularly sensitive to neuraminidase in venular endothelium, especially in old animals; as a result of neuraminidase treatment to remove sialoconjugates, a twofold greater number (~25%) of plasmalemmal vesicles (as compared with normal) are labeled by cationic ferritin.[110,111]

3. Glycoproteins

As revealed by studies with specific lectins, oligosaccharide residues such as mannosyl, N-acetylglycosaminyl, N-acetylgalactosyl, fucosyl, and galactosyl are present in a patchy distribution on endothelial plasmalemma of capillaries[108] and large vessels.[67] Plasmalemmal vesicles, channels, and their diaphragms are particularly rich in β-D-galactosyl and β-N-acetylglucosaminyl moieties, whereas fenestral diaphragms almost lack such residues. The oligosaccharide microdomains appear to a degree complementary to the anionic microdomains, but their exact involvement in transport has not been clarified.

B. Cell-Surface-Associated Plasma Proteins

Plasma molecules approach the endothelium within the thin unstirred layer of plasma intermixed with the luminal glycocalyx. Some molecules are taken up (permeant molecules) whereas others become temporarily associated with the molecules of the cell coat and are commonly designated as adsorbed molecules. Because of the technical problems in the chemical characterization of the plasma protein–glycocalyx interactions *in vivo*, the distinct molecular configuration and nature of this association are unknown.

Cytochemical techniques used to identify plasma proteins *in situ* have revealed the existence at the level of blood–endothelial interface of several molecular species such as fibrinogen–fibrin,[153,155] immunoglobulins,[89,90,98] α_2-macroglobulin,[7] and albumin.[9,89,90,98,155] The exact quantity and distribution of the immunoperoxidase-detected plasma proteins are difficult to assess because of the unavoidable enzymatic amplification effect and diffusion of the reaction product. A possible role of the adsorbed plasma proteins in various endothelial functions including transport was implied.[23]

1. Fibrinogen–Fibrin

An endo-endothelial lining of fibrin(ogen) was postulated as an anticoagulant insulation.[20] Fibrinogen and fibronectin injected intravenously were found to accumulate nonhomogeneously on the ECS, especially in venules, whereas γ-globulin and albumin did not show any affinity for the vessel wall.[153]

2. Albumin

As other plasma proteins detected immunocytochemically via a peroxidase reaction, endogenous as well as exogenous (perfused) albumin was visualized on the ECS.[9,98,155] The adsorbed albumin layer was completely removed from the endothelial cell membrane by simple perfusion with a physiological buffer or fluorocarbon emulsion.[98] This indicates that a weak, unstable interaction may exist between albumin and molecules of the glycocalyx. Moreover, recent observations showed that in numerous vascular endothelia examined, an albumin–gold conjugate does not bind significantly to endothelial plasmalemma. However, in some capillaries (heart, lung, skeletal muscle, and adipose tissue), this ligand was adsorbed with high affinity to restricted domains of the luminal plasma membrane, namely to uncoated pits and plasmalemmal vesicles.[36,38] Thus, the endothelial glycocalyx interacts with many plasma proteins, and obviously with all permeant macromolecules approaching the endothelial structures involved in their transport. There is no conclusive evidence that albumin in particular participates in the control of permeability as postulated by the fiber matrix theory.[23,136] Moreover, some investigators reported that, in rabbit heart, albumin has no significant effects on permeability of sucrose and inulin,[149] whereas others have demonstrated that serum factors other than albumin are required for the maintenance of normal capillary permselectivity of hindlimb muscle.[44]

IV. SORTING OF MOLECULES: ENDOCYTOSIS AND TRANSCYTOSIS

Under the special chemical and hemorheological conditions occurring at the blood–vessel wall interface, a fraction of circulating molecules retarded within the unstirred juxtaendothelial layer are actively sorted and gated by the endothelial cells, according to their molecular size, charge, and chemistry. The routes of sorted molecules vary, being either internalized and used for the cell itself (endocytosis) or transported across the endothelial cell (transcytosis) to the interstitial fluid. Since the size sorting has been extensively presented in other reviews,[84,93,115,133] we shall discuss the electrostatic and chemical sorting.

A. Electrostatic Sorting

The existence on the ECS of microdomains of different electrical charge or charge densities corresponding to structures involved in transport suggests that endothelium may discriminate permeant molecules by their charge in addition to size. The endothelial surface charge was detected indirectly, using electronopaque probes of various pI values. Plasma membrane does not bind significantly molecules of pI 4.5 to 7.0, but is labeled by probes of pI outside of this range.[4,33,34,110] In most endothelia, coated pits bind preferentially cationic tracers, except in liver sinusoids where they are marked by both anionic and cationic probes.[35]

Most plasmalemmal vesicles, transendothelial channels, and their diaphragms are not labeled by polycations of pI 7.2 to 9.0 which strongly bind to fenestral diaphragms. Probes of pI higher than 9.0 (e.g., cationized hemeundecapeptide, cationic ferritin pI > 9.2, cytochrome C pI 10.6) decorate the entire ECS including open plasmalemmal vesicles. The membrane of plasmalemmal vesicles is devoid of strong anionic sites, but expresses very weak acidic residues of high pK_a values; thus, vesicles discriminate against cationic macromolecules up to pI 9.0 only. This upper charge limit for electrostatic restriction was identified in the fenestrated endothelium of mouse pancreatic capillaries.[34] The surface charge and chemistry of vesicle membrane may show significant modulations in other endothelia.[110] Thus, it appears that, at least in some capillary beds, plasmalemmal vesicles and channels are designed to facilitate the access of anionic molecules (that is the case with most plasma proteins). In visceral capillaries, the fenestrae provided with a diaphragm rich in heparan sulfate-generated anionic sites may repel anionic proteins. This presumption is supported by physiological experiments which show that the presence of fenestrae is not associated with a higher permeability to proteins (as compared to continuous endothelia) but with a greater permeability to water and small solutes.[93,133] It was demonstrated that the pI of ferritin significantly influences its transport through capillaries with continuous[6,136] or fenestrated endothelium[33] as well as across arterial and vasa vasorum endothelia.[4,5] Cationized IgG was found to be more extensively transcytosed across isolated capillaries *in vitro* as compared to native IgG.[27]

Considering the local variations in charge and charge densities and the preferential occurrence in some endothelial vesicles of specific binding sites for plasma proteins, one realizes the wide modulation and biochemical heterogeneity in the otherwise morphologically similar population of plasmalemmal vesicles. The effect of this nonhomogeneity on the transport of macromolecules in various endothelia remains to be further elucidated. Taken together, these data support the idea that the endothelial cell is a refined charge barrier. The uptake and transport across the endothelium are probably influenced not only by the charge and size of permeant molecules, but also by the size and net charge of endothelial structures involved in transport. The findings suggest that anionic plasma proteins may be to a degree repeled by the negatively charged plasmalemma; the repulsion can generate concentration gradients at the level of the introits of plasmalemmal vesicles and channels, a condition that may favor diffusion of anionic molecules in these structures.

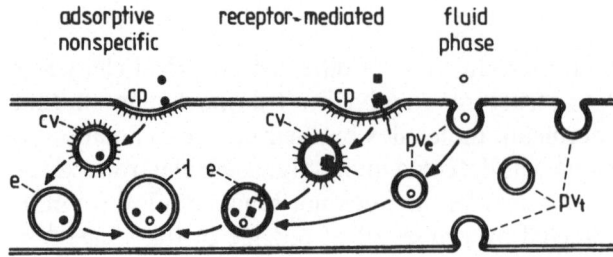

adsorptive   receptor-mediated   fluid
nonspecific   phase

Figure 2. Diagrammatic illustration of the mechanisms and pathways of endocytosis in endothelial cells. cp, coated pit; cv, coated vesicle; e, endosome; l, lysosome; pv_e, plasmalemmal vesicles involved in endocytosis; pv_t, plasmalemmal vesicles performing transcytosis.

B. Chemical Sorting

There is increasing evidence that in some endothelia, certain plasma constituents can be selectively taken up by specific binding sites (or receptors) and either internalized into, or transported across, the cell. Some of these binding sites are of high affinity and are localized either to coated pits (e.g., LDL receptors and β-VLDL receptors) or to uncoated pits and plasmalemmal vesicles (e.g., LDL and albumin binding sites). Plasmalemmal vesicles appear to have also low-affinity binding sites for LDL and β-VLDL.[137–140] Other plasma molecules (insulin, transferrin, ceruloplasmin) have been shown to bind to specific sites on the ECS and to be transported by a receptor-mediated mechanism. It was also demonstrated that in microvessels isolated from rat epididymal fat vesicle ingestion of glycosylated albumin was more pronounced than that of albumin.[150–152] These observations strongly suggest that chemical specificity may play an important role in endothelial uptake of macromolecules.

C. Endocytosis

Molecules to be delivered to the endothelial cell itself can be taken up by adsorption or by fluid phase. Adsorption usually occurs restrictively on coated pits either as a nonspecific process (nonspecific adsorptive endocytosis) or as binding to a cognate receptor (receptor-mediated endocytosis) (Fig. 2). The first case is exemplified by any polycation bound to and internalized by coated pits–coated vesicles, to endosomes, multivesicular bodies, and lysosomes. Receptor-mediated endocytosis is represented by the internalization of LDL or β-VLDL particles.[138,139] The LDL high-affinity pathway is saturable and temperature-dependent and presumably provides the endothelial cell with cholesterol for membrane biogenesis. The cell can also take up chemically modified lipoproteins and albumin via distinct scavenger receptors.[3,10,48–50,91]

Fluid-phase endocytosis or pinocytosis may be performed by a small fraction of plasmalemmal vesicles; such is the case with endocytosis of the albumin–gold complex especially in the capillaries with a continuous endothelium. In other endothelia (e.g., aortic, arterial, venous), the same ligand (albumin–gold) appears to be endocytosed by both coated pits and some plasmalemmal vesicles.[39,105]

Pulmonary endothelial cells in culture are able to phagocytose 5-μm poly-

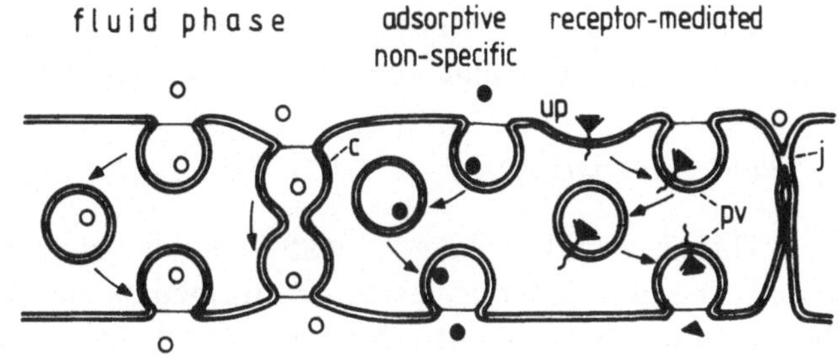

fluid phase adsorptive receptor-mediated
non-specific

Figure 3. Diagram of mechanisms and pathways for transcytosis of macromolecules across endothelial cell. The carriers are plasmalemmal vesicles (pv) which either can shuttle between the two endothelial fronts or can fuse to form transient transendothelial channels (c). Upon ligand binding, uncoated pits (up) of the luminal membrane may form vesicles carrying the ligand. j, interendothelial junction.

styrene beads, a process associated with the expression of Fc receptors on the ECS.[96]

D. Transcytosis

In eukaryotic cells, carrier vesicles are used either to discharge their secretory products by exocytosis, to ingest macromolecules by endocytosis, or to transport macromolecules across the cell by transcytosis. The possible role of plasmalemmal vesicles as carriers through the endothelial cell was first postulated by Palade.[82,83] This vesicular transport has received along the years various names (for review see Refs. 114, 115, 117). In the last decade, new parameters and mechanisms of endothelial transport have been identified: e.g., passage through transient channels formed by fused vesicles,[120] receptor-mediated transport.[105] Moreover, it was found that structures other than plasmalemmal vesicles can carry large molecules and particles across the cell, i.e., coated vesicles, vacuoles. This broader concept covering all these possible mechanisms has been included in the term *transcytosis* introduced in 1979 by N. Simionescu.[112] Transport in quanta via discrete vesicles represents dissipative transcytosis, whereas that through patent channels and fenestrae represents convective transcytosis. Both the concept and the terms, *transcytosis* and *receptor-mediated transcytosis*, have been extended to similar processes occurring in other endothelial or nonepithelial cells.[1,47,72,75,86]

Transcytosis employs the same basic uptake mechanisms that operate in endocytosis: fluid-phase ingestion, nonspecific adsorption, and specific receptor-mediated adsorption (Fig. 3). Unlike endocytosis, vesicle carriers which are instrumental in transcytosis bypass the lysosomal compartment and transport the molecules directly across the cell. Particles, macromolecular complexes, macromolecules, small solutes, and fluid can be transported by transcytosis. Some molecular species (e.g., LDL, β-VLDL, IgG, albumin, insulin, transferrin) can be concomitantly endocytosed and transcytosed. For fluid-phase transcytosis the rate of uptake depends on the size and number of vesicles, rate of their formation, solute concentration, steric exclusion; for receptor-mediated transcytosis the deciding factors are the number and affinity of binding sites. Frequently, *in vivo* transcytosis is a dynamic combination of fluid and adsorptive uptake.

1. Fluid Transcytosis

This is a bulk uptake of water-soluble molecules that do not bind to the membrane of the carrier vesicle. Performed by smooth vesicles, this process represents a major mechanism for transendothelial exchanges.

2. Nonspecific Adsorptive Transcytosis

Molecules can bind by electrostatic forces to the membrane of a carrier vesicle: hemeundecapeptide pI 4.85, anionized hemeundecapeptide pI 3.5, cationized hemeundecapeptide pI 10.6,[34] and cytochrome C pI 11 are some examples.

3. Receptor-Mediated Transcytosis

Since the early 1980s it has been intimated that the endothelial cell (at least *in vitro*) can recognize selectively some molecules. The vesicular ingestion exhibits preference for anionic proteins as compared to cationic species, for carbohydrate-conjugated albumin versus native albumin.[152] Some molecules can bind specifically to high-affinity receptors. In epithelia, such binding sites have been found concentrated in most cases in coated pits: e.g., hepatocyte receptors for IgG, asialoglycoproteins, galactosylated albumin, asialotransferrin, IgA; fetal yolk sac receptors for IgG; carcinoma cell receptors for transferrin (for brief review see Ref. 75). In vascular endothelium, high-affinity binding sites have been identified for albumin using an albumin–gold complex; they are localized restrictively in some uncoated pits of the luminal plasmalemma and on the membrane of plasmalemmal vesicles. Such putative albumin receptors exist in the capillary endothelium of lung, heart, skeletal muscle, and adipose tissue.[38,39] Since albumin–gold complex binding is competed by monomeric albumin, it is assumed that both polymeric and monomeric albumin share the same binding sites. Receptor-mediated transcytosis in endothelia can also involve low-affinity binding sites such as those for LDL and β-VLDL that have been localized also to uncoated pits and some plasmalemmal vesicles.[137–139] The involvement of plasmalemmal vesicles in receptor-mediated transcytosis adds a new line of evidence for the active participation of these organelles in transport. Based on three-dimensional reconstructions of a tiny fragment of a chemically fixed capillary endothelial cell, the almost total absence of free vesicles in the cytoplasm was taken as proof of vesicle immobility. Thus, it was implied that vesicles do not participate in transport.[14,30] These results were contradicted by other investigations using a comparable stereological approach[15,143,145] as well as by recent tracer experiments in which the probe molecules were injected *in vivo* before the endothelium was chemically fixed.[11,18,19,57,58,73,88,92] Of special relevance is the evidence for vesicular transport obtained with physiological probes, i.e., plasma molecules, in time-course experiments.[27,28,36–40,65,74,79,89,90,97,98,118,131,132,137–142,146,150–152,155]

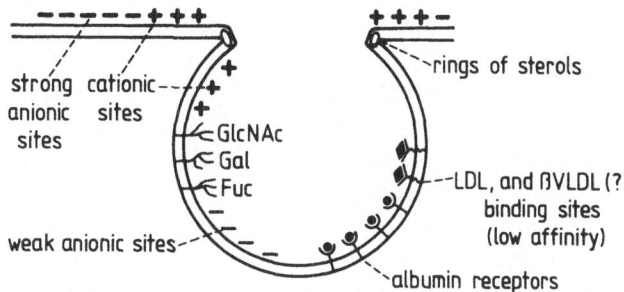

Figure 4. Diagrammatic illustration of some biochemical characteristics of the membrane of plasmalemmal vesicles so far detected in some capillary endothelia.

size and charge sorting:
FLUID PHASE,
and ADSORPTIVE NONSPECIFIC
TRANSCYTOSIS

chemical sorting:
RECEPTOR-MEDIATED
TRANSCYTOSIS

4. Some Characteristics of Transcytosis in the Endothelial Cell

The discovery of biochemically differentiated microdomains of the ECS revealed that the membrane of the plasmalemmal vesicle is chemically distinct from the neighboring cell membrane (Fig. 4). In most endothelia examined, the vesicle membrane appears to be poor or devoid of strong anionic sites of low pK_a value, sialoconjugates, and proteoglycans; other moieties occur in characteristic high density, i.e., weak anionic groups of high pK_a values, cationic residues (essentially amino groups), galactosyl-, and N-acetylglucosaminyl-rich glycoproteins, and low-affinity binding sites for LDL and β-VLDL. Moreover, on this chemical background, plasmalemmal vesicles of some continuous capillaries (heart, lung, skeletal muscle, adipose tissue) contain albumin receptors (Fig. 4).

The finding that albumin–gold conjugate perfused through the vasculature reaches rapidly multivesicular bodies while coated pits and coated vesicles are not marked by the tracer, suggests that endocytosis of this ligand is performed by a fraction of plasmalemmal vesicles that carry the complex by fluid phase. Thus, it is apparent that the structurally uniform plasmalemmal vesicles are represented (at least) by two functionally distinct types: pinocytotic (or endocytotic) vesicles and transcytotic vesicles. The variety in behavior of plasmalemmal vesicles in different vascular beds also strongly suggests that these structures may be very heterogeneous in physiological properties and roles.

The observations obtained on various endothelia examined *in vivo*, *in situ*, or *in vitro* revealed that transcytosis can be, at least in part, regulated by temperature, pressure,[17] molecular composition of the surrounding medium (being enhanced by anionic proteins), divalent cations,[135] serotonin, glucose concentration, and so forth (for review see Ref. 114). Vesicle movement through the endothelial cell is considered to be essentially random for the bulk fluid transport. During receptor-mediated transcytosis—by analogy with epithelial transcytosis of IgA—the transit might be strictly controlled by an assumed conductor–receptor couple targeting the vesicle to cognate site of fusion on the opposite plasma membrane.

Depending on the endothelium involved and on the experimental conditions used, transcytosis was demonstrated to occur, in various degrees, either bidirectionally[57,120] or unidirectionally (as in the retinal and iris vessels[92]).

Data on the energy requirements for transcytosis are still controversial. Since ATP inhibitors usually do not depress endothelial transport *in vitro*, it was assumed that ATP is not an energy source for fluid transport.[144,146] However, by analogy with endocytosis and exocytosis, vesicle attachment and detachment from the plasmalemma may be energy dependent.

Since fluid transcytosis seems to be constitutive,[114] it can be presumed that *de novo* vesicle formation may be induced by specific ligands (i.e., LDL, β-VLDL, albumin), by invagination of a restricted microdomain of the plasmalemma appearing as an uncoated pit. The vesicles may require a certain recycling mechanism (not yet demonstrated), with subsequent recovery of the transcytotic receptor.

A rather rare form of transcytosis occurring in abnormal conditions is what was called "clearing transcytosis" by which aggregates of polycations (such as cationized ferritin clusters formed *in vivo*) can be cleared from plasma by an extensive extravasation involving smooth vesicles and vacuoles as well as coated vesicles. This process unique to the endothelium and different from endothelial phagocytosis[96] was observed in most capillary beds except brain microvessels.[117]

The membrane traffic and the extent of recycling of membranes and receptors during transcytosis in endothelium remain a largely unknown and very exciting field of research.

V. RECEPTOR-MEDIATED TRANSCYTOSIS

It has been demonstrated that on the luminal front of endothelium there are specific binding sites (putative receptors) for several plasma components such as histamine,[46] insulin,[25,26,62–64] transferrin,[56,124] LDL,[45,137,139] and β-VLDL.[3,138] Recently, evidence was obtained that some of these specific binding sites are instrumental in a chemically selective transport of plasma proteins across endothelium. The concept of receptor-mediated transcytosis was increasingly substantiated by cell biological, cytological, and physiological data on the preferential specific transendothelial transport of important molecules such as albumin, LDL, β-VLDL, transferrin, insulin, ceruloplasmin, and probably transcobalamin. The cellular events taking place during receptor-mediated transcytosis have been most extensively investigated for albumin.

A. Albumin

Serum albumin is reputably known to be not only largely responsible for the oncotic pressure of plasma and interstitial fluid, but also for providing a reservoir of bound (and rapidly dissociating) ions, and to serve as a carrier for important molecules such as fatty acids, amino acids, bilirubin, cholesterol, steroid and thyroid hormones.[29,87,127,148] Receptors for albumin were demonstrated on liver cells and on HB virus.[68,69,134] In hepatocytes and myocardium, it was demonstrated that the uptake of fatty acids is mediated by an albumin receptor.[53,54,80,147] Estimates of lymph/plasma ratio revealed that the amount of albumin passing through endothelium varies largely from one organ to another (for references see Ref. 38). However, no firm correlation could be established between the endothelial structure (continuous, fenestrated, discontinuous) and the rate of albumin transport.[133] Thus, it became apparent that other properties of the endothelial cell may account for the difference in rate, speed, and kinetics of albumin uptake by various tissues.

Recent evidence indicates that in the endothelium of capillaries and small pericytic venules of the lung, heart, skeletal muscle, and adipose tissue, the transport of albumin occurs by adsorption on specific binding sites restrictively located on plasmalemmal vesicles followed by the transport of ligand-bound vesicles across the cell. On the contrary, in all other endothelia examined (arterial, arteriolar, venular, venous, endocardiac, fenestrated, and sinusoid capillaries) albumin is transported at a low rate by nonspecific fluid-phase transcytosis. These findings emerged from studies *in vivo* and *in situ* with radiolabeled monomeric albumin or albumin–gold complex and other tagged proteins used as controls.[36–40,102,103]

1. Chemically Specific Transcytosis

After blood was removed by perfusion with Dulbecco's phosphate-buffered saline (PBS), homologous or heterologous serum albumin adsorbed on 5-nm gold particles (Alb-Au) was perfused *in situ* in mouse and rabbit. After 3 to 35 min the unbound conjugate was washed out, the vasculature was fixed by perfusion, and various vessels and tissues were processed for electron microscopy. In the endothelium of capillaries and small postcapillary pericytic venules (smaller than 20 μm in diameter) of the lung, heart, skeletal muscle, diaphragm, and adipose tissue, Alb-Au bound restrictively to the membrane of plasmalemmal vesicles open to the luminal front and to some uncoated pits of plasma membrane (Figs. 5, 6). This binding pattern of Alb-Au was absent on the brain cortex capillaries, fenestrated and sinusoid capillaries. The scarcity of Alb-Au binding to plasma membrane does not support the postulate that *in vivo* albumin is electrostatically attached to the glycocalyx of the endothelium.[23] Morphometric analysis showed that, per unit volume, the concentration of particles within a vesicle (mostly adsorbed on membrane) was 1000 times higher than in the perfusate, and was much greater than expected from a Poisson distribution. These data indicate that the vesicle loading with Alb-Au does not result solely from diffusion from the lumen but from strong attractive forces exposed on the vesicle membrane. The findings also suggest that some of the ligand-carrying vesicles may be formed upon Alb-Au binding to their receptors followed by clustering in uncoated pits. A similar observation emerged from the experiments with LDL and β-VLDL. In vesicles

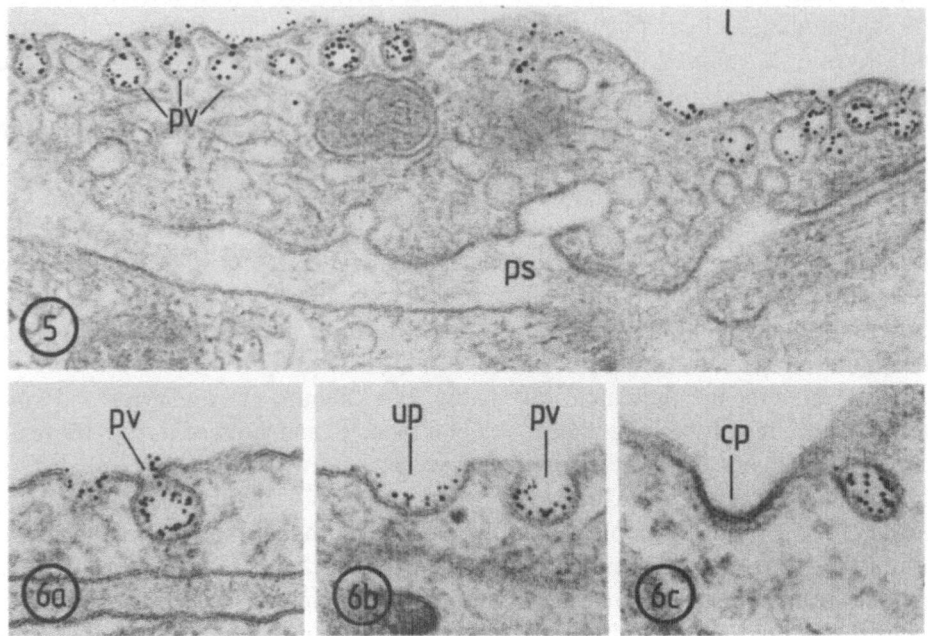

Figures 5 and 6. Binding of albumin–gold conjugate (Alb-Au) to specific domains of the endothelial luminal surface in capillaries of adipose tissue (Fig. 5) and lung (Fig. 6). At 3 min of perfusion, Alb-Au characteristically labels the membrane of plasmalemmal vesicles (pv) open to the vessel lumen (l) (Figs. 5, 6a) and uncoated pits (up) (Fig. 6b), whereas coated pits (cp) are not decorated (Fig. 6c). ps, pericapillary space. Fig. 5, × 74,000; Fig. 6, × 108,000.

with higher load, in addition to Alb-Au particles adsorbed on the vesicle membrane, other particles may occur in the middle of vesicles apparently in fluid phase.

a. Albumin Binding Is Saturable. Using increasing ligand concentrations, morphometric analysis revealed that binding of Alb-Au on alveolar capillaries was saturable at very low concentrations and short exposure. Similar results were obtained with radioiodinated monomeric albumin ($[^{125}I]$-Alb) or albumin–gold complex ($[^{125}I]$-Alb-Au) perfused in the lung and analyzed by radioassay of the pulmonary parenchyma (Fig. 7).

b. Albumin Binding Is Specifically Competed. When Alb-Au was injected *in vivo* (in the presence of homologous serum albumin) or its binding *in situ* was competed with free albumin, no significant binding was detected (Fig. 8). These observations suggest that both Alb-Au and monomeric albumin compete for the same binding sites on vesicle membrane. The fact that neither heparin nor high ionic strength could displace the Alb-Au attached to vesicle membrane, indicates that the binding was not electrostatic in nature. Gold complexes with other plasma proteins (fibronectin, fibrinogen) or lipoproteins (LDL, β-VLDL), or nonplas-

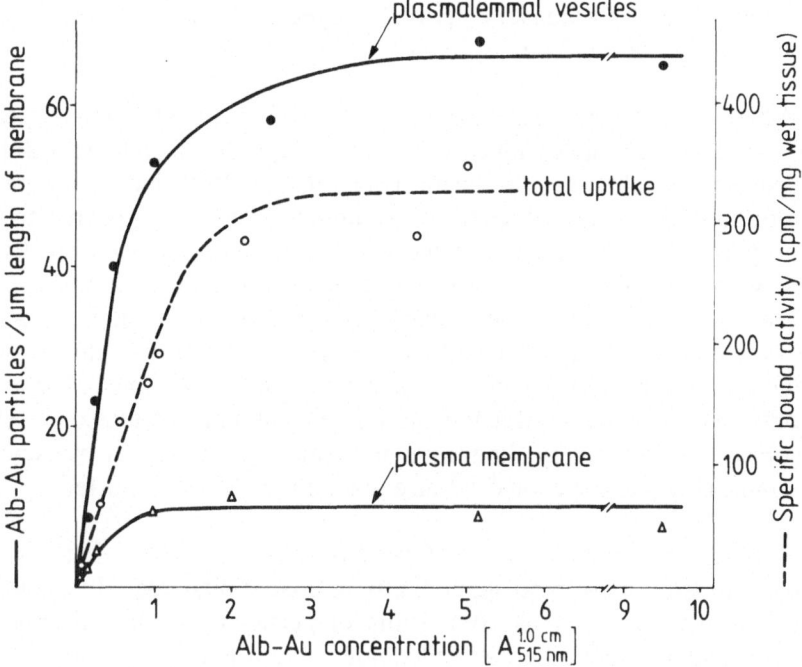

Figure 7. Saturation curve of Alb-Au binding and uptake by the mouse lung. Solid lines represent the number of Alb-Au particles bound per micrometer length of membrane as a function of tracer concentration. Note that the binding of Alb-Au is about six times higher on the plasmalemmal vesicles than on the plasma membrane proper. Dashed line shows the total uptake of radiolabeled A!b-Au perfused *in situ* for 3 min at various ligand concentrations.

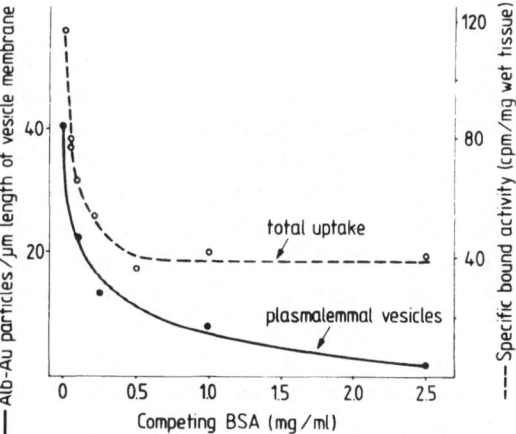

Figure 8. Competition by monomeric bovine serum albumin (BSA) of the uptake of [^{125}I]-Alb-Au (dashed line) and the binding of Alb-Au to plasmalemmal vesicle membrane (solid line).

matic proteins (glucose oxidase) or polyethyleneglycol did not give a decoration comparable to that of albumin.

c. *Alb-Au Is Transcytosed Exclusively by Vesicles.* After 3 to 5 min of perfusion with Alb-Au, labeled vesicles appeared within the cytoplasm and on the abluminal front apparently discharging the ligand (Figs. 9, 10). With increasing perfusion time, more labeled vesicles reached the abluminal front and appeared to release the ligand particles in the subendothelial space (Figs. 11, 12). Occasionally ligand particles were detected along transendothelial channels and beyond their abluminal stomata. Transport was particularly prominent in heart and lung capillaries (but this might be due to a better perfusion of these organs). No Alb-Au particles were seen in the intercellular junctions of the endothelia examined. These results were confirmed with radioiodinated monomeric albumin analyzed by electron microscopic autoradiography. Immunocytochemical evidence for transcytosis of free albumin, either endogenous[9,155] or exogenous,[74,98] was reported.

d. *Alb-Au, Though Endocytosed, Does Not Label Coated Pits.* At any time of perfusion, Alb-Au did not bind significantly to coated pits–coated vesicles (Figs. 6c, 13). However, as early as after 5 min of perfusion the ligand appeared in structures tentatively identified as endosomes and in multivesicular bodies (Fig. 14). The plasmalemmal vesicles which took up Alb-Au in fluid phase ($\sim$10%) seemed to be the carriers for the endocytosed particles (Fig. 14). The routing to the lysosomal compartment may either represent a normal endocytotic pathway for a certain amount of albumin or may be the consequence of the multivalent character of the Alb-Au complex. The adsorption of Alb-Au to coated pits was totally different in endothelia where the ligand is transcytosed in fluid phase only.

Modulation in the density and affinity of endothelial albumin binding sites in different microvascular beds may explain, at least in part, the difference in the lymph/plasma concentration of albumin detected in various organs and tissues.[133]

The results obtained with homologous (mouse) and heterologous (bovine) albumin were similar although the binding affinity of the former seems to be higher. This suggests that the endothelial albumin receptor has been phylogenetically well preserved with no major species differences in its expression.

e. *Alb-Au Does Not Bind Specifically to Alveolar Epithelium.* To determine whether in the lung this ligand also has affinity for alveolar type I or type II epithelial cells, experiments were conducted in the mouse, in which under general anesthesia without or with previous bronchoalveolar lavage (with 0.5 ml of PBS, five times at 37°C) 0.3 ml of Alb-Au at a concentration corresponding to $A_{515\ nm}^{1.0\ cm} = 1.0$ was infused intratracheally. After 7 min, the unbound tracer was removed with PBS, followed by intratracheal administration of a triple fixative;[119] specimens were then collected and processed for standard electron microscopy. It was found that Alb-Au did not bind significantly to alveolar epithelial cells, but the ligand particles were, as expected, avidly ingested by the alveolar macrophages, wherever these were present. This indicates that, at the level of the alveolo-capillary unit, among the luminal surfaces exposed to the blood or to the

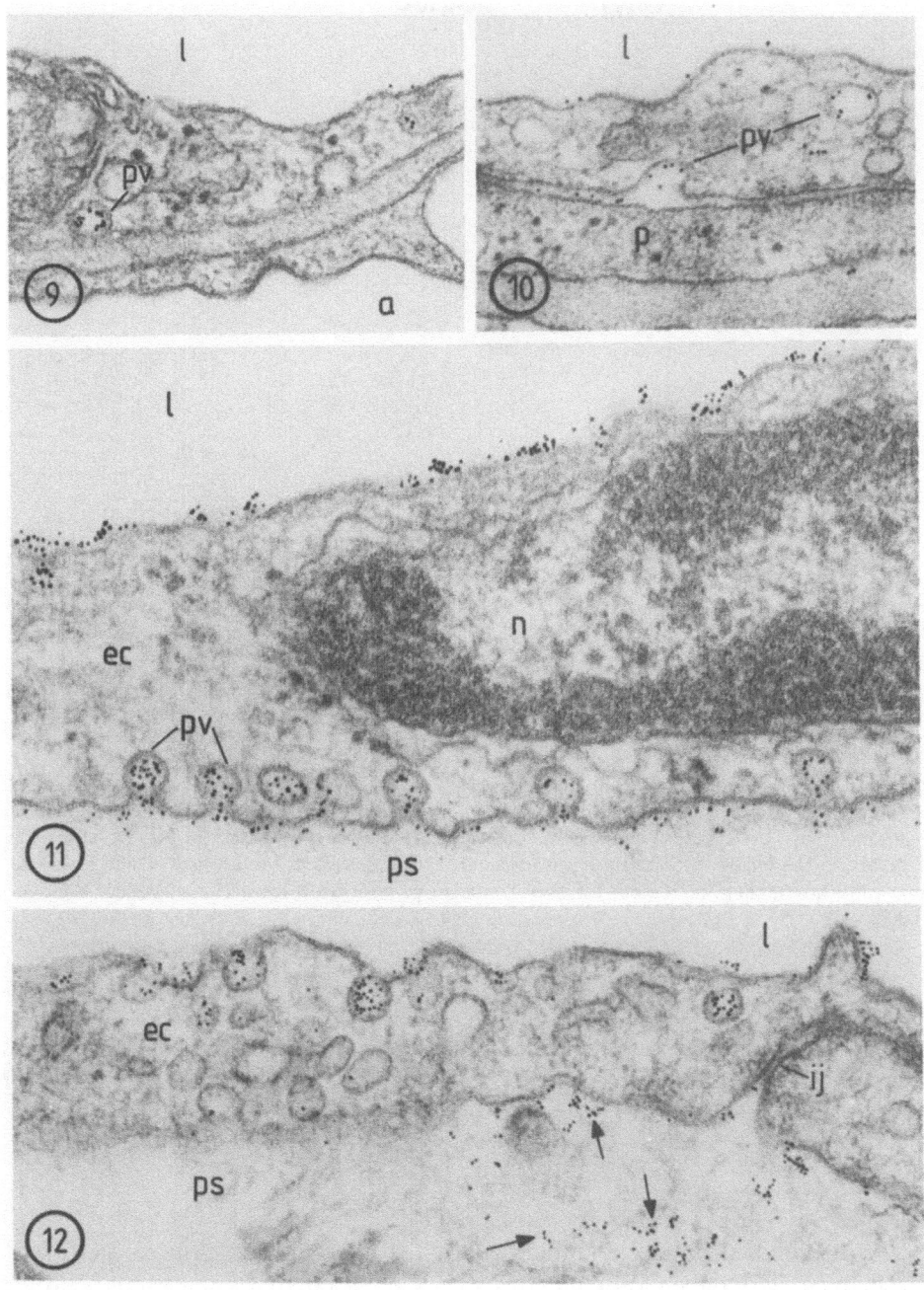

Figures 9–12. Transcytosis of albumin–gold complex (Alb-Au) by plasmalemmal vesicles (pv). After 5 min of perfusion, Alb-Au-labeled vesicles approach the abluminal front of endothelium (Fig. 9), and apparently discharge the ligand in the pericapillary space (ps) (Figs. 10 and 11). At 35 min, Alb-Au appears free in the interstitia (Fig. 12, arrows). l, lumen; ec, endothelial cell; n, nucleus; ij, intercellular junction; a, alveolar space; p, pericyte. Figures 9 and 10, × 87,000; Figs. 11 and 12, × 105,000.

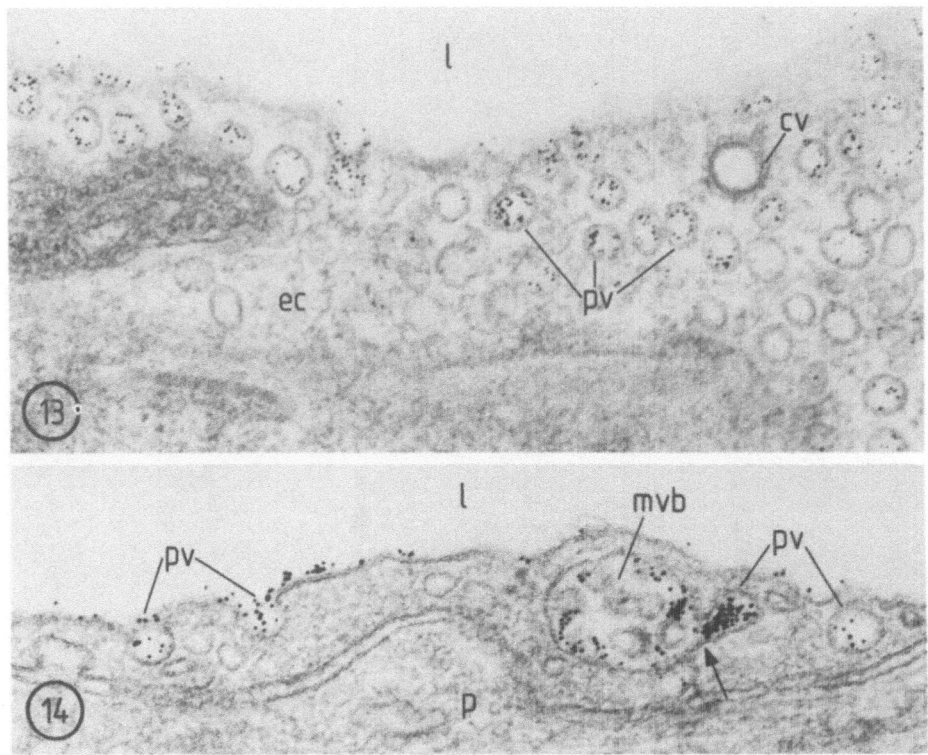

Figures 13 and 14. Endocytosis of albumin–gold complex. Figure 13 shows an obliquely cut endothelial cell (ec); 5 min after perfusion through the mouse lung vasculature, Alb-Au decorates a large number of plasmalemmal vesicles (pv) whereas the coated vesicle (cv) is not labeled. In Fig. 14, after 15 min of perfusion the ligand is found in a multivesicular body (mvb), at the level of which a vesicle is apparently discharging its content (arrow). l, lumen; p, pericyte. Figure 13, × 74,000; Fig. 14, × 90,000.

air, only endothelium expresses receptors for albumin (unpublished observations). As reported by others, more than 92% of resistance to albumin flux across the alveolo-capillary partition lies in the epithelial barrier.[41]

2. Nonspecific Transcytosis

We examined by electron microscopy the interaction of Alb-Au perfused *in situ* in mice, with the endothelium of successive vascular segments: endocardium, aorta abdominalis, phrenic artery, arterioles, capillaries, and venules (in bipolar microvascular fields of the diaphragm), phrenic vein, and vena cava abdominalis. All these vessels are lined by continuous endothelium. For comparative studies, additional samples were collected from brain, pancreas, jejunum, heart, lung, trapezius muscle, white adipose tissue (retroperitoneal and periaortic), and brown

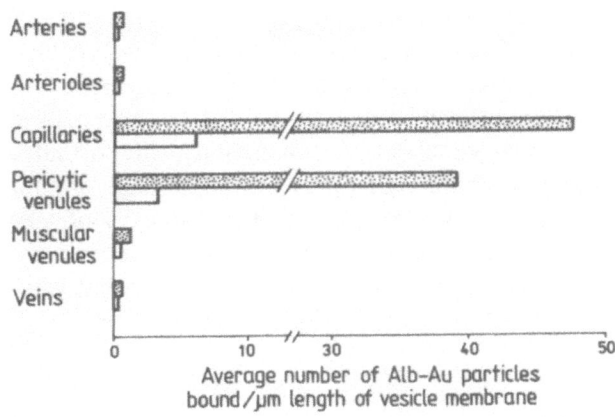

Figure 15. Morphometric data on the density of Alb-Au binding to the membrane of plasmalemmal vesicles (stippled columns) and its competition by monomeric albumin (white columns) in the endothelium of successive vascular segments. Note the high density and the competition of binding of Alb-Au in capillaries and pericytic (postcapillary) venules.

fat (interscapular and periadrenal). Competition experiments were carried out with free bovine serum albumin. Morphometric analysis showed that, except for brain cortex, in the continuous endothelium of capillaries and postcapillary pericytic venules (diameter <20 μm) of the organs examined (lung, heart, skeletal muscle, adipose tissue), Alb-Au bound to specific binding sites and was transported by receptor-mediated transcytosis. Commonly, in these vessels, coated pits and coated vesicles did not bind Alb-Au particles, but the endosomal and lysosomal compartments were labeled. Conversely, in all other endothelia examined, relatively few plasmalemmal vesicles were marked by Alb-Au particles (Fig. 15) in concentration comparable to that in the vascular lumen. The ligand was transported at a very low rate by nonspecific fluid-phase transcytosis. Fluid-phase transport as well as endocytosis of Alb-Au were not competed by monomeric albumin. In these endothelia, coated pits and coated vesicles were frequently marked by Alb-Au which also decorated endosomes and lysosomes.

This remarkable differentiation in the mechanism and magnitude of albumin transport in various endothelia was confirmed with radioiodinated monomeric albumin analyzed by light microscopy and electron microscopy radioautography. Morphometric analysis on specimens, especially lung, diaphragm, and myocardium, collected from experiments using [125I]-Alb, [125I]-Alb-Au, or [198Au]-Alb gave similar values for the saturation of binding sites and for 50% competition of ligand binding by free albumin (0.2 mg/ml). In comparative studies in which [125I]-IgG was perfused instead of radiolabeled albumin, morphometric analysis revealed that the net uptake and transcytosis of albumin in the heart and diaphragm capillaries was 20–30 times higher than that of IgG, and also an order of magnitude higher than at the level of the large vessels.[39,40]

These observations indicate that the mechanism and rate of endothelial transport of albumin are characteristically differentiated in various vascular beds. In the capillaries and postcapillary (pericytic) venules of the lung, heart, skeletal muscle, and adipose tissue, albumin is specifically and extensively transported by receptor-mediated transcytosis. The rest of the endothelia examined performed a rather nonspecific low rate of uptake and fluid-phase transcytosis.

It appears that receptor-mediated transcytosis of albumin operates restrictively in capillaries with continuous endothelium, except brain cortex, whereas fluid-phase transcytosis governs the transport of albumin in fenestrated and discontinuous capillaries as well as in the endothelium of large vessels including muscular venules larger than 20 μm. Reportedly, in some visceral capillaries the fenestral diaphragms prevent the passage of endogenous albumin.[90]

An active transendothelial transport of albumin was demonstrated for the monolayers of cultured porcine pulmonary artery endothelium, but in that system the preferential transit occurred from, what was considered in culture, the interstitial to the luminal aspect of endothelium.[99]

3. Transcytosis of Fatty Acid–Albumin

It is known that free fatty acids (FFA) constitute a major source of lipids for tissues providing the substrate for energy production and complex lipid biosynthesis. Because of their very low hydrosolubility in the circulation, FFA are transported mostly tightly bound to albumin in anionic form (one or a few fatty acid anions per molecule of albumin). Since albumin does not cross the cell membrane, in order to enter cells FFA must dissociate from the albumin carrier. Although this may not be the only mechanism for the cellular uptake of FFA,[127] it has been postulated that some cells, e.g., hepatocytes,[80,128,129,147] adipocytes,[12] and myocardium,[53,54] take up FFA via a membrane albumin receptor. Fatty acids are extensively used by the striated muscle for oxidative phosphorylation, by the lung for surfactant production, and by the adipose tissue for lipid synthesis. The existence of albumin receptors in the capillary endothelium of these tissues may provide a mechanism for securing the constant amount of FFA required for their metabolism. This constant minimal supply is controlled first at the level of their capillary endothelium endowed with high-affinity receptors for albumin. Yet, we do not know whether during the transendothelial passage of the albumin–FFA complex (Alb-FFA) a dissociation and reassociation occur before reaching the target cell. To this intent, we tried to get an insight into some steps and aspects of this complex process. First we investigated whether the transcytotic mechanism and magnitude for Alb-FFA are the same as for the fat-free albumin.[32] For this preliminary inquiry we used oleic acid-loaded bovine serum albumin (OA-Alb) adsorbed on 5-nm gold particles [(OA-Alb)-Au], its radiolabeled derivative {(OA-[^{125}I]-Alb)-Au}, and for comparison Alb-Au and [^{125}I]-Alb-Au. Using [^{14}C]-OA and [^{125}I]-BSA it was found that in the (OA-Alb)-Au complex the OA/BSA molar ratio was ~2.5 with 4.85 ± 0.95 ($n = 4$) molecules of OA-loaded BSA adsorbed onto one gold particle. Each tracer was perfused in mice *in situ* through the pulmonary vasculature at a flow rate of 3 ml/min for 3 to 30 min; at the end of perfusion the unbound ligand was flushed out with PBS. In experiments with radiolabeled conjugates the total uptake by the lung was measured; for the other experiments tissues were fixed *in situ* and processed for electron microscopy. Either Alb-Au or (OA-Alb)-Au was used to compete the [^{125}I]-Alb-Au binding. The total uptake by the lung of (OA-[^{125}I]-Alb)Au was ~1.5 times higher than the uptake of fatty acid-free counterpart (Fig. 16). By electron microscopy the (OA-

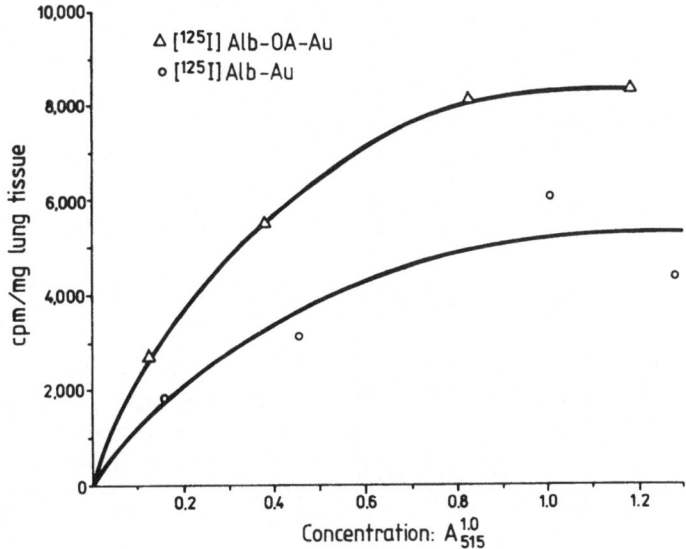

Figure 16. Uptake of radioiodinated albumin–gold ([^{125}I]-Alb-Au) and oleic acid–albumin–gold complexes ([^{125}I]-Alb-OA-Au) by the mouse lung perfused *in situ* for 3 min (9 ml) at various concentrations of the ligands. Note the increased uptake of oleic acid-carrying albumin at all concentrations used.

Alb)-Au complex displayed the same general pattern of binding as Alb-Au, with preferential location on plasmalemmal vesicles and uncoated pits; however, the labeling density of (OA-Alb)-Au was 2.5-fold higher than the values obtained for fatty acid-free albumin. The concentration at which (OA-Alb)-Au binding was achieved, was similar to that recorded for Alb-Au. Both Alb-Au and (OA-Alb)-Au were effective in competing the [^{125}I]-Alb-Au binding, suggesting that the defatted as well as the fatty acid-loaded albumin conjugates bound specifically to the same binding sites. Autoradiography showed a prevalent uptake in capillaries and small postcapillary pericytic venules of both [^{125}I]-Alb-Au and (OA-[^{125}I]-Alb)-Au. The magnitude of transcytosis of (OA-Alb)-Au was significantly faster and greater than that observed with Alb-Au. These data suggest that in mouse lung capillaries the endothelial specific binding sites for albumin express a higher affinity for this protein when it carries fatty acids (i.e., OA).[32] As in the case of Alb-Au, the (OA-Alb)-Au conjugate is taken up nonspecifically and transported at a low rate across the endothelium of large vessels, arterioles, and muscular venules. Tentatively, one can hypothesize that the endothelial albumin receptor senses the conformationally modified albumin molecule (i.e., upon binding of free fatty acids), and reacts with different affinities.

4. Transcytosis of Glycosylated Albumin

Capillary endothelial cells isolated from rat epididymal fat have been shown to take up glycosylated proteins such as albumin, myoglobin, and ovalbumin at a greater rate than their nonglycosylated forms. Although no ultrastructural evi-

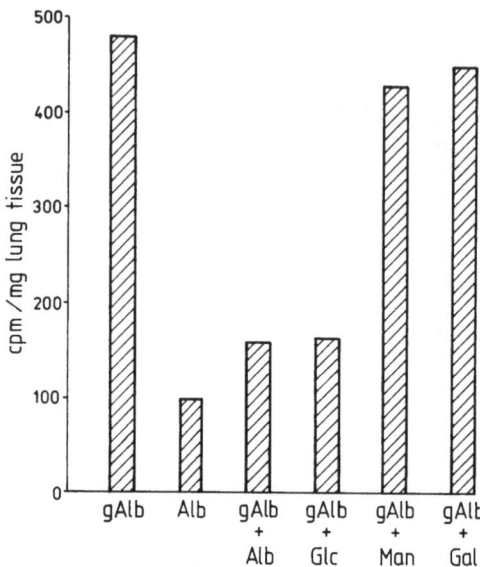

Figure 17. The uptake of [^{125}I]-glycosylated albumin (gAlb) by the mouse lung perfused *in situ* is about four times higher than that of native [^{125}I]albumin (Alb). Uptake is partially (about 60%) competed by albumin and glucose (Glc) but not by D-mannose (Man) or D-galactose (Gal).

dence was provided, it was assumed that all these proteins were ingested by micropinocytotic vesicles.[150–152] In perfused rat lungs it was shown that binding, internalization, and transport of the albumin–gold complex were enhanced by nonenzymatic glycosylation of the protein.[141]

Preliminary experiments conducted in our laboratory have revealed that radiolabeled nonenzymatically glycosylated albumin ([^{125}I]-gAlb) perfused *in situ* in the mouse lung, is taken up ~4 times more intensely than its native counterpart. The uptake is reduced by 60% by competition with either native albumin or D-glucose, but none of these competitors was able to abolish completely the [^{125}I]-gAlb binding. Competition with D-mannose and D-galactose was ineffective (Fig. 17).[91]

Glycosylated albumin conjugated with 5-nm gold particles (gAlb-Au) perfused *in situ* for 3 min was found to be adsorbed on some uncoated pits and the membrane of plasmalemmal vesicles open on the luminal surface of alveolar capillaries. The binding pattern was generally similar to that reported for the gold complex with native albumin (Alb-Au). However, unlike the latter, gAlb-Au also bound as scattered, heterogeneously distributed particles on plasma membrane proper. Perfusion of gAlb-Au for 10 to 30 min was followed by internalization of vesicles and transcytosis of the tracer in the subendothelial space. Specific adsorptive transcytosis of gAlb-Au across capillary endothelium was ~30% higher than that of Alb-Au. As indicated by morphometric analysis, binding was partially competed by either albumin (mostly on vesicles) or D-glucose (mainly on plasma membrane). These preliminary observations suggest that gAlb behaves as a bifunctional ligand with affinities for plasmalemmal vesicle albumin receptor and for a lectinlike membrane component. Its higher uptake by the microvascular walls should be rationalized and further elucidated in terms of the dual endothelial interaction with the protein moiety and the carbohydrate moiety of this modified serum albumin.[91]

B. Low-Density Lipoprotein

Electron microscopic examination of blood vessels of spontaneously hyper-lipoproteinemic rats revealed that *in vivo* lipoprotein particles within the size range of LDL (~20–25 nm) and VLDL (~30–50 nm) were consistently associated with endothelial structures involved in transcytosis (i.e., plasmalemmal vesicles and transendothelial channels) and to a lesser degree to those participating in endocytosis (i.e., coated pits–vesicles, endosomes, lysosomes).[118] Similar but more detailed information came from experiments in which exogenous LDL was perfused in rat *in situ* for up to 10 min and detected by immunoperoxidase cytochemistry. It was found that both homologous and heterologous LDL particles are taken up and transported by endocytosis (a relatively small amount) and transcytosis (the prevalent process). At early time points, LDL bound to some uncoated pits of plasma membrane and labeled some plasmalemmal vesicles open to the lumen. At later intervals, labeled vesicles appeared on the abluminal front and apparently discharged their content.[139] LDL transcytosis by arterial endothelium appeared not to be influenced by temperature but was augmented at high concentrations of LDL. Heparin was able to displace most of LDL from coated pits, and reduced to a lower degree LDL uptake by plasmalemmal vesicles. These findings were interpreted as suggesting the existence on the endothelial cell *in vivo* of two types of LDL-binding sites: high-affinity binding sites on coated pits, and low-affinity binding sites on plasmalemmal vesicles. The possibility that heparin did not have full access to plasmalemmal vesicles cannot be ruled out. Experiments in parallel with native LDL and methylated LDL, each of them double-labeled with [^{125}I]- and [^{14}C]-oleate cholesterol, perfused *in situ* in guinea pigs for 60 min, followed by radioassay of the aortic intima, showed that the amounts of the two probes accumulated in the arterial wall were similar.[138,140] Based on the assumption that reductively methylated LDL is not recognized by the high-affinity LDL receptor, it may be speculated that methyl-LDL was transcytosed by a low-affinity or receptor-independent mechanism. For a more detailed account see Chapter 18.

Because of the difficulties in estimating the LDL receptor affinities *in vivo*, some investigators have applied an *in vitro* system and found that LDL transcytosis across porcine arterial endothelial cells in culture was saturable and temperature- and energy-dependent.[45] Recent work on the hamster lung perfused *in situ* with [^{3}H]cholesterol LDL, [^{125}I]-LDL, or LDL-Au conjugate demonstrated that ~50% of LDL uptake was receptor-mediated.[79]

In no experiments examined by electron microscopy could LDL particles be detected at the level of endothelial intercellular spaces or junctions.

For transport of β-VLDL across endothelium, see Chapter 18.

C. Transferrin

Experiments on capillaries isolated from rat epididymal fat incubated with either [^{14}C]sucrose or fluorescein-labeled transferrin showed that endocytosis of transferrin was six to seven times greater than that of sucrose suggesting a mechanism of selection for the protein. Although no ultrastructural investigation was performed, it was speculated that transferrin might be adsorbed to the vesicle membrane from which it was partially released by exocytosis.[146]

A monoclonal antibody against rat and human transferrin receptors was applied on cryostat sections from various rat tissues followed by peroxidase-conjugated rabbit F(ab')$_2$ anti-mouse IgG antibody. A positive reaction was detected only on brain capillaries. When the monoclonal antibody was injected into the blood, followed by the same second antibody, again only the brain capillaries were marked by the reaction product indicating a restrictive localization of transferrin receptors on the endothelium of the blood–brain barrier.[56] These observations prompted experiments in which after washing the blood, rat brains were perfused with iron-saturated [^{125}I]transferrin and microvessels were isolated. Pulse–chase experiments and competition experiments showed that within 3–5 min after its addition to the perfusate, transferrin moved into endothelium by a receptor-mediated mechanism and in 5–10 min reached the brain gray matter. No specific uptake of radioiodinated transferrin was observed in the hindlimb. It was concluded that transferrin crosses the blood–brain barrier by receptor-mediated transcytosis.[28] A similar conclusion was reached by studies on the transcytosis of iron–transferrin–gold complex in the bone marrow. These studies considered that such transport occurs via a system of tubules and endosomal vesicles.[123] It has also been reported that in the liver transferrin receptors are limited to endothelial cells.[124] These take up transferrin by a receptor-mediated mechanism, to be presumably desialylated, then released and recognized by the galactosyl receptors on the hepatocyte surface.[131] Though the data so far available indicate that transport of transferrin in restricted vascular beds is receptor-mediated, the cellular localization of the endothelial transferrin receptors, their route and kinetics remain to be elucidated.

D. Insulin

There is a relatively extensive literature on the endothelial insulin receptors and their role in the metabolic activities of microvascular and macrovascular endothelial cells. However, only recently has it been revealed that, at least under culture conditions, the processing of cell-bound insulin by the endothelial cells is rather complex. It was found that degradation in endothelial cells is low; the cell-bound insulin is rapidly internalized, transcytosed, and dissociated from the cells, primarily as biologically intact insulin, to be delivered to the target cells.[25,26] It was demonstrated that the vectorial transport of intact [^{125}I]insulin across the endothelium is blocked by both unlabeled insulin and anti-insulin receptor antibodies,[62–64] and that occupying the endothelial insulin binding sites with unlabeled insulin prevents almost completely the transport of [^{125}I]insulin across the endothelium. It is assumed that transendothelial transport of insulin is performed by vesicles, using the insulin receptors as transporters.[63] Vesicular ingestion was observed in the uptake of ferritin-conjugated insulin by the endothelium of capillaries isolated from rat epididymal fat.[126] However, more work is needed for a high-resolution localization of the transcytotic receptors for insulin in various types of endothelia. Down-regulation of insulin receptors in endothelial cells could provide an additional mechanism for regulation of hormone action.[63]

E. Ceruloplasmin

Ceruloplasmin is a 130,000- to 160,000-dalton multifunctional copper-containing glycoprotein carrying more than 90% of plasma copper (seven cupric ions per molecule). In addition to copper transport, ceruloplasmin catalyzes several oxidative reactions and expresses superoxide dismutase activity. It is largely removed from the circulation by the liver upon binding to ceruloplasmin receptors which are exclusively located on endothelium.[60] Results with ceruloplasmin–gold conjugates showed that this ligand is transcytosed across sinusoid endothelium via coated vesicles by a receptor-mediated mechanism.[61] The liver endothelium desialylates ceruloplasmin[55] which is subsequently released and recognized by the galactosyl receptors of the underlying hepatocytes.[132] A comparable process was reported to take place with transcobalamin II, a B_{12}-binding serum polypeptide.[125]

VI. ENDOTHELIAL CELLS OF CONTINUOUS CAPILLARIES EXPRESS ALBUMIN-BINDING PROTEINS

Based on the occurrence of specific albumin-binding sites (ABS) in the capillaries of the lung, heart, diaphragm, and adipose tissue, biochemical investigations were carried out to identify the protein(s) responsible for this binding.[156,158] Extracts were prepared from: (a) cells and tissues which *in situ* express ABS, e.g. microvascular endothelia (ME) freshly isolated from rat epididymal fat and organs which contain ABS-positive capillaries, i.e., lung, heart, adipose tissue, and (b) cells which *in situ* do not express ABS, e.g., arterial endothelia. The extracts were subjected to solubilization, SDS-gel electrophoresis (in various conditions), electroblotting on nitrocellulose paper, and blot incubation with either albumin–gold complex, [^{125}I]-monomeric or polymeric albumin, or [^{125}I]-albumin–gold. Controls included incubation of transfer strips with other plasma proteins adsorbed on gold particles, or competition for albumin-binding by the putative albumin-binding protein(s) (ABP) with free albumin or other plasmatic and nonplasmatic proteins.

Findings revealed that extracts of capillary endothelial cells isolated from rat adipose tissue, as well as the lung and heart homogenates, contain ABPs, represented by two pairs of polypeptides with major components of molecular mass of 31 kDa and 18 kDa. The ABP peptides have a pI of 8.05 to 8.75. Extracts of arterial endothelia do not express such proteins. ABP bind specifically both monomeric and polymeric albumin: the binding is saturable and specifically competed with albumin. Sulfhydryl reducing agents such as dithiothreitol and β-mercaptoethanol did not significantly affect ABPs' electrophoretic mobility and binding activity.

Radioiodination of the cell surface of isolated capillary endothelia, indicated that ABPs are membrane-associated, with the binding sites exposed in their ectodomains.

The 31 kDa and 18 kDa polypeptides, resolved in the electrophoresed SDS gels, may represent either components of one or more ABP, or the latter may be represented by an oligomer of nonidentical subunits. The possible identity between ABS and ABP remains to be determined.

VII. RECEPTOR-MEDIATED TRANSCYTOSIS IS A BASIC PROCESS SHARED BY MOST EPITHELIA

A growing body of evidence indicates that the controlled transfer of molecules between compartments separated by an epithelium is a fundamental cellular process. Data reported in the last decade substantiate (or suggest) a receptor-mediated mechanism for the transcellular transfer of several macromolecules. The best-characterized examples are the transport of IgG by the intestinal epithelium of the neonatal rat,[95] human placenta and rabbit yolk sac, and the receptor-mediated transcytosis of IgA across various epithelia (for a brief review see Ref. 75). Using inside-out thyroid follicles, it was demonstrated that upon stimulation with TSH, thyroglobulin–gold conjugate is transcytosed by small vesicles from the apical to the basolateral plasma membranes.[47] Transcytosis contributes substantially to the blood-to-bile transport of substances including biliary secretion[66] and transfer of G protein of vesicular stomatitis virus between two apposed cell membrane surfaces.[86] Endocytosis, exocytosis, and transcytosis are interconnected pathways in a network of membrane traffic, the sorting of which has only begun to be unraveled. Receptor-mediated transcytosis in various tissues secures vital functions such as nutrition, blood–tissue exchanges, blood–biliary circulation, hormone transport, immunologic reactions, and so forth.

VIII. CONCLUDING REMARKS

Permeant plasma macromolecules actively selected by vascular endothelium according to their size, charge, and chemistry can be transported either into the cell (endocytosis) or across the cell (transcytosis). The latter process can occur by three mechanisms of uptake: fluid phase, adsorptive nonspecific, and adsorptive specific (or receptor-mediated). The endothelial cell is capable of regulating vascular permeability by three groups of receptors or specific binding sites: vasomediator receptors (mediating the impulses to the endothelial cytoskeleton), endocytotic receptors (securing the metabolic needs of the cell), and transcytotic receptors (acting as transporters across the cell). In certain tissues, endothelial cells express transcytotic receptors of low affinity (for LDL and β-VLDL), or/ and high affinity (for albumin, transferrin, insulin, ceruloplasmin). Specific binding sites for albumin were demonstrated to be limited to the continuous endothelium of capillaries and postcapillary venules of lung, heart, skeletal muscle, and adipose tissue. These binding sites are characteristically restricted to plasmalemmal vesicles which carry albumin across endothelium by receptor-mediated transcytosis; the process is more pronounced for fatty acid-carrying albumin. It may be assumed, yet not proved, that receptor-mediated transcytosis of albumin represents a selective mechanism for site-specific delivery of important molecules carried by albumin (e.g., free fatty acids, steroids, thyroid hormones). Receptor-mediated transcytosis is a basic cellular process shared by most polarized epithelia, aimed to secure major physiologic functions such as blood–tissue exchanges, nutrition, and hormone transport. It is reasonable to assume that as a function of local conditions in certain vascular beds, other plasma constituents are specifically selected for transport, a possibility worth exploring.

ACKNOWLEDGMENTS. The work presented in this review was largely supported by the Ministry of Education, Romania, and by National Institutes of Health (USA) Grant HL-26343. The excellent word processing by D. Neacsu, graphics by C. Neacsu, and photographic work by E. Stefan and V. G. Ionescu are gratefully acknowledged.

REFERENCES

1. Adams, C. J., Maurcy, K. M., and Storrie, B., 1982, Exocytosis of pinocytic contents by Chinese hamster ovary cells, *J. Cell Biol.* **93**:632–637.
2. Antohe, F., Heltianu, C., and Simionescu, N., 1987, Further evidence for the distribution and nature of histamine receptors on microvascular endothelium, *Microcirc. Endothel. Lymph.* **3**:163–185.
3. Baker, D. P., Van Lenten, B. J., Fogelman, A. M., Edwards, P. A., Kean, C., and Berliner, J. A., 1984, LDL, scavenger, β-VLDL receptors on aortic endothelial cells, *Arteriosclerosis* **4**:248–255.
4. Baldwin, A. L., and Chien, S., 1984, Endothelial transport of anionized and cationized ferritin in the rabbit thoracic aorta and vasa vasorum, *Arteriosclerosis* **4**:372–382.
5. Baldwin, A. L., and Chien, S., 1985, Effect of plasma proteins on endothelial binding and vesicle loading of anionized ferritin in rabbit aorta, *Arteriosclerosis* **5**:451–458.
6. Baldwin, A. L., and Chien, S., 1985, Regulation of aortic endothelial vesicular uptake of cationized ferritin by plasmalemmal binding, *Atherosclerosis* **55**:233–245.
7. Becker, C. G., and Harpel, C. P., 1976, Alpha-2-macroglobulin on human vascular endothelium, *J. Exp. Med.* **144**:1–9.
8. Betz, A. L., 1985, Epithelial properties of brain capillary endothelium, *Fed. Proc.* **44**:2614–2615.
9. Bignon, J., Chahinian, P., Feldman, G., and Sapin, C., 1975, Ultrastructural immunoperoxidase demonstration of autologous albumin in the alveolar capillary membrane and in the alveolar lining material in normal rats, *J. Cell Biol.* **64**:503–509.
10. Blomhoff, R., Eskild, W., and Berg, T., 1984, Endocytosis of formaldehyde-treated serum albumin via scavenger pathway in liver endothelial cells, *Biochem. J.* **218**:81–86.
11. Boyles, J., L'Hernault, N., Laks, H., and Palade, G. E., 1981, Evidence for vesicular shuttle in heart capillaries, *J. Cell Biol.* **91**:418a.
12. Brandes, R., Okner, R. K., Weisiger, R. A., and Lysenko, N., 1982, Specific and saturable binding of albumin to rat adipocytes: Modulation by epinephrine and possible role in free fatty acid transfer, *Biochem. Biophys. Res. Commun.* **105**:821–827.
13. Bundgaard, M., 1984, The three dimensional organization of tight junctions in a capillary endothelium revealed by serial-section electron microscopy, *J. Ultrastruct. Res.* **88**:1–17.

14. Bundgaard, M., Hagman, P., and Crone, C., 1983, The three dimensional organization of plas-malemmal vesicular profiles in the endothelium of rat heart capillaries, *Microvasc. Res.* **25**:358–368.

15. Casley-Smith, J. R., 1985, Vesicular form and fusion as revealed by freeze-immobilization and stereoscopy of semi-thin sections: Implications for permeation via these structures, *Prog. Appl. Microcirc.* **9**:6–20.

16. Chang, R. L. S., Dean, W. M., Robertson, C. R., and Brenner, B. M., 1975, Permselectivity of the glomerular capillary wall. III. Restricted transport of polyanions, *Kidney Int.* **8**:212–218.

17. Chien, S., Fan, F-C., Lee, M. A. L., and Handley, D. A., 1984, Effects of arterial pressure on endothelial transport of macromolecules, *Biorheology* **21**:631–641.

18. Clough, G., 1982, The steady state transport of cationized ferritin by endothelial cell vesicles, *J. Physiol. (London)* **328**:389–401.

19. Clough, G., and Michel, C. C., 1981, The role of vesicles in the transport of ferritin through frog endothelium, *J. Physiol. (London)* **315**:127–148.

20. Copley, A. L., 1983, The endo-endothelial fibrin lining: A historical account, *Thromb. Res.* **5**(Suppl.):1–26.

21. Crone, C., 1986, Modulation of solute permeability in microvascular endothelium, *Fed. Proc.* **45**:77–83.

22. Curry, F.-R. E., 1982, The effect of charge on the transport of intermediate sized protein probes across the capillary wall, in: *The Pathogenicity of Cationic Proteins* (P. P. Lambert, P. Bergmann, and R. Beauwens, eds.), Raven Press, New York, pp. 120–124.

23. Curry, F.-R. E., and Michel, C. C., 1980, A fiber matrix theory of capillary permeability, *Microvasc. Res.* **20**:96–99.

24. Danon, D., and Skutelsky, E., 1979, Endothelial surface charge and its positive relationship to thrombogenesis, *Ann. N.Y. Acad. Sci.* **275**:47–63.

25. Dernovsek, K. D., and Bar, R. S., 1985, Processing of cell bound insulin by capillary and ma-crovascular endothelial cells in culture, *Am. J. Physiol.* **248**:E244–E251.

26. Dernovsek, K. D., Bar, R. S., Ginsberg, B. H., and Lioubin, M. N., 1984, Rapid transport of biologically intact insulin through cultured endothelial cells, *J. Clin. Endocrinol. Metab.* **58**:761–763.

27. Devenny, J. J., and Wagner, R. C., 1985, Transport of immunoglobulin G by endothelial vesicles in isolated capillaries, *Microcirc. Endothel. Lymph.* **2**:15–26.

28. Fishman, J. B., Andrahan, J. V., Connor, J., Dickey, B. F., and Fine, R. E., 1985, Receptor-mediated transcytosis of transferrin across the blood brain barrier, *J. Cell Biol.* **101**:423a.

29. Forker, E. L., and Luxon, B. A., 1983, Albumin-mediated transport of rose bengal by perfused rat liver, *J. Clin. Invest.* **72**:1764–1771.

30. Frøkjaer-Jensen, J., 1980, Three-dimensional organization of plasmalemmal vesicles in endo-thelial cells: An analysis by serial sectioning of frog mesenteric capillaries, *J. Ultrastruct. Res.* **73**:9–20.

31. Fujimoto, T., and Singer, S. J., 1986, Immunocytochemical studies of endothelial cells in vivo. I. The presence of desmin only, or of desmin plus vimentin, or vimentin only, in the endothelial cells of different capillaries of the adult chicken, *J. Cell Biol.* **103**:2775–2787,

32. Galis, Z., Ghitescu, L., Simionescu, M., and Simionescu, N., 1987, Fatty acid binding to albumin increases its uptake and transcytosis by the lung capillary endothelium, *J. Cell Biol.* **105**:612.

33. Ghinea, N., and Hasu, M., 1986, Charge effect on binding, uptake and transport of ferritin through fenestrated endothelium, *J. Submicrosc. Cytol.* **18**:647–659.

34. Ghinea, N., and Simionescu, N., 1985, Anionized and cationized hemeundecapeptides as probes for cell surface charge and permeability studies: Differentiated labeling of endothelial plasma-lemmal vesicles, *J. Cell Biol.* **100**:606–612.

35. Ghitescu, L., and Fixman, A., 1984, Surface charge distribution on the endothelial cell of liver sinusoids, *J. Cell Biol.* **99**:639–647.

36. Ghitescu, L., Fixman, A., Simionescu, M., and Simionescu, N., 1985, Albumin is transported through capillaries with continuous endothelium by receptor-mediated transcytosis, *J. Cell Biol.* **101**:424a.

37. Ghitescu, L., Fixman, A., Simionescu, M., and Simionescu, N., 1986, Transendothelial transport

of serum albumin in the aorta, coronaries and heart valves of experimental hypercholesterolemic hamsters, in: *XVIth International Congress of the International Academy of Pathology, Vienna*, Abstracts Volume, p. 69.

38. Ghitescu, L., Fixman, A., Simionescu, M., and Simionescu, N., 1986, Specific binding sites for albumin restricted to plasmalemmal vesicles of continuous capillary endothelium: Receptor-mediated transcytosis, *J. Cell Biol.* **102**:1304–1311.

39. Ghitescu, L., Fixman, A., Simionescu, M., and Simionescu, N., 1986, Differentiated uptake and transcytosis of albumin by the continuous endothelium of successive vascular segments, *J. Cell Biol.* **103**:449a.

40. Ghitescu, L., Galis, Z., Simionescu, M., and Simionescu, N., 1987, Diferentiated uptake and transcytosis of albumin in successive vascular segments, *Circ. Res.* (in press).

41. Gorin, A. B., and Stewart, P. A., 1979, Differential permeability of endothelial and epithelial barriers to albumin flux, *J. Appl. Physiol.* **47**:1315–1324.

42. Gorog, P., and Born, G. V. R., 1982, Increased uptake of circulating low density lipoproteins and fibrinogen by arterial walls after removal of sialic acid from the endothelial surface, *Br. J. Exp. Pathol.* **63**:447–451.

43. Grotte, G., 1956, Passage of dextran molecules across the blood–lymph barrier, *Acta Clin. Scand. Suppl.* **211**:1–84.

44. Haraldsson, H., and Rippe, B., 1985, Serum factors other than albumin are needed for the maintenance of normal capillary permselectivity in rat hindlimb muscle, *Acta Physiol. Scand.* **123**:427–436.

45. Hashida, R., Anamizu, C., Kimura, J., Ohkuma, S., Yoshida, Y., and Takano, T., 1986, Transcellular transport of lipoprotein through arterial endothelial cells in monolayer culture, *Cell Struct. Funct.* **11**:31–42.

46. Heltianu, C., Simionescu, M., and Simionescu, N., 1982, Histamine receptors of the microvascular endothelium revealed in situ with a histamine–ferritin conjugate: Characteristic high affinity binding sites in venules, *J. Cell Biol.* **93**:357–364.

47. Herzog, V., 1983, Transcytosis in thyroid follicular cells, *J. Cell Biol.* **97**:607–617.

48. Horiuchi, S., Murakami, M., Takata, K., and Morino, Y., 1986, Scavenger receptor for aldehyde-modified proteins, *J. Biol. Chem.* **261**:4962–4966.

49. Horiuchi, S., Takata, K., and Morino, Y., 1985, Characterization of a membrane associated receptor from rat sinusoidal liver cells that binds formaldehyde-treated serum albumin, *J. Biol. Chem.* **260**:475–481.

50. Horiuchi, S., Takata, K., and Morino, Y., 1985, Purification of a receptor for formaldehyde-treated serum albumin from rat liver, *J. Biol. Chem.* **260**:482–488.

51. Horvat, R., Hovorka, A., Dekan, G., Poczewski, H., and Kerjaschki, D., 1986, Endothelial cell membranes contain podocalyxin—the major sialoprotein of visceral glomerular epithelial cells, *J. Cell Biol.* **102**:484–491.

52. Huang, T. W., and Langlois, J. C., 1985, Podoendin—a new cell surface protein of the podocyte and endothelium, *J. Exp. Med.* **162**:245–267.

53. Hutter, J. F., Piper, H. M., and Spieckermann, P. G., 1984, Kinetic analysis of myocardial fatty acid oxidation suggesting an albumin receptor mediated uptake process, *J. Mol. Cell. Cardiol.* **16**:219–226.

54. Hutter, J. F., Piper, H. M., and Spieckermann, P. G., 1984, Myocardial fatty acid oxidation: Evidence for an albumin-receptor-mediated membrane transfer of fatty acids, *Basic Res. Cardiol.* **79**:274–282.

55. Irie, S., and Tavassoli, M., 1986, Liver endothelium desialates ceruloplasmin, *Biochem. Biophys. Res. Commun.* **140**:94–100.

56. Jefferies, W. A., Brandon, M. R., Hunt, S. V., Williams, A. F., Gatter, K. C., and Mason, D. Y., 1984, Transferrin receptor on endothelium of brain capillaries, *Nature* **312**:162–163.

57. Johansson, B. R., 1979, Capillary permeability to interstitial microinjections of macromolecules and influence of capillary hydrostatic pressure on endothelial ultrastructure, *Acta Physiol. Scand.* **463**:45–50.

58. Johansson, B. R., Karlsson, R., and Bagge, U., 1983, Ultrastructural observations on endothelial

cell vesicles after horseradish peroxidase microinjection into blood and dextran-perfused microvessels of rat mesentery, *Int. J. Microcirc. Clin. Exp.* **2**:157–169.

59. Joris, I., Majno, G., and Ryan, G. B., 1972, Endothelial contraction in vivo: A study of the rat mesentery, *Virchows Arch. B.* **12**:73–83.
60. Kataoka, M., and Tavassoli, M., 1984, Ceruloplasmin receptors in liver cell suspensions are limited to the endothelium, *Exp. Cell Res.* **155**:232–240.
61. Kataoka, M., and Tavassoli, M., 1985, The role of liver endothelium in the binding and uptake of ceruloplasmin: Studies with colloidal gold probe, *J. Ultrastruct. Res.* **90**:194–202.
62. King, G. L., Jialal, I., Buchwald, S., and Johnson, S., 1984, Receptor-mediated uptake and transport of insulin by endothelial cells, *Diabetes* **33**(Suppl. 1):9A.
63. King, G. L., and Johnson, S. M., 1985, Receptor-mediated transport of insulin across endothelial cells, *Science* **227**:1583–1586.
64. King, G. L., Johnson, S. M., and Jialal, I., 1985, Processing and transport of insulin by vascular endothelial cells: Effects of sulfonylurease on insulin receptors, *Am. J. Med.* **79**:43–47.
65. Kurozumi, T., Imamura, T., Tanaka, K., Yae, Y., and Koga, S., 1984, Permeation and deposition of fibrinogen and low density lipoprotein in the aorta and cerebral artery of rabbits—Immune-electron microscopic study, *Br. J. Exp. Pathol.* **65**:355–364.
66. Lake, J. R., Licko, V., van Dyke, R. W., and Scharschmidt, B. F., 1985, Biliary secretion of fluid-phase markers by the isolated perfused rat liver: Role of transcellular vesicular transport, *J. Clin. Invest.* **76**:676–684.
67. Leabu, M., Ghinea, M., Muresan, V., Colceag, J., Hasu, M., and Simionescu, N., 1987, Cell surface chemistry of arterial endothelium and blood monocytes in the normolipidemic rabbit, *J. Submicrosc. Cytol.* **19**:193–208.
68. Lenkei, R., Onica, D., and Ghetie, V., 1977, Receptors for polymerized albumin on liver cells, *Experientia* **33**:1046–1047.
69. Machida, A., Shimoto, S., Ohnuma, H., Miyamoto, H., Baba, K., Oda, K., Nakamura, T., Miyakawa, Y., and Mayumi, M., 1983, A hepatitis B surface antigen polypeptide (P31) with the receptor for chimpanzee albumins, *Gastroenterology* **85**:268–274.
70. Majno, G., Gilmore, V., and Leventhal, M., 1967, On the mechanism of vascular leakage caused by histamine-type mediators, *Circ. Res.* **21**:833–847.
71. Majno, G., and Palade, G. E., 1961, Studies on inflammation. I. The effect of histamine and serotonin on vascular permeability: An electron microscopic study, *J. Biophys. Biochem. Cytol.* **11**:571–606.
72. Maratos-Flier, E., Yang-Kao, C. Y., and King, G. L., 1986, Receptor-mediated transcytosis of epidermal growth factor in MDCK cells, *J. Cell Biol.* **103**:449a.
73. McGuire, P. C., and Twietmayer, T. A., 1985, Transcytosis of ferritin and increased production of subendothelial matrix components by aortic endothelial cells during the development of hypertension, *Microcirc. Endothel. Lymph.* **2**:129–149.
74. Milici, A. J., Watrous, N. E., and Palade, G. E., 1986, Immunogold localization of exogenous albumin in murine myocardial capillaries, *J. Cell Biol.* **103**:194a.
75. Mostov, K. E., and Simister, N. E., 1985, Transcytosis, *Cell* **43**:389–390.
76. Muller, W. A., and Gimbrone, M. A., Jr., 1986, Plasmalemmal proteins of cultured vascular endothelial cells exhibit apical–basal polarity: Analysis by surface-selective iodination, *J. Cell Biol.* **103**:2389–2402.
77. Muresan, V., and Constantinescu, M. C., 1985, Distribution of sialoconjugates on the luminal surface of the endothelial cell in fenestrated capillaries of the pancreas, *J. Histochem. Cytochem.* **33**:474–476.
78. Muresan, V., and Simionescu, N., 1984, Intracellular fate of a multivalent ligand covalently bound to cell surface components, in: *International Cell Biology, 1984* (S. Seno and Y. Okada, eds.), Academic Press Japan, Inc., Tokyo, p. 334.
79. Nistor, A., and Simionescu, M., 1986, Uptake of low density lipoproteins by the hamster lung: Interactions with capillary endothelium, *Am. Rev. Respir. Dis.* **134**:1266–1272.
80. Ockner, R. K., Weisiger, R. A., and Gollan, J. L., 1983, Hepatic uptake of albumin-bound substances: Albumin receptor concept, *Am. J. Physiol.* **245**:613–619.

81. Olesen, S.-P., and Crone, C., 1984, Serotonin increases microvascular permeability in the brain, *Int. J. Microcirc. Clin. Exp.* **3**:466.

82. Palade, G. E., 1960, Transport in quanta across the endothelium of blood capillaries, *Anat. Rec.* **136**:254.

83. Palade, G. E., 1961, Blood capillaries of the heart and other organs, *Circulation* **24**:368–384.

84. Palade, G. E., Simionescu, M., and Simionescu, N., 1979, Structural aspects of the permeability of the microvascular endothelium, *Acta Physiol. Scand. Suppl.* **463**:11–32.

85. Pappenheimer, J. R., 1953, Passage of molecules through capillary walls, *Physiol. Rev.* **33**:387–423.

86. Pesonen, M., Ansorge, W., and Simons, K., 1984, Transcytosis of the G protein of vesicular stomatitis virus after implantation into the apical membrane of Madin–Darby canine kidney cells. I. Involvement of endosomes and lysosomes, *J. Cell Biol.* **99**:796–802.

87. Peters, T. Jr., 1975, Serum albumin, *Adv. Prot. Chem.* **37**:161–245.

88. Pietra, G. G., Sampson, P., Lanken, P. N., Hansen-Flaschen, J., and Fishman, A. P., 1984, Transcapillary movement of cationized ferritin in the isolated perfused rat lung, *Lab. Invest.* **49**:54–61.

89. Pino, R. M., 1985, Restriction to endogenous plasma proteins by a fenestrated capillary endothelium: An ultrastructural immunocytochemical study of the choriocapillary endothelium, *Am. J. Anat.* **172**:279–289.

90. Pino, R. M., and Thouron, C. L., 1983, Vascular permeability in the rat eye to endogenous albumin and immunoglobulin (IgG) examined by immunohistochemical methods, *J. Histochem. Cytochem.* **31**:411–416.

91. Predescu, D., Simionescu, M., Simionescu, N., and Palade, G. E., 1987, The bifunctional nature of glycated-albumin may account for its enhanced binding and transport by capillary endothelium, in situ, *J. Cell Biol.* **105**:362a.

92. Raviola, G., and Butler, J. M., 1983, Unidirectional vesicular transport mechanism in retinal vessels, *Invest. Ophthalmol. Vis. Sci.* **24**:1465–1474.

93. Renkin, E. M., 1985, Capillary transport of macromolecules: Pores and other endothelial pathways, *J. Appl. Physiol.* **58**:315–325.

94. Rennke, H. G., Patel, Y., and Venkatachalam, M. A., 1978, Glomerular filtration of proteins: Clearance of anionic neutral and cationic horseradish peroxidase in the rat, *Kidney Int.* **13**:324–328.

95. Rodewald, R., and Kraehenbuhl, J. P., 1984, Receptor-mediated transport of IgG, *J. Cell Biol.* **99**:159s–164s.

96. Ryan, U. S., 1986, The endothelial surface and response to injury, *Fed. Proc.* **45**:101–108.

97. Schneeberger, E. E., 1983, Proteins and vesicular transport in capillary endothelium, *Fed. Proc.* **42**:2419–2424.

98. Schneeberger, E. E., and Hamelin, M., 1984, Interaction of serum proteins with lung endothelial glycocalyx: Its effect on endothelial permeability, *Am. J. Physiol.* **246**:H206–H217.

99. Shasby, D. M., and Shasby, S. S., 1985, Active transendothelial transport of albumin: Interstitium to lumen, *Circ. Res.* **57**:903–908.

100. Simionescu, D., and Simionescu, M., 1983, Differentiated distribution of the cell surface charge on the alveolar–capillary unit: Characteristic paucity of anionic sites on the air–blood barrier, *Microvasc. Res.* **25**:85–100.

101. Simionescu, M., 1985, Regional differentiation of the surface distribution in the continuous endothelium of the microvasculature, in: *Glomerular Dysfunction and Biopathology of the Vascular Wall* (A. L. Copley, Y. Hamashima, S. Seno, and M. A. Venkatachalam, eds.), Academic Press Japan, Inc., Tokyo, pp. 3–11.

102. Simionescu, M., Ghitescu, L., Fixman, A., and Simionescu, N., 1986, Receptor-mediated transcytosis of albumin in vascular endothelium, *Acta Biol. Acad. Hung. Sci.* **37**:104.

103. Simionescu, M., Ghitescu, L., Fixman, A., and Simionescu, N., 1987, How plasma macromolecules are transported by vascular endothelium, *News Physiol. Sci.* **2**:97–100.

104. Simionescu, M., and Simionescu, N., 1984, Ultrastructure of the microvascular wall: Functional correlations, in: *Handbook of Physiology*, Sect. 2, Vol. IV (E. M. Renkin and C. C. Michel, eds.), American Physiological Society, Bethesda, pp. 41–101.

105. Simionescu, M., and Simionescu, N., 1986, Receptor-mediated transcytosis of plasma molecules by vascular endothelium, in: *Fourth International Symposium on the Biology of Vascular Endothelial Cell, Noordwijkerhout,* Abstract Volume, p. 21.

106. Simionescu, M., and Simionescu, N., 1986, Functions of the endothelial cell surface, *Annu. Rev. Physiol.* **48:**279–293.

107. Simionescu, M., Simionescu, N., and Palade, G. E., 1982, Biochemically differentiated microdomains of the cell surface of capillary endothelium, *Ann. N.Y. Acad. Sci.* **401:**9–24.

108. Simionescu, M., Simionescu, N., and Palade, G. E., 1982, Differentiated microdomains on the luminal surface of capillary endothelium: Distribution of lectin receptors, *J. Cell Biol.* **94:**406–413.

109. Simionescu, M., Simionescu, N., and Palade, G. E., 1982, Preferential distribution of anionic sites on the basement membrane and the abluminal aspect of the endothelium in fenestrated capillaries, *J. Cell Biol.* **95:**425–434.

110. Simionescu, M., Simionescu, N., Santoro, F., and Palade, G. E., 1985, Differentiated microdomains of the luminal plasmalemma of murine muscle capillaries: Segmental variations in young and old animals, *J. Cell Biol.* **100:**1396–1407.

111. Simionescu, M., Simionescu, N., Silbert, J., and Palade, G. E., 1981, Differentiated microdomains on the luminal surface of capillary endothelium. II. Partial characterization of the anionic sites, *J. Cell Biol.* **90:**614–621.

112. Simionescu, N., 1979, The microvascular endothelium: Segmental differentiations; transcytosis, selective distribution of anionic sites, in: *Advances in Inflammation Research,* Vol. 1 (G. Weissmann, B. Samuelson, and R. Paoletti, eds.), Raven Press, New York, pp. 61–70.

113. Simionescu, N., 1980, Transcytosis and endocytosis in the endothelial cell, *Eur. J. Cell Biol.* **22:**180.

114. Simionescu, N., 1981, Transcytosis and traffic of membranes in the endothelial cell, in: *International Cell Biology 1980–1981* (H. Schweiger, ed.), Springer-Verlag, Berlin, pp. 657–672.

115. Simionescu, N., 1983, Cellular aspects of transcapillary exchange, *Physiol. Rev.* **63:**1536–1579.

116. Simionescu, N., and Simionescu, M., 1981, Hydrophilic pathways of capillary endothelium, a dynamic system, in: *Water Transport across Epithelia* (H. H. Ussing, N. B. Bindslev, and O. Sten-Knudsen, eds.), Munksgaard, Copenhagen, pp. 228–247.

117. Simionescu, N., and Simionescu, M., 1984, Fluid-phase and adsorptive transcytosis in the endothelial cell, in: *International Symposium on Membrane Biogenesis and Recycling, Kanvami,* Abstracts Volume, pp. VI–I.

118. Simionescu, N., and Simionescu, M., 1985, Interactions of endogenous lipoproteins with capillary endothelium in spontaneously hyperlipoproteinemic rats, *Microvasc. Res.* **30:**314–332.

119. Simionescu, N., Simionescu, M., and Palade, G. E., 1972, Permeability of intestinal capillaries: Pathway followed by dextrans and glycogens, *J. Cell Biol.* **53:**365–392.

120. Simionescu, N., Simionescu, M., and Palade, G. E., 1976, Structural–functional correlates in the transendothelial exchange of water-soluble macromolecules, *Thromb. Res.* **8**(Suppl. 2):257–289.

121. Simionescu, N., Simionescu, M., and Palade, G. E., 1978, Open junctions in the endothelium of the postcapillary venules of the diaphragm, *J. Cell Biol.* **79:**27–44.

122. Simionescu, N., Simionescu, M., and Palade, G. E., 1981, Differentiated microdomains on the luminal surface of capillary endothelium. I. Preferential distribution of anionic sites, *J. Cell Biol.* **90:**605–613.

123. Soda, R., and Tavassoli, M., 1984, Transendothelial transport (transcytosis) of iron–transferrin complex in the bone marrow, *J. Ultrastruct. Res.* **88:**18–29.

124. Soda, R., and Tavassoli, M., 1984, Liver endothelium and not hepatocytes or Kupffer cells have transferrin receptors, *Blood* **63:**270–276.

125. Soda, R., Tavassoli, M., and Jacobsen, D. W., 1985, Receptor distribution and the endothelial uptake of transcobalamin II in liver cell suspensions, *Blood* **65:**795–802.

126. Solenski, N. J., and Williams, S. K., 1985, Insulin binding and vesicular ingestion in capillary endothelium, *J. Cell. Physiol.* **124:**87–95.

127. Spector, A., 1986, Plasma albumin as a lipoprotein, in: *Biochemistry and Biology of Plasma Lipoproteins* (A. M. Scann and A. A. Spector, eds.), Dekker, New York, pp. 247–279.

128. Stremmel, W., Potter, B. J., and Berk, P. D., 1983, Studies of albumin binding to rat liver plasma membranes, *Biochim. Biophys. Acta* **756**:20–27.

129. Stremmel, W., Strohmeyer, G., and Berk, P. D., 1986, Hepatocellular uptake of oleate is energy dependent, sodium linked and inhibited by an antibody to a hepatocyte plasma membrane fatty acid binding protein, *Proc. Natl. Acad. Sci. USA* **83**:3584–3588.

130. Svensjo, E., and Grega, G. J., 1986, Evidence for endothelial cell-mediated regulation of macromolecular permeability by postcapillary venules, *Fed. Proc.* **45**:89–95.

131. Tavassoli, M., Kishimoto, T., Soda, R., Kataoka, M., and Harjes, K., 1986, Liver endothelium mediates the uptake of iron–transferrin complex by hepatocytes, *Exp. Cell Res.* **165**:369–379.

132. Tavassoli, M., Kishimoto, T., and Kataoka, M., 1986, Liver endothelium mediates the hepatocyte's uptake of ceruloplasmin, *J. Cell Biol.* **102**:1298–1303.

133. Taylor, A. E., and Granger, D. N., 1984, Exchange of macromolecules across the microcirculation, in: *Handbook of Physiology,* Sect. 2, Vol. IV (E. M. Renkin and C. C. Michel, eds.), American Physiological Society, Bethesda, pp. 467–520.

134. Trevisan, A., Gudat, F., Luoud, M., Guggenheim, R., Krey, G., Durmuller, G., Duggelin, M., Landmann, J., Tondelli, P., and Bianchi, L., 1982, Demonstration of albumin receptors on isolated human hepatocytes by light and scanning electron microscopy, *Hepatology* **2**:823–835.

135. Trout, J. J., Koenig, H., Lu, C. Y., and Goldstone, A. D., 1986, Acute stimulation of transcytosis in rat heart capillaries following a calcium depletion–repletion cycle is Ca^{2+} and polyamine dependent, *J. Cell Biol.* **103**:60a.

136. Turner, M. R., Clough, G., and Michel, C. C., 1983, The effects of cationized ferritin upon the filtration coefficient of single frog capillaries: Evidence that proteins in the endothelial cell coat influence permeability, *Microvasc. Res.* **25**:205–222.

137. Vasile, E., Nistor, A., Nedelcu, S., Simionescu, M., and Simionescu, N., 1980, Dual pathway of low density lipoprotein transport through aortic endothelium and vasa vasorum, in situ, *Eur. J. Cell Biol.* **22**:181.

138. Vasile, E., Popescu, G., Simionescu, M., and Simionescu, N., 1986, Enhanced transcytosis and accumulation of β-very low density lipoproteins in the aorta of rabbits with experimental hyperlipidemia, in: *XVIth International Congress of the International Academy of Pathology, Vienna,* Abstracts Volume, p. 68.

139. Vasile, E., Simionescu, M., and Simionescu, N., 1983, Visualization of the binding, endocytosis and transcytosis of low-density lipoprotein in the arterial endothelium in situ, *J. Cell Biol.* **96**:1677–1689.

140. Vasile, E., and Simionescu, N., 1985, Transcytosis of low density lipoprotein through vascular endothelium, in: *Glomerular Dysfunction and Biopathology of Vascular Wall* (A. L. Copley, Y. Hamashima, S. Seno, and M. A. Venkatachalam, eds.), Academic Press Japan, Inc., Tokyo, pp. 87–101.

141. Villaschi, S., Johns, L., Cirigliano, M., and Pietra, G. G., 1986, Binding and uptake of native and glycosylated albumin–gold complexes in perfused rat lungs, *Microvasc. Res.* **32**:190–199.

142. Vorbrodt, A. W., and Lossinsky, A. S., 1986, Transport of homologous albumin through various capillary endothelia in mice, *J. Cell Biol.* **103**:193a.

143. Wagner, R. C., and Andrews, S. B., 1985, Ultrastructure of the vesicular system in rapidly frozen capillary endothelium of the rete mirabile, *J. Ultrastruct. Res.* **90**:172–182.

144. Wagner, R. C., and Casley-Smith, J. R., 1981, Endothelial vesicles—Review, *Microvasc. Res.* **21**:267–298.

145. Wagner, R. C., and Robinson, C. S., 1984, High-voltage electron microscopy of capillary endothelial vesicles, *Microvasc. Res.* **28**:197–205.

146. Wagner, R. C., Robinson, C. S., Cross, P. J., and Devenny, J. J., 1983, Endocytosis and exocytosis of transferrin by isolated capillary endothelium, *Microvasc. Res.* **25**:387–396.

147. Weisiger, R., Gollan, J., and Ockner, R., 1981, Receptor for albumin on the liver cell surface may mediate uptake of fatty acids and other albumin-bound substances, *Science* **211**:1048–1050.

148. Weisiger, R. A., Zacks, C. M., Smith, N. D., and Boyer, J. L., 1984, Effect of albumin binding on extraction of sulfobromphtalein by perfused elasmobranch liver: Evidence for dissociation-limited uptake, *Hepatology* **4**:492–501.

149. Wesselcouch, E. O., Luneau, C. J., Williams, K. J., and Gosselin, R. E., 1984, The failure of

serum albumin to affect capillary permeability in the isolated rabbit heart, *Microvasc. Res.* **28**:373–386.

150. Williams, S. K., 1983, Vesicular transport of proteins by capillary endothelium, *Ann. N.Y. Acad. Sci.* **416**:457–467.

151. Williams, S. K., Devenny, J. J., and Bittensky, M. W., 1981, Micropinocytotic ingestion of glycosylated albumin by isolated microvessels: Possible role in pathogenesis of diabetic microangiopathy, *Proc. Natl. Acad. Sci. USA* **78**:2393–2397.

152. Williams, S. K., and Solenski, N. J., 1984, Enhanced vesicular ingestion of nonenzymatically glycosylated proteins by capillary endothelium, *Microvasc. Res.* **28**:311–321.

153. Witte, S., 1983, The endothelial lining as studied by fluorescent labeling technique in situ, *Thromb. Res.* **5**(Suppl.):93–104.

154. Wolinsky, H., 1980, A proposal linking clearing of circulating lipoproteins to tissue metabolic activity as a basis for understanding atherogenesis, Circ. Res. **47**:301–311.

155. Yokota, S., 1983, Immunocytochemical evidence for transendothelial transport of albumin and fibrinogen in rat heart and diaphragm, *Biomed. Res.* **4**:577–586.

156. Ghinea, N., Fixman, A., Alexandru, D., Popov, D., Hasu, M., Ghitescu, L., Eskenasy, M., Simionescu, M., and Simionescu, N., 1988, Identification of albumin binding proteins in capillary endothelial cells. *J. Cell Biol.* (in press).

157. Simionescu, M., Ghinea, N., Fixman, A., Lasser, M., Kukes, L., Simionescu, N., and Palade, G. E., 1988, The cerebral microvasculature of the rat: Structure and luminal surface properties during early development. J. Submicrosc. Cytol. (in press)

158. Simionescu, N., and Simionescu, M., 1987. Receptor mediated transcytosis of albumin: Identification of albumin binding proteins in the plasma membrane of capillary endothelium, in: *Proceedings of the IVth World Congress on Microcirculation* Tokyo, Japan (M. Tsuchiya, M. Asano, Y. Mishima, M. Oda, eds.), Elsevier, Amsterdam and New York, pp. 67–82.

5

Studies of Pulmonary Endothelial Permeability Using Tritiated Dextrans

John H. Hansen-Flaschen and Alfred P. Fishman

I. INTRODUCTION

For many years physiologists have sought to understand the movement of plasma proteins across the endothelial surface of small blood vessels by applying the physical principles of diffusion and convection to geometric models of small particles and semipermeable membranes. The endothelium has been viewed as a continuous surface containing water-filled channels or "pores" large enough to allow the passage of plasma proteins but small enough to restrict their passage relative to water. The plasma proteins, in turn, have been viewed as inert particles that move through the pores at a rate proportionate to their size and the hydrostatic and oncotic pressure gradients across the endothelium.

Theoretical equivalent pore models based upon this conceptual framework have provided many useful insights into the function of the endothelium, both under normal conditions and in a variety of disease states. However, the equivalent pore models have sometimes yielded results that are internally inconsistent when applied to experimental data. For example, Renkin recently reanalyzed previous measurements of transendothelial protein flux in the cat ileum and the dog paw, and found that the equivalent pore models that best fit the data obtained at high transcapillary flow rates systematically underestimated the transendothelial protein flux measured at lower volume flows.[25] Renkin concluded that either the estimated equivalent pore sizes in the existing models were wrong or that mechanisms other than diffusion and convection contribute importantly to the transcapillary transport of macromolecules.

Other mathematical models have yielded results that are difficult to reconcile with known ultrastructural features of the endothelium. For example, Harris and Roselli reported that experimental measurements of protein transport across pulmonary capillary walls in healthy awake sheep are best explained by the presence of three distinct populations of transendothelial pathways: small pores with an

John H. Hansen-Flaschen and Alfred P. Fishman • Cardiovascular–Pulmonary Division, University of Pennsylvania, Philadelphia, Pennsylvania 19104.

equivalent radius of 2.8 nm, intermediate pores with an equivalent radius of 18.0 nm, and large pores with an equivalent radius of 100 nm.[13] The small and intermediate pores in this theoretical model may correspond respectively to the endothelial intercellular junctions and the plasmalemmal vesicles; however, no ultrastructural analogue has yet been identified for the large pore equivalent pathways.

Discrepancies between theoretical analyses and ultrastructural findings have also emerged from studies of increased vascular permeability in acute lung injury. Mathematical models of injury-induced changes in blood-to-lymph transport of plasma proteins have led to the conclusion that increased-permeability pulmonary edema is due to subtle changes in normal transendothelial pathways,[27,31] whereas morphological studies have shown leakage of macromolecular tracers through large endothelial discontinuities.[15] These discrepancies suggest that the concept of inert molecules traversing fluid-filled channels by convection and diffusion is incomplete.

We questioned whether the movement of plasma proteins across the pulmonary endothelium may be influenced to an important extent by specific chemical or charge interactions between the proteins and one or more structural components of the endothelial barrier. To explore this possibility, we compared the movement of plasma proteins with that of tritiated neutral dextrans across the fluid-exchanging vessels of the lungs in awake sheep. We chose dextran—the macromolecular polymer of glucose—for our experiments because these molecules have no electrically charged sites and because dextrans are similar in shape, flexibility, and physicochemical properties over a broad continuous range of molecular sizes. Although dextrans have been widely used in physiological and morphological studies of endothelial permeability, they have not previously been employed in lymphatic studies of pulmonary vascular permeability. In addition, tritium labeling of the dextrans enabled us to use considerably lower concentrations of the tracer than have been used in most previous studies.

After our initial experiments using awake sheep showed striking differences in the transport of plasma proteins and neutral dextrans from blood to lung lymph, we repeated the dextran experiments in sheep that had breathed 100% O_2 for 3 to 4 days in order to examine the effects of acute pulmonary vascular injury on transendothelial transport of dextrans and proteins. This review summarizes our studies on healthy and oxygen-toxic sheep and discusses possible implications of these results in light of other recent observations on transendothelial transport of macromolecules.

II. EXPERIMENTAL METHODS

A. Preparation of [³H]-Dextran Tracers

Using a modification of techniques originally described by Chang and coworkers,[5] Sampson and others in our laboratory developed a tritium-labeled neutral dextran tracer that has a high specific activity (40–113 μCi/mg) for studies of pulmonary endothelial permeability.[17] This polydisperse probe contains [³H]dextrans which range in effective molecular (Stokes–Einstein) radius from approximately 1 to 9 nm; they are also electrically uncharged as determined by cellulose acetate electrophoresis. The molecular size distribution of the [³H]dextran tracer is the same whether it is dissolved in saline or in sheep plasma, indicating that the tracer does not bind to plasma proteins. Prior to use, each dose of the tracer is tested using the *Limulus* assay to exclude the presence of endotoxin.

B. Infusion of [³H]-Dextran in Sheep

The passage of [³H]dextran from blood to lung lymph was studied using male yearling sheep that were prepared with chronic lung lymph fistulas and hemodynamic catheters using a modification of the surgical techniques developed by Staub *et al.*[30] We sought to minimize contamination of the lymph from extrapulmonary sources by resecting a portion of the caudal mediastinal node below the right pulmonary ligament and by surgically interrupting lymph vessels joining the node above the point of resection from both hemidiaphragms. When properly performed, this approach effectively eliminates lymph from nonpulmonary sources except for a small contribution from the upper esophagus.[4]

Our experiments were performed 5 or more days after completion of surgery while the animals stood unrestrained in metabolic cages. We infused [³H]dextran dissolved in normal saline through a 0.22-μm bacteria filter into the superior vena cava using a syringe infusion pump. A loading dose of 13.5 μCi/kg was followed by a continuous infusion of the tracer for $7\frac{1}{2}$–30 hr at a rate of 0.15 μCi/kg per min. At this infusion rate, the quantity of [³H]dextran administered to the sheep was less than 0.5 mg/kg per hr, and the total volume infused during the experiments was less than 100 ml.

We recorded left atrial, pulmonary arterial, and systemic arterial pressures for 2–3 hr before each infusion was begun, and at frequent intervals during the infusions. Cardiac output was measured using the thermodilution technique, and arterial blood samples were obtained for determination of pH and blood gas tensions. Lung lymph was collected in centrifuge tubes containing dried sodium heparin, and lymph volumes were recorded at 30-min intervals. Sixty-minute pooled lymph samples and hourly plasma samples were analyzed for albumin and total proteins.

C. Pulmonary Oxygen Toxicity in Sheep

To examine the effects of increased pulmonary vascular permeability on the passage of [³H]dextran from blood to lung lymph, we prepared sheep with chronic tracheostomies, lung lymph fistulas, and indwelling hemodynamic catheters.[11] After a $7\frac{1}{2}$-hr baseline infusion of [³H]dextran was completed, five of the animals breathed humidified 100% O_2 via the tracheostomy. The animals were allowed free access to food and water during exposure to oxygen. Inhalation of 100% O_2

was discontinued when lymph flow increased to approximately three times the baseline flow rate. A second $7\frac{1}{2}$-hr infusion of [^{3}H]dextran was then performed while the animals breathed either ambient air or 40% O_2. Other sheep breathed compressed air through the same gas delivery system for 96 hr to serve as controls. After completion of the second infusion of [^{3}H]dextran, the animals were killed and the lungs removed for gravimetric determination of total lung water content and for morphological studies as described below.

D. [^{3}H]-Dextran Analysis

The [^{3}H]dextrans contained in lymph and plasma samples from the sheep were separated into approximately 25 fractions according to molecular size by gel chromatography. The chromatographic column was calibrated using blue dextran and a series of eight globular proteins that ranged from 1.6 to 8.4 nm in effective molecular radius. Total radioactivity (counts/min per ml) was measured in consecutive chromatography fractions using an automated liquid scintillation spectrometer. To determine lymph-to-plasma concentration ratios (L/P) for the [^{3}H]dextran fractions, we divided the total radioactivity of each lymph fraction by that of the corresponding plasma fraction after correcting for background counts.

E. Morphological Studies

We examined postmortem lung samples from the sheep that breathed 100% O_2 by light and electron microscopy to relate the [^{3}H]dextran data with concurrent morphological changes in the alveolar capillary membranes. Lung slices were stained with hematoxylin and eosin and assessed semiquantitatively for the severity of lung injury. For electron microscopy, the cranial lobe of the right lung was fixed immediately after the terminal thoracotomy by installation of formaldehyde–glutaraldehyde into the bronchus. Electron micrographs of cross sections of randomly selected alveolar capillaries from the five sheep that breathed oxygen were compared with similar electron micrographs from postsurgical control sheep.

III. RESULTS

A. Transport of [^{3}H]-Dextran from Blood to Lung Lymph in Normal Sheep

In normal sheep, continuous intravenous infusion of [^{3}H]dextran for up to 30 hr had no detectable effect on hemodynamic pressures, arterial blood gas tensions, cardiac output, or lymph flow rates. The concentrations of endogenous albumin and total protein in lung lymph and plasma also remained unchanged throughout the infusions. L/P for the [^{3}H]dextran fractions reached stable values within 5 hr and remained unchanged for up to 30 hr during continuous infusion of the tracer.

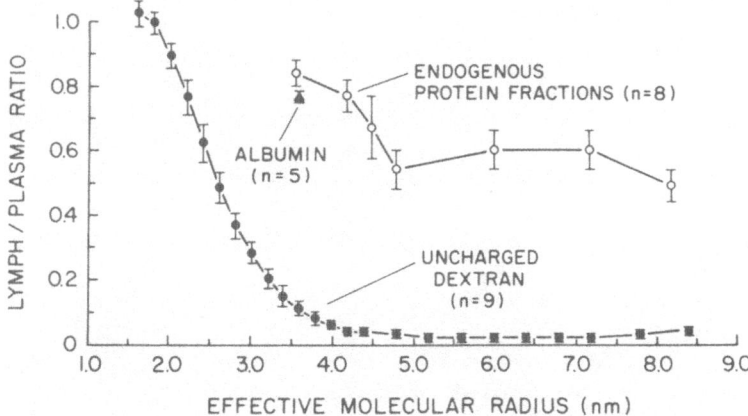

Figure 1. Steady-state lymph-to-plasma concentration ratios (L/P) for fractions of [³H]dextrans and endogenous albumin measured in healthy awake sheep under conditions of normal lymph flow (mean ± S.E.). Data for [³H]dextrans (●) and albumin (▲) from Lanken *et al.*[17] Data for endogenous protein fractions (○) from Brigham *et al.*[2] Reprinted with permission from Lanken *et al.*[17]

In the steady state, the blood–lymph barrier of the lungs exhibited marked selectivity for [³H]dextran on the basis of molecular size (Fig. 1).[17] The average L/P for the [³H]dextrans fell sharply with increasing molecular size from near 1.0 for [³H]dextrans with an effective molecular radius of 1.6 nm to 0.03 for [³H]dextrans with a radius of 4.8 nm. Between 5.0 and 8.4 nm radius, the L/P for the [³H]dextran fractions were not significantly different from zero.

For comparison, Fig. 1 also illustrates the steady-state L/P for endogenous albumin measured in our experiments, and the steady-state L/P for eight endogenous protein fractions previously measured by Brigham *et al.* in healthy unanesthetized sheep.[2] It can be seen that the L/P for the [³H]dextran fractions were markedly lower than those for the endogenous proteins of the same molecular radius throughout the molecular size range of the dextran tracer. Moreover, whereas plasma proteins behave as though the blood–lymph barrier is a hetero-porous semipermeable membrane,[13] we found that the L/P for the [³H]dextrans closely approximated theoretical predictions for a homoporous membrane. Figure 2 compares steady-state L/P for [³H]dextrans with expected values for transport of inert particles across a membrane containing uniform cylindrical pores with a radius of 5.0 nm.

B. Pulmonary Oxygen Toxicity in Sheep

After 24–36 hr, continuous inhalation of 100% O_2 causes progressive diffuse lung injury in adult unanesthetized sheep.[11,22] The flow of lymph from the lungs begins to increase after 50–70 hr of oxygen breathing and often reaches four to five times baseline flow rates before respiratory failure develops. In the present experiments, administration of 100% O_2 was discontinued when lymph flow

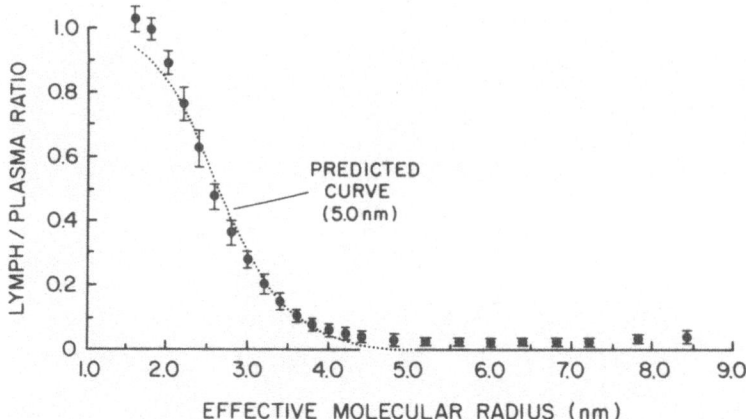

Figure 2. Comparison of mean steady-state L/P (± S.E.) for [³H]dextran fractions with theoretical prediction for molecular sieving of spherical particles across an ideal membrane through cylindrical pores with a radius of 5.0 nm. Reprinted with permission from Lanken *et al.*[17]

reached approximately three times the prehyperoxia baseline rate (66–84 hr, mean 74 hr). After 100% O_2 was replaced with air or 40% O_2, lymph flow tended to stabilize and then remained relatively constant during the posthyperoxia infusions of [³H]dextran.

Arterial blood gases, measured while sheep breathed ambient air, indicated that inhalation of 100% O_2 for 66–84 hr caused moderately severe abnormalities in gas exchange; the arterial PO_2 fell to an average of 42 torr after hyperoxia while the average PCO_2 increased mildly to 52 torr and the pH fell slightly to 7.36. Possibly because of mild dehydration, the mean left atrial and pulmonary artery pressures both decreased an average of 3 torr from baseline values while cardiac output fell approximately 16%.

Figure 3 compares the average L/P for the [³H]dextran fractions from the last 2 hr of the posthyperoxia infusions with corresponding L/P from the baseline infusions. Also shown are the average L/P for endogenous albumin from the same periods. The L/P for albumin and total protein tended to be higher than prehyperoxia baseline values after 100% O_2 exposure; however, the differences failed to reach statistical significance. In contrast, hyperoxia caused a significant increase in L/P for all of the fractions containing [³H]dextrans with radii larger than 2.0 nm. The L/P for the [³H]dextran fraction with the same effective molecular radius as albumin (3.6 nm) increased on the average by 2.6-fold above baseline. Changes in L/P were proportionately greater for the larger [³H]dextran fractions; average L/P increased from 0.02 to 0.09 at 8.4 nm radius. Inhalation of compressed air for 96 hr had no effect on lymph flow rates or L/P for the [³H]dextran fractions in the control sheep.

At autopsy, lungs from the sheep that breathed 100% O_2 appeared congested and focally atelectatic. The gravimetrically determined total water content of the lungs [(wet weight − dry weight)/dry weight] was significantly increased in com-

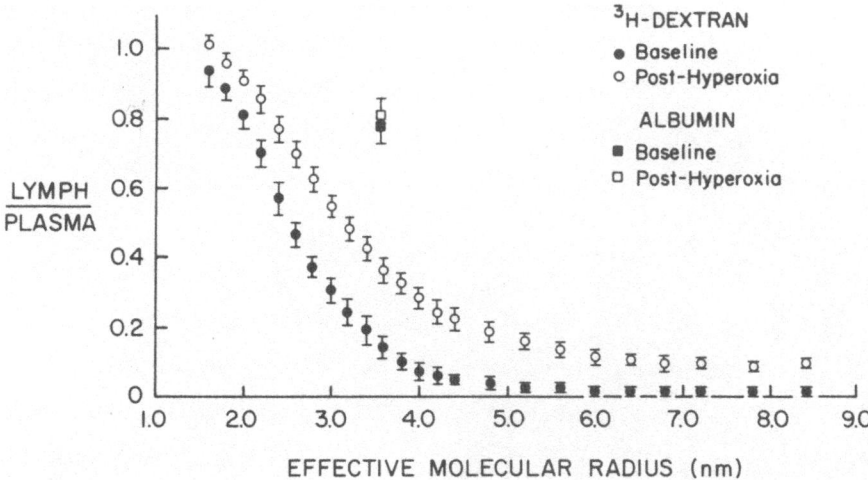

Figure 3. Comparison of average posthyperoxia L/P for [³H]dextran fractions and endogenous albumin with corresponding values from pre-hyperoxia baseline infusions (mean ± S.E. for five sheep). After hyperoxia, L/P were significantly above baseline values for every [³H]dextran fraction greater than 2.0 nm in radius ($P < 0.05$, Student's t test). Reprinted with permission from Hansen-Flaschen *et al.*[11]

parison with values obtained from postsurgical control sheep, indicating the presence of pulmonary edema. Light microscopy showed widespread thickening and increased cellularity of the alveolar walls. Numerous neutrophils were present within the interstitial spaces and occasionally within the alveoli as well. Hyaline membranes were seen in some areas but alveolar edema was rarely present.

Electron microscopy disclosed a variety of changes within the alveolar walls. In addition to increased cellularity, the interstitial spaces of the alveolar capillary membranes contained many electronlucent spaces that were thought to represent accumulation of edema fluid. Many epithelial type I cells exhibited intracellular changes, and in some areas the epithelial cell surface was fragmented or lost. Capillary endothelial cells also showed widespread intracellular changes, including focal swelling of the cytoplasm, abnormal cytoplasmic vacuoles, and swollen mitochondria. In contrast to the epithelial surface, the endothelial cell layer generally appeared continuous; however, three discrete endothelial gaps were seen in 200 random electron micrographs of alveolar capillaries (Fig. 4). Normal-appearing basement membrane could be seen beneath each of these gaps. Endothelial gaps were not seen in lung samples from control sheep.

IV. DISCUSSION

A. [³H]-Dextran Studies in Healthy Sheep

Our studies of pulmonary vascular permeability in healthy awake sheep showed that the blood-to-lymph barrier of the lungs is much more restrictive to

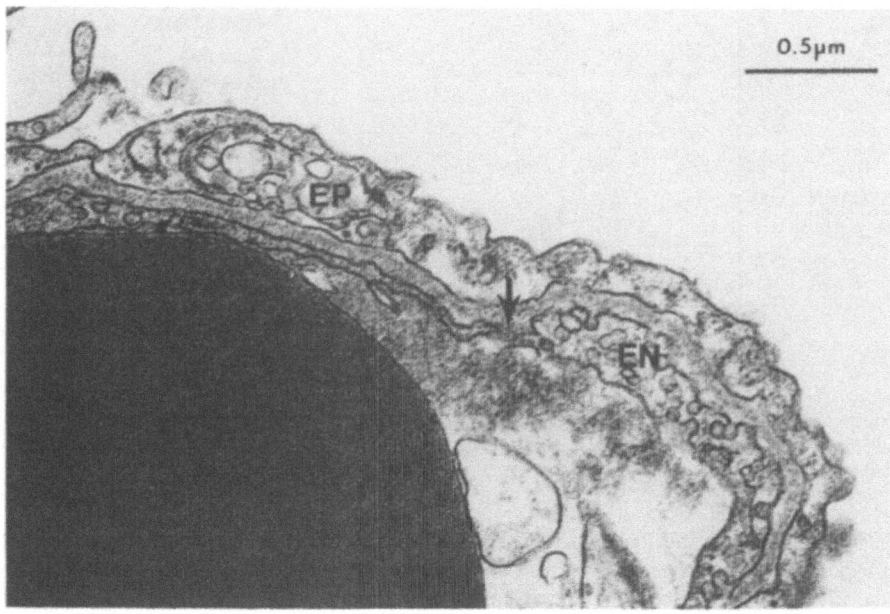

Figure 4. Electron micrograph showing part of an alveolar wall after a sheep inhaled 100% O_2 for 70 hr. A gap (arrow) is present in endothelium (EN) on thin side of alveolar capillary membrane. Basement membrane beneath gap appears intact. EP, epithelial cell. Reprinted with permission from Hansen-Flaschen *et al.*[11]

transport of neutral dextrans than plasma proteins under conditions of normal lymph flow. Other investigators have also found that L/P for neutral dextrans or polyvinylpyrrolidones are lower than those for plasma proteins in the dog paw and the rabbit paw, although the magnitude of the differences is somewhat smaller.[8,9] These observations are difficult to reconcile with theoretical models of the blood-to-lymph barrier that envision transport of inert spherical particles through hydraulically conductive channels entirely by convection and diffusion. Perhaps physicochemical properties of the molecules other than size are important determinants of transport through the channels, or alternatively, mechanisms other than convection and diffusion may contribute to the movement of at least some macromolecules across the barrier, especially in the lung.

It is unlikely that reduced transport from blood to lymph of dextrans relative to proteins is due to differences in the shape or flexibility of the molecules. In aqueous solution, dextrans form loosely coiled, hydrated spheres that are highly deformable in comparison with the relatively rigid plasma proteins.[3,23] In fact, dextrans move more readily through biological gels and across the capillary walls of the renal glomerulus than do negatively charged or neutral proteins of similar molecular size, probably because the dextrans deform into elongated structures that can pass through narrow channels and around obstacles.[7,26] Nor is it likely that differences in blood-to-lymph transport of plasma proteins and neutral dextrans result entirely from electrostatic attraction or repulsion of the proteins by electrically charged sites lining the transendothelial pathways. Both the luminal surface of the pulmonary endothelium and the endothelial basement membrane are richly endowed with negatively charged polysaccharides that should repel anionic proteins and restrict their passage relative to neutral macromolecules, as has been shown for the renal glomerulus.[33]

Differences in the shape of the steady-state sieving profiles for proteins and neutral dextrans (Fig. 1) may provide insight into the mechanism of transport of these molecules across pulmonary endothelium. In the steady state and under conditions of normal lymph flow, the concentration of endogenous proteins in lung lymph relative to plasma declines gradually as a function of molecular size, reaching ratios of 0.3 to 0.4 for proteins as large as 10 to 11 nm in radius; in contrast, L/P for neutral dextrans fall sharply with increasing molecular size and approach 0 above 5 nm radius. Mathematical models of protein sieving data have suggested that plasma proteins move across pulmonary capillary walls through two or three distinct pathways that differ in selectivity with respect to molecular size.[13] However, the sieving profile for neutral dextrans can be modeled by a single population of pathways that is highly selective on the basis of size (Fig. 2). This striking discrepancy suggests that dextrans and proteins may not have equal access to the less selective pathways through the barrier.

Considerable support for this possibility can be found in other recent ultrastructural and physiological studies of endothelial interactions with circulating macromolecules. Bignon et al. examined the composition of the endothelial surface of alveolar capillaries in the rat lung using ultrastructural immunocytochemical techniques.[1] In 1976 they reported that albumin and other plasma proteins coat the glycocalyx of the endothelial cell membranes, forming a dense protein layer that can be washed away during fixation. Schneeberger and Hamelin extended these observations and showed that removal of plasma proteins from the endothelial surface of the rat lung greatly increases transendothelial transport of native ferritin.[28] This effect was rapidly reversed when albumin was replaced in the perfusion medium. Ghitescu et al. studied the interactions between plasma proteins and capillary endothelium in mouse lung, heart, and diaphragm using gold-labeled ultrastructural tracers.[10] They found that gold-labeled albumin binds preferentially to the membrane surfaces of plasmalemmal vesicles engaged in transcytosis, and that this binding is more pronounced in the capillaries of the lung than in the heart or diaphragm. Binding of albumin to the endothelial membrane was saturable, but was not prevented by heparin or high ionic strength, suggesting a specific binding mechanism that is not electrostatic in nature. These authors concluded that albumin moves across the endothelium, at least in part, by receptor-mediated transcytosis in plasmalemmal vesicles. Villaschi et al. have also studied interactions between proteins and pulmonary endothelium using gold-labeled albumin.[34] They found that endothelial binding and transendothelial transport of gold–albumin complexes were not inhibited by competition with dextran but were greatly enhanced by glycosylation of the albumin.

Using physiological techniques, several laboratories have demonstrated that the presence of plasma proteins has an important influence on capillary permeability to nonprotein macromolecules.[6,12,16,19] Using the isolated perfused hind-

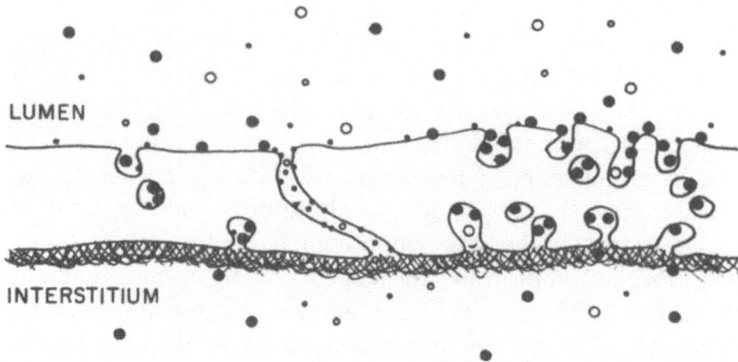

Figure 5. Schematic drawing of pulmonary capillary endothelium illustrating a possible explanation for differences in transport of [^{3}H]dextrans from blood to lung lymph. Smaller proteins (●) and dextrans (○) may both move through intracellular junctions, whereas vesicular transport of proteins may be enhanced relative to dextrans by adsorption of proteins to vesicular membranes.

limb of the cat, Kinter and Pappenheimer found that capillary permeability to dextrans increased greatly when plasma proteins were removed from the perfusion fluid; this effect could be reversed by addition of 1% protein to the fluid.[16] Recently, Michel and Phillips studied the effects of bovine serum albumin on the transport of the neutral macromolecular polymer Ficoll 70 in isolated perfused capillaries of the frog mesentery.[19] They found that the reflection coefficient for Ficoll was greatly reduced when albumin was added in small concentrations to the medium, and proposed that albumin exerted this effect by binding to the endothelial cell surface and excluding Ficoll from the transcellular pathways. In another study, Michel *et al.* found that the effects of albumin on capillary permeability could be reversed by chemically modifying the cationic arginine groups of the albumin molecule but not by modifying the cationic lysine groups, suggesting that the interaction between albumin and the endothelium involves a specific site on the albumin molecule rather than net electrical charge.[20] In studies using isolated capillary segments from the rat epididymal fat pad, Wagner *et al.* found that the rate of vesicular endocytosis of transferrin was six to seven times that for sucrose.[35] Kinetic analysis of the data suggested that transferrin was taken up by adsorption as well as simple fluid endocytosis while sucrose was taken up by fluid endocytosis alone. Wagner *et al.* proposed that transferrin binds to the membrane of endothelial vesicles so that the vesicles contain more transferrin than sucrose after internalization.

The results of these studies provide compelling evidence for binding of certain proteins to the surface of the capillary endothelium, and strongly suggest that this binding has different effects on transendothelial transport of proteins and non-protein macromolecules such as dextran and Ficoll. Against this background, a possible explanation emerges for the differences we observed in the transport of plasma proteins and neutral dextrans of various sizes from blood to lung lymph (Fig. 5). Smaller dextrans and plasma proteins may both move through pulmonary endothelial intracellular junctions by convection and diffusion in proportion to molecular size, while transendothelial transport of proteins via plasmalemmal vesicles may be facilitated relative to neutral dextrans by adsorption of the proteins onto the membranes of the vesicles. Membrane adsorption may effectively "pack" the vesicles with a higher concentration of proteins than is present in the plasma and, at the same time, reduce access of circulating dextrans to luminal vesicles or reduce the space available within the vesicles for fluid-phase endocytosis of dextrans. Moreover, the magnitude of this effect may be particularly important in the pulmonary circulation. Since dextrans can be visualized by electron microscopy,[29] it should be possible to test this hypothesis directly using ultrastructural techniques.

B. Permeability Studies in Pulmonary Oxygen Toxicity

The virtual exclusion of larger [³H]dextrans from the lymph of healthy sheep suggested that this tracer might be particularly useful for studies of permeability changes in acute lung injury. In addition, we wondered whether increased pulmonary vascular permeability might reduce or eliminate differences in the blood-to-lymph transport of dextrans and proteins. We chose to study oxygen toxicity because, in contrast to most other experimental models of acute lung injury in the sheep, continuous inhalation of 100% O_2 causes progressive diffuse lung damage with no associated increase in pulmonary vascular pressures or cardiac output.[11,22]

We found that exposure to oxygen of sufficient duration to cause a threefold increase in lymph flow rates (66–84 hr) consistently produces diffuse interstitial edema with minimal alveolar flooding. At this stage in the evolution of pulmonary oxygen toxicity, L/P were increased significantly from prehyperoxia baseline values for each [³H]dextran fraction that exceeded 2.0 nm in radius. In addition, oxygen toxicity allowed larger dextrans to enter the lymph; whereas L/P approached zero for [³H]dextrans greater than 5 nm in radius before hyperoxia, [³H]dextrans as large as 8–9 nm were present in the lymph after hyperoxia. The shape of the posthyperoxia sieving profile for [³H]dextrans is also of interest (Fig. 3). After hyperoxia the L/P declined little as a function of molecular radius for [³H]dextrans between 5 and 8.4 nm as if these larger dextrans gained access to the lymph through relatively nonselective pathways or "leaks" that had equivalent pore radii substantially larger than 8.4 nm. Either some of the normal fluid pathways were greatly altered by hyperoxia, or new pathways for macromolecules had appeared in the barrier.

The development of nonselective "leaks" might be expected to have similar effects on blood-to-lymph transport of dextrans and proteins. However, in contrast to [³H]dextran, the average L/P for endogenous albumin was unchanged from baseline after hyperoxia. The L/P for albumin may not be as sensitive to the appearance of leaks in the barrier as the L/P for larger plasma proteins because the L/P for albumin is relatively close to 1 under baseline conditions (~ 0.8). In fact, Newman et al. reported that L/P for the largest endogenous plasma proteins do increase from baseline values after hyperoxia in sheep, suggesting "partial dissolution" of the barrier.[22]

The finding of larger dextrans in the lymph after hyperoxia is also consistent with results of recent ultrastructural studies by Michel of acute pulmonary vascular injury induced by α-naphthylthiourea in dogs.[21] He found that intravenously administered dextran 75 (average molecular radius 12.5 nm) appeared to remain within the vasculature of the lungs in control animals. After α-naphthylthiourea, dextran was found in the interstitial spaces surrounding the alveolar capillaries and within the collecting lymphatic ducts. Comparatively little dextran was observed in the interstitium around arterioles or venules and virtually none was found around the airways. These observations localize the major site of leakage of dextran to the alveolar capillaries.

The nature of the capillary defect that results in leakage of dextrans after acute lung injury cannot be determined with certainty from our studies or from those of Michel. However, our ultrastructural studies did reveal occasional gaps in the alveolar capillary endothelium at the time that our posthyperoxia measurements were obtained. It appears likely that the "leaks" suggested by the [³H]dextran studies correspond to the endothelial gaps that were observed by electron microscopy.

Despite occasional discontinuities in the capillary endothelium, neither the selectivity of the barrier with respect to the molecular size of uncharged dextrans nor the difference in transport of dextrans and plasma proteins was abolished by hyperoxia. This observation suggests that the movement of larger dextrans from blood to lymph may have been restricted by the basement membrane beneath the damaged endothelium. In support of this possibility is the finding that the glomerular basement is primarily responsible for the sieving of dextrans in the kidney[33] and that this sieving effect is largely preserved in acute glomerular injury.[24]

In addition to the pulmonary endothelium, circulating macromolecules must traverse a number of other structures en route from the bloodstream to the postnodal collection site in the sheep lymph fistula preparation. These structures include the endothelial basement membrane, pulmonary interstitium, the lymphatic vessels, and the caudal mediastinal lymph node. If dextrans are removed from the vascular filtrate anywhere along this route, then L/P may not accurately reflect transendothelial transport of these molecules. Recent studies have suggested that uncharged dextrans do not bind to structures within the interstitium of the lung *in vivo*[21] or to isolated lung interstitial constituents *in vitro* (P. Sampson, unpublished observations). However, dextrans are taken up to some extent by phagocytic cells, including those that reside within lymph nodes.[32] Perfusion studies of the popliteal node in dogs,[18] and lymphoscintigraphic studies using dextran in dogs and humans,[14] have suggested that the amount of tracer taken up from the lymph during a single passage through lymph nodes is very small. Nevertheless, it is possible that our studies using intact sheep underestimated pulmonary endothelial permeability to dextrans because of uptake of the exogenous tracer within the node. To examine this possibility, Lanken and others in our laboratory have begun quantitative perfusion studies of the caudal mediastinal node in the sheep using [³H]dextrans in tracer concentrations.

C. Perspective

Much remains to be learned about the mechanisms involved in transport of various macromolecules from the bloodstream into interstitial spaces of the lungs and other organs. The conventional view that macromolecules move across the endothelium entirely by convection and diffusion through fluid-filled channels must be modified by recent observations that certain proteins bind to the luminal surface of capillary endothelium and to the membranes of endothelial vesicles. However, the quantitative significance of this binding with respect to overall transport of proteins across the endothelium remains to be determined. Until techniques are developed to obtain endothelial monolayers with physiological structure and function for *in vitro* study, the quantitative significance of protein binding to endothelium might best be examined by comparing rates of transport of endogenous proteins from blood to lymph with that of macromolecules that do not bind to the endothelium. Our studies in sheep with lung lymph fistulas suggest that plasma proteins enter the pulmonary interstitium much more readily than neutral dextrans having the same molecular size. However, because of the possibility that dextrans may be taken up to some extent by phagocytosis as they traverse the pulmonary interstitium or the lymphatic system in the sheep, additional studies will be required to substantiate this conclusion. Ultrastructural studies of dextran interactions with pulmonary endothelium in the presence and absence of plasma proteins are also needed.

The binding of plasma proteins to the endothelial surface may have important implications for studies of increased vascular permeability in acute lung injury. The effects of endothelial cell injury on the binding of plasma proteins have not yet been determined. Moreover, little is known about the effect of acute injury on the rate or direction of macromolecular transport through vesicular pathways. In view of these uncertainties, interpretation of injury-induced changes in blood-to-lymph transport of endogenous proteins of various sizes may be considerably more complex than previously envisioned. Our studies of pulmonary oxygen toxicity, as well as the studies of α-naphthylthiourea-induced lung edema by Michel suggest that nonprotein macromolecular tracers may be particularly useful in detecting and in characterizing increased capillary permeability in the lung.

V. CONCLUDING REMARKS

Despite intensive investigation, the mechanisms involved in transport of macromolecules across the endothelium of fluid-exchanging blood vessels remain incompletely understood. We have examined molecular determinants of transendothelial transport by comparing the passage of plasma proteins with that of tritiated neutral dextrans from blood to lung lymph in healthy awake sheep. Our studies show that the blood-to-lymph barrier of the lungs in sheep is much more restrictive to transport of neutral dextrans than endogenous proteins of the same molecular size. These findings are not readily explained by differences in shape, flexibility, or net electrical charge of the molecules but are consistent with the hypothesis that vesicular transport of plasma proteins is facilitated relative to neutral dextrans by adsorption of the proteins to the membranes of the endothelial vesicles engaged in transcytosis.

We have also used tritium-labeled dextrans of graded molecular sizes to study increased pulmonary vascular permeability in awake sheep with acute oxygen toxicity. After the sheep breathe 100% oxygen for 3 to 4 days, the blood-to-lymph barrier of the lungs becomes less selective for neutral dextrans on the basis of molecular size, as if nonselective pathways or "leaks" have appeared in the barrier. These "leaks" may correspond to the occasional gaps in the endothelium of the alveolar capillaries that can be seen by electron microscopy in acute oxygen toxicity.

ACKNOWLEDGMENTS. The authors thank P. N. Lanken, S. M. Albelda, P. M. Sampson, G. G. Pietra, and R. Haselton for their contributions to these studies and for their critical review of the manuscript. The work summarized herein was supported by National Heart Lung Blood Program Project Grant H-708805, and by a grant from the Southeastern Pennsylvania Chapter of the American Heart Association.

REFERENCES

1. Bignon, J., Jaubert, F., and Jaurand, M. C., 1976, Plasma protein immunocytochemistry and polysaccharide cytochemistry at the surface of alveolar and endothelial cells in the rat lung, *J. Histochem. Cytochem.* **24**:1076–1084.
2. Brigham, K. L., Bowers, R. E., and Haynes, J., 1979, Increased sheep lung vascular permeability caused by *Escherichia coli* endotoxin, *Circ. Res.* **45**:292–297.
3. Cerf, R., and Scheraga, H. A., 1952, Flow birefringence in solutions of macromolecules, *Chem. Rev.* **51**:185–261.
4. Chanana, A. D., and Darrel, D. J., 1986, Contamination of lung lymph following standard and modified procedures in sheep, *J. Appl. Physiol.* **60**:809–816.
5. Chang, R. L. S., Dean, W. P. M., Robertson, C. P. R., and Brenner, B. M., 1975, Permeability of the glomerular capillary wall. III. Restricted transport of polyanions, *Kidney Int.* **8**:212–218.
6. Danielli, J. F., 1940, Capillary permeability and edema in the perfused frog, *J. Physiol. (London)* **98**:109–129.
7. DeGennes, P. G., 1971, Reptation of a polymer chain in the presence of fixed obstacles, *J. Chem. Phys.* **55**:572–579.
8. Firrell, J. C., Lewis, G. P., and Youlten, L. J. F., 1982, Vascular permeability to macromolecules in rabbit paw and skeletal muscle: A lymphatic study with mathematical interpretation of transport processes, *Microvasc. Res.* **23**:294–310.
9. Garlick, D. G., and Renkin, E. M., 1970, Transport of large molecules from plasma to interstitial fluid and lymph in dogs, *Am. J. Physiol.* **219**:1595–1605.
10. Ghitescu, L., Fixman, A., Simionescu, M., and Simionescu, N., 1986, Specific binding sites for albumin restricted to plasmalemmal vesicles of continuous capillary endothelium: Receptor-mediated transcytosis, *J. Cell Biol.* **102**:1304–1311.

11. Hansen-Flaschen, J. H., Lanken, P. N., Pietra, G. G., Sampson, P. M., Johns, L. P., and Fishman, A. P., 1986, Effect of 100% O_2 on passage of uncharged dextrans from blood to lung lymph, *J. Appl. Physiol.* **60:**1797–1809.

12. Haraldsson, B., and Rippe, B., 1985, Serum factors other than albumin for the maintenance of normal capillary permselectivity in rat hindlimb muscle, *Acta Physiol. Scand.* **123:**427–436.

13. Harris, T. R., and Roselli, A., 1981, A theoretical model of protein, fluid, and small molecule transport in the lung, *J. Appl. Physiol.* **50:**1–14.

14. Henze, E., Schelbert, H. R., Collins, J. D., Najafi, A., Barrio, J. P. R., and Bennett, L. R., 1982, Lymphoscintigraphy with ^{99m}Tc-labeled dextran, *J. Nucl. Med.* **23:**923–929.

15. Hurley, J. V., 1978, Current views on mechanisms of pulmonary edema, *J. Pathol.* **125:**59–79.

16. Landis, E. M., and Pappenheimer, J. R., 1963, Exchange of substances through capillary walls, in: *Handbook of Physiology,* Section 2, Vol. II (W. F. Hamilton and P. Dow, eds.), American Physiological Society, Washington, D.C., pp. 961–1034.

17. Lanken, P. N., Hansen-Flaschen, J. H., Sampson, P. M., Pietra, G. G., Haselton, F. R., and Fishman, A. P., 1985, Passage of uncharged dextrans from blood to lung lymph in awake sheep, *J. Appl. Physiol.* **59:**580–591.

18. Mayerson, H. S., Patterson, R. M., McKee, A., LeBrie, S. J., and Mayerson, P., 1962, Permeability of lymphatic vessels, *Am. J. Physiol.* **203:**98–106.

19. Michel, C. C., and Phillips, M. E., 1985, The effects of bovine serum albumin and a form of cationised ferritin upon the molecular selectivity of the walls of single frog capillaries, *Microvasc. Res.* **29:**190–203.

20. Michel, C. C., Phillips, M. E., and Turner, M. R., 1982, The effects of chemically modified albumin on the filtration coefficient of single frog mesenteric capillaries, *J. Physiol. (London)* **332:**111P–112P.

21. Michel, R. P., 1985, Lung microvascular permeability to dextran in alpha-naphthylthiourea-induced edema, *Am. J. Pathol.* **119:**474–484.

22. Newman, J. H., Loyd, J. E., English, D. K., Ogletree, M. L., Fulkerson, W. J., and Brighan, K. L., 1983, Effects of 100% oxygen on lung vascular function in awake sheep, *J. Appl. Physiol.* **54:**1379–1386.

23. Ogston, A. G., Preston, B. N., and Wells, J. D., 1973, On the transport of compact particles through solutions of chain-polymers, *Proc. R. Soc. London Ser. B* **A333:**297–309.

24. Olson, J. L., Rennke, H. G., and Venkatachalam, M. A., 1981, Alterations in the charge and size selectivity barrier of the glomerular filter in aminonucleoside nephrosis in rats, *Lab. Invest.* **44:**271–279.

25. Renkin, E. M., 1985, Capillary transport of macromolecules: Pores and other endothelial pathways, *J. Appl. Physiol.* **58:**315–325.

26. Rennke, H. G., and Venkatachalam, M. A., 1979, Glomerular permeability of macromolecules, *J. Clin. Invest.* **63:**713–717.

27. Rutili, G., Kvietys, D., Martin, J. C., and Taylor, A. E., 1982, Increased pulmonary microvascular permeability induced by alpha-naphthylthiourea, *J. Appl. Physiol.* **52:**1316–1323.

28. Schneeberger, E. E., and Hamelin, M., 1984, Interaction of serum proteins with lung endothelial glycocalyx: Its effect on endothelial permeability, *Am. J. Physiol.* **247:**H206–H217.

29. Simionescu, N., and Palade, G. E., 1971, Dextrans and glycogens as particulate tracers for studying capillary permeability, *J. Cell Biol.* **50:**616–624.

30. Staub, N. C., Bland, R. D., Brigham, K. L., Demling, R., Erdman, A. J., and Woolverton, W. C., 1975, Preparation of chronic lung lymph fistulas in sheep, *J. Surg. Res.* **19:**315–320.

31. Staub, N. C., 1979, Pathways for fluid and solute fluxes in pulmonary edema, in: *Pulmonary Edema* (A. P. Fishman and E. M. Renkin, eds.), American Physiological Society, Washington, D.C., pp. 113–124.

32. Thorball, N., 1981, FITC-dextran tracers in microcirculatory and permeability studies using combined fluorescence stereomicroscopy, fluorescence light microscopy and electron microscopy, *Histochemistry* **71:**209–233.

33. Venkatachalam, M. A., and Rennke, H. G., 1978, The structural and molecular basis of glomerular filtration, *Circ. Res.* **43:**337–347.
34. Villaschi, S., Johns, L., Cirigliano, M., and Pietra, G. G., 1986, Binding and uptake of native and glycosylated albumin–gold complexes in perfused rat lungs, *Microvasc. Res.* **32:**190–199.
35. Wagner, R. C., Robinson, C. S., Cross, P. J., and Derenny, J. J., 1983, Endocytosis and exocytosis of transferrin by isolated capillary endothelium, *Microvasc. Res.* **25:**387–396.

III

Endothelial Cell Growth and Differentiation

Endothelial Morphogenesis

Ronald L. Heimark and Stephen M. Schwartz

This review will discuss evidence for common mechanisms controlling development of the vascular system and pathological growth responses of the endothelium.

I. DEVELOPMENTAL BIOLOGY OF THE ENDOTHELIUM

The morphologic sequence of developmental events in early vertebrate blood vessels suggests a pattern which may be fundamental to all further vascular growth. The initial blood vessels appear in the mesoderm as small sacs lined by endothelial cells. These sacs, arising from localized aggregation and proliferation of angioblasts, are among the first recognizable organs to appear during differentiation.[52,136] Even the earliest of these blood vessels shows hematopoiesis. It is somewhat surprising to note that current evidence supports the hypothesis that all hematopoietic tissues, including liver, spleen, and bone marrow, develop hematopoiesis as a result of seeding of primitive stem cells derived from these very early embryonic vessels.[19,27]

The network of blood vessels develops as a result of two further changes in these sacs. The first is an extension of the sacs to form a set of tubes connected to one another and to the primitive heart. Smooth muscle cells appear, coincidentally, with the connection to the heart, apparently derived from the local mesenchyme surrounding the endothelial tube. It is of some interest to note that smooth muscle cells may be derived from more than one germ layer.[16,77] LeDouarin has suggested that some smooth muscle cells associated with branches of thoracic aorta are derived from the "mesectoderm" rather than from the mesoderm.

Nothing is known about the mechanisms that trigger the differentiation of endothelial cells from primitive mesenchyme to form the vascular tree. *In vitro* development of vascular structures can be induced from blastocyst-derived embryonic mouse stem cell lines by addition of an as yet undefined factor in human

Ronald L. Heimark and Stephen M. Schwartz • Department of Pathology, University of Washington, Seattle, Washington 98195.

cord serum.[28] A number of embryonic tissues have been shown to attract capillary invasion when grown on the chick chorioallantoic membrane.[33,68,122] From both embryonic brain and kidney, heparin-binding angiogenic factors have been purified and characterized.[104,105] In the developing mouse embryo, day 7 to day 9 is a critical period in the development of the vasculature with proliferation of cords extending from the yolk sac into the embryo body.[52] Based on a recent report by Schreiber et al.[109] describing the angiogenic properties of transforming growth factor-α (TGF-α), it is worth speculating on the possibility that TGF-α is a fetal developmental hormone involved in vascular formation. Two peaks of TGF-α expression have been described during development of the mouse.[131] The first of these peaks occurs at day 7 and the second smaller peak occurs at day 13.

Once the primitive system of tubes develops, they become invested with smooth muscle and interconnected to form a complete circuit.[44] Further development of blood vessels occurs either by fusion of preexisting tubes or by extension of new tubes. This process may be thought of as analogous to development of the viscera by branching and rebranching of the primitive endodermal tube.[74] Once the primitive tube is defined by these differentiated cell layers, it appears unlikely that any new cells are recruited from undifferentiated precursors to form more endoderm. Similarly, once the vascular network is defined, it is likely that new vessels develop only by extension and replication of preexisting endothelium.

This assertion that new vessels only develop by extension from preexisting vessels implies that further change in blood vessels requires some modulation of the mechanisms maintaining the tubular structure. Perhaps the most studied example of this would be the formation of new blood vessels as seen in angiogenesis. A number of factors derived from lymphocytes,[3] normal tissues, and many tumors[37] have been described as having the ability to induce the formation of new blood vessels. The sequence of events as described by Ausprunk and Folkman[4] is quite striking. At some focus along the vessel an outpouching occurs such that endothelial cells invade the normal investing basement membrane forming a cord of cells and penetrating the surrounding connective tissue. Continuity is maintained within that cord between the newly forming lumen and the lumen of the existing blood vessels. It is important to note that this growth process resembles the morphogenic sequence which occurs during extension of the primitive blood sacs. The chick chorioallantoic membrane, a structure undergoing normal vascular neogenesis, has been traditionally used as an *in vivo* assay of angiogenesis factors. Results on this system, despite the obvious species barriers, have generally been consistent with results using rabbit cornea. The similarity in results across such a wide species barrier and in an embryonic versus an adult system suggests that the underlying sequence of events may be fundamental processes during vertebrate development.

This restriction of newly formed blood vessels to tubular structures is maintained even during tumorigenesis. All but the most widely undifferentiated hemangioendothelial sarcomas, for example, maintain a tubular–vascular structure. In turn, this suggests that endothelial growth can be stimulated without the cells losing their ability to maintain themselves in a strict monolayer.

This separation between morphogenic changes resulting in proliferation of new vascular tubes and stimulation of endothelial DNA synthesis is also supported by observations of angiogenesis in response to different stimuli. For example, Smeds and Wollman[119] have shown a centrifugal pattern of stimulation of endothelial DNA synthesis surrounding thyroid tumors. The result of this stimulation is a local increase of endothelial density without the apparent formation of new blood vessels. The mechanism underlying this stimulation is not known. In contrast, among the many factors that have been described as angiogenic, there is no obvious requirement for direct stimulation of endothelial DNA synthesis. Banda et al.[6] and Vallee and co-workers[35] have described angiogenic factors which stimulate endothelial cell migration in vitro but do not stimulate endothelial cell replication in vitro.[35] Sholley et al.[117] showed that the initial stages of migration of endothelial cells to form new vascular structures in vivo occur even when vessels have been irradiated to prevent cell replication. It seems likely that the initiation of new blood vessels in response to angiogenic factors occurs by stimulation of a morphogenic change, i.e., the formation of new branch points, without stimulation of endothelial replication. Replication follows, perhaps as a result of a localized breakdown of cell–cell interactions. This is well illustrated by the observation of Ausprunk and Folkman[4] that replication in newly formed blood vessels occurs in the cells behind the leading tip.

II. ENDOTHELIAL REPLICATION IN VIVO

To this point we have concentrated on stimulation of endothelial cell replication within *intact* cell sheets. At birth the aortic endothelium of the rat has a replication rate of 13%.[111] As shown in Fig. 1, these cells are typical of embryonic or regenerating cells, with large amounts of cytoplasm and an extensive network of endoplasmic reticulum.

In contrast, data on endothelial cell turnover in uninjured vessels show that the overall rate of cell replication in normal adult animals is phenomenally low.[111] Estimates vary from a few cells per thousand to a few cells per ten thousand per day. Morphologically, the cells show a highly attenuated cytoplasm, as thin as 300 nm, with little evidence of active protein synthesis. This quiescent picture suggests an extraordinarily stable tissue. Analysis of the distribution of replicating cells, however, shows something quite different. While most endothelium shows no replication, focal areas, called high-turnover regions, may show as much as a few percent replication per day. Despite the fact that cell replication may be quite high in focal areas, no cell denudation is observed in these sites. There are suggestions that high-turnover regions have altered permeability and that the cells show evidence of cell injury or adaptation.[53,76]

Cell replication in these focal areas of large vessels apears to represent a true form of high turnover. Evidence for this comes from two sources. First, these focal areas are seen in adult animals that no longer show any increase in the total amount of endothelium or cell density. Replication must represent turnover, not growth. Second, there is a high correlation between the location of replicating

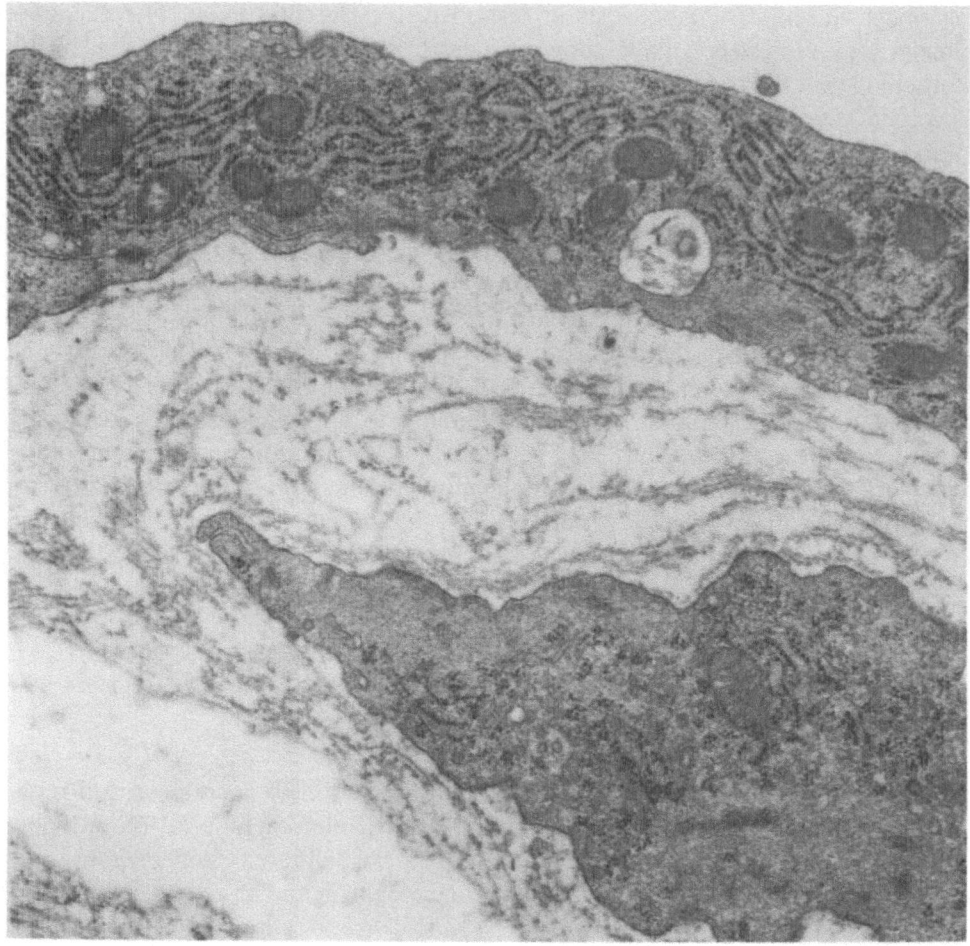

Figure 1. Aortic intima of newborn rat. The newborn aortic endothelium shows a more extensive endoplasmic reticulum and fewer vesicles compared with adult endothelium. × 25,000.

endothelial cells in the adult animal and the location of dead cells.[54] Time-lapse video microscopy of dying cells observed *in vitro* as well as cell kinetic analysis following administration of endotoxin to animals agree that the normal process of cell death includes undermining of the detaching cell by intact neighbor cells.[53] This results in the maintenance of continuity even in the presence of extensive cell death and is comparable to phenomena that have long been established as occurring at the tip of intestinal villi. In summary, we believe that a normal process of cell loss and replacement accounts for the basal rate of cell replication.

A somewhat different set of observations has been made in regard to stimulation of endothelial cell replication in large vessels subjected to different kinds of injury. Endotoxin causes a dramatic increase in the frequency of cell replication.[43,101] The basis for this change is not known; however, the increase in replication correlates well with an increase in cell death and there is no net change in the number of cells present.[54] Thus, endotoxin appears to cause a form of cell death with compensatory replacement. The results of treatment of animals with endotoxin should be contrasted with the response of neighboring vessels to the presence of a thyroid carcinoma cited above. In the case of thyroid carcinoma, cell replication is stimulated and cell density changes.[119] Nonetheless, in both cases increased replication occurs without any loss of the normal monolayer configuration. This implies a distinction between morphogenic changes underlying angiogenesis and mitogenic changes in response to growth factors.

So far we have only considered initiation of replication in intact monolayers. An extensive effort has been made to study regeneration of endothelium following mechanical abrasion of portions of the cell layer.[7] It was quickly noted, however, that the denudation of vessels *in vivo* was rapidly followed by the migration and proliferation of smooth muscle cells to form an intimal cell mass. Observations in several laboratories that smooth muscle cells form the principal noninflammatory component of the atherosclerotic lesion led to an extensive use of balloon catheter denudation as a model for the smooth muscle cell proliferative response. The hypothesis was that removal of the endothelium resulted in attachment of platelets, release of platelet growth factor, and stimulation of smooth muscle cell migration and proliferation.[106] The first attempt to study this process in a kinetic fashion showed a correlation between the distribution of smooth muscle proliferation and the distribution of regenerated endothelium.[56] Areas that became rapidly re-covered by endothelium did not show smooth muscle proliferation. The regenerating endothelium exhibits large amounts of rough endoplasmic reticulum reminiscent of endothelial cells in the developing aorta (Fig. 2). Junctional structures in the regenerating cells are poorly formed and characteristically occluded only at spotlike regions of membrane fusion.

While these observations supported the reasonable teleologic expectation that endothelial continuity would inhibit smooth muscle growth, a number of observations interfered with this conclusion. Minick et al.[85] found that the greatest area of intimal thickening after balloon denudation was at the edge of the regenerated sheet of endothelium. The model of extensive denudation with the balloon catheter is obviously not similar to the spontaneous denudation likely to occur in the progress of atherosclerotic lesions, particularly in the early stages. A more defined approach to removing endothelium has been developed.[100] Even extensive denudation with this approach did not produce smooth muscle proliferation unless there was some injury or alteration deeper to the wall. The nature of that injury and the mechanism leading to smooth muscle proliferation remains an open question. Endothelial cell replication is initiated rapidly by the removal of neighboring endothelial cells. If, however, the cells removed constitute only a very narrow strip, no endothelial replication occurs. Apparently, endothelial cell movement has to continue for a minimum of 6 to 8 hr before cell replication will be stimulated. When extensive areas were denuded, the endothelium would only grow out 10 mm at 3 months and failed to regenerate further.[102] This cessation of endothelial growth complicated the interpretation of Minick and colleagues' experiment. It

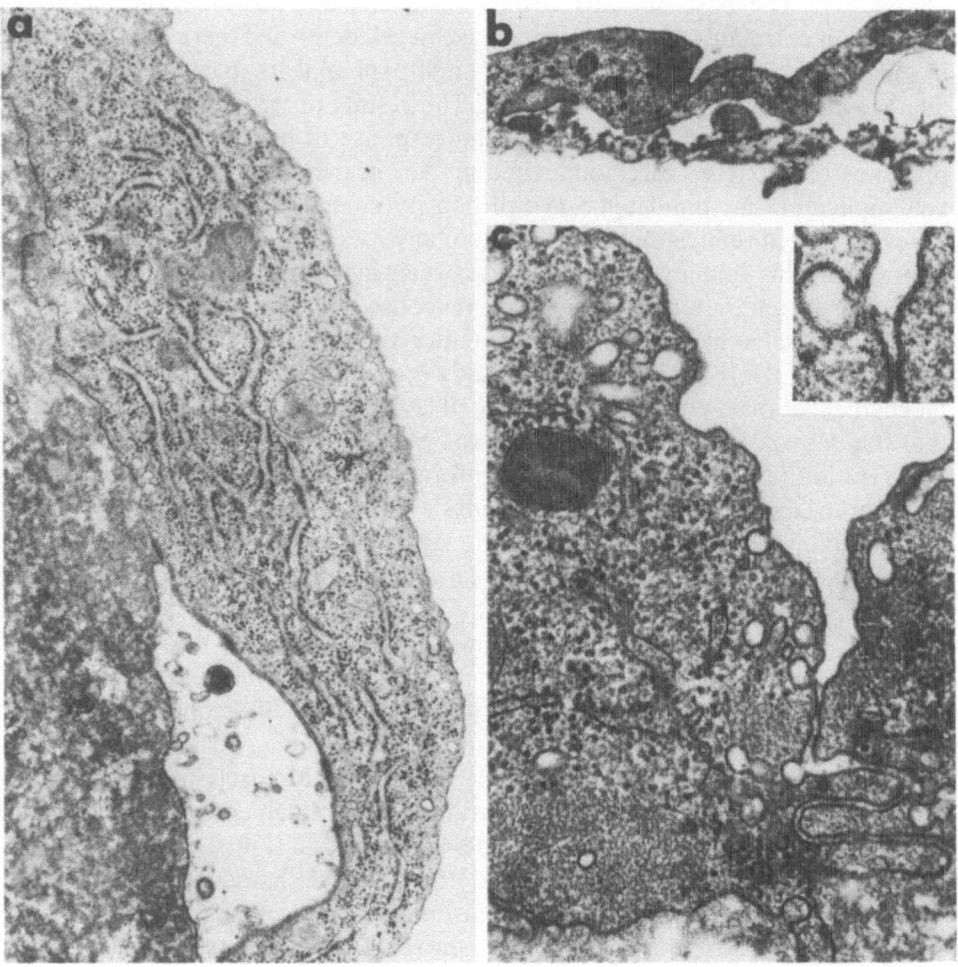

Figure 2. Endothelial cell during regeneration. (a) Four weeks after injury ($\times$ 25,000); (b) 1 week after injury with loose attachment to subendothelium; (c) 2 weeks after injury; note that cell junction is "open" except for a small occluded region (inset, $\times$ 105,000) ($\times$ 45,000).

now appears likely that the thickened intima occurred at the edge of a sheet of endothelium which itself was no longer able to replicate. The failure of endothelium to replicate is as important a mystery as is the increased intimal thickening at the same location.

Two alternative hypotheses have been offered to explain this limited ability of endothelium to regenerate.[99] The first is that the limit is intrinsic to the endothelial cells themselves. In this hypothesis, the endothelial cell, perhaps like the human diploid fibroblasts studied by Hayflick and others, is only able to undergo a finite number of generations.[58] According to this hypothesis, the cells at the edge would lose the ability to divide and perhaps the ability to migrate, resulting in a sessile fence able to inhibit replication of cells farther back. We have shown that the cytoskeleton in these cells at the regenerated edge, rather than being oriented consistently along the long axis of the vessel, oriented cells at right angles to the direction of regeneration.[39] A second hypothesis is that the limited ability of endothelial cells to regenerate is dependent upon some kind of interaction between endothelial cells and the denuded surface. At the time endothelial cell regeneration ceases, the surface is coated with pseudoendothelial cell derived from smooth muscle cells. These altered smooth muscle cells have many of the properties of endothelium including nonthrombogenicity[50,102] and apparently the ability to produce PGI_2.[26,34] No further information is available on the cellular interactions between pseudoendothelial cells and endothelial cells at a biochemical level. Obviously, it is difficult to study such interactions when the only known way of producing pseudoendothelial cells requires an intact animal.

In summary, the current view of the endothelium is as a dynamic population with large stable areas but having focal areas with considerable cell turnover. The rate of that turnover is sufficiently high that one should not expect to see morphological evidence of discontinuities. This can be shown quantitatively, as follows.

Studies in the rat show that endothelium can recover an area one cell wide within about 3 hr.[100] Turnover studies in the same species imply a rate of cell area loss of approximately 1×10^{-3} per day.[110,111] These values may be combined to estimate the total average area of denudation present at one time:

If

$$E = A \times r \times t$$

where E = denuded area (μm^2/cell), A = area of each cell $\simeq 10^3 \ \mu m^2$, r = rate of turnover $\simeq 10^{-3} \ day^{-1}$, and t = time required to replace each cell $\simeq 3$ hr, then

$$E \simeq 0.125 \ \mu m^2/cell$$

The assumptions used in this calculation should result in a maximal estimate of the amount of denudation present during normal turnover. Nonetheless, 0.125 μm^2 is a very small area, which might be difficult to detect at the usual level of resolution available by scanning electron microscopy.

In other words, endothelial cells of large vessels maintain themselves as a continuous cell layer both during normal turnover and during regeneration. This raises an important set of questions: What are the mechanisms that maintain cell–cell interactions and how do these molecules interact with the process or processes controlling cell growth?

III. INHIBITION OF ENDOTHELIAL REPLICATION BY CELL–CELL INTERACTION

As discussed previously, endothelial cells are capable of regenerating only a finite extent of denuded surface. Cell replication also fails to occur *in vivo* if the wound is sufficiently small. This happens when wounds are smaller than the extent required to be covered by about 6 to 8 hr. In these cases, endothelial cells either spread or move rapidly over the denuded surface, but there is no initiation of DNA synthesis. Possibly this failure to initiate DNA synthesis is a result either of the rapid reestablishment of cell–cell contacts across the wound or, alternatively, the failure to break cell–cell adhesion in the advancing cell sheet.

The hypothesis that replication in a small wound *in vivo* is prevented by maintenance of some kind of cell–cell interaction is also supported by studies *in vitro* of various kinds of wounds. For example, despite the substantial evidence that various growth factors stimulate or assist endothelial cell replication when these cells are plated sparsely, no growth factor to our knowledge has been described which stimulates growth once the cells have reached stationary density.[57,113] This contrasts significantly with the general observation from experiments with fibroblasts and with smooth muscle cells. Using these cell types, saturation density of cultured cells is dependent upon the concentration of serum.[129] Using bovine aortic endothelial cells, we have examined cell densities over a wide range of serum concentrations. While the rate of replication required to reach saturation density may be serum dependent, the final cell density appears to be independent of serum concentration.[113,114] There are factors which reinitiate replication in this stationary density cell sheet. These factors cover a diverse range of chemistry including collagen,[24] fibrin,[69] and colchicine.[116] The common feature among all of these "mitogens" is their ability to disrupt the continuity of the monolayer. Even colchicine, while able to inhibit migration from a wound edge, is able to stimulate DNA synthesis in a confluent layer concomitant with the ability of colchicine to cause cell–cell retraction.[116] In contrast, once endothelial cells have been separated from an extensive wound edge, replication can be readily inhibited by agents that inhibit cell movement, including cytochalasins B and D.[115] The cytochalasins are also able to inhibit cell replication initiated by colchicine. This suggests that either cell locomotion or retraction (also inhibited by cytochalasin) is required for the initiation of growth. These observations suggest that at the very least, endothelial replication can be inhibited by cell–cell contact.

The concept of contact inhibition of replication is not new. It has, however, been controversial because of the lack of identification of molecules able to mimic the effects of cell–cell interaction at saturation density. Perhaps the most direct evidence against the cell–cell interaction hypothesis comes from studies of 3T3 fibroblasts. Martz and Steinberg[84] conducted a careful longitudinal time-lapse image cinematographic study of the cell replication in subconfluent cultures. Their hypothesis was that the incidence of mitosis would correlate with some measure of percent periphery occupied by neighbors or frequency of contact between neighbors. This was, however, not observed. Failure to demonstrate contact inhibition of growth contrasts with a generally accepted observation that cell–cell contact does inhibit movement as noted by Abercrombie[1] and Albrecht-Buehler.[2] As a result, Martz and Steinberg and others have tried to separate the concept of contact inhibition of movement from the concept of density-dependent inhibition of replication.

Studies of growth control of 3T3 cells under conditions where factors present in conditioned medium might be limiting have also shown inconsistent results. The saturation density of 3T3 cells can be increased by the simple expedient of increasing the amount of growth factor available in the culture medium. Most investigators today would accept the hypothesis that density-dependent control of growth of the 3T3 cells is not simply a matter of the availability of cell contact. Instead, population density would appear to be regulated by the availability of mitogens either added exogenously or synthesized by the cells, as well as the availability of growth inhibitors. Two growth inhibitors that the cells themselves can synthesize are β-interferon and TGF-β.[25,32]

Despite the evidence for soluble factors controlling the density of 3T3 cell culture,[121] there remain substantial reasons for believing that cell contact is also important in this system. First, the Martz and Steinberg[84] experiments are potentially flawed by the relatively short time period studied in the time-lapse videos. The time period was barely more than a single cell cycle. It is conceivable that inhibition of replication by cell–cell interaction represents either an integral part of intercellular adhesions over a longer period of time or that the inhibition of replication requires more than one cell generation. More directly to the point, there is evidence for cell surface membrane fragments from 3T3 cells that are able to mimic the effects of contact inhibition of growth.[88,137] While studies have not led to identification of a single molecule, they have shown that membrane preparations from these cells contain a protease-sensitive activity that is able to inhibit replication in a reversible fashion. Furthermore, the inhibition has been shown not to be due merely to the depletion of mitogens from the medium.[133] We would suggest that it is important to consider at least two kinds of growth control mechanisms for 3T3 cells: (1) density-dependent mechanisms involving paracrine or endocrine factors available in the surrounding medium and (2) controls representing direct cell–cell interaction.

Similar considerations need to be given to control of endothelial growth. For example, a number of workers have described growth factors that support endothelial cell replication in culture.[11,18,20,46,79] Several of these now turn out to be one of two factors originally described by Gospodarowicz et al. as "fibroblast growth factor" (FGF).[45] Today we know that FGF consists of two related proteins. Acidic FGF, also known as endothelial cell growth factor (ECGF), has been shown to have clear mitogenic effects for human endothelium.[82,128] It is interesting to note, however, that ECGF prolonged the life span of cells that had already been grown in its absence. This is similar to observations in other systems where the major action of growth factors may be to prolong replicative life span, possibly by preventing terminal differentiation rather than by simply stimulating DNA synthesis. Examples include the ability of basic FGF to prevent skeletal muscle differentiation[78] and EGF to prevent epidermal cell differentiation.[103] Interestingly, Maciag and his collaborators note a marked change in endothelial structure upon withdrawal of ECGF, perhaps implying a "differentiation" into vascular tube structures.[67,83] The ability of ECGF or basic FGF to stimulate endothelial replication in a more general sense is suggested by the recent discovery that these factors are angiogenic.[70] This, however, is complicated by the possibility that other cells, e.g., pericytes or smooth muscle cells, might be the in vivo target of these growth factors. It has also been observed that α-interferon inhibits the mitogenic activity of FGF.[30] In contrast, to our knowledge, no growth factor has been shown to stimulate endothelial cell replication in confluent endothelium *in vitro*.

There is also evidence for other soluble factors able to inhibit endothelial replication. With respect to inflammatory cells, it has been shown that soluble factors from activated lymphocytes are able to inhibit endothelial migration.[17] The nature of this factor has yet to be defined. There have been reports of inhibition of endothelial replication by tumor necrosis factor and γ-interferon,[123] factors likely to be present at sites of inflammation. In addition, γ-interferon and interleukin 1 induce a number of changes in endothelial cells including reorganization of actin filaments and expression of new cell surface molecules.[10,89,97,123] TGF-β released from platelets inhibits both migration and replication in regenerating endothelial cell sheets.[5,38,62] It is certainly possible that substantial amounts of TGF-β might be released from platelets, perhaps inhibiting or delaying endothelial replication. We have shown that purified TGF-β from platelets blocked cells in a wound-edge assay from entering S phase and the fraction of cells in G_1 increased. The inhibition was observed 24 hr but not 48 hr after addition of TGF-β. In at least one culture system, a soluble TGF-β-like protein has been proposed as an important mediator of density-dependent inhibition of growth.[130] In other cell types, TGF-β has been shown to enhance plasminogen activator activity[72] and stimulate production of major extracellular matrix proteins.[66]

Of interest is the possibility that cell–cell interactions play a definite role in regulation of endothelial cell growth. Schwartz *et al.*[112] and Spagnoli *et al.*[120] have shown that regenerating *in vivo* endothelial cells lose their gap junctions. This is a common feature of cell regeneration in epithelia and has been suggested as a major controlling factor in the stimulation of growth by release from quiescence.[80] There is also evidence *in vitro* that composition of cell surface proteins in the endothelium is dependent on cell–cell contact.[60,134] Reappearance of the cell surface protein CSP-60 after trypsinization and replating correlates re-formation of the monolayer. Finally, we have shown that addition of a cell surface membrane fraction from confluent cells to growing cells inhibits DNA synthesis.[59] Both migration and replication were blocked by the cell surface membranes. Apparently, the membranes have arrested the cells in the G_0/G_1 phase of the cell cycle. The effect was not seen with membrane preparations from smooth muscle cells. The inhibition of replication is nontoxic and reversible without altering the composition of the medium. These data imply the presence of a cell–cell recognition molecule (Fig. 3). The activity is labile to proteases, heat treatment, and reduction, suggesting a protein. This is in contrast to the heparinlike inhibitor of smooth muscle cell growth described by Castellot *et al.*[13] It is solubilized by the detergent octyl-D-glucoside but is not dissociated by treatment which removes extrinsic membrane proteins. Analysis of the cell surface membrane fraction by SDS gel electrophoresis shows that a number of proteins are present including CSP-60. The octyl-D-glucoside-solubilized activity binds to wheat germ agglutinin–agarose and is eluted with *N*-acetylglucosamine. Cell surface glycoproteins have also been implicated in the inhibition of growth in fibroblasts.[138]

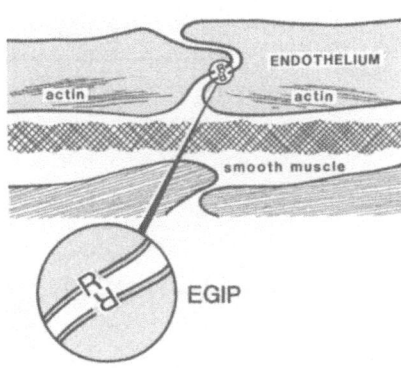

Figure 3. Interaction of the endothelial growth inhibitory protein (EGIP) at saturation density. EGIP is represented as the integral transmembrane protein (R). Occupancy of a few receptors apparently is not enough to stop growth but could inform interacting cells to change direction of migration. Only when a critical number of receptors become occupied would this be expected to stop growth.

In line with the results of inhibition of DNA synthesis in endothelial cells, Pereira-Smith *et al.*[92] have described inhibition of fibroblast growth by surface membrane-enriched preparations from senescent cells. Interestingly, polyadenylated mRNA isolated from senescent fibroblasts inhibited DNA synthesis in growing cells after microinjection,[81] whereas polyadenylated mRNA from cells made quiescent by removal of serum growth factors had only a slight inhibitory effect. The authors suggest that the production of this inhibitory activity is part of the mechanism of cellular quiescence and senescence. It is intriguing to consider the possibility that the mRNA species is coding for the membrane growth inhibitor.

IV. CELL–CELL AND CELL–SUBSTRATE ADHESION MOLECULES

At this point we need to consider the relationship between endothelial cell movement and replication. As already noted, agents that cause cell–cell separation also stimulate endothelial cell replication. Inhibitors of cell movement, and therefore the ability of cells to move relative to one another, also inhibit endothelial cell replication.[115] These findings suggest we have to begin to consider how endothelial cells assemble into a cell sheet.

In addition to the classical morphological junctional structures in the endothelium,[73,118] there are extensive areas of apposition with less specific morphology between membranes of adjacent cells. Several cell surface proteins have been identified as involved in cell adhesion.[21,40,125,140] In epithelial cells a protein, termed uvomorulin, has been localized in a similar intercellular region to intermediate junctions.[12] Monoclonal antibodies to uvomorulin prevent re-formation of tight junctions and dissociate cell–cell contact in a manner similar to removal of calcium.[8,51] Edelman and others have speculated on the role of cell adhesion molecules in regulating temporal and spatial organization during development.[31,55,75,127] It is surprising that none of the antibodies to cell adhesion molecules identified to date cross-react with vascular cells.

We have found that vascular smooth muscle cells and endothelial cells also

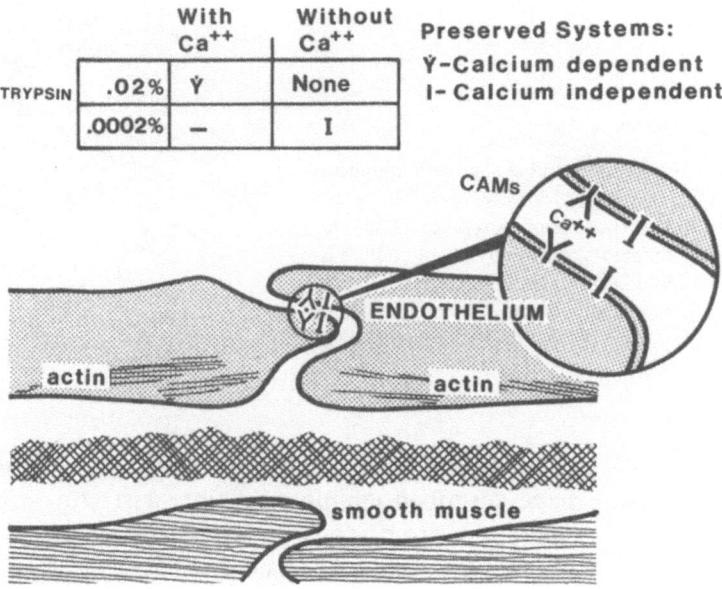

TRYPSIN		With Ca^{++}	Without Ca^{++}
	.02%	Ÿ	None
	.0002%	−	I

Preserved Systems:

Ÿ-Calcium dependent

I- Calcium independent

Figure 4. Intercellular adhesion mechanisms. The endothelium shows both a Ca^{2+}-dependent and a Ca^{2+}-independent cell adhesion mechanism which are distinguished by their sensitivity to trypsin. Dissociation of confluent monolayers of endothelial cells with 0.002% trypsin in the presence of EGTA abolishes Ca^{2+}-dependent aggregation but not the Ca^{2+}-independent mechanism. Raising the trypsin concentration to 0.02% with EGTA abolishes both mechanisms. If calcium is present with the high trypsin concentration, the Ca^{2+}-dependent mechanism is preserved and Ca^{2+}-independent aggregation is destroyed.

show multiple cell adhesion mechanisms.[61] These fall into two classes of cell adhesion mechanisms: either Ca^{2+}-dependent or Ca^{2+}-independent, the difference lying in the sensitivity to proteolysis in the presence or absence of calcium (Fig. 4). Both endothelial and smooth muscle cells contain a related Ca^{2+}-dependent aggregation system. For this reason we shall refer to this Ca^{2+}-dependent molecule as V-CAD (vascular cadherin). Photomicrographs of Ca^{2+}-dependent aggregated endothelial cells show that the cells aggregate as sheets. Fab' fragments of antibodies raised against intact endothelial cells inhibit only the Ca^{2+}-dependent mechanism and further suggest that molecular mechanisms of the two aggregation systems are different. Gel filtration on Sephacryl S200 of the trypsin-releasable fragments of the Ca^{2+}-dependent activity showed that a 22,000-molecular-weight fragment was able to reverse the inhibition of aggregation by antiendothelial Fab' fragments. Immunoblot analysis of plasma membranes shows that the endothelial cell surface antibodies recognize several proteins ranging in molecular weight from 29,000 to 230,000. Membranes from cells trypsinized under conditions that remove the Ca^{2+}-dependent adhesion mechanism show loss of proteins with a molecular weight of 135,000. Membrane proteins were separated by SDS gel electrophoresis and transferred to nitrocellulose. The membranes were cut into ten equal pieces of molecular weight classes and incubated in a solid-phase neutralization assay with the antiendothelial cell Fab' fragment. One region with a molecular weight range of 130,000 to 145,000 contained all of the neutralizing activity.

The existence of this phenomenon, even before the inhibitory activity is purified, has had a number of important effects on our thinking about mechanisms of cell–cell interaction. For example, we would predict that a set of monoclonal antibodies binding to the endothelial surface would be mitogenic by neutralization of function required for integrity. During screening for hybridomas that recognize the endothelial cell surface, we identified a set of clones that are mitogenic when added to confluent endothelial cells. As a result of these studies, we have identified a monoclonal antibody (designated Ec4.10a) that disrupts cell–substrate adhesive processes by probably binding to an adhesion-related molecule on the cell surface.[141] Antibodies with similar activity have been reported in the literature; however, none of these antibodies recognize the endothelium.[22,90] Given the sensitivity of endothelial cells to mitogenesis in response to agents that disrupt the monolayer, it is not surprising that antibodies directed at cell–substrate adhesive molecules or cell–cell adhesive molecules would be mitogenic.

In this light, we should also give thought to a class of molecules recognized for their role in cell–substrate adhesion. By a similar rationale, we might expect any alteration in cell–substrate interaction to have marked effects on cell–cell interactions responsible for controlling growth. Geiger et al.[42] have shown that the subcellular location of a 135,000-molecular-weight adherens junction protein is at sites of intercellular contact and that another junction protein, talin, is restricted to sites of focal contacts. This suggests that information could be uniquely transmitted through these proteins that are associated with various types of adherens junctions. The importance of the extracellular matrix in cell proliferation has been previously recognized.[47] Madri and co-workers[96] have further shown that matrix components can affect the degree of migration and proliferation of endothelial cells when cultured as a fibronectin versus a type I/III collagen substrate. The cells showed a marked preference for type I/III collagen. These results suggest that extracellular matrix components have a role in regulation of endothelial cell growth, which has yet to be clearly defined. One could speculate that endothelial cells in a developing capillary could "sense" some stable interaction between membrane proteins and fixed extracellular structures to promote or inhibit growth.

To take this further, we need to define the specific cell and substrate molecules involved in assembly of the endothelium on the extracellular matrix. Two monoclonal antibodies, CSAT and JG22, dissociate myoblasts and fibroblasts and prevent cell–matrix adhesion.[22,48] They both recognize a complex of three glycoproteins with molecular weights of 160,000, 140,000, and 110,000 which constitute the fibronectin receptor in avian cells. A number of laboratories have implicated the 140,000-molecular-weight surface glycoprotein in binding to fibronectin[15,71,98] in addition to a 47,000-molecular-weight glycoprotein.[64,91,132] Synthetic peptides of the cell binding region in fibronectin containing the sequence Arg-Gly-Asp-Ser (RGDS) have been used to promote cell attachment.[93] The RGDS region of this fragment also blocks cell attachment.[93,139]

We have found that the RGDS fragment readily detaches confluent endothelial cells from their matrix, implying that some molecule in the endothelial cell membrane, perhaps analogous to the determinant of the CSAT antibody, is important in substrate adhesion. Obvious candidates are the fibronectin receptor or the GpIIb/IIIa-like proteins.[36,124] Glycoproteins IIb/IIIa have been most extensively described in the platelet where they appear to account for aggregation via interactions with fibrinogen and fibronectin.[9,87,94] Fibrinogen and von Willebrand factor also contain RGD sequences and binding of these proteins to platelets is inhibited by RGD-containing peptides.[41,95] The platelet proteins GpIIb/IIIa which bind to RGDS-containing proteins have also been identified in endothelial cells.[36,124] Addition of Fab' fragments from IgG isolated from GpIIb/IIIa antiserum also detaches cells from their matrix but much more slowly.[14] Apparently, alterations in cell–substrate interaction can have effects on cell–cell interactions. The activity of the GpIIb/IIIa complex, however, does not appear to be simply mediated by cell–substrate interaction. RGDS or monoclonal antibodies to cytoadhesion proteins will cause detachment of cells. The cells separate not only from the substrate but from each other, suggesting that the cell–substrate interaction plays an important role in maintaining stability of the cell–cell junction as well.[65] Depending on the substrate, monoclonal antibodies will prevent reattachment of trypsinized cells. In contrast, Fab' fragments to the GpIIb/IIIa complex do not alter reattachment. These data suggest to us that the role of GpIIb/IIIa in attachment is complex and indirect. It is interesting to note that several laboratories have observed that fibrin or fibrinopeptides added to the top of the cell layer alter endothelial cell shape.[23,69,107,108] Perhaps the GpIIb/IIIa complex acts as a ''sensor'' of the abnormal presence of such molecules on the luminal surface of the cells, stimulating a change in cell shape rather than causing detachment by direct interaction with a cell–substrate complex.

V. MORPHOGENIC CONTROL OF GROWTH: A HYPOTHESIS

A summary of the mechanisms maintaining a confluent monolayer is shown in Fig. 5. We would like to propose that there are at least three different fundamental mechanisms responsible for controlling endothelial cell replication.

The primary one is the maintenance of quiescence mediated by cell–cell interactions and perhaps cell–substrate interactions which in some manner inform the endothelial cell that there is no need to replicate. As long as these interactions are intact, endothelial cells will not respond to mitogenic stimuli, even when those stimuli are plentifully available. This would be consistent with the general observation that the final density obtained by endothelial cells in cultured systems is independent of the availability of growth-controlling substances in the medium.

The secondary mechanism would be stimulation of migration and replication by disruption of these stable structures. This could occur in a number of ways. First, one might imagine the availability of factors able to modulate the cell's expression of these critical cell surface molecules. This would be directly anal-

Figure 5. Sites of cell surface molecules which regulate endothelial morphogenesis. The cell surface contains receptors for protein factors which regulate growth. In addition, there are structural transmembrane proteins which communicate with the exterior and interior of the cell and function in maintenance of integrity of the endothelium. These include EGIP; the fibronectin receptor, cytoadherence molecules, such as GpIIb/IIIa; and the CAMS which include calcium-dependent and independent cell adhesion molecules.

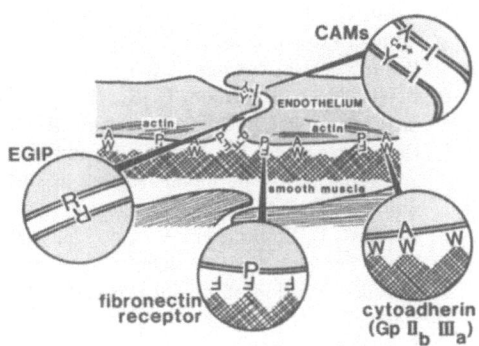

ogous to the morphogenic processes proposed by Edelman and others for control of the morphogenesis of structures derived from the neural tube.[31,75] In that system, development of new neural structures occurs when cells cease to express intercellular adherence molecules and begin to express molecules associated with the attachment and migration along extracellular matrix.[29,126,135] It is intriguing to consider the possibility that such sequential expression of cell adherence molecules might be found in the early developmental stages of endothelial embryology. As yet, we have no indication of factors which might modulate such expression. It is intriguing, however, to consider the possibility that angiogenic factors, particularly those which seem to act by stimulating migration rather than directly stimulating replication, may function in this way. Interestingly, such factors have been described by Banda *et al.*[6] and by Vallee and co-workers.[35] A placental protein of the basic FGF family has recently been purified and, in addition to stimulating DNA synthesis, it promotes production of plasminogen activator and latent collagenase in capillary endothelial cells.[86] It is also intriguing to speculate on the relationship between growth and protease production in neovascularization.[49] Release and activation of protease may also be necessary for remodeling of the matrix breakdown of cell–cell or cell–substrate interactions, releasing the cells from growth inhibition.

Finally, growth factors might act on the endothelium by interfering directly with the cell surface molecules. In this hypothesis, one might imagine a growth factor specifically altering an intercellular adherence system, perhaps disrupting the endothelial growth inhibitory protein, or the complex interactions between endothelial cells and their substrate.

REFERENCES

1. Abercrombie, M., 1970, Contact inhibition in tissue culture, *In Vitro* **6**:128.
2. Albrecht-Buehler, G., 1977, The phagokinetic tracks of 3T3 cells, *Cell* **17**:299.
3. Auerbach, R., and Sidky, Y. A., 1979, Nature of the stimulus leading to lymphocyte-induced angiogenesis, *J. Immunol.* **123**:751.
4. Ausprunk, D. H., and Folkman, J., 1977, Migration and proliferation of endothelial cells in preformed and newly formed blood vessels during tumor angiogenesis, *Microvasc. Res.* **14**:53.

5. Baird, A., and Durkin, T., 1986, Inhibition of endothelial cell proliferation by type beta-trans-forming growth factor: Interactions with acidic and basic fibroblast growth factors, *Biochem. Biophys. Res. Commun.* **138:**476.
6. Banda, M. J., Knighton, D. R., Hunt, T. K., and Werb, Z., 1982, Isolation of a nonmitogenic factor from wound fluid, *Proc. Natl. Acad. Sci. USA* **79:**7773.
7. Baumgartner, H. R., Muggli, R., and Tschoff, T. B., 1976, Platelet adhesion, release and ag-gregation in flowing blood: Effects of surface properties and platelet function, *Thromb. Hae-mostasis* **35:**124.
8. Behrens, J., Birchmeier, W., Goodman, S. L., and Imhof, B. A., 1985, Dissociation of Mandin-Darby canine kidney epithelial cells by the monoclonal antibody anti Arc-1: Mechanistic aspects and identification of the antigen as a component related to uvomorulin, *J. Cell Biol.* **101:**1307.
9. Bennett, J. S., Vilaire, G., and Cines, D. B., 1982, Identification of the fibrinogen receptor on human platelets by photoaffinity labeling, *J. Biol. Chem.* **257:**8049.
10. Bevilacqua, M. P., Pober, J. S., Majeau, G. R., Cotran, R. S., and Gimbrone, M. A., Jr., 1984, Interleukin 1 (IL-1) induces biosynthesis and cell surface expression of procoagulant activity in human vascular endothelial cells, *J. Exp. Med.* **160:**618.
11. Bohlen, P., Baird, A., Esch, F., Ling, N., and Gospodarowicz, A., 1984, Isolation and partial molecular characterization of pituitary fibroblast growth factor, *Proc. Natl. Acad. Sci. USA* **81:**5364.
12. Boller, K., Vestweber, D., and Kemler, R., 1985, Cell-adhesion molecule uvomorulin is localized in intermediate junctions of adult intestinal epithelial cells, *J. Cell Biol.* **100:**327.
13. Castellot, J. J., Addonizio, M. L., Rosenberg, R., and Karnovsky, M. J., 1981, Cultured en-dothelial cells produce a heparin-like inhibitor of smooth muscle cell growth, *J. Cell Biol.* **90:**372.
14. Chen, C. S., Thiagarajan, P., Schwartz, S. M., Harlan, J. M., and Heimark, R. L., 1987, The platelet glycoprotein IIb/IIIa-like protein on human endothelial cells promotes adhesion not at-tachment, *J. Cell Biol.* **105:**1885.
15. Chen, W.-T., Greve, J. M., Gottlieb, D. I., and Singer, S. J., 1985, Immunocytochemical lo-calization of 140 kd cell adhesion molecules in cultured chicken fibroblasts, and in chicken smooth muscle and intestinal epithelial tissues, *J. Histochem. Cytochem.* **33:**576.
16. Ciment, G., and Weston, J. A., 1985, Segregation of developmental abilities in neural-crest-derived cells: Identification of partially restricted intermediate cell type in the branchial arches of avian embryos, *Dev. Biol.* **111:**73.
17. Cohen, M. C., Picciano, P. T., Douglas, W. J., Yoshida, T., Kreutzer, D. L., and Cohen, S., 1982, Migration inhibition of endothelial cells by lymphokine-containing supernatants, *Science* **215:**301.
18. Conn, G., and Hatcher, V. B., 1984, The isolation and purification of two anionic endothelial cell growth factors from human brain, *Biochem. Biophys. Res. Commun.* **124:**262.
19. Cudennec, C. A., Thiery, J.-P., and LeDouarin, N. M., 1981, *In vitro* induction of adult eryth-ropoiesis in early mouse yolk sac, *Proc. Natl. Acad. Sci. USA* **78:**2412.
20. D'Amore, P. A., Glaser, B., Brunson, S. K., and Fenselau, A. H., 1981, Angiogenic activity from bovine retina: Partial purification and characterization, *Proc. Natl. Acad. Sci. USA* **78:**3068.
21. Damsky, C. H., Richa, J., Solter, D., Knudsen, K., and Buck, C. A., 1984, Identification of a cell surface glycoprotein mediating intercellular adhesion in embryonic and adult tissue, *Cell* **34:**455.
22. Damsky, C. H., Knudsen, K. A., Bradley, D., Buck, C. A., and Horowitz, A. F., 1985, Dis-tribution of the cell substratum attachment (CSAT) antigen on myogenic and fibroblastic cells in culture, *J. Cell Biol.* **100:**1528.
23. Dejana, E., Languino, L. R., Polentarutti, N., Balconi, G., Ryckewaert, J. J., Larrieu, M. J., Donat, M. B., Mantovani, A., and Marguerie, G., 1985, Interaction between fibrinogen and cultured endothelial cells: Induction of migration and specific binding, *J. Clin. Invest.* **75:**11.
24. Delvos, U., Gajdusek, C., Sage, H., Harker, L. A., and Schwartz, S. M., 1982, Interactions of vascular wall cells with collagen gels, *Lab. Invest.* **46:**61.
25. Dernck, R., Jarrett, J. A., Chen, E. Y., Eaton, D. H., Bell, J. R., Assoian, R. K., Roberts, A. B., Sporn, M. B., and Goeddel, D. V., 1985, Human transforming growth factor-beta: Comple-mentary DNA sequence and expression in normal and transformed cells. *Nature* **316:**701.

26. DeWitt, D. L., Day, J. S., Sonnenburg, W. K., and Smith, W. L., 1983, Concentrations of prostaglandin endoperoxide synthase and prostaglandin I_2 synthase in the endothelium and smooth muscle of bovine aorta, *J. Clin. Invest.* **72:**1882.

27. Dieterlen-Lievre, F., and Martin, C., 1981, Diffuse intraembryonic hemopoiesis in normal and chimeric avian development, *Dev. Biol.* **88:**180.

28. Doetschman, T. C., Eistetter, H., Katz, M., Schmidt, W., and Kemler, R., 1985, The *in vitro* development of blastocyst-derived embryonic stem cell lines: formation of visceral yolk sac, blood islands and myocardium, *J. Embryol. Exp. Morphol.* **87:**27.

29. Duband, J. L., and Thiery, J. P., 1982, Distribution of fibronectin in the early phase of avian cephalic neural crest cell migration, *Dev. Biol.* **93:**308.

30. du Heynes, A. D., Eldor, A., Vlodavsky, I., Fridman, R., and Panet, A., 1985, The antiproliferative effect of interferon and the mitogenic activity of growth factors are independent cell cycle events: Studies with vascular smooth muscle cells and endothelial cells, *Exp. Cell Res.* **161:**297.

31. Edelman, G. M., 1985, Cell adhesion and the molecular process of morphogenesis, *Annu. Rev. Biochem.* **54:**135.

32. Einhorn, S., Eldor, A., Vlodavsky, I., Fuks, Z., and Panet, A., 1985, Production and characterization of interferon from endothelial cells, *J. Cell. Physiol.* **122:**200.

33. Ekblom, P., Sariola, H., Karkinen, M., and Saxen, L., 1982, The origin of the glomerular endothelium, *Cell Diff.* **11:**35.

34. Eldor, A., Falcone, D. J., Hajjar, D. P., Minick, C. R., and Weksler, B. B., 1981, Recovery of prostaglandin production by de-endothelialized rabbit aorta: Critical role of neointimal smooth muscle cells, *J. Clin. Invest.* **67:**735.

35. Fett, J. W., Strydan, D. J., Lobb, R. R., Alderman, E. M., Bethune, J. L., Riordan, J. F., and Vallee, B. L., 1985, Isolation and characterization of angiogenin, an anionic protein from human carinoma cells, *Biochemistry* **24:**5480.

36. Fitzgerald, L. A., Charo, I. F., and Phillips, D. R., 1985, Human and bovine endothelial cells synthesize membrane proteins similar to human platelet glycoproteins IIb and IIIa, *J. Biol. Chem.* **260:**10893.

37. Folkman, J., and Cotran, R., 1976, Relation of vascular proliferation to tumor growth, *Int. Rev. Exp. Pathol.* **16:**207.

38. Frater-Schröder, M., Müller, G., Birchmeier, W., and Böhlen, P., 1986, Transforming growth factor-beta inhibits endothelial cell proliferation, *Biochem. Biophys. Res. Commun.* **137:**295.

39. Gabbiani, G., Gabbiani, F., Lombardi, D., and Schwartz, S. M., 1983, Organization of actin cytoskeleton in normal and regenerating arterial endothelial cells, *Proc. Natl. Acad. Sci. USA* **80:**2361.

40. Gallin, W. J., Edelman, G. M., and Cunningham, B. A., 1983, Characterization of L-CAM, a major cell adhesion molecule from embryonic liver cells, *Proc. Natl. Acad. Sci. USA* **80:**1038.

41. Gartner, T. K., and Bennett, J. S., 1984, The tetrapeptide analog of the cell attachment site of fibronectin inhibits platelet aggregation and fibrinogen binding to activated platelets, *J. Biol. Chem.* **260:**11891.

42. Geiger, B., Volk, T., and Volberg, T., 1985, Molecular heterogeneity of adherens junctions, *J. Cell Biol.* **101:**1523.

43. Gerrity, R. G., Caplan, B. A., Richardson, M., Cade, J. F., Hirsh, J., and Schwartz, C. J., 1975, Endotoxin-induced endothelial injury. I. Endothelial cell turnover in the aorta of the rabbit, *Exp. Mol. Pathol.* **23:**379.

44. Girard, H., 1973, Arterial pressure in the chick embryo, *Am. J. Physiol.* **224:**454.

45. Gospodarowicz, D., Moran, J., Braun, D., and Birdwell, C., 1976, Clonal growth of bovine vascular endothelial cells: Fibroblast growth factor as a survival agent, *Proc. Natl. Acad. Sci. USA* **73:**4120.

46. Gospodarowicz, D., Cheng, J., Lui, G. M., Baird, A., and Bohlen, A., 1984, Isolation of brain fibroblast growth factor by heparin–Sepharose affinity chromatography: Identity with pituitary fibroblast growth factor, *Proc. Natl. Acad. Sci. USA* **81:**6963.

47. Greenburg, G., and Gospodarowicz, D., 1982, Inactivation of a basement membrane component responsible for cell proliferation but not cell attachment, *Exp. Cell Res.* **140:**1.

48. Greve, J. M., and Gottlieb, D. L., 1982, Monoclonal antibodies which alter the morphology of cultured chick myogenic cells, *J. Cell. Biochem.* **18:**221.

49. Gross, J. L., Moscatelli, D., and Rifkin, D. B., 1983, Increased capillary endothelial protease activity in response to angiogenic stimuli in vitro, *Proc. Natl. Acad. Sci. USA* **80:**2623.

50. Groves, H. M., Kinlough-Rathbone, R. L., Richardson, M., Moore, S., and Mustard, J. F., 1983, Platelet interaction with damaged rabbit aorta, *Lab. Invest.* **40:**194.

51. Gumbiner, B., and Simons, K., 1986, A functional assay for proteins involved in establishing an epithelial occluding barrier: Identification of a uvomorulin-like polypeptide, *J. Cell Biol.* **102:**457.

52. Haar, J. L., and Ackerman, G. A., 1971, A phase and electron microscopic study of vasculogenesis and erythropoiesis in the yolk sac of the mouse, *Anat. Rec.* **170:**199.

53. Hansson, G. K., and Schwartz, S. M., 1983, Endothelial cell dysfunction without cell loss, in: *Biochemical Interactions at the Endothelium* (A. Cryer, ed.), Elsevier/North-Holland, Amsterdam, p. 343.

54. Hansson, G., Chao, S., Schwartz, S. M., and Reidy, M. A., 1985, Aortic endothelial cell death and replication in normal and lipopolysaccharide-treated rats, *Am. J. Pathol.* **121:**123.

55. Hatta, K., Okada, T. S., and Takeichi, M., 1985, A monoclonal antibody disrupting calcium-dependent cell–cell adhesion of brain tissues: Possible role of its target antigen in animal pattern formation, *Proc. Natl. Acad. Sci. USA* **82:**2789.

56. Haudenschild, C., and Schwartz, S. M., 1979, Endothelial regeneration. II. Restitution of endothelial continuity, *Lab. Invest.* **41:**407.

57. Haudenschild, C. C., Zahniser, D., Folkman, J., and Klagsbrun, M., 1976, Human vascular endothelial cells in culture: Lack of response to serum growth factors, *Exp. Cell Res.* **98:**175.

58. Hayflick, L., and Moorhead, P. S., 1961, The serial cultivation of human diploid cell strains, *Exp. Cell Res.* **25:**585.

59. Heimark, R. L., and Schwartz, S. M., 1985, The role of membrane–membrane interactions in the regulations of endothelial cell growth, *J. Cell Biol.* **100:**1934.

60. Heimark, R. L., and Schwartz, S. M., 1987, Characterization of the luminal surface domain of aortic endothelial cells by restrictive iodination, *J. Cell. Physiol.* (in press).

61. Heimark, R. L., and Schwartz, S. M., 1987, Cell adhesion mechanisms in endothelial and smooth muscle cells: Characterization of a Ca^{2+}-dependent cell adhesion mechanism, *J. Cell Biol.* (submitted).

62. Heimark, R. L., Twardzik, D., and Schwartz, S. M., 1986, Inhibition of endothelial regeneration by type beta transforming growth factor from platelets, *Science* **233:**1078.

63. Hirsch, E., and Robertson, A. L., 1977, Selective acute arterial endothelial injury and repair. I. Methodology and surface characteristics, *Atherosclerosis* **282:**271.

64. Hughes, R. C., Butters, T. D., and Aplin, J. D., 1981, Cell surface molecules involved in fibronectin-mediated adhesion: A study using specific antisera, *Eur. J. Cell Biol.* **26:**198.

65. Hynes, R. O., 1981, Relationships between fibronectin and the cytoskeleton, in: *Cell Surface Reviews*, Vol. 7 (G. Poste and G. L. Nicolson, eds.), Elsevier/North-Holland, Amsterdam, p. 97.

66. Ignotz, R. A., and Massague, J., 1986, Transforming growth factor-β stimulates the expression of fibronectin and collagen and their incorporation into the extracellular matrix, *J. Biol. Chem.* **261:**4337.

67. Jaye, M., McConathy, E., Drohan, W., Tong, B., Duel, T., and Maciag, T., 1985, Modulation of the sis gene transcript during endothelial cell differentiation in vitro, Science **228:**882.

68. Jotereau, F. V., and LeDouarin, N. M., 1978, The developmental relationship between osteocytes and osteoclasts: A study using the quail–chick nuclear marker in endochondral ossification, *Dev. Biol.* **63:**253.

69. Kadish, J. L., Butterfield, C. E., and Folkman, J., 1979, The effect of fibrin on cultured vascular endothelial cells, *Tissue Cell* **11:**99.

70. Klagsburn, M., and Shing, Y., 1985, Heparin affinity of anionic and cationic capillary endothelial cell growth factors: Analysis of hypothalamus-derived growth factors and fibroblast growth factors, *Proc. Natl. Acad. Sci. USA* **82:**805.

71. Knudsen, K. A., Horwitz, A. F., and Buck, C. A., 1985, A monoclonal antibody identifies a glycoprotein complex involved in cell–substratum adhesion, *Exp. Cell Res.* **157:**218.

72. Laiho, M., Saksela, O., and Keski-Oja, J., 1986, Transforming growth factor β alters plasminogen activator activity in human skin fibroblasts, *Exp. Cell Res.* **164**:399.
73. Larson, D. M., and Sheridan, J. D., 1982, Intercellular junctions and transfer of small molecules in primary vascular endothelial cultures, *J. Cell Biol.* **92**:183.
74. LeDouarin, N. M., 1975, An experimental analysis of liver development, *Med. Biol.* **53**:427.
75. LeDouarin, N. M., 1984, Cell migration in embryos, *Cell* **38**:353.
76. Stemmerman, M. B., Spaet, T. H., Pitlick, K. F., Cintron, J., Lejnieks, I., and Tiell, M. L., 1977, Intimal healing: The pattern of reendothelialization and intimal thickening, *Am. J. Pathol.* **87**:125.
77. LeLievre, C. S., and LeDouarin, N. M., 1975, Mesenchymal derivatives of neural crest: Analysis of chimaeric quail and chick embryos, *J. Embryol. Exp. Morphol.* **34**:125.
78. Lim, R. W., and Hauschka, S., 1984, A rapid decrease in epidermal growth factor-binding capacity accompanies terminal differentiation of mouse myoblasts in vitro, *J. Cell Biol.* **98**:739.
79. Lobb, R. R., and Fett, J. W., 1984, Purification of two distinct growth factors from bovine neural tissue by heparin affinity chromatography, *Biochemistry* **23**:6295.
80. Loewenstein, W. R., 1979, Junctional intercellular communication and the control of growth, *Biochim. Biophys. Acta* **560**:1.
81. Lumpkin, C. K., Jr., McClung, J. K., Pereira-Smith, O. M., and Smith, J. R., 1986, Existence of high abundance antiproliferative mRNA's in senescent human diploid fibroblasts, *Science* **232**:393.
82. Maciag, T., Hoover, G. A., Stemerman, M. B., and Weinstein, R., 1981, Serial propagation of human endothelial cells *in vitro*, *J. Cell Biol.* **91**:420.
83. Maciag, T., Kadish, J., Wilkins, L., Stemerman, M. B., and Weinstein, R., 1982, Organizational behavior of human umbilical vein endothelial cells, *J. Cell Biol.* **94**:511.
84. Martz, E., and Steinberg, M. S., 1972, The role of cell–cell contact in "contact" inhibition of cell division: A review and new evidence, *J. Cell. Physiol.* **79**:189.
85. Minick, C. R., Stemerman, M. B., and Insull, W., 1977, Effect of regenerated endothelium on lipid accumulation in the arterial wall, *Proc. Natl. Acad. Sci. USA* **74**:1724.
86. Moscatelli, D., Presta, M., and Rifkin, D. B., 1986, Purification of a factor from human placenta that stimulates capillary endothelial cell protease production, DNA synthesis and migration, *Proc. Natl. Acad. Sci. USA* **83**:2091.
87. Nachman, R. L., and Leung, L. L., 1982, Complex formation of platelet membrane GpIIb and IIIa with fibrinogen, *J. Clin. Invest.* **69**:263.
88. Natraj, C. V., and Datta, P., 1978, Control of DNA synthesis in growing Balb/c 3T3 mouse cells by a fibroblast growth regulatory factor, *Proc. Natl. Acad. Sci. USA* **75**:6115.
89. Nawroth, P. P., Handley, D. A., Esman, C. T., and Stern, D. M., 1986, Interleukin 1 induces endothelial cell procoagulant while suppressing cell-surface anticoagulant activity, *Proc. Natl. Acad. Sci. USA* **83**:3460.
90. Oesch, B., and Birchmeier, W., 1982, New surface component of fibroblast's focal contacts identified by a monoclonal antibody, *Cell* **31**:671.
91. Oppenheimer-Marks, N., and Grinnell, F., 1984, Calcium ions protect cell–substratum adhesion receptors against proteolysis, *Exp. Cell Res.* **152**:467.
92. Pereira-Smith, O. M., Fisher, S. F., and Smith, J. R., 1985, Senescent and quiescent cell inhibitors of DNA synthesis, *Exp. Cell Res.* **160**:297.
93. Pierschbacher, M. D., and Ruoslahti, E., 1984, Variants of the cell recognition site of fibronectin that retain attachment-promoting activity, *Proc. Natl. Acad. Sci. USA* **81**:5985.
94. Plow, E. F., and Ginsberg, M. H., 1981, Specific and saturable binding of plasma fibronectin to thrombin stimulated platelets, *J. Biol. Chem.* **256**:9477.
95. Plow, E. F., Pierschbacher, M. D., Ruoslahti, E., Marguerie, G. A., and Ginsberg, M. H., 1985, The effect of Arg-Gly-Asp-containing peptides on fibrinogen and von Willebrand factor binding to platelets, *Proc. Natl. Acad. Sci. USA* **82**:8057.
96. Pratt, B. M., Harris, A. S., Morrow, J. S., and Madri, J. A., 1984, Mechanisms of cytoskeletal regulation: Modulation of aortic endothelial cell spectrin by the extracellular matrix, *Am. J. Pathol.* **117**:349.
97. Prober, J. S., Gimbrone, M. A., Cotran, R. S., Reiss, C. S., Burkoff, S. J., Fiers, W., and Ault,

K. A., 1983, Ia expression by vascular endothelium is inducible by activated T cells and by human γ interferon, *J. Exp. Med.* **157**:1339.

98. Pytela, R., Pierschbacher, M. D., and Ruoslahti, E., 1985, Identification and isolation of a 140 kd cell surface glycoprotein with properties expected of a fibronectin receptor, *Cell* **40**:191.

99. Reidy, M. A., 1985, A reassessment of endothelial injury and arterial lesion formation, *Lab. Invest.* **53**:513.

100. Reidy, M. A., and Schwartz, S. M., 1981, Endothelial regeneration. III. Time course of intimal changes after small defined injury to rat aortic endothelium, *Lab. Invest.* **44**:301.

101. Reidy, M. A., and Schwartz, S. M., 1983, Endothelial injury and regeneration. IV. Endotoxin: A nondenuding injury to aortic endothelium, *Lab. Invest.* **48**:25.

102. Reidy, M. A., Clowes, A. W., and Schwartz, S. M., 1983, Endothelial regeneration. V. Inhibition of endothelial regrowth in arteries of rat and rabbit, *Lab. Invest.* **49**:569.

103. Rheinwald, J. G., and Green, H., 1975, Serial cultivation of strains of human epidermal keratinocytes: The formation of keratinizing colonies from single cells, *Cell* **6**:331.

104. Risau, W., 1986, Developing brain produces an angiogenesis factor, *Proc. Natl. Acad. Sci. USA* **83**:3855.

105. Risau, W., and Ekblom, P., 1986, Production of a heparin-binding angiogenesis factor by embryonic kidney, *J. Cell Biol.* **103**:1101.

106. Ross, R., 1986, The pathogenesis of atherosclerosis: An update, *N. Engl. J. Med.* **314**:488.

107. Rowland, F. N., Donovan, M. J., Pieciano, P. T., Wilner, G. D., and Kreutzer, D. L., 1984, Fibrin-mediated vascular injury: Identification of fibrin peptides that mediate endothelial cell retraction, *Am. J. Pathol.* **117**:418.

108. Schleef, R. R., and Birdwell, C. R., 1984, Biochemical changes in endothelial cell monolayers induced by fibrin deposition in vitro, Arteriosclerosis **4**:14.

109. Schreiber, A. B., Winkler, M. E., and Derynck, R., 1986, Transforming growth factor-α: A more potent angiogenic mediator than epidermal growth factor, *Science* **232**:1250.

110. Schwartz, S. M., and Benditt, E. P., 1973, Cell replication in the aortic endothelium: A new method for study of the problem, *Lab. Invest.* **28**:699.

111. Schwartz, S. M., and Benditt, E. P., 1977, Aortic endothelial cell replication. I. Effects of age and hypertension in the rat, *Circ. Res.* **41**:248.

112. Schwartz, S. M., Stemerman, M. B., and Benditt, E. P., 1975, The aortic intima. II. Repair of the aortic lining after mechanical denudation, *Am. J. Pathol.* **81**:1.

113. Schwartz, S. M., Selden, S. C., III, and Bowman, P., 1979, Growth control in aortic endothelium at wound edges, in: *Hormones and Cell Culture*, Vol. 6 (R. Ross and G. Sato, eds.), Cold Spring Harbor Laboratory, Cold Spring Harbor, N.Y., p. 593.

114. Schwartz, S. M., Gajdusek, C. M., and Selden, S. C., III, 1981, Vascular wall growth control: The role of the endothelium, *Arteriosclerosis* **1**:107.

115. Selden, S. C., III, and Schwartz, S. M., 1979, Cytochalasin B inhibition of endothelial proliferation at wound edges in vitro, J. Cell Biol. **81**:348.

116. Selden, S. C., III, Rabinovitch, P. S., and Schwartz, S. M., 1981, Effects of cytoskeletal disrupting agents on replication of bovine endothelium, *J. Cell. Physiol.* **108**:195.

117. Sholley, M. M., Ferguson, G. P., Seibel, H. R., Montour, J. L., and Wilson, J. D., 1984, Mechanism of neovascularization: Vascular sprouting can occur without proliferation of endothelial cells, *Lab. Invest.* **51**:624.

118. Simionescu, M., Simionescu, N., and Palade, G, E., 1975, Segmental differentiations of cell junctions in the vascular endothelium: The microvasculature, *J. Cell Biol.* **67**:863.

119. Smeds, S., and Wollman, S. H., 1983, ^{3}H-Thymidine labeling of endothelial cells in thyroid arteries, veins and lymphatics during thyroid stimulation, *Lab. Invest.* **48**:285.

120. Spagnoli, L. G., Pietra, G. G., Villaschi, S., and Johns, L. W., 1982, Morphometric analysis of gap junctions in regenerating arterial endothelium, *Lab. Invest.* **46**:139.

121. Steck, P. A., Blenis, J., Voss, P. G., and Wang, J. L., 1982, Growth control in cultured 3T3 fibroblasts. II. Molecular properties of a fraction enriched in growth inhibitory activity, *J. Cell Biol.* **92**:523.

122. Stewart, P. A., and Willey, M. J., 1981, Developing nervous tissue induces formation of blood–

brain barrier characteristics in invading endothelial cells: A study using quail–chick transplantation, *Dev. Biol.* **84**:183.

123. Stolpen, A. H., Guinan, E. C., Fiers, W., and Prober, J. S., 1986, Recombinant tumor necrosis factor and immune interferon act singly and in combination to reorganize human vascular endothelial cell monolayers, *Am. J. Pathol.* **123**:16.

124. Thiagarajan, P., Shapiro, S. S., Levine, E., DeMarco, L., and Yalcin, A., 1985, Monoclonal antibody to human platelet glycoprotein IIIa detects a related protein in cultured human endothelial cells, *J. Clin. Invest.* **75**:896.

125. Thiery, J.-P., Brackenbury, R., Rutishauser, U., and Edelman, G. M., 1977, Adhesion among neural cells of the chick embryo. II. Purification and characterization of a cell adhesion molecule from neural retina, *J. Biol. Chem.* **252**:6841.

126. Thiery, J.-P., Duband, J. L., and Delouvee, A., 1982, Pathways and mechanisms of avian trunk neural crest cell migration and localization, *Dev. Biol.* **93**:324.

127. Thiery, J.-P., Delouvee, A., Gallin, W. J., Cunningham, B. A., and Edelman, G. M., 1984, Ontogenetic expression of cell adhesion molecules: L-CAM is found in epithelia derived from the three primary germ layers, *Dev. Biol.* **102**:61.

128. Thornton, S. C., Mueller, S. N., and Levine, E. M., 1983, Human endothelial cells: Use of heparin in cloning and long-term serial cultivation, *Science* **222**:623.

129. Todaro, G. J., Lazar, G. K., and Green, H., 1965, The initiation of cell division in a contact-inhibited mammalian cell line, *J. Cell. Comp. Physiol.* **56**:325.

130. Tucker, R. F., Shipley, G. D., Moses, H. L., and Holley, R. W., 1984, Growth inhibitor from BSC-1 cells closely related to platelet type β transforming growth factor, *Science* **226**:705.

131. Twardzik, D. R., 1985, Differential expression of transforming growth factor-α during prenatal development of the mouse, *Cancer Res.* **45**:5413.

132. Urushihara, H., and Yamada, K. M., 1986, Evidence for involvement of more than one class of glycoprotein in cell interaction with fibronectin, *J. Cell. Physiol.* **126**:323.

133. Vale, R. D., Peterson, S. W., Matiuck, N. V., and Fox, C. F., 1984, Purified plasma membranes inhibit polypeptide growth factor-induced DNA synthesis in subconfluent 3T3 cells, *J. Cell Biol.* **98**:1129.

134. Vlodavsky, I., Johnson, L. K., and Gospodarowicz, D., 1979, Appearance in confluent vascular endothelial cell monolayers of a specific cell-surface protein (CSP-60) not detected in actively growing endothelial cells or in cell types growing in multiple layers, *Proc. Natl. Acad. Sci. USA* **76**:2306.

135. Volinsky, J. E., and LeDouarin, N. M., 1985, Production of plasminogen activator by migrating cephalic neural crest cells, *EMBO J.* **4**:1403.

136. Wagner, R. C., 1980, Endothelial cell embryology and growth, *Adv. Microcirc.* **9**:45.

137. Whittenberger, B., and Glaser, L., 1977, Inhibition of DNA synthesis in culture of 3T3 cells by isolated surface membranes, *Proc. Natl. Acad. Sci. USA* **74**:2251.

138. Wieser, R., and Brunner, G., 1983, Imitation of contact inhibition by substrate-bound plasma membrane glycoproteins and lectins in serum-free hormone-supplemented cultures of GH₃ cells, *Exp. Cell Res.* **147**:23.

139. Yamada, K. M., and Kennedy, D. W., 1984, Dualistic nature of adhesive protein function: Fibronectin and its biologically active peptide fragments can autoinhibit fibronectin in function, *J. Cell Biol.* **99**:29.

140. Yoshida, C., and Takeichi, M., 1982, Teratocarcinoma cell adhesion: Identification of a cell-surface protein involved in calcium-dependent cell aggregation, *Cell* **28**:217.

141. Heimark, R. L., Cheme, and Schwartz, S. M., 1987, Stimulation of endothelial cell growth by a monoclonal antibody disrupting cell-substrate interaction, *J. Biol. Chem.* (submitted).

IV

Endothelial Cytoskeleton and Matrix and Their Interactions

7

The Endothelial Junction
The Plaque and Its Components

Werner W. Franke, Pamela Cowin, Christine Grund, Caecilia Kuhn, and Hans-Peter Kapprell

I. INTRODUCTION

The endothelium of most blood vessels is a single layer of tightly packed cells which line the vascular lumen and border on the basal lamina and, in some arteries and arterioles, on the processes of vascular smooth muscle cells. Like single-layered epithelia, the endothelial cells are polar, with an apical, i.e., adluminal, and a basal, i.e., abluminal, plasma membrane region which appear to be segregated from each other by a special membrane region containing occluding, i.e., "tight," junctions (for reviews see Ref. 89). Again similarly to polar epithelia, the endothelium is capable of vectorial sorting, secretion, and virus budding as well as endocytotic and transcytotic processes (for examples see Refs. 47, 66, 72). Obviously, tight sealing of endothelial cells to each other is a prerequisite for their physiological functioning, and situations in which the coherence of the endothelial layer is locally and/or transiently interrupted are usually associated with pathological processes (e.g., Refs. 16, 17, 52, 54, 77, 86, 92).

The elucidation of the molecular organization of the "endothelial junction" is crucial to our understanding of the structures and molecules involved in the intercellular adhesion of the endothelial cells. This junctional zone is characterized by a relatively close apposition of the adjacent plasma membranes which are flanked by a so-called "parajunctional zone" of cortical cytoplasm (for definition see Ref. 90). So far only two specific junctions have been identified within this junctional complex. These are the occluding (tight) and the communicating (gap) junctions which are morphologically similar to those present in other tissues (reviewed in Refs. 89–92; for certain freeze-cleave aspects see also Refs. 98, 110). Typical desmosomes or other adhering junction structures have not been demonstrated in higher vertebrates. However, Fawcett[29,30] has attracted attention to

Werner W. Franke, Pamela Cowin, Christine Grund, Caecilia Kuhn, and Hans-Peter Kapprell • Division of Membrane Biology and Biochemistry, Institute of Cell and Tumor Biology, German Cancer Research Center, D-6900 Heidelberg, Federal Republic of Germany. *Present address of P.C.:* Department of Cell Biology, New York University, School of Medicine, New York, New York 10016.

the presence of an extensive, "desmosome-like" plaque structure in the endothelium of the blood vessels of the fish swim bladder.

Most electron micrographs of endothelial junctions available in the literature
do not show a typical dense plaque and the anchorage of bundles of actin-containing 5-nm microfilaments and/or 7- to 12-nm intermediate-sized filaments (IFs).
Only in a few micrographs published can a loosely woven filamentous web be
recognized at the cytoplasmic aspect of the endothelial junctions (e.g., Refs. 30,
47, 54, 59, 60, 111). However, considerable ambiguity concerning the nature of
this plaque and this type of junction has reigned until recently when marker proteins capable of distinguishing between the various kinds of plaque-bearing junctions have been used (cf. Refs. 13, 33, 34, 44).

Intercellular junctions of the *adhaerens* category are generally characterized
by the presence of a pair of dense cytoplasmic plaques on either plasma membrane. This category contains two major types of junction, the desmosome (*ma-
cula adhaerens*) and various nondesmosomal junctions (*zonulae* and *fasciae adhaerentes, puncta adhaerentia*). While the latter have loosely matted plaques
associated with bundles of microfilaments,[28,30,44,45,61,70,75,96] the typical desmosome has a relatively thick (15–30 nm), densely stained, and sharply contoured
plaque, usually associated with bundles of IFs.[12,23,28,44,63,95] Asymmetric forms
showing some relationship to these major intercellular junction types also exist:
Microfilament-associated plaques are characteristic of special plasma membrane
domains bordering on extracellular matrix material and are referred to as "focal
adhesions," "subplasmalemmal densities," or, in certain nerve and muscle cells,
as "pre- or postsynaptic densities." IF-associated forms of asymmetric plaque
domains in basal plasma membrane regions bordering on extracellular matrix are
usually collectively referred to as "hemidesmosomes."

While the electron microscopic identification and classification of junctions
of typical appearance presents no difficulty, this is more complicated, and often
impossible, in cases where the ultrastructural features are less distinct, including
all formations with small, thin, or very loosely matted plaque structures. This has
resulted in numerous examples of misclassification in the literature, a problem
also of considerable importance in the differential diagnosis of tumors (e.g., for
reviews see Refs. 26, 46, 69). Moreover, many junctions have been given am-
biguous names such as "desmosome-like" or are referred to by the name of the
cell type in which they occur such as "Sertoli cell junctions" of testis or "endothelial junctions." However, recent immunocytochemical studies have brought
new information on the endothelial junction, specifically its plaque, so that now
its classification as an adhering junction related to the *zonula adhaerens* of epithelial cells seems to be justified.

II. ELECTRON MICROSCOPY OF ENDOTHELIAL JUNCTIONS

In preliminary experiments we noticed that postfixational soaking of the specimens in uranyl acetate or the inclusion of tannic acid in the initial fixative[13,38,39]
increased the electron contrast of the plaque material of endothelial junctions.
With such fixation and staining procedures, typical endothelial junctions of con-

tinuous and fenestrated capillaries (Fig. 1a) as well as larger blood vessels (Fig. 1b,c) are characterized by regions of closely parallel, though not directly apposed plasma membrane profiles and by a pair of cytoplasmic plaques of loosely packed fibrillogranular material. This plaque is less densely stained and sharply demarcated than typical desmosomal plaques but rather resembles the plaque arrangements of nondesmosomal adhering junctions such as those of the *zonula adhaerens* of polar epithelial cells (compare, e.g., Refs. 28, 44, 45, 61, 70, 95). As with the latter, the plaques of the endothelial junctions are associated, to variable extents, with webs of microfilaments which are best recognized in oblique or tangential sections to the plaque (e.g., Fig. 1c). IFs sometimes approach these plaque-associated microfilamentous webs but do not appear to be attached to the plaque itself.

The sizes and the detailed membrane morphology of these plaque-covered endothelial junction domains vary considerably as does the plaque thickness (Fig. 2). However, some plaque structure has been recognized in all cross-sectioned endothelia of capillaries and larger blood vessels, indicating that the endothelial junction and its plaque form a continuous belt system around the entire cell circumference (for a discussion of the absence of communicating junctions in certain capillaries and pericytic venules, see Refs. 89, 91, 92). Frequently, these plaque-covered junctional membrane domains are interspersed with regions lacking a well-developed subplasmalemmal coat, and here the intercellular region is often wider and displays variously-sized intercellular cavities (e.g., Figs. 1b and 2d). Formations of coated pits have also been occasionally observed within these plaque-free regions.

For most of the endothelial junction the membranes are parallel with an intercellular cleft of 10–20 nm (Figs. 1c and 2a–d). However, in certain places the intermembranous cleft is narrowed (Fig. 2c) or absent as in the communicating (gap) junctions (Figs. 1c and 2a,e) and the sites corresponding to occluding junctions (e.g., see Fig. 4d of Ref. 13). While at some of the communicating junctions the plaque material appears to be reduced, other junctions of this type appear continually coated by plaque material (Fig. 2a,e).

In the endothelia of some larger blood vessels such as the pulmonary artery, two special modifications of the endothelial junction are often seen. One is represented by multiple and extensive interdigitations which leave an approximately 15-nm intercellular cleft and are filled with loosely arranged filamentous material (Fig. 3a,b). The other special formations observed are large aggregates of microfilaments of up to 0.5-μm diameter associated with the junctional plaques (Fig. 3c).

Dense plaques associated with microfilament bundles also occur, at highly variable frequencies and probably in response to differences of hemodynamics, at the abluminal plasma membrane bordering on the basal lamina (Fig. 3d,e). Besides their common association with microfilaments, the relationship of these asymmetric plaques, which have been repeatedly described at the basal[18,22,59,60,65,79,106,111] and, occasionally, at the adluminal (e.g., Ref. 48) plasma membrane, to the plaque of the intercellular junctions of the same cells is not clear (see below).

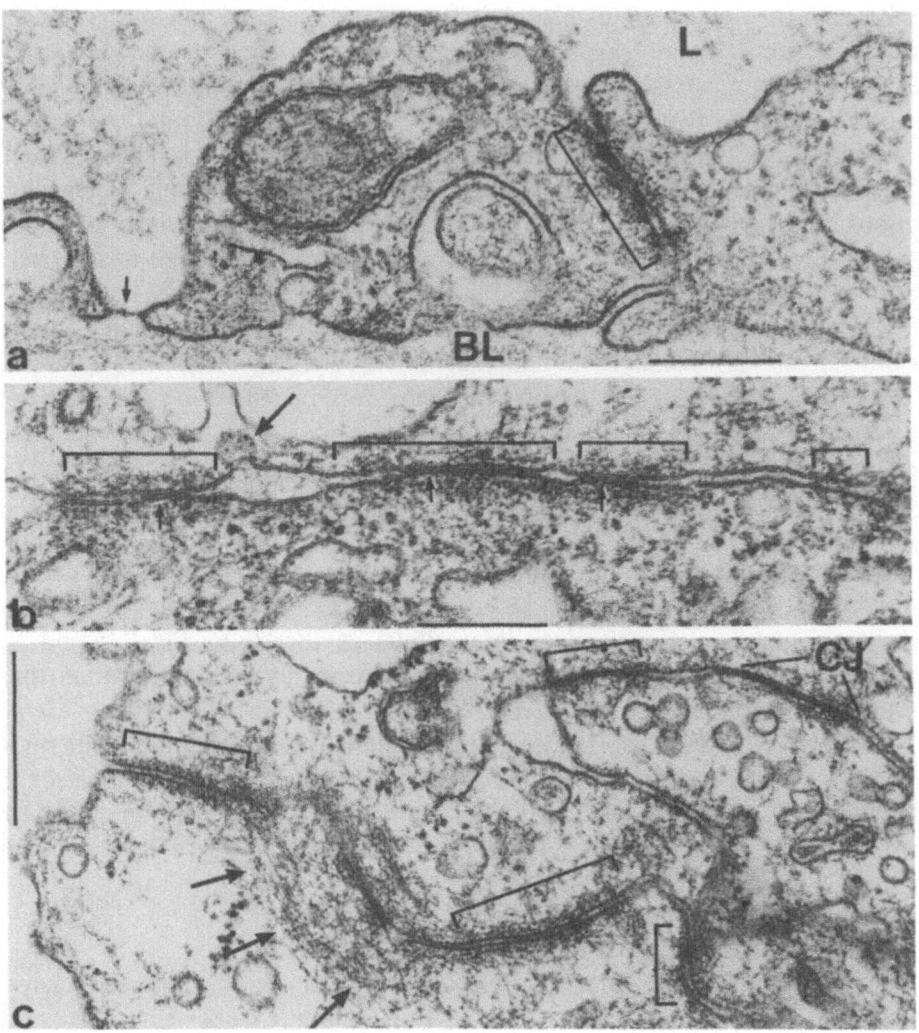

Figure 1. Electron microscopy of endothelial junction complexes as revealed in cross (*a, b*) and oblique (*c*) sections of the endothelium of fenestrated capillary of bovine kidney (*a*) and pulmonary artery (*b, c*). For details of fixation see Ref. 13. Note that the region in which cells border on each other is characterized by a cytoplasmic coat of densely stained granulofibrillar material, i.e., the plaque (brackets), and an apparent intercellular cleft of ~ 20 nm, which is interrupted in some places by loci of direct membrane contact at communicating (CJ in *c*) and occluding (cf. Ref. 13) junctions and by regions of wider intercellular cavities at which sometimes vesicles seem to form (e.g., arrow in *b*). Microfilamentous structures are sometimes resolved within the parajunctional plaque-associated fibrillar webs, specifically in grazing sections (such a region is demarcated by arrows in *c*). L, lumen; BL, basal lamina. The short arrow at the left of panel *a* denotes a fenestra. Bars = 0.25 μm (*a, b*) and 0.5 μm (*c*).

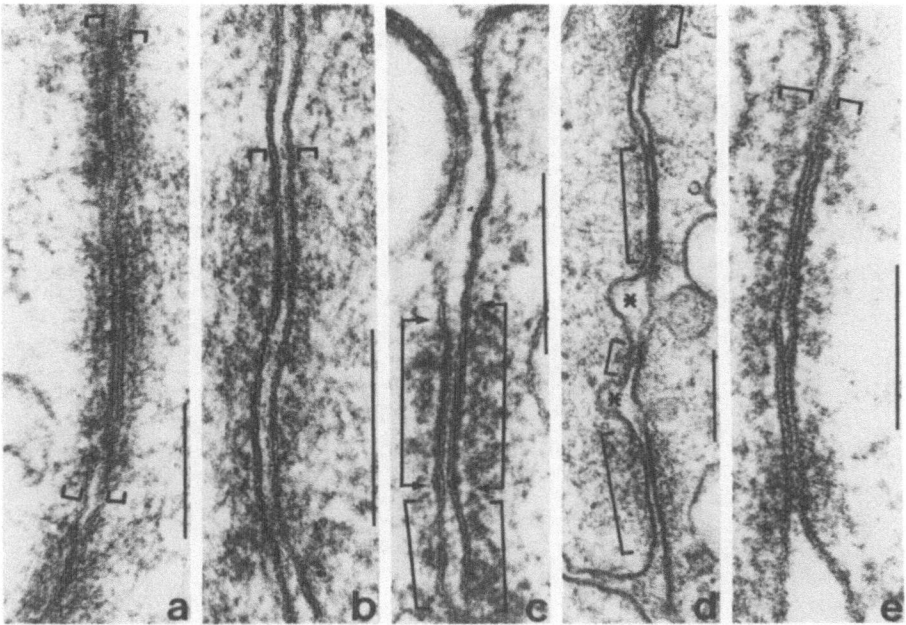

Figure 2. Higher-resolution electron micrographs showing ultrastructural details of plaques of bovine endothelial junctions in near-vertical sections through pulmonary artery (*a, d, e*) and renal capillaries (*b, c*). These adhering junctions and their plaques (demarcated by brackets) can be continuous for relatively longer (*a, b*) or shorter (*c, d*) distances and interspersed with regions of wider intercellular distance (*d*, asterisk). Both communicating (*e*) and tight (cf. Ref. 13) junctions also occur within such plaque-covered domains. Note also differences of intermembrane distances in adjacent junctional domains (the bracket with arrows denotes a relatively close-spaced junction, the lower bracket one with a wider cleft). Bars = 0.2 μm.

III. CHARACTERIZATION OF JUNCTIONS BY MARKER PROTEINS

Information about the chemical composition of the endothelial junction is essential for their identification, classification, and the understanding of their functions. Recently, the different types of junctions have been distinguished by typical constituents which may be used as markers for the typing of junctions. For example, a major component of the communicating (gap) junction has been identified in the form of a transmembrane protein (e.g., Refs. 19, 56, 64, 74; for review see Ref. 78). A protein of M ~ 225,000 daltons has recently been reported by Stevenson *et al.*[97] to be associated with—and characteristic for—occluding junctions of epithelia and endothelia.

In the case of the two major types of plaque-bearing junctions, biochemical analyses and immunocytochemical studies have shown that most of the proteins are unique to one type or the other. Desmosomes, for example, are characterized by the presence of plaque proteins such as desmoplakin I (M_r 250,000) and the "D1 antigen" (a minor component) as well as the cytoplasmic part of desmoglein, a glycosylated polypeptide of M_r 165,000 (also referred to as "band 3 polypeptide"

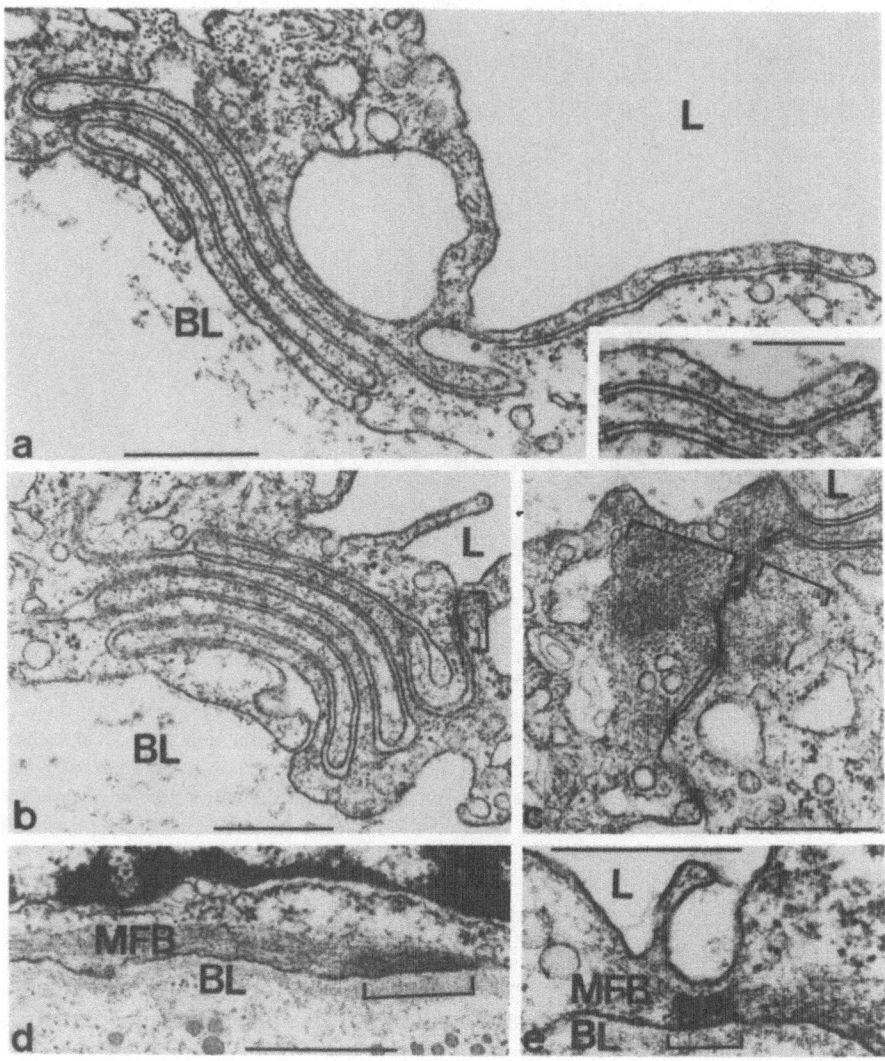

Figure 3. Electron micrographs of cross sections through endothelium of bovine pulmonary artery (*a*–*c*; cells shown here have been collected in endothelial monolayer sheets by scraping from the internal surface; cf. Ref. 13) and renal capillaries (*d, e*; only basal cell portions are shown). Note extended zones of interdigitations of special regions of endothelial junctions (*a, b*). These interdigitating cell processes contain finely filamentous material (see also arrow in insert in lower right corner of *a*) and reveal a relatively narrow and constant (∼ 20 nm) intercellular cleft. In some cells the adhering junctions are associated with large masses of actin-containing microfilaments in the parajunctional zone (*c*, bracket). Densely stained plasma membrane-associated plaques are not restricted to intercellular junctions but also occur at basal cell surfaces bordering on the basal lamina (BL; *d, e*). These asymmetrical plaques are also attachment sites for microfilament bundles (MFB). BL, basal lamina; L, lumen. Bars = 0.5 μm, except for *a* (0.25 μm).

of isolated bovine snout epidermal desmosomes) which may be accompanied, at least in certain stratified epithelia, by desmoplakin II (M_r 215,000) and desmocalmin, a less tightly bound calmodulin-binding protein of M_r 240,000, and a basic polypeptide of M_r 75,000 (also referred to as "band 6 polypeptide" of isolated bovine snout epidermal desmosomes; for references see Refs. 9–12, 33–35, 44, 50, 71, 83, 84, 96, 100). Desmocollins, a pair of glycosylated polypeptides of M_r ~ 115,000 and ~ 130,000, have also been described in desmosomes from various tissues and species.[9,10,33,50,68,71,93,96]

Conversely, vinculin (M_r ~ 130,000) and α-actinin (M_r ~ 100,000) have been identified as major components of the plaques of nondesmosomal intercellular junctions (*zonulae* and *fasciae adhaerentes, puncta adhaerentia,* "type II plaques") and to the asymmetric "focal adhesions" where they appear to be tightly associated with the actin microfilaments adhering to these sides.[2,3,6,14,21,42–45,73,87,88,94] In certain forms of nondesmosomal adhering junctions, the location of these plaque proteins appears to correspond to the localization of certain membrane-bound glycoproteins such as uvomorulin and A-CAM, but it is not clear whether the latter proteins interact directly with the plaque proteins.[1,5,15,24,45,51,53,101,105] Different from these mutually exclusive junction proteins, plakoglobin, a polypeptide of M_r 83,000 (previously also referred to as "band 5 polypeptide" of isolated bovine snout epidermal desmosomes[10–12,20,33,37–39]) is common to desmosomal and diverse kinds of nondesmosomal types of adhering junctions[13] and can therefore be used as a general marker for a broader group of adhering junctions.

IV. IMMUNOLOCALIZATION OF CYTOSKELETAL PROTEINS IN ENDOTHELIAL CELLS

In discussing the nature of the endothelial junctions, the composition of the IF cytoskeleton present in endothelial cells is also relevant. Although endothelial cells are often subsumed under "epithelia", most mammalian endothelia differ from true epithelia by the absence of cytokeratin IFs (for the occurrence of cytokeratin IFs in amphibian endothelia see Ref. 61). The only type of IF protein usually detected in mammalian endothelial cells is vimentin [4,22,32,49,57,82,85] (for a cytokeratin-positive subtype of endothelium in human synovia and certain mesenchymal tissues see Ref. 61). Fujimoto and Singer[40] have recently reported that in chicken certain renal and hepatic endothelia contain only desmin and pancreatic capillary endothelia coexpress desmin and vimentin; this type of coexpression has also been reported for the high endothelium venules of lymph nodes in rat but not in man.[99]

When plakoglobin antibodies are used for immunocytochemistry on complex tissues, positive staining along intercellular boundaries is seen not only in cells known to contain desmosomes, as demonstrable by staining for desmosome-specific markers such as desmoplakins, but also in cells of vascular tracts (Fig. 4 shows human uterine tissue as an example). This positive reaction for plakoglobin

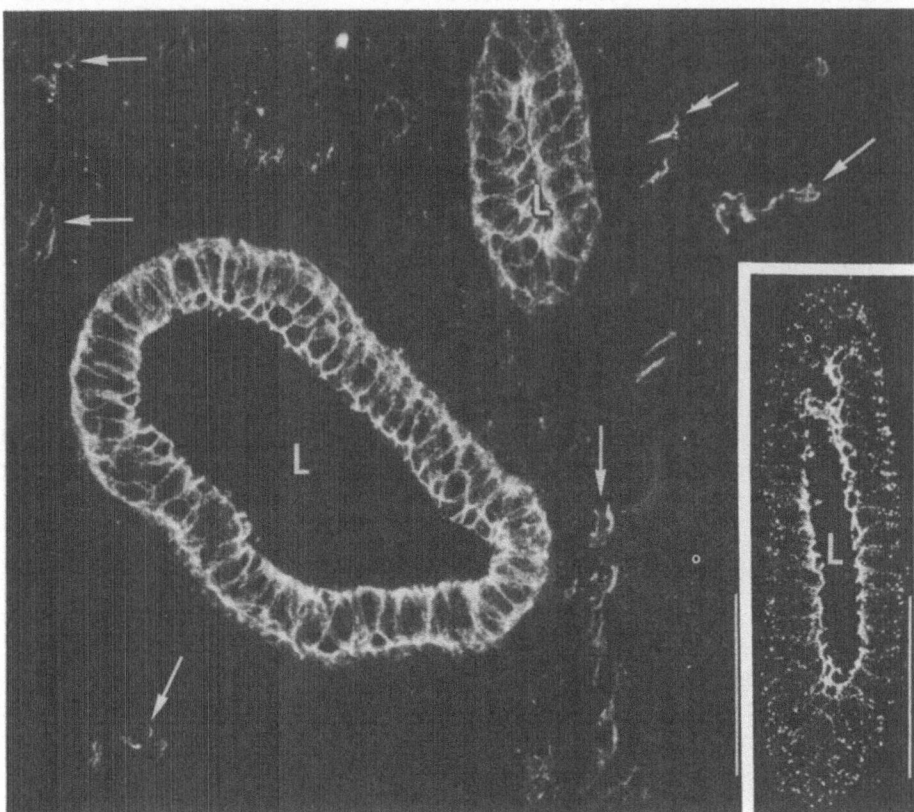

Figure 4. Immunofluorescence microscopy showing the reaction of plakoglobin in endometrium ot human uterus with antibodies to plakoglobin, in comparison with monoclonal antibodies (DP1&2-2.15 and DP1-2.17) to desmoplakins (insert). Antibodies to plakoglobin decorate the lateral cell borders of the epithelial cells (L, lumen). In some places, the staining appears as a punctate pattern which is much denser than that found with antibodies to desmoplakins (such a figure should replace Fig. 6c of Ref. 11, which has been incorrectly labeled; the correct legend of that figure is presented in Ref. 36). In addition to their reaction on epithelial cells, the plakoglobin antibodies stain the endothelia of blood vessels (arrows) within the uterine stroma which are negative for antibodies to desmoplakins. Bars = 50 μm.

is seen in both continuous and fenestrated capillaries, including those of the renal glomeruli (cf. Ref. 39) and in veins and arteries of all calibers (Figs. 4–6). In double-label immunofluorescence experiments, vinculin antibodies colocalize with plakoglobin but also react with other regions of endothelial cells (for a detailed study see Ref. 22) and, very intensely, with the smooth muscle tissue of the vascular walls.[39] Vinculin antibodies also react with sites corresponding to the subplasmalemmal densities of neuromuscular junctions (see Refs. 3, 87) which are not stained by plakoglobin antibodies. The absence of plakoglobin staining at neuromuscular junctions is best demonstrated by double-label immunofluorescence microscopy for plakoglobin and synaptophysin, a marker for neurosecretory vesicles and, therefore, neuronal structures, including presynaptic endings (Fig. 5a,b; cf. Ref. 108).

The distribution of the plaque material of adhering junctions in endothelia can also be demonstrated in whole mount preparations of detached endothelial monolayers ("Häutchen-Präparate"). Figure 6 presents immunofluorescence micrographs of vimentin and plakoglobin in whole mount preparations of endothelial cell sheets scraped off from the internal surface of bovine pulmonary arteries. While the masses of vimentin IFs are located mostly in the central part of the cytoplasm (Fig. 6a), plakoglobin is clearly concentrated along the cell circumference (Fig. 6b,c). In such preparations the intensity of the plakoglobin reaction appears rather uniform in certain groups of cells but shows fluctuations and local inhomogeneities in other regions of the same cell sheet (Fig. 6c). It is possible that such local differences are correlated with differences in plaque sizes and/or the presence of the interdigitations of endothelial junctions observed in the electron microscope (see above). However, artifacts during the preparations cannot be excluded at present. Concentrations of plakoglobin at sites of intercellular junctions have also been shown in cultured bovine endothelial cells, although with the culture conditions and the cell lines used the junctional zones do not encircle the entire cell circumferences (e.g., Ref. 13).

When such whole mount preparations are labeled by immunofluorescence microscopy with antibodies to vinculin, α-actinin, myosin, and actin, staining of the cell periphery can also be observed to variable degrees, depending on the specific endothelium and its pretreatment.[41,55,60,80,106,107,109] A similar observation has been made in cultured endothelial cells.[58] However, in contrast to plakoglobin, these proteins are also located in other structures. Notably, actin, myosin, and α-actinin occur in the microfilament bundles extending through the cytoplasm ("stress fibers"[41,55,60,80,106,107,109]; for cultured endothelial cells see also Ref. 58) that often show a preferential orientation in parallel to the direction of the blood flow in the vessel from which they have been obtained.[31,41,60,106,107,109] On the other hand, vinculin is most highly concentrated in the basal plaques to which the microfilament bundles are attached.[22]

It is not clear whether proteins of the spectrin family, which colocalize with plakoglobin at intercellular boundaries of lens tissue,[38] are excluded from the endothelial junctions. The occurrence of some spectrin-related proteins along cell-to-cell boundaries has been described[76] in certain cultures of bovine endothelial cells growing on specific collagen substrates. Another component of the subplasmalemmal cytoskeletal meshwork, i.e., a protein related to erythroid protein 4.1 of $M_r \sim 78,000$ which colocalizes with spectrin proteins in a variety of cells and tissues (for references see Ref. 38), has not been identified at intercellular junctions between cultured endothelial cells.[8]

The concentration of plakoglobin in the plaques of the endothelial junctions has been confirmed by immunoelectron microscopy (Fig. 7a,b; cf. Refs. 13, 37, 38). Moreover, the biochemical identity of the protein recognized by the plakoglobin antibodies in endothelial cells has been examined by one- and two-dimensional gel electrophoresis of structure-bound, i.e., pelletable, material, followed by immunoblotting (Fig. 8). The reactive polypeptide spot comigrates with

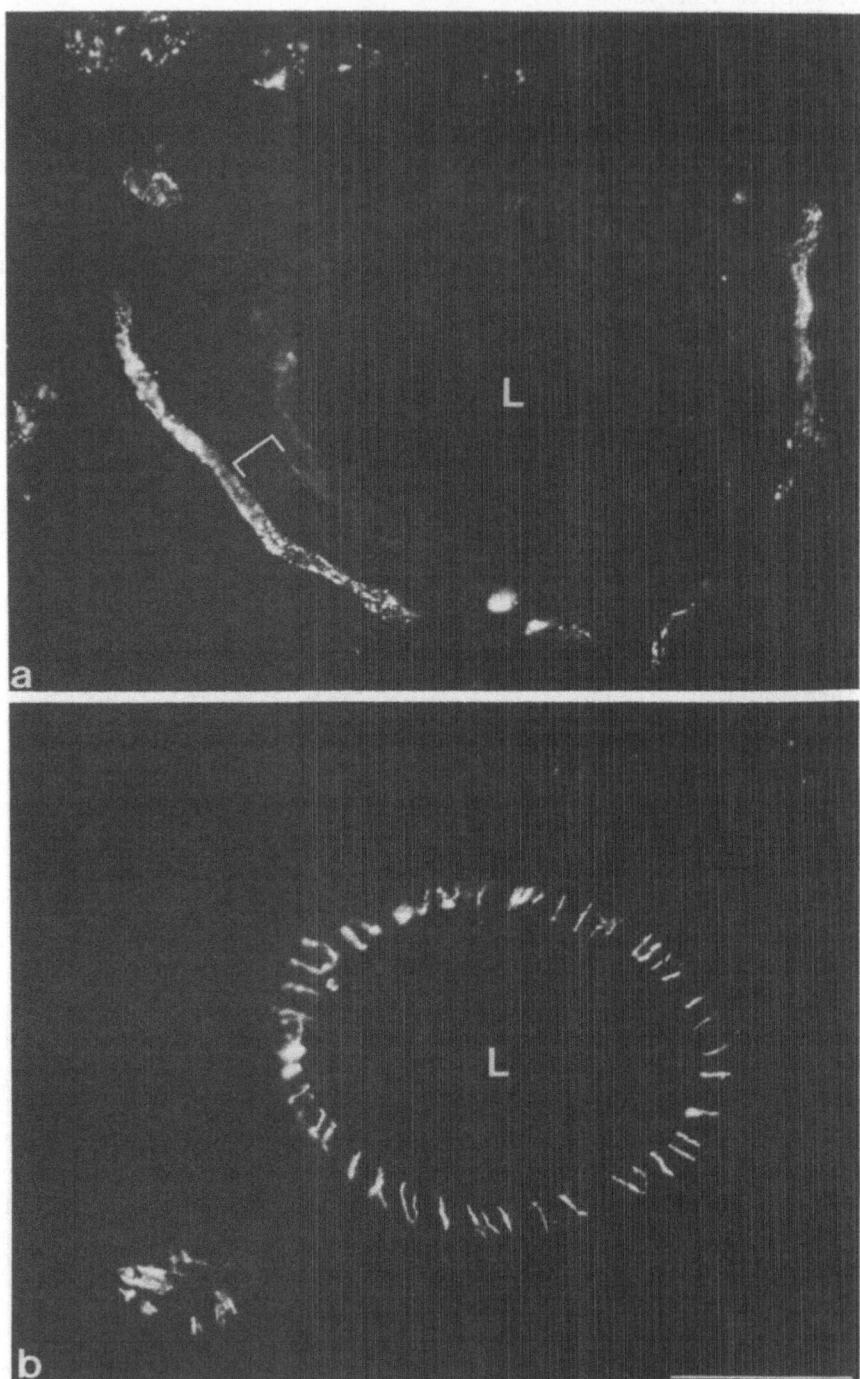

plakoglobin isolated from desmosomes of bovine snout epidermis. The conclusion that endothelial and epithelial plakoglobin are identical is also supported by Northern blot hybridizations using plakoglobin cDNA probes.[13]

All vascular endothelia tested, i.e., human, bovine, and murine (rat and mouse) ones, are negative for desmoplakins, desmoglein, and D1 antigen.[33,37,39,83,84] This is in agreement with the electron microscopic appearance of the endothelial junctions of mammals which lack typical desmosomal features (see above). Whether fish endothelial junctions, which show a more developed plaque and even a structure resembling the "midline" of desmosomes,[30] essentially differ in their composition from mammalian endothelial junctions remains to be examined. In contrast, junctions with plaques positive for desmoplakins and other desmosomal markers can be found to interconnect the stellate endothelial cells extending through the sinusoidal spaces of bovine and human lymph nodes.

V. STRUCTURE-BOUND AND SOLUBLE FORMS OF JUNCTION PLAQUE PROTEINS

The occurrence of typical adhering plaque proteins such as vinculin and plakoglobin is not restricted to the plaque structures. The existence of a soluble form of vinculin has been indicated by several experiments (e.g., Refs. 27, 67, 81, and our unpublished results), and an ~ 7 S form of plakoglobin has been identified in cell lysates from various tissues and cultured cell lines, endothelial cells included.[13,62] In several tissues and cultured cells, 20–30% of the total plakoglobin can be recovered in this solubilized form. If these findings can be supported by direct demonstration of a pool of diffusible vinculin and plakoglobin in the living cell, they would indicate that the major components of the endothelial plaque exist in a dynamic equilibrium between a soluble and a plaque-bound form. The existence of such a pool of diffusible vinculin and plakoglobin in the cytoplasm of the endothelial cell is not inconsistent with the predominant immunolocalization of these proteins at the plaque structures, as the local concentration of the dispersed soluble plakoglobin is probably very low and a considerable part of soluble proteins may be lost during the incubation and washing procedures applied during immunolocalization experiments.

The existence of an equilibrium between the plaque-bound and the soluble

Figure 5. Double-label immunofluorescence microscopy of cross section through bovine tongue musculature, showing the reaction of synaptophysin (a) as a marker for neuronal cells and neuromuscular junctions (a; antibody SY38; cf. Ref. 108) in comparison with plakoglobin (b; guinea pig antibodies; cf. Refs. 11, 13). Note that in this tissue plakoglobin stains exclusively the endothelial junctions (b) of an arteriole (L, lumen) and a smaller vessel (lower left part) but does not react with the neuronal cells and the plaques of neuromuscular junctions (identified by SY38 in a) and with the subplasmalemmal plaque densities of the smooth muscle cells of the vascular wall (bracket in a). In this specimen the endothelial junctions are conspicuously oriented due to partial agonist vasoconstriction, resulting in a scalloped appearance of the internal elastic lamina to the ridges of which the intercellular junction-containing regions of the endothelial cell are attached. Bar = 25 μm.

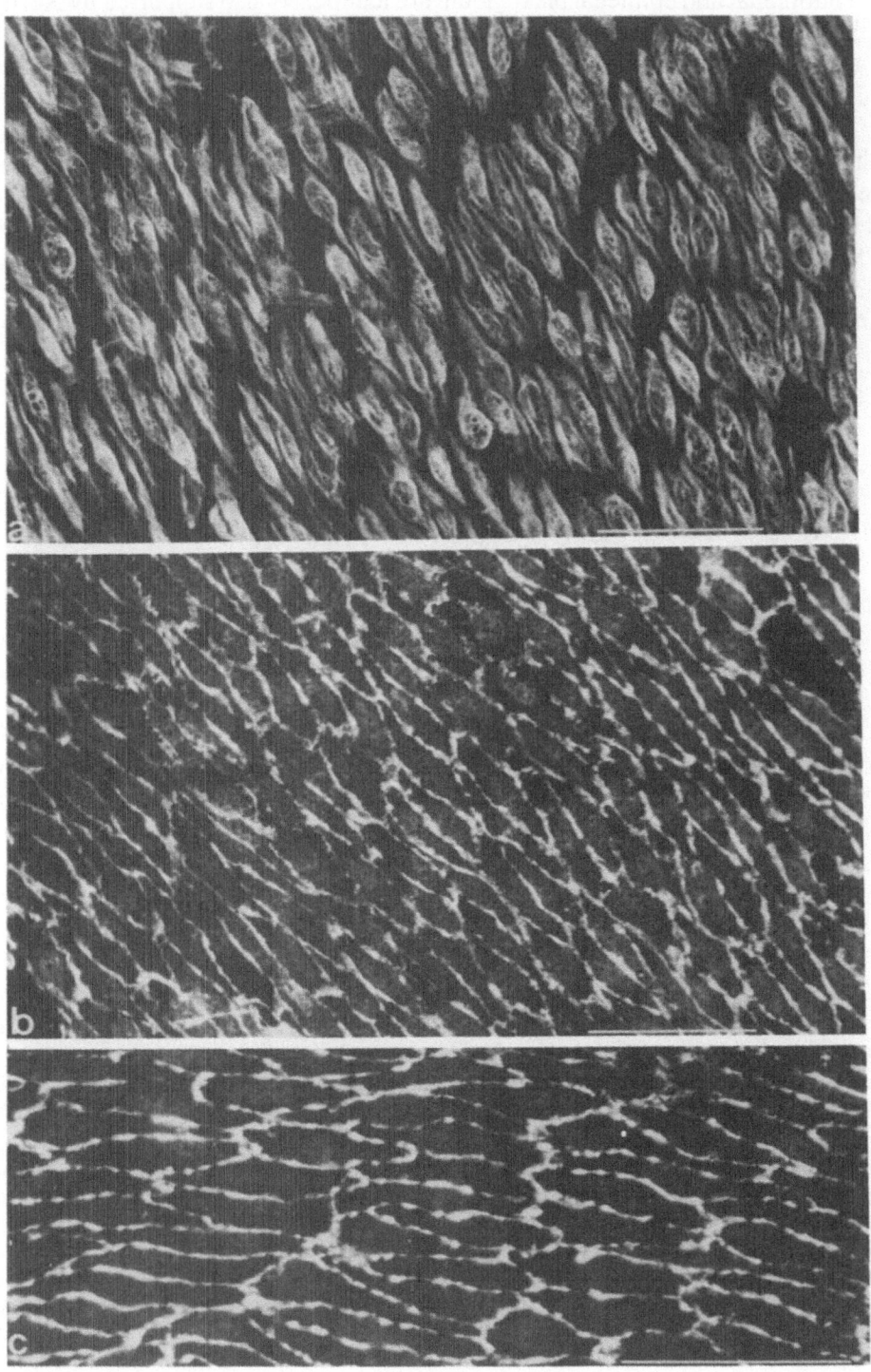

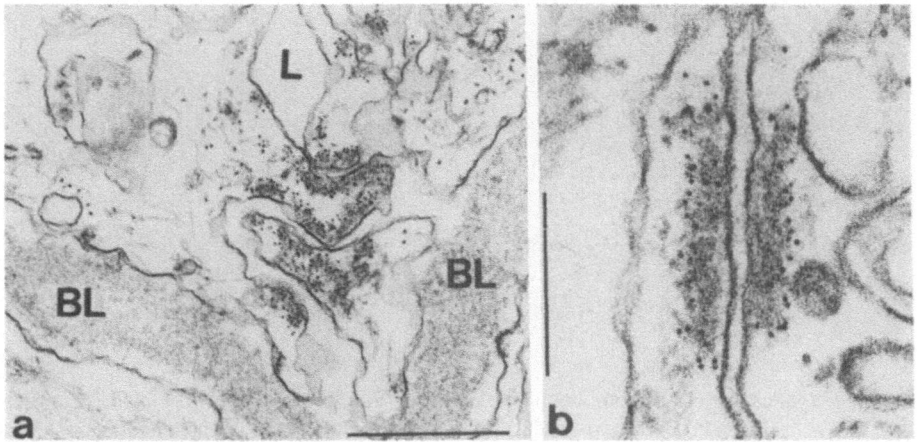

Figure 7. Immunoelectron micrographs of cross sections through bovine endothelium (cardiac capillary; L, lumen) as revealed after reaction with monoclonal plakoglobin antibodies, using a preembedding procedure involving partial cell lysis by brief treatment with saponin and secondary antibodies coupled to 5-nm gold particles (for details see Ref. 13). Note that most gold particles are associated with the plaque structures, and no reaction is seen on the basal lamina (BL), plaque-free membranes, vimentin filaments, and microfilament bundles (a, survey; b, higher magnification). Bars = 0.5 μm (a) and 0.2 μm (b).

forms of vinculin and plakoglobin may be important in the regulation of assembly and disassembly of the plaque components. Clearly the regulation of plaque formation must be different for plakoglobin and vinculin, as vinculin but not plakoglobin also occurs in the asymmetric plaques (subplasmalemmal densities) at the basal cell surface bordering on the extracellular matrix (e.g., Ref. 22).

VI. THE ENDOTHELIAL JUNCTION AS A ZONULA ADHAERENS: CONCLUSIONS AND PERSPECTIVES

All electron microscopic and immunocytochemical data available support the conclusion that the endothelial junction of mammalian blood vessels represents a continuous belt system related to the *zonula adhaerens* of polar epithelial cells and the plaque-bearing "ectoplasmic specializations" of Sertoli cells of testis.[13,102]

←───

Figure 6. Immunofluorescence micrographs of monolayer sheets of endothelial cells prepared by scraping from the luminal surface of bovine pulmonary artery (for details see Ref. 13) with antibodies to vimentin (a) and plakoglobin (b, c). Note that vimentin antibodies stain fibrillar arrays which extend throughout the cytoplasm but do not seem to be intimately associated with intercellular boundaries (a) whereas plakoglobin appears concentrated along cell-to-cell borders (b). This peripheral decoration of plakoglobin is of rather uniform intensity (e.g., in some cells in the lower left part of c) or shows regional fluctuations which may reflect local differences of plaque thickness or frequencies of interdigitations, but an artificial cause of this phenomenon cannot be excluded at present. Bars = 50 μm.

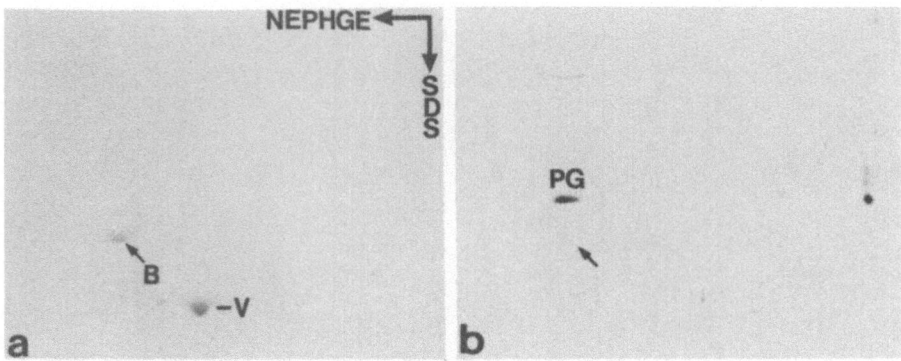

Figure 8. Identification of structure-bound plakoglobin (PG) from cultured calf pulmonary arterial endothelial cells (line CPAE) by two-dimensional gel electrophoresis and immunoblotting using monoclonal antibody PG 5.1. Nonequilibrium pH gradient electrophoresis (NEPHGE) was used in the first dimension (basic polypeptides to the left) and electrophoresis in the presence of SDS in the second dimension. (a) Ponceau S staining of polypeptides of a 100,000g pellet from CPAE cells after transfer to nitrocellulose paper. V denotes vimentin, B and the arrow coelectrophoresed bovine serum albumin. (b) Autoradiograph corresponding to a showing the immunoblot reaction of monoclonal antibody PG 5.1 to plakoglobin. Note that plakoglobin of endothelial cells has the same electrophoretic mobility in NEPHGE as that from bovine snout epidermal cells (cf. Refs. 11, 13).

These junctions are characterized by a plaque containing plakoglobin, vinculin, and α-actinin which is associated with actin microfilaments. At present the corresponding transmembranous and intermembranous components involved in the formation of this membrane cytoskeleton-domain are not known, and systematic development of procedures for isolating fractions enriched in endothelial junction material as well as antibodies against junctional membrane components are needed. Besides the striking similarity in plaque composition, the plaque-bearing zonula of the endothelial junction differs from the *zonula adhaerens* of polarized epithelial cells in that it is more pleomorphic and topologically less restricted. While the epithelial *zonula adhaerens* is a sharply defined, 0.2- to 0.5-μm belt structure in a specific position within the subapical junctional complex and shows a rather uniform, well-developed plaque,[28] the endothelial *zonula adhaerens* is less regular. In different subtypes of endothelia it varies considerably in depth and in the thickness of the plaque, and often it does not seem to form a single uniform belt but rather a complex, sometimes anastomosing zonular system. Future detailed studies will have to show whether certain morphological subtypes of the endothelial *zonula* are related to special forms of endothelial cells and/or blood vessels.

The topological and functional relationship of the endothelial *zonula adhaerens* and its plaque to the adjacent occluding and communicating junctions also remains to be clarified. Moreover, the patterns of the sparser intercellular connections and junctions in the discontinuous endothelium of the hepatic sinusoids need to be elucidated, and it is hoped that the newly defined junctional markers will be valuable tools in such studies.

The IF cytoskeleton in vascular endothelial cells also appears to be different from that of epithelial cells, not only by the absence of cytokeratin IFs and desmosomes but also by the fact that the vimentin IFs present are not anchored at junctional plasma membrane domains, unlike the situation in those cells in which vimentin IFs attach to desmosomal plaques such as in the meningothelial cells of the *arachnoidea* layer of some mammalian species.[63,84]

The information available, albeit incomplete, tends to indicate that the asymmetric dense plaques at the basal cell membrane of the endothelial cells ("subplasmalemmal densities," "ring fibers," "microfilament condensations") are only partly related to the symmetric plaques of the intercellular junctions between endothelial cells. Here the situation may be similar to that in certain cultured epithelial cells containing both a subapical *zonula adhaerens* and basal "focal adhesions."[13] While the plaques of both "junctions" are associated with vinculin, α-actinin, and actin microfilaments (for endothelia see also Refs. 22, 60), plakoglobin has not been detected in "focal adhesions."[13] Vice versa, talin seems to be a protein that coexists with vinculin in asymmetric plaques such as subplasmalemmal densities at sites bordering on extracellular matrix but is absent from the zonula plaque.[7,45,67,104] On the other hand, the third type of subplasmalemmal junction plaque occurring, although less frequently, in endothelial cells, i.e., the fuzzy cytoplasmic coat at the heterologous intercellular junctions between endothelial cells and subjacent smooth muscle cells at sites of basal lamina interruption ("myoendothelial junctions"), might represent a plakoglobin-positive symmetric *adhaerens* junction although a specific study on this structure is lacking.

Although the biological functions of the endothelial junction and its plaque are not fully clear, it is reasonable to assume that this junctional complex is involved in the sealing of the endothelial layer and the formation and maintenance of a semipermeable barrier which also protects the subendothelial tissue. The formation of the broad endothelial junction zone may also be an important prerequisite for the formation of occluding and communicating junctions within this zone, although the sequence of appearance of the diverse types of junctions in endothelia has not been studied.

The endothelial literature presents numerous examples of pathological phenomena accompanied by changes interpreted as "loosening," "opening," or "splitting" of endothelial junctions, notably those of the postcapillary venules. These observations suggest that various physiological influences and experimental treatments, from changes of temperature and air oxygen to the effects of a wide range of pharmacologically active compounds (e.g., prostaglandins, angiotensins, serotonin, histamine, epinephrine, nicotine), lead to the splitting of endothelial junctions along their symmetry plane, followed by cell separation and the appearance of transient or long-lived relatively large "gaps" between endothelial cells, thus allowing the passage not only of large particles but even of whole cells (for references see Refs. 54, 89, 92). The findings of Volberg *et al.*[103] that in polar epithelial cells the *zonula adhaerens* plaque material detaches as an entire ring from the membrane upon reduction of the concentration of extracellular Ca^{2+} and that its components are subsequently dispersed in the cytoplasm raise the question whether Ca^{2+}; ions are also important in the formation and maintenance of the *zonula adhaerens*-like plaque-belt of the endothelial junctions and hence in the regulation of the semiselective barrier and gate function of the endothelium as a whole. The availability of marker proteins for the endothelial junction, specifically its plaque proteins, should stimulate detailed cell biological experiments to answer these questions.

REFERENCES

1. Behrens, J., Birchmeier, W., Goodman, S. L., and Imhof, B. A., 1985, Dissociation of Madin–Darby canine kidney epithelial cells by the monoclonal antibody anti-Arc-1: Mechanistic aspects and identification of the antigen as a component related to uvomorulin, *J. Cell Biol.* **101:**1307.
2. Bloch, R. J., 1986, Actin at receptor-rich domains of isolated acetylcholine receptor clusters, *J. Cell Biol.* **102:**1447.
3. Bloch, R. J., and Hall, Z. W., 1983, Cytoskeletal components of the vertebrate neuromuscular junction: Vinculin, α-actinin, and filamin, *J. Cell Biol.* **97:**217.
4. Blose, S. H., and Meltzer, D. I., 1981, Visualization of the 10-nm filament vimentin rings in vascular endothelial cells in situ, *Exp. Cell Res.* **135:**299.
5. Boller, K., Vestweber, D., and Kemler, R., 1985, Cell-adhesion molecule uvomorulin is localized in the intermediate junctions of adult intestinal epithelial cells, *J. Cell Biol.* **100:**327.
6. Bretscher, A., and Weber, K., 1978, Localization of actin and microfilament associated proteins in the microvilli and terminal web of the intestinal brush border by immunofluorescence microscopy, *J. Cell Biol.* **79:**839.
7. Burridge, K., and Connell, L., 1983, A new protein of adhesion plaques and ruffling membranes, *J. Cell Biol.* **97:**359.
8. Constantinescu, E., Heltianu, C., and Simionescu, M., 1986, Immunological detection of an analogue of the erythroid protein 4.1 in endothelial cells, *Cell Biol. Int. Rep.* **10:**861.
9. Cowin, P., and Garrod, D. R., 1983, Antibodies to epithelial desmosomes show wide tissue and species cross-reactivity, *Nature* **302:**148.
10. Cowin, P., Mattey, D., and Garrod, D., 1984, Distribution of desmosomal components in the tissues of vertebrates, studied by fluorescent antibody staining, *J. Cell Sci.* **66:**119.
11. Cowin, P., Kapprell, H.-P., and Franke, W. W., 1985a, The complement of desmosomal plaque proteins in different cell types, *J. Cell Biol.* **101:**1442.
12. Cowin, P., Kapprell, H.-P., and Franke, W. W., 1985b, The desmosome–intermediate filament complex, in: *The Cell in Contact* (G. Edelman and J.-P. Thiery, eds.), Wiley, New York, pp. 427–460.
13. Cowin, P., Kapprell, H.-P., Franke, W. W., Tamkun, J., and Hynes, R. O., 1986, Plakoglobin: A protein common to different kinds of intercellular adhering junctions, *Cell* **46:**1063.
14. Craig, S. W., and Pardo, J. V., 1979, Alpha-actinin localization in the junctional complex of intestinal epithelial cells, *J. Cell Biol.* **80:**203.
15. Damsky, C. H., Richa, J., Solter, D., Knudsen, K., and Buck, C. A., 1983, Identification and purification of a cell surface glycoprotein mediating intercellular adhesion in embryonic and adult tissue, *Cell* **34:**455.
16. Davies, P. F., 1986, Vascular cell interactions with special reference to the pathogenesis of atherosclerosis, *Lab. Invest.* **55:**5.
17. Davies, P. F., Remuzzi, A., Gordon, E. J., Dewey, C. F., Jr., and Gimbrone, M. A., Jr., 1986, Turbulent fluid shear stress induces vascular endothelial cell turnover in vitro, *Proc. Natl. Acad. Sci. USA* **83:**2114.
18. De Bruyn, P. P. H., and Cho, Y., 1974, Contractile structures in endothelial cells of splenic sinusoids, *J. Ultrastruct. Res.* **49:**24.

19. Dermietzel, R., Liebstein, A., Frixen, U., Janssen-Timmen, U., Traub, O., and Willecke, K., 1984, Gap junctions in several tissues share antigenic determinants with liver gap junctions, *EMBO J.* **3**:2261.

20. Docherty, R. J., Edwards, J. G., Garrod, D. R., and Mattey, D. L., 1984, Chick embryonic pigmented retina is one of the group of epithelioid tissues that lack cytokeratins and desmosomes and have intermediate filaments composed of vimentin, *J. Cell Sci.* **71**:61.

21. Drenckhahn, D., and Franz, H., 1986, Identification of actin-, α-actinin-, and vinculin-containing plaques at the lateral membrane of epithelial cells, *J. Cell Biol.* **102**:1843.

22. Drenckhahn, D., and Wagner, J., 1986, Stress fibers in the splenic sinus endothelium in situ: Molecular structure, relationship to the extracellular matrix, and contractility, *J. Cell Biol.* **102**:1738.

23. Drochmans, P., Freudenstein, C., Wanson, J.-C., Laurent, L., Keenan, T. W., Stadler, J., Leloup, R., and Franke, W. W., 1978, Structure and biochemical composition of desmosomes and tonofilaments isolated from calf muzzle epidermis, *J. Cell Biol.* **79**:427.

24. Edelman, G. M., 1984, Expression of cell adhesion molecules during embryogenesis and regeneration, *Exp. Cell Res.* **161**:1.

25. Eriksson, A., and Thornell, L.-E., 1979, Intermediate (skeletin) filaments in heart Purkinje fibers, *J. Cell Biol.* **80**:231.

26. Erlandson, R. A., 1981, *Diagnostic Transmission Electron Microscopy of Human Tumors*, Masson Publishing, New York.

27. Evans, R. R., Robson, R. M., and Stromer, M. H., 1984, Properties of smooth muscle vimentin, *J. Biol. Chem.* **259**:3916.

28. Farquhar, M. G., and Palade, G. E., 1963, Junctional complexes in various epithelia, *J. Cell Biol.* **17**:375.

29. Fawcett, D. W., 1961, Intercellular bridges, *Exp. Cell Res. Suppl.* **8**:174.

30. Fawcett, D. W., 1981, *The Cell*, Saunders, Philadelphia.

31. Franke, R.-P., Gräfe, M., Schnittler, H., Seiffge, D., Mittermayer, C., and Drenckhahn, D., 1984, Induction of human vascular endothelial stress fibres by fluid shear stress, *Nature* **307**:648.

32. Franke, W. W., Schmid, E., Osborn, M., and Weber, K., 1979, Intermediate-sized filaments of human endothelial cells, *J. Cell Biol.* **81**:570.

33. Franke, W. W., Schmid, E., Grund, C., Müller, H., Engelbrecht, I., Moll, R., Stadler, J., and Jarasch, E.-D., 1981, Antibodies to high molecular weight polypeptides of desmosomes: Specific localization of a class of junctional proteins in cells and tissues, *Differentiation* **20**:217.

34. Franke, W. W., Moll, R., Schiller, D. L., Schmid, E., Kartenbeck, J., and Müller, H., 1982, Desmoplakins of epithelial and myocardial desmosomes are immunologically and biochemically related, *Differentiation* **23**:115.

35. Franke, W. W., Moll, R., Mueller, H., Schmid, E., Kuhn, C., Krepler, R., Artlieb, U., and Denk, H., 1983, Immunocytochemical identification of epithelium-derived human tumours with antibodies to desmosomal plaque proteins, *Proc. Natl. Acad. Sci. USA* **80**:543.

36. Franke, W. W., Moll, R., Achtstätter, T., and Kuhn, C., 1986, Cell typing of epithelia and carcinomas of the female genital tract using cytoskeletal proteins as markers, in: *Banbury Report 21: Viral Etiology of Cervical Cancer* (R. Peto and H. zur Hausen, eds.), Cold Spring Harbor Laboratory, Cold Spring Harbor, N.Y., pp. 121–148.

37. Franke, W. W., Cowin, P., Schmelz, M., and Kapprell, H.-P., 1987a, The desmosomal plaque and the cytoskeleton, *Ciba Found. Symp.* **125**:26–44.

38. Franke, W. W., Kapprell, H.-P., and Cowin, P., 1987b, Plakoglobin is a component of the filamentous subplasmalemmal coat of lens cells, *Eur. J. Cell Biol.* **43**:301.

39. Franke, W. W., Kapprell, H.-P., and Cowin, P., 1987c, Immunolocalization of plakoglobin in endothelial junctions: Identification as *zonulae adhaerentes*, *Biol. Cell* **59**:205.

40. Fujimoto, T., and Singer, S. J., 1986, Immunocytochemical studies of endothelial cells in vivo. I. The presence of desmin only, or of desmin plus vimentin, or vimentin only, in the endothelial cells of different capillaries of the adult chicken, *J. Cell Biol.* **103**:2775.

41. Gabbiani, G., Gabbiani, F., Lombardi, D., and Schwartz, S. M., 1983, Organization of actin cytoskeleton in normal and regenerating arterial endothelial cells, *Proc. Natl. Acad. Sci. USA* **80**:2361.

42. Geiger, B., 1979, A 130-K protein from chicken gizzard: Its localization at the termini of micro-filament bundles in cultured chicken cells, *Cell* **18**:193.
43. Geiger, B., Dutton, A. H., Tokuyasu, K. T., and Singer, S. J., 1981, Immunoelectron microscope studies of membrane–microfilament interactions: Distribution of α-actinin, tropomyosin and vin-culin in intestinal epithelial brush border and in chicken gizzard smooth muscle cells, *J. Cell Biol.* **91**:614.
44. Geiger, B., Schmid, E., and Franke, W. W., 1983, Spatial distribution of proteins specific for desmosomes and adhaerens junctions in epithelial cells demonstrated by double immunofluo-rescence microscopy, *Differentiation* **23**:189.
45. Geiger, B., Avnur, Z., Volberg, T., and Volk, T., 1985, Molecular domains of adhaerens junc-tions, *in*: *The Cell in Contact* (G. Edelman and J.-P. Thiery, eds.), Wiley, New York, pp. 461–489.
46. Ghadially, F. N., 1982, *Ultrastructural Pathology of the Cell and Matrix*, 2nd ed., Butterworths, London.
47. Ghitescu, L., Fixman, A., Simionescu, M., and Simionescu, N., 1986, Specific binding sites for albumin restricted to plasmalemmal vesicles of continuous capillary endothelium: Receptor-me-diated transcytosis, *J. Cell Biol.* **102**:1304.
48. Giamcomelli, F., Wiener, J., and Spiro, D., 1970, Cross-striated arrays of filaments in endothe-lium, *J. Cell Biol.* **45**:188.
49. Giorno, R., 1984, Unusual structure of human splenic sinusoids revealed by monoclonal anti-bodies, *Histochemistry* **81**:505.
50. Gorbsky, G., and Steinberg, M. S., 1981, Isolation of the intercellular glycoproteins of desmo-somes, *J. Cell Biol.* **90**:243.
51. Gumbiner, B., and Simons, K., 1987, The role of uvomorulin in the formation of epithelial oc-cluding junctions, *Ciba Found. Symp.* **125**:168–180.
52. Hansson, G. K., and Schwartz, S. M., 1983, Evidence for cell death in the vascular endothelium in vivo and in vitro. *Am. J. Pathol.* **112**:278.
53. Hatta, K., and Takeichi, M., 1986, Expression of N-cadherin adhesion molecules associated with early morphogenetic events in chick development, *Nature* **320**:447.
54. Heltianu, C., Simionescu, M., and Simionescu, N., 1982, Histamine receptors of the microvas-cular endothelium revealed in situ with a histamine–ferritin conjugate: Characteristic high-affinity binding sites in venules, *J. Cell Biol.* **93**:357.
55. Herman, I. M., Pollard, T. D., and Wong, A. J., 1982, Contractile proteins in endothelial cells, *in*: *Endothelium* (A.P. Fishman, ed.), New York Academy of Sciences, New York, pp. 50–60.
56. Hertzberg, E. L., and Skibbens, R. V., 1984, A protein homologous to the 27,000 dalton liver gap junction protein is present in a wide variety of species and tissues, *Cell* **39**:61.
57. Hormia, M., Linder, E., Lehto, V.-P., Vartio, T., Badley, R. A., and Virtanen, I., 1982, Vimentin filaments in cultured endothelial cells from butyrate-sensitive juxtanuclear masses after repeated subculture, *Exp. Cell Res.* **138**:159.
58. Hormia, M., Badley, R. A., Lehto, V.-P., and Virtanen, I., 1985, Actomyosin organization in stationary and migrating sheets of cultured human endothelial cells, *Exp. Cell Res.* **157**:116.
59. Huettner, I., Costabella, P. M., de Chastonay, C., and Gabbiani, G., 1982, Volume, surface and junctions of rat aortic endothelium during experimental hypertension: A morphometric and freeze fracture study, *Lab. Invest.* **46**:489.
60. Huettner, I., Walker, C., and Gabbiani, G., 1985, Aortic endothelial cell during regeneration: Remodelling of cell junctions, stress fibers, and stress fiber–membrane attachment domains, *Lab. Invest.* **53**:287.
61. Jahn, L., Fouquet, B., Rohe, K., and Franke, W. W., 1987, Cytokeratins in certain endothelial and smooth muscle cells of two taxonomically distant vertebrate species, *Xenopus laevis* and man, *Differentiation* **36**:234.
62. Kapprell, H.-P., Cowin, P., and Franke, W. W., 1987, Biochemical characterization of the soluble form of the junctional plaque protein, plakoglobin, from different cell types, *Eur. J. Biochem.* **166**:505.
63. Kartenbeck, J., Schwechheimer, K., Moll, R., and Franke, W. W., 1984, Attachment of vimentin

filaments to desmosomal plaques in human meningiomal cells and arachnoidal tissue, *J. Cell Biol.* **98**:1072.

64. Kumar, N. M., and Gilula, N. B., 1986, Cloning and characterzation of human and rat liver cDNAs coding for a gap junction protein, *J. Cell Biol.* **103**:767.

65. Le Beux, Y. J., and Willemot, J., 1978, Actin-like filaments in the endothelial cells of adult rat brain capillaries, *Exp. Neurol.* **58**:446.

66. Lombardi, T., Montesano, R., and Orci, L., 1985, Polarized plasma membrane domains in cultured endothelial cells, *Exp. Cell Res.* **161**:242.

67. Mangeat, P., and Burridge, K., 1984, Actin–membrane interaction in fibroblasts: What proteins are involved in this association? *J. Cell Biol.* **99**:95s.

68. Mattey, D. L., Suhrbier, A., Parrish, E., and Garrod, D. R., 1987, Recognition, calcium and the control of desmosome formation, *Ciba Found. Symp.* **125**:49–60.

69. Moll, R., Cowin, P., Kapprell, H.-P., and Franke, W. W., 1986, Desmosomal proteins: New markers for identification and classification of tumours, *Lab. Invest.* **54**:4.

70. Mooseker, M. S., 1985, Organization, chemistry and assembly of the cytoskeletal apparatus of the intestinal brush border, *Annu. Rev. Cell Biol.* **1**:209.

71. Müller, H., and Franke, W. W., 1983, Biochemical and immunological characterization of desmoplakins I and II, the major polypeptides of the desmosomal plaque, *J. Mol. Biol.* **163**:647.

72. Muresan, V., 1986, Pathways of transcytosis in the fenestrated endothelium of pancreatic capillaries, *J. Submicrosc. Cytol.* **18**:691.

73. Pardo, J. V., D'Angelo Siciliano, J., and Craig, S. W., 1983, Vinculin is a component of an extensive network of myofibril–sarcolemma attachment regions in cardiac muscle fibers, *J. Cell Biol.* **97**:1081.

74. Paul, D. L., 1986, Molecular cloning of cDNA for rat liver gap junction protein, *J. Cell Biol.* **103**:123.

75. Peters, A., Palay, S. L., and Webster, H. F., 1976, *The Fine Structure of the Nervous System: The Neurons and Supporting Cells*, Saunders, Philadelphia, pp. 1–406.

76. Pratt, B. M., Harris, A. S., Morrow, J. S., and Madri, J. A., 1984, Mechanisms of cytoskeletal regulation: Modulation of aortic endothelial cell spectrin by the extracellular matrix, *Am. J. Pathol.* **117**:349.

77. Reidy, M. A., 1985, A reassessment of endothelial injury and arterial lesion formation, *Lab. Invest.* **53**:513.

78. Revel, J. P., Yancey, S. B., Nicholson, B., and Hoh, J., 1987, Sequence diversity of gap junction proteins, *Ciba Found. Symp.* **125**:108–121.

79. Röhlich, P., and Oláh, I., 1967, Cross-striated fibrils in the endothelium of the rat myometrial arterioles, *J. Ultrastruct. Res.* **18**:667.

80. Rogers, K. A., and Kalnins, V. I., 1983, A method for examining the endothelial cytoskeleton in situ using immunofluorescence, *J. Histochem. Cytochem.* **31**:1317.

81. Rosenfeld, G. C., Hou, D. C., Dingus, J., Meza, I., and Bryan, J., 1985, Isolation and partial characterization of human platelet vinculin, *J. Cell Biol.* **100**:669.

82. Savion, N., Vlodavsky, I., Greenburg, G., and Gospodarowicz, D., 1982, Synthesis and distribution of cytoskeletal elements in endothelial cells as a function of cell growth and organization, *J. Cell. Physiol.* **110**:129.

83. Schmelz, M., Duden, R., Cowin, P., and Franke, W. W., 1986a, A constitutive transmembrane glycoprotein of M_r 165,000 (desmoglein) in epidermal and non-epidermal desmosomes. I. Biochemical identification of the polypeptide, *Eur. J. Cell Biol.* **42**:177.

84. Schmelz, M., Duden, R., Cowin, P., and Franke, W. W., 1986b, A constitutive transmembrane glycoprotein of M_r 165,000 (desmoglein) in epidermal and non-epidermal desmosomes. II. Immunolocalization and microinjection studies, *Eur. J. Cell Biol.* **42**:184.

85. Schmid, E., Osborn, M., Rungger-Brändle, E., Gabbiani, G., Weber, K., and Franke, W. W., 1982, Distribution of vimentin and desmin filaments in smooth muscle tissue of mammalian and avian aorta, *Exp. Cell Res.* **137**:329.

86. Schwartz, S. M., Gajdusek, C. M., Reidy, M. A., Selden, S. C., III, and Haudenschild, C. C., 1980, Maintenance of integrity in aortic endothelium, *Fed. Proc.* **39**:2618.

87. Sealock, R., Paschal, B., Beckerle, M., and Burridge, K., 1986, Talin is a post-synaptic component of the rat neuromuscular junction *Exp. Cell Res.* **163**:143.

88. Shear, C. R., and Bloch, R. J., 1985, Vinculin in subsarcolemmal densities in chicken skeletal muscle: Localization and relationship to intracellular and extracellular structures, *J. Cell Biol.* **101**:240.

89. Simionescu, N., and Simionescu, M., 1983, The cardiovascular system, in: *Histology* (L. Weiss, ed.), Elsevier, Amsterdam, pp. 371–433.

90. Simionescu, M., Simionescu, N., and Palade, G. E., 1974, Morphometric data on the endothelium of blood capillaries, *J. Cell Biol.* **60**:128.

91. Simionescu, M., Simionescu, N., and Palade, G. E., 1975, Segmental differentiations of cell junctions in the vascular endothelium, *J. Cell Biol.* **67**:863.

92. Simionescu, N., Simionescu, M., and Palade, G. E., 1978, Open junctions in the endothelium of the postcapillary venules of the diaphragm, *J. Cell Biol.* **79**:27.

93. Skerrow, C. J., and Matoltsy, A. G., 1974a, Isolation of epidermal desmosomes, *J. Cell Biol.* **63**:515.

94. Small, J. V., 1985, Geometry of actin–membrane attachments in the smooth muscle cell: The localisation of vinculin and α-actinin, *EMBO J.* **4**:45.

95. Staehelin, L. A., 1974, Structure and function of intercellular junctions, *Int. Rev. Cytol.* **39**:191.

96. Steinberg, M. S., Shida, H., Giudice, G. J., Shida, M., Patel, N. H., and Blaschuk, O. W., 1987, On the molecular organization, diversity and functions of desmosomal proteins, *Ciba Found. Symp* **125**:3–19.

97. Stevenson, B. R., Siliciano, J. D., Mooseker, M. S., and Goodenough, D. A., 1986, Identification of ZO-1: A high molecular weight polypeptide associated with the tight junction (zonula occludens) in a variety of epithelia, *J. Cell Biol.* **103**:755.

98. Tani, E., Yamagata, S., and Ito, Y., 1977, Freeze-fracture of capillary endothelium in rat brain, *Cell Tissue Res.* **176**:157.

99. Toccanier-Pelte, M. F., Skalli, O., Kapanci, Y., and Gabbiani, G., 1987, Characterization of stromal cells with myoid features in lymph nodes and spleen in normal and pathologic conditions, *Amer. J. Pathol.* **129**:109.

100. Tsukita, S., and Tsukita, S., 1985, Desmocalmin: A calmodulin-binding high molecular weight protein isolated from desmosomes, *J. Cell Biol.* **101**:2070.

101. Vestweber, D., Ocklind, C., Gossler, A., Odin, P., Öbrink, B., and Kemler, R., 1985, Comparison of two cell-adhesion molecules, uvomorulin and cell-CAM105, *Exp. Cell Res.* **157**:451.

102. Vogl, A. W., and Soucy, L. J., 1985, Arrangement and possible function of actin filament bundles in ectoplasmic specializations of ground squirrel Sertoli cells, *J. Cell Biol.* **100**:814.

103. Volberg, T., Geiger, B., Kartenbeck, J., and Franke, W. W., 1986a. Changes in membrane–microfilament interaction in intercellular adhaerens junctions upon removal of extracellular Ca^{2+} ions, *J. Cell Biol.* **102**:1832.

104. Volberg, T., Sabanay, H., and Geiger, B., 1986b, Spatial and temporal relationships between vinculin and talin in the developing chicken gizzard smooth muscle, *Differentiation* **32**:34.

105. Volk, T., and Geiger, B., 1984, A-CAM: A 135-kD receptor of intercellular adherens junctions. I. Immunoelectron microscopic localization and biochemical studies, *J. Cell Biol.* **103**:1441.

106. White, G. E., and Fujiwara, K., 1986, Expression and intracellular distribution of stress fibers in aortic endothelium, *J. Cell Biol.* **103**:63.

107. White, G. E., Gimbrone, M. A., Jr., and Fujiwara, K., 1983, Factors influencing the expression of stress fibers in vascular endothelial cells in situ, *J. Cell Biol.* **977**:416.

108. Wiedenmann, B., and Franke, W. W., 1985, Identification and localization of synaptophysin, an integral membrane glycoprotein of M_r 38,000 characteristic of presynaptic vesicles, *Cell* **41**:1017.

109. Wong, A. J., Pollard, T. D., and Herman, I. M., 1983, Actin filament stress fibers in vascular endothelial cells in vitro, *Science* **219**:867.

110. Yee, A. G., and Revel, J. P., 1975, Endothelial cell junctions, *J. Cell Biol.* **66**:200.

111. Yohro, T., and Burnstock, G., 1973, Filament bundles and contractility of endothelial cells in coronary arteries, *Z. Zellforsch.* **138**:85.

8

Endothelial Cell–Extracellular Matrix Interactions
Matrix as a Modulator of Cell Function

Joseph A. Madri, Bruce M. Pratt, and Judith Yannariello-Brown

I. INTRODUCTION

The role of extracellular matrix as a modulator of cell behavior is widely accepted and is currently under intensive study by a number of investigators using diverse cell, tissue, and organ culture systems.[10,34] Several general findings have emerged from this work, namely, that cell behavior is dramatically different when cells are grown and maintained on extracellular matrix as compared to tissue culture plastic or glass; and that cell behavior can be modulated, depending on the composition and organization of the matrix component(s) or tissue used.[35] For example, in recent studies Lwebuga-Mukasa *et al.* have demonstrated that cultured type II pneumocytes exhibit variable behavior patterns, depending on the nature of the underlying matrix on which they are cultured. When cultured on the stromal aspect of the acellular amnionic membrane, these cells appear flattened, having few cytoplasmic lamellar bodies, basolateral junctional complexes, and apical microvilli. In contrast, when cultured on the basement membrane surface of the amnion, they maintain their cuboidal morphology and have abundant microvilli, basolateral junctional complexes, and cytoplasmic lamellar bodies.[21,22] Additionally, in both instances the cultured cells produce a basal lamina-like structure. Conversely, when cultured and maintained on an acellular pulmonary basement membrane, these cells do not synthesize a basal lamina. With time they become flattened and attenuated, losing their cytoplasmic lamellar bodies and apical microvilli but maintaining their basolateral junctional complexes, suggesting a "differentiation" into type I pneumocytes.[22] As a further example, Ingber *et al.* have studied the effects of extracellular matrix on the behavior of rat pancreatic adenocarcinoma cells. This tumor exhibits cytodifferentiation in vivo when closely associated with vascular or peritoneal connective tissue, suggesting the possibility

Joseph A. Madri, Bruce M. Pratt, and Judith Yannariello-Brown • Department of Pathology, Yale University School of Medicine, New Haven, Connecticut 06510.

of specific cell–matrix interactions.[14] In later studies it has been demonstrated that these cells exhibit unique attachment and spreading kinetics and cytodifferentiation including specific organelle localization, cell surface component organization, and cytoskeletal rearrangement in response to culture on the basement membrane surface of the amnion and the basement membrane component laminin.[15,16]

Over the past several years there has been intensive study in the area of endothelial cell cytoskeleton organization and dynamics *in vivo* and *in vitro*.[8,9] Parallel to these studies has been a growing appreciation of the complex interactions of cells with their surrounding matrix, which, in part, has led to a discovery of a variety of collagen-, fibronectin-, and laminin-binding proteins.[31,37,38] Continued work in the above two areas has led several investigators to consider the extracellular matrix as an important potential modulator of endothelial cell behavior.[27,35] In these studies, changes in cell attachment, motility, proliferation, multicellular organization, and biosynthesis have been documented.[25,28,29]

In this chapter we will discuss vascular endothelial cell–extracellular matrix interactions, namely, evidence for endothelial cell surface matrix binding proteins, organization of selected endothelial cell cytoskeletal components by matrix, and correlation of endothelial cell behavior with cytoskeletal organization and extracellular matrix composition and organization.

II. ENDOTHELIAL CELL–SUBENDOTHELIAL MATRIX INTERACTIONS: IN VIVO CONSIDERATIONS

Endothelial cells are polar cells, having one surface (luminal) in contact with the blood and the other (abluminal) in close contact with the subendothelial matrix. In the normal, quiescent state both large and small vessel endothelial cells usually are in close apposition to a subendothelial matrix composed of basal lamina components. In previous studies we have determined that the matrix underlying both microvascular and large vessel endothelium is composed, in part, of collagen types IV and V and laminin and appears to be devoid of significant amounts of the interstitial collagen types I and III.[24,25] Since organisms having vascular systems are subjected to vascular and soft tissue injury, their endothelial cells must have the ability to interact with the wide variety of extracellular components that are encountered during development and repair. Indeed, following some denudation injuries in large vessels, the endothelial cells at the wound margins migrate and proliferate over a matrix composed of significant amounts of interstitial collagens. Similarly, following soft tissue injury, microvascular endothelial cells are released from the constraints of their investing basement membrane and migrate and proliferate into a three-dimensional interstitial matrix. In both instances, following initial migratory and proliferative responses the endothelial cells synthesize a new basal lamina and eventually reexpress a differentiated, mitotically quiescent phenotype.[26] The effects of extracellular matrix on endothelial cell behavior in normal and pathological states, while incompletely understood, are thought to be important. Changes in matrix composition and organization have been correlated with changes in endothelial cell size and shape,[19,39] proliferative rate,[29] and multicelluular organization.[29,35] While informative, *in vivo* models are complex and manipulation and control of variables is difficult to accomplish. Because of these and many other complex problems encountered in the use of *in vivo* models, many investigators are developing and are using *in vitro* culture systems to study endothelial cell interactions with extracellular matrix.[26] Although useful, tissue and organ culture experiments cannot provide a complete understanding of complex biological phenomena and where possible, an approach utilizing both *in vitro* and *in vivo* studies should be used.

III. ENDOTHELIAL CELL–EXTRACELLULAR MATRIX INTERACTIONS: IN VITRO CONSIDERATIONS

A. Large Vessel Endothelial Cells

Arterial endothelial cells *in vivo* form a continuous, nonthrombogenic, metabolically active lining for the vascular system. To maintain this lining they have a complex, effective response to injury. Following denudation injury, the cells at the wound margins are stimulated to spread, migrate, and if necessary proliferate into the denuded area to reconstitute a continuous lining. Over the past several years the roles of underlying matrix and the extracellular matrix components synthesized by the stimulated endothelial cells have been implicated as being important modulators of this repair response.[24,26,35] In addition to the finding that cultured large vessel endothelial cells are capable of attaching to a wide variety of matrix components,[33] we have demonstrated that these cells require continual collagen synthesis and secretion for migration following release from contact inhibition and that the cells synthesize and secrete specific matrix components in a temporally ordered sequence.[28] In a similar fashion to epidermal cells, migrating large vessel endothelial cells initially express laminin and only at later times express type IV collagen in the underlying matrix. Following reconstitution of the continuous monolayer, the cells appear to synthesize a complete subendothelial matrix.[12,28]

1. Morphology/Behavior

The above studies prompted our investigation into the mechanisms responsible for the observed changes in cell behavior noted when large vessel endothelial cells interact with various matrix components. Using a migration assay developed in the laboratory, we noted dramatic differences in the areas covered by cultured bovine aortic endothelial cells depending on what matrix component the cells were plated.[36] When examined by light microscopy the cells exhibited marked differences in size when plated on a substrate of plasma fibronectin as compared to collagen types I and III (Fig. 1). On the fibronectin substrate the cells appeared as flattened, polygonal cells having irregular, intermittent cell–cell contacts. In

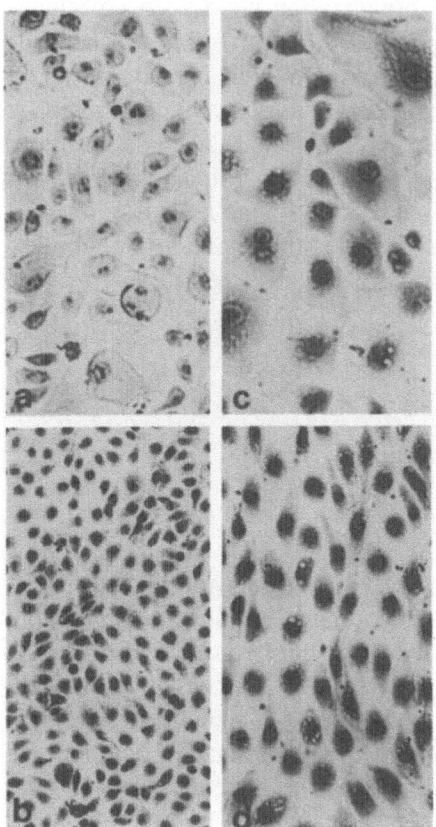

Figure 1. Morphology of fixed and hematoxylin-stained bovine aortic endothelial cells (BAEC) in nonmigrating and migrating states on fibronectin (Fn) and collagen type I/III (I/III) matrices. Original magnification, × 200. (a) A typical field of nonmigrating BAEC on a Fn matrix illustrating the large, flattened, polygonal cells forming a loose cobblestone pattern. (b) A representative field of nonmigrating BAEC on a I/III matrix. The cells appear small and polygonal in a characteristic cobblestone pattern. (c) A typical field of BAEC released from contact inhibition on a Fn substrate. The cells appear large, flat, and polygonal. (d) A representative field of migrating BAEC on a I/III matrix. The cells appear fusiform, being larger than those observed on this matrix in the nonmigrating state, and oriented toward the stimulus of migration. (Data from Leto *et al.*[19] with permission.)

contrast, the cells on the collagenous substrate were smaller polygonal cells having extensive cell–cell contact. Following release from contact inhibition, the cells on the fibronectin substrate maintained their large, flattened polygonal shape and exhibited minimal migration; while the cells on the collagenous substrates and laminin became fusiform, aligning their long axes with the direction of migration and exhibiting considerable migration. When examined on the electron microscopic level in the confluent, nonmigrating state, the cells on both matrices exhibited similar interactions with the substratum, namely, extensive close contact sites. In contrast, following the stimulus to migrate, the cells on the collagenous substrate exhibited intermittent apposition with the substratum with these areas exhibiting extensive cytoskeletal arrays.[35] These data suggest that matrix composition and organization may exert their effects on cell behavior, in part, by modulation of the cytoskeleton.

2. Matrix Receptors

Over the past several years there has been a growing appreciation of complex, specific interactions of cells with the surrounding matrix, leading to the discovery

Figure 2. Demonstration of a type IV collagen binding protein (CB-48) in bovine aortic endothelial cells (BAEC). Cell membranes were prepared from BAEC grown to confluency on tissue culture trays (Nunc). Cell membranes were solubilized in 1% NP-40 in Tris-buffered saline and passed over a type IV collagen–Sepharose column. After washing, the bound proteins were eluted with 4 M urea and run on a 10% Laemmli SDS-PAGE system. Lane a, solubilized BAEC membranes stained with Coomassie blue and silver nitrate. Lane b, Coomassie blue stain of the 48k band specifically eluted from the type IV collagen–Sepharose column and run under reducing conditions. Lane c, silver-stained sample of CB-48 run under

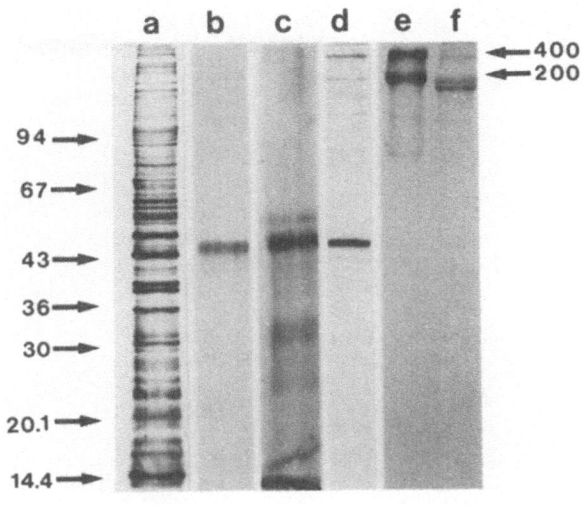

nonreducing conditions. Lane d, fluorograph of eluate from the affinity column obtained after metabolic labeling of BAEC with [³⁵S]-Met. Lanes e and f, Coomassie blue staining of laminin and type IV collagen, respectively.

of collagen-, fibronectin-, and laminin-binding cell surface proteins in a variety of cell types.[20] Recently, we have accrued preliminary evidence suggesting the existence of laminin and type IV collagen cell surface binding proteins on large and microvascular endothelial cells.[43] Specifically, we have demonstrated saturable and specific binding of soluble laminin and type IV collagen to bovine aortic endothelial cells in suspension and on confluent monolayers that is time and temperature dependent. In addition, we have isolated specific cell surface proteins from detergent-solubilized endothelial plasma membranes by affinity chromatography on laminin– and type IV collagen–Sepharose. As illustrated in Fig. 2, a 48,000-dalton type IV collagen binding protein (CB-48) can be isolated by elution with 4 M urea from a type IV collagen affinity column. CB-48 migrates in a Laemmli SDS-PAGE system with similar mobility under reducing and nonreducing conditions (Fig. 2b,c) and is a metabolic product of endothelial cells as evidenced by the incorporation of [³⁵S]methionine (Fig. 2d). This protein is similar in molecular weight to "colligin," a recently discovered collagen binding protein of cultured teratocarcinoma cells.[18] In contrast to the single molecular weight species endothelial cell type IV collagen binding protein, we have found several molecular weight species of laminin binding proteins in bovine aortic endothelial cells, the major species having molecular weights of 30, 50, 75–80, 90, and 120–140k. Notably, CB-48 has never been observed in the eluant, suggesting that CB-48 is not involved in laminin binding. The apparent molecular heterogeneity observed in laminin–Sepharose eluates may reflect multiple mechanisms of endothelial cell interaction with laminin. Each of these proteins may bind to specific epitopes on the laminin molecule which may transmit a specific signal to the endothelial cell. Some of the moieties we have reported correspond in molecular

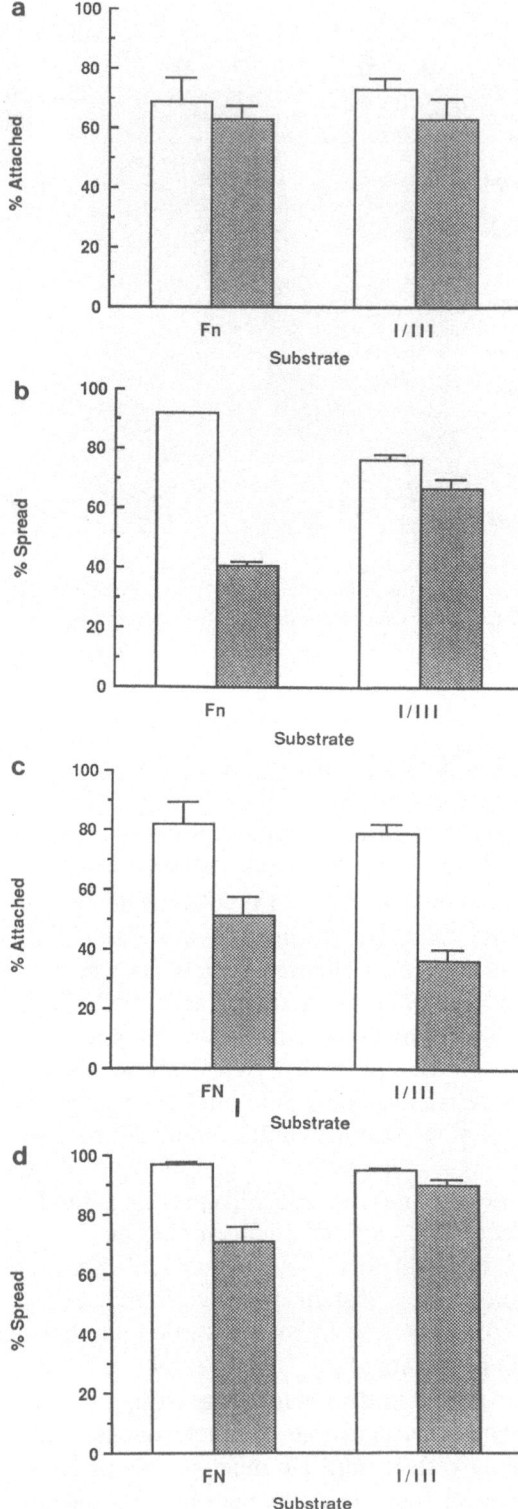

Figure 3. Synthetic tetrapeptide (RGDS) effects on bovine aortic endothelial cell (BAEC) attachment and spreading. At 400 µg/ml the RGDS peptide had no observable effect on BAEC attachment to either fibronectin (Fn) or collagen type I/III (I/III) substrates (a). However, at this concentration there was a preferential effect on cell spreading on the Fn matrix (b). At higher concentration (500 µg/ml), BAEC attachment on both matrices was affected, being reduced to 50 and 40% on Fn and I/III substrates, respectively (c), while only cell spreading on the Fn substrate was affected (d). The error bars represent the results of triplicate samples.

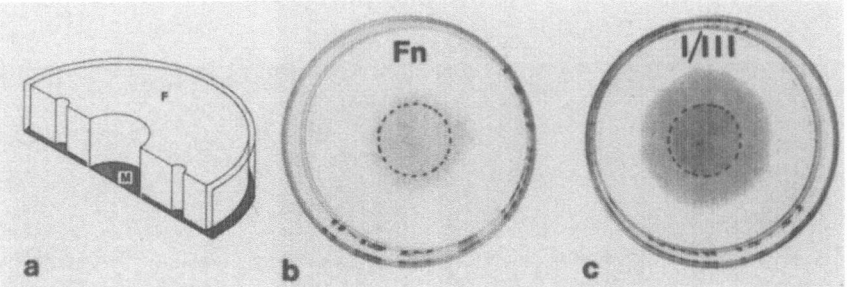

Figure 4. Schematic drawing of the Teflon fence used for migration studies and morphology of BAEC cultured and allowed to migrate on fibronectin and collagen type I/III substrates. (a) The Teflon fence (F) is placed in a bacteriologic plastic petri dish coated with a matrix substrate (M). Cells are plated into the center well of the fence at high density. After 24 hr the BAEC have formed a confluent monolayer. The fence is then removed and the migrating BAEC are cultured for 6 days for assessment of the effects of various matrix substrates on cell migration/proliferation. (b) After 6 days in culture the cells on Fn display minimal increase in net surface area covered; whereas (c) cells on I/III display greater increase in net surface area covered. Dashed lines indicate areas covered by BAEC seeded within the central well of the fence. (Data from Pratt *et al.*[36] with permission.)

weight to other laminin binding proteins reported in the literature. For example, the CSAT antigen complex isolated from chick embryo fibroblasts has a molecular weight range of approximately 150k and has been reported to bind to laminin and fibronectin, while the most commonly reported laminin binding protein has a molecular weight of approximately 65–72k. However, it remains to be demonstrated if the endothelial cell-derived laminin binding proteins are related to the described CSAT antigen and laminin receptor.[11,13,20] In addition to laminin and type IV collagen binding proteins, several groups have presented evidence for fibronectin receptors on endothelial cell surfaces (GP IIb/IIIa and possibly the CSAT antigen).[4,11] In Fig. 3 we have demonstrated the effects of the synthetic tetrapeptide Arg-Gly-Asp-Ser (RGDS) on the attachment and spreading of bovine aortic endothelial cells plated on substrates of fibronectin or collagen types I/III. At lower concentrations of the RGDS tetrapeptide (400 µg/ml), cell attachment appears unaffected on either substrate; while cell spreading on the fibronectin substrate is affected preferentially. At higher concentrations (500 µg/ml), cell attachment on both matrices is reduced; however, only cell spreading on the fibronectin substrate is affected. These observations (see Fig. 5a,c) are consistent with the finding of RGDS sequences in both fibronectin and type I collagen.[40] The lack of an RGDS effect on BAEC spreading on collagen types I/III suggests the presence of an additional collagen binding protein which does not utilize the RGDS sequence. The role(s) of the RGD sequence found in the various proteins (matrix and other) that endothelial cells come into contact with are still largely unknown and should provide a fruitful area of investigation.

3. Cytoskeletal Dynamics

Several investigators have illustrated dramatic changes in the organization of selected cytoskeletal elements following *in vivo* and *in vitro* endothelial cell

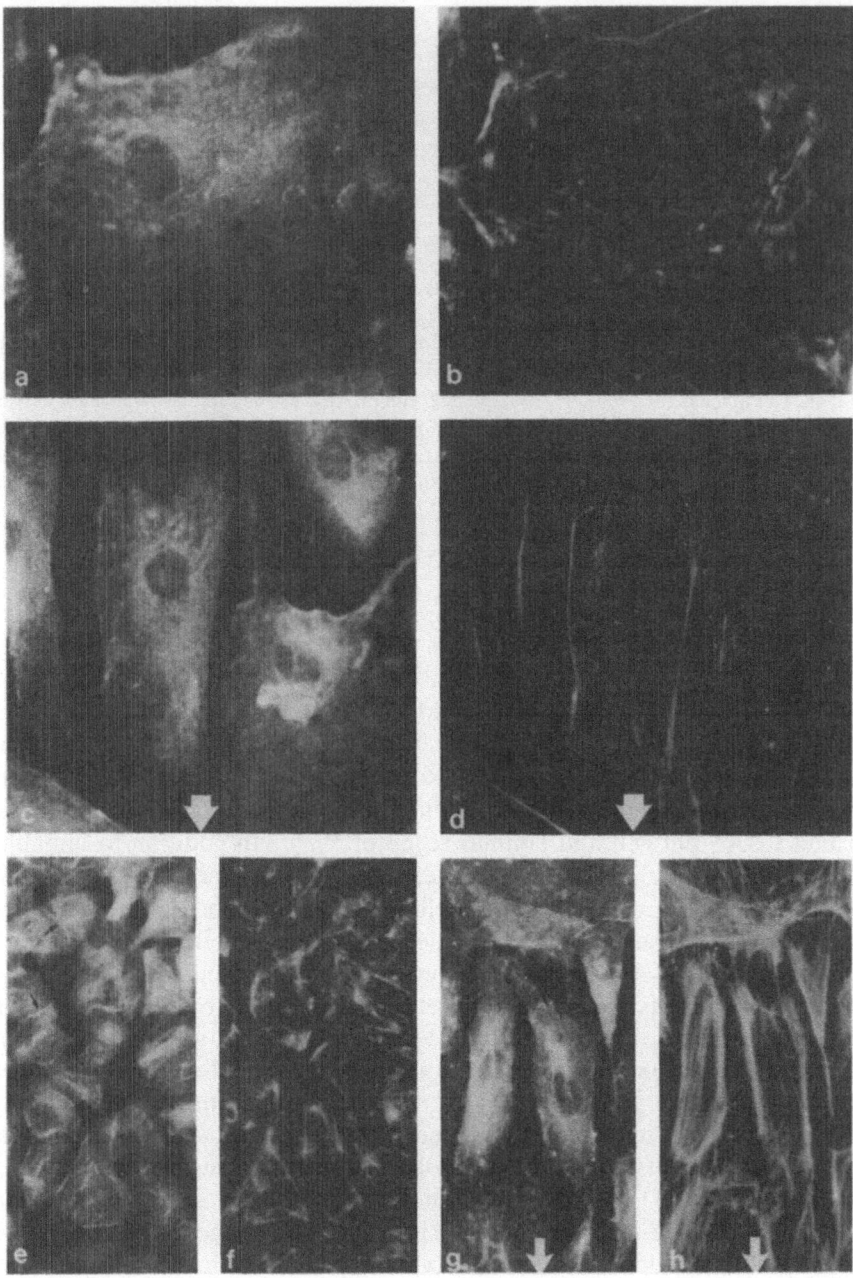

Figure 5. Immunofluorescence micrographs of BAEC in nonmigrating and migrating states on fibronectin (Fn) and collagen type I/III (I/III) matrices stained using antibodies to fodrin and NBD-phallacidin. Original magnification (a–h), × 250. (a) Nonmigrating BAEC on a Fn substratum labeled with
antifodrin. The large, flattened cells exhibit a general, diffuse, finely granular cytoplasmic fluorescence.
The nuclear area is negative and there is no appreciable peripheral band staining noted. (b) The same
cells as in (a) decorated with NBD-phallacidin revealing a complex cytoplasmic filamentous actin

injury.[8,9,41] The evidence supporting the concept that extracellular matrix composition and organization affects endothelial cell behavior, in part, by modulation of the cytoskeleton has come from immunolocalization studies. As noted above, when bovine aortic endothelial cells are grown to confluency on various matrix components, cell and nuclear size varies considerably. Furthermore, in cultures released from contact inhibition, markedly different migration rates are observed, collagen types I and III permitting the most and fibronectin the least (Fig. 4). When these cultures are stained with antibodies to fodrin and protein band 4.1, two cortical cytoskeletal proteins, dramatic differences in fluorescence patterns are noted, depending on what matrix component the cells are plated on and whether the cells are at confluence or stimulated to migrate.[19,36] Figure 5 illustrates fodrin localization in confluent bovine aortic endothelial cells plated on fibronectin. The random granular pattern of fodrin localization in the large, polygonal cells does not appear to colocalize with actin filaments visualized with a phalloidin probe in the nonmigrating, confluent cultures. Following release from contact inhibition there is little apparent migration and the localization pattern of fodrin does not change although the actin filaments align along the long axis of the cell and the dense peripheral bands are no longer apparent. In contrast, the cells plated on a collagen I/III matrix are considerably smaller and exhibit a dense peripheral staining pattern with fodrin antibodies which partially colocalizes with actin filaments. When permitted to migrate the cells do so rapidly, exhibiting dramatic changes in fodrin localization, namely, a loss of the dense peripheral staining and the appearance of filamentous staining along the long axis of the migrating cells which partially colocalizes with actin filaments. Changes in protein band 4.1 localization on the different matrices and in response to migratory stimulus were also observed (Fig. 6). Protein 4.1 localization in confluent cells on a fibronectin substrate was complex, being nuclear and filamentous, exhibiting partial colocalization with actin filaments. Stimulus to migrate caused a realignment of the protein 4.1 in filamentous structures along the long axis of the cells and partial colocalization with actin filaments. In contrast, cells on the collagen I/III matrix exhibited dense peripheral protein 4.1 localization as well as nuclear staining. Following migratory stimulus, the peripheral staining was lost and a delicate fi-

---

network, having no particular orientation. (c) Migrating BAEC on a Fn substrate labeled with anti-fodrin. The large, flattened cells orient toward the migratory stimulus (large arrow) and exhibit an unchanged fluorescence pattern as compared to those noted in (a), namely, a general, diffuse granular cytoplasmic stain, negative in nuclear areas and in the areas of peripheral bands. (d) The same cells as in (c) decorated with NBD-phallacidin revealing a cytoplasmic filamentous network oriented in the direction of migration (large arrow). (e) Nonmigrating BAEC on a I/III substrate labeled with antifodrin. The smaller, rounded cells exhibit a diffuse granular cytoplasmic fluorescence, absent from the nuclear areas. In addition, there are intense linear fluorescence patterns in areas of apparent cell–cell contact (small arrows). (f) The same cells as in (e) decorated with NBD-phallacidin revealing dense peripheral banding patterns with variable cytoplasmic filamentous labeling. (g) Migrating BAEC on a I/III substrate labeled with antifodrin. The fusiform cells exhibit a delicate filamentous fluorescence pattern aligned along the long axes of the cells in the direction of migration (large arrow). In addition, the intense peripheral staining along areas of cell–cell contact is no longer detectable. (h) The same cells as in (g) decorated with NBD-phallacidin revealing an intense cytoplasmic filamentous labeling pattern aligned along the direction of migration (large arrow).

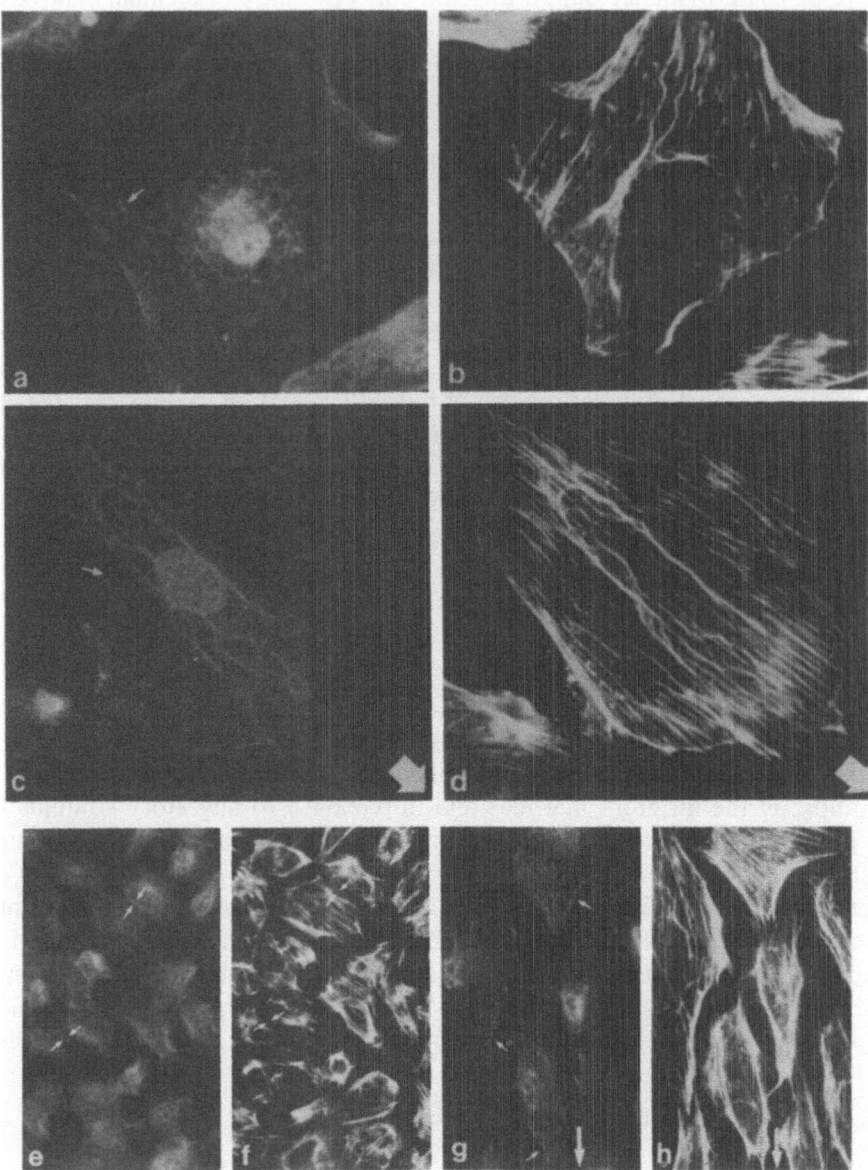

Figure 6. Immunofluorescence micrographs of BAEC in nonmigrating and migrating states on fibronectin (Fn) and collagen type I/III (I/III) matrices stained using antibodies to protein 4.1 and NBD-
phallacidin. Original magnification (a–h), × 250. (a) Nonmigrating BAEC on a Fn substrate labeled
with anti-4.1. The flattened, polygonal cells reveal a nuclear and perinuclear labeling with a radiating,
delicate filamentous staining pattern throughout the cytoplasm. No appreciable peripheral staining is
noted. Arrow indicates a representative cluster of particulate cytoplasmic structures which label with
anti-4.1. (b) The same cell as in (a) decorated with NBD-phallacidin revealing an intense filamentous
actin network in the cell cytoplasm. No orientation of the filaments is apparent and there is no detectable nuclear labeling. (c) Migrating BAEC on a Fn substratum labeled with anti-4.1. The flattened
cells have a diffuse nuclear and a cytoplasmic filamentous labeling pattern with the filaments aligned

lamentous staining, colocalizing with major actin filaments along the long axis of the cells, became apparent. The observed nuclear protein 4.1 staining may be indicative of potential interactions with other filamentous proteins associated with the nucleus such as vimentin and tubulin, suggesting that protein 4.1 (being localized to nuclear and plasma membrane regions as well as to cytoplasmic filamentous structures) may serve as a "link" protein in the transduction of information from the extracellular matrix to the nucleus.[19] Although morphological and chemical evidence supporting connections between the cortical and membranous cytoskeleton and the extracellular matrix via matrix receptors is far from convincing at this time, preliminary immunofluorescence evidence supports this concept.[44] In collaboration with Dr. Lance Liotta of the NIH, we have begun to investigate the cellular distribution and cytoskeletal associations of a putative endothelial cell laminin receptor utilizing the monoclonal antibody LR2 which specifically recognizes a 72k protein isolated from human breast carcinoma tissues.[20,44] Utilizing the Teflon fence migration assay (Fig. 4), we have examined the effects of matrix and migratory behavior on laminin receptor and actin localization in bovine aortic endothelial cells. While the large vessel cells were stationary, LR2 distribution was diffuse with occasional staining around the periphery of the cells. During migration, LR2 staining was noted in linear arrays and as an intense peripheral banding pattern which appeared to colocalize with actin microfilaments detected with NBD-phallacidin.[44] These data demonstrate that the distribution of this laminin binding protein is not static in this culture system and that there may be associations (direct and/or through other associated proteins) with the cytoskeletal framework. Evidence for a connection between matrix binding proteins and the cortical cytoskeleton has been reported by others at the morphological and biochemical levels.[1,3,42] However, a great deal of work has yet to be done in order to substantiate these concepts and preliminary observations in endothelial cells.

In addition to its function as a dynamic modulator of cell behavior by directing

---

in the direction of the migratory stimulus (indicated by the large arrow). There is also some peripheral labeling at the leading edge of the cells. In addition to the filamentous patterns, there is labeling of particulate structures in the cytoplasm (small arrow). (d) The same cells as in (c) decorated with NBD-phallacidin, revealing an intense filamentous actin network oriented in the direction of the migratory stimulus (large arrow) and in the periphery along the leading edge of the cells. In addition, the labeled spherical structures observed in (c) are closely associated with actin filaments. (e) Nonmigrating BAEC on a I/III substratum labeled with anti-4.1. The smaller, rounded polygonal cells reveal a diffuse granular cytoplasmic labeling which is more intense in nuclear regions. In addition, there is peripheral staining in areas of apparent cell–cell contact (pairs of small arrows). (f) The same cells as in (e) decorated with NBD-phallacidin revealing a dense peripheral banding pattern with variable cytoplasmic filamentous labeling. The pairs of small arrows delimit the regions of the linear 4.1 staining. Note the absence of phallacidin decoration in these areas. (g) Migrating BAEC on a I/III substrate labeled with anti-4.1. The elongated, fusiform cells exhibit a complex pattern of staining. There is nuclear staining and a granular cytoplasmic pattern with delicate filamentous cytoplasmic staining emanating from the nuclear areas. In addition, irregular peripheral staining is noted on the migrating cells (small arrows). Large arrow denotes the direction of migration. (h) The same cells as in (g) decorated with NBD-phallacidin revealing a delicate filamentous cytoplasmic pattern and an intense fluorescence in peripheral areas. Large arrow denotes the direction of migration. (Data from Leto et al.[19] with permission.)

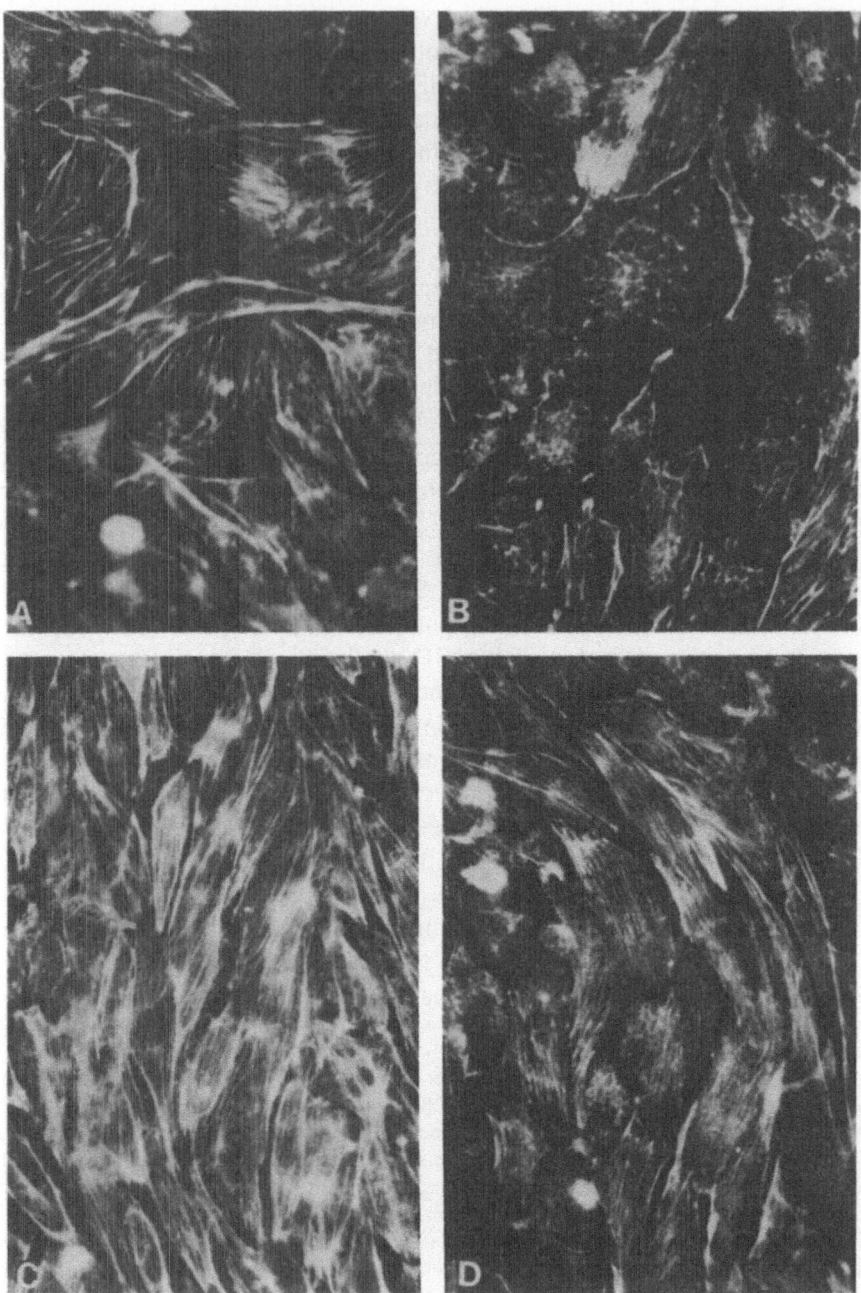

Figure 7. Rhodamine-phalloidin decoration of BAEC grown on tissue culture plastic (TCP) or type I/III collagen (I/III) and treated with 12-*O*-tetradecanoyl phorbol 13-acetate (TPA). (A) Cells grown on TCP appear as roughly cobblestone, being in close apposition with one another and having an intense linear cytoplasmic filamentous actin network. (B) In contrast, cultures treated with 20 nM TPA appear markedly rounded and retracted from one another. Their actin network has also dramatically changed,

cytoskeleton organization, the matrix may also serve as a stabilizer of the cytoskeleton, thereby acting to maintain certain aspects of cell behavior. For example, extracellular matrix may act as an organizer of a variety of cell surface receptors whose organization may, in turn, cause the stabilization of selected cytoskeletal elements which may preserve cell shape thus affecting a variety of cellular processes such as polarity, proliferation and migration rates, and biosynthesis.[17] Recent preliminary data from our laboratory are consistent with this concept. When various concentrations of TPA (12-O-tetradecanoyl phorbol 13-acetate) were added to bovine aortic endothelial cell cultures grown to confluency on tissue culture plastic, fibronectin and collagen type I/III substrates, the cells plated on the fibronectin and type I/III collagen substrates were noted to be more resistant to cell shape change (cell retraction) and actin filament rearrangement, while cells plated on tissue culture plastic were less resistant to the actin filament rearrangement and morphological changes induced by TPA (Fig. 7). The presence of various extracellular components appears to stabilize some components of the cytoskeleton in the face of soluble mediators that are known to cause cytoskeletal reorganization. Thus, matrix-mediated stabilization of cytoskeleton may exert profound influence on cell behavior, e.g., maintenance of differentiated phenotype.

B. Microvascular Endothelial Cells

1. In Vitro Culture Models

Capillary endothelial cells, unlike their large vessel counterparts, exhibit a significant arc of curvature and single cells are surrounded by an investing basement membrane. Their response to injury is also very different from that of large vessel endothelial cells in that they migrate and proliferate *through* a three-dimensional matrix, not *over* it. The importance of extracellular matrix in microvascular endothelial biology became apparent during early attempts at tissue culture of this cell type. Folkman found that a gelatin substratum facilitated the culture of capillary endothelial cells and also permitted occasional "tube formation" in long-term cultures.[5] Many laboratories have devoted considerable efforts in attempting to better understand the roles of extracellular matrix during the complex process of angiogenesis and its presumed *in vitro* correlates.[23,32] Our efforts have concentrated on the roles of specific matrix components in the various phases of the angiogenic response: activation, migration, proliferation, differentiation/stabilization, and regression.[27] When grown on substrates consisting of the interstitial collagen types I and III, cultured microvasculature endothelial cells

becoming less intense and forming multiple "polygonal" structures throughout the cytoplasm. (C) Cells grown on a I/III matrix appear as somewhat fusiform cells in close apposition with one another, having an intense linear cytoplasmic filamentous actin network. (D) Unlike cells cultured on TCP, cells cultured on I/III and treated with 20 nM TPA exhibited only minimal changes in cell shape and cell–cell interaction, for the most part maintaining their fusiform shape and cell–cell appositions. Similarly, their actin network is also essentially unchanged, exhibiting linear filamentous arrays as did the untreated cultures. Original magnification, × 250.

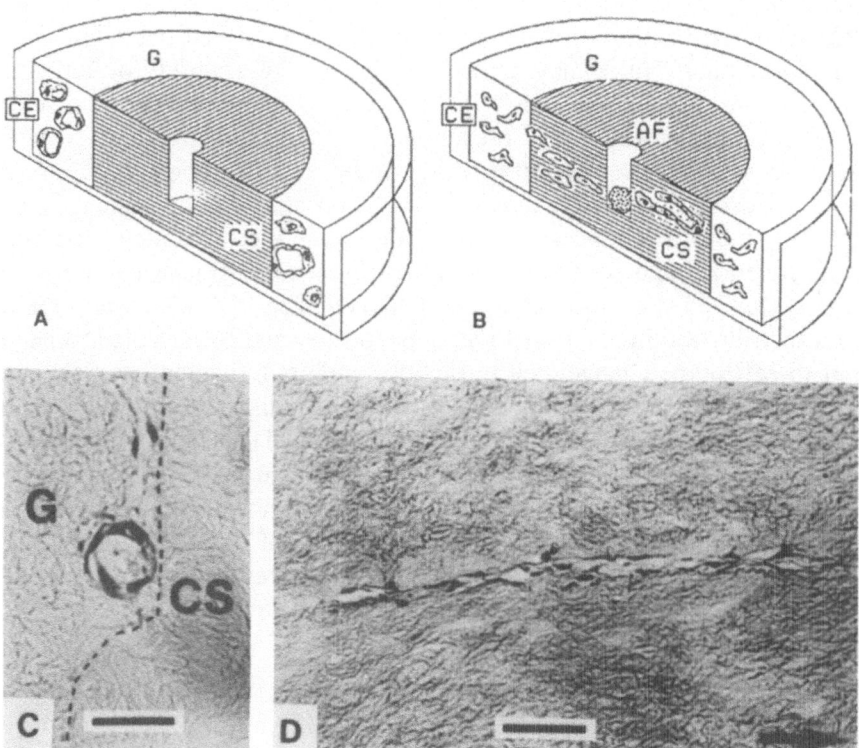

Figure 8. Schematic diagram of capillary endothelial cells (CE) seeded into a collagen gel (G) surrounding a plug of bovine corneal stromal tissue (CS). (A) In the absence of an angiogenic factor the
cells form luminal structures within the collagen gel but do not enter the corneal stroma. (B) In the
presence of an angiogenic factor (AF) in a slow-release polymer pellet placed in the center of the
corneal stroma, the capillary endothelial cells migrate into the corneal stroma and form tubelike structures parallel to the collagen lamellae of the corneal stroma. (C) Light micrograph of a hematoxylin-
and-eosin-stained 6-μm section of cultured rat epididymal fat pad capillary endothelial cells (RFC)
seeded in a three-dimensional collagen gel surrounding a 5-mm plug of bovine corneal stromal tissue.
In the absence of angiogenic factor, RFC form luminal structures in the collagen gel (G) but do not
enter the corneal stroma (CS) delimited by the dashed line. Following implantation of angiogenic factor,
the RFC can be observed migrating into the corneal stroma from the periphery in planes parallel to
the orthogonal arrays of collagen bundles. RFC are also observed to form tubelike structures having
lumina. Arrow indicates the direction of migration. Bars = 50 μm.

exhibited a higher proliferation rate than similar cultures plated on the basement
membrane components type IV and V collagen.[29] However, the cells grown on
the basement membrane components formed characteristic "tubelike structures"
at early time periods in culture as compared to similar cells grown on the interstitial
collagens. In addition to the above-mentioned differences in proliferation rate and
multicellular organization, the cultured capillary endothelial cells exhibited markedly different collagen synthetic profiles when cultured on the two classes of
matrices. Namely, the cells were observed to produce more basement membrane

collagens when they were grown on heterologous basement membrane components and were observed to be differentiated into tubelike structures.[29,35]

In other studies, Milici and Carley have demonstrated that composition/organization of the extracellular matrix can modulate number of fenestrae and transendothelial channels produced by cloned bovine adrenal cortical capillary endothelial cells (a fenestrated capillary bed *in vivo*). In addition, they have demonstrated that cells derived from a continuous capillary bed (epididymal fat pad) do not form fenestrae in culture even when grown on the extracellular matrix (MDCK cell matrix) that elicits maximal fenestra formation in the adrenal capillary endothelial cells.[2,30] These studies demonstrate that a vascular bed specific differentiated phenotype, e.g., fenestrae, can be selectively elicited by particular matrix composition/organization.

Since capillary endothelial cells normally exist in a three-dimensional environment, two-dimensional tissue culture systems may not adequately approximate the *in vivo* condition. Therefore, several three-dimensional tissue culture systems have been developed which allow evaluation of the roles of extracellular matrix components in angiogenesis.[27,32,35] One tissue culture system that has been used with some success which approximates an "intact biological matrix" is the acellular amnion. Capillary endothelial cells grown on the basement membrane aspect of the amnion form tubelike structures at short time periods in culture.[29] The cells comprising these structures exhibit abluminal and luminal plasma membrane domain specializations as well as a highly organized cytoskeleton. In contrast, cells grown on the stromal aspect of the amnion achieve a high cell density and migrate into the stroma, where, with time, they are observed to form occasional tubelike structures having complex plasma membrane junctional complexes, and eventually produce an abluminal basal lamina-like material.[35] Another three-dimen-

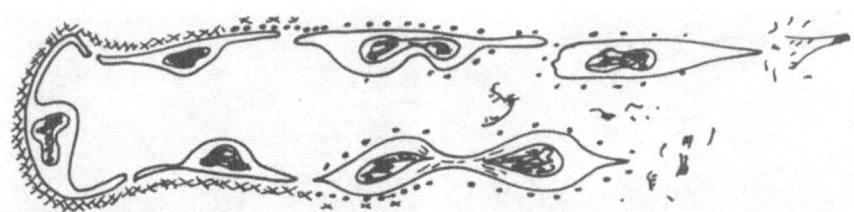

Figure 9. Schematic diagram of the temporally ordered appearance of laminin (Ln) and type IV collagen (IV) noted during the angiogenic response. Following stimulation, endothelial cells lining locally affected microvessels degrade their investing basement membrane and migrate into the surrounding stroma toward the injury site in response to a multitude of factors. Following the migratory response the cells undergo a proliferative response. Migration and proliferation are observed at and near the distal tip of the newly forming vessels, while nearer the original vessels the endothelial cells have a lower proliferative rate and organize into tubelike structures having luminal and abluminal differentiation. Endothelial cells in the more distal regions have been noted to express Ln (dots) but not IV and no morphologically identifiable basement membrane is noted in this region. In contrast, endothelial cells closer to the parent vessels express Ln *and* IV (cross-hatched design), and form a morphologically identifiable basement membrane. Thus, the appearance of Ln alone can be correlated with a high proliferative and migratory rate while the presence of IV and the appearance of a basement membrane correlates with the expression of a differentiated phenotype (tube formation).

sional matrix culture system that has been used successfully by several investigators in culturing capillary endothelial cells is collagen gel.[32] Capillary endothelial cells cultured in such gels form tubelike structures rapidly. As observed in the cultures on the amnion, the cells exhibit specialized luminal and abluminal features including the formation of a morphologically identifiable basal lamina. While useful, these culture systems have several shortcomings including difficulty in manipulation and eliciting directed angiogenesis. A system that has been used in our laboratory to investigate the effects of angiogenic factors as well as to study the effects of preexisting and newly synthesized matrix during angiogenesis involves the use of bovine cornea and collagen gels. As illustrated in Fig. 8, acellular bovine corneal plugs (5 mm diameter) are placed in the bottoms of the 16-mm diameter wells of 24-well cluster plates. Following this, a suspension of capillary endothelial cells in a type I collagen solution is added around the corneal plug and the collagen

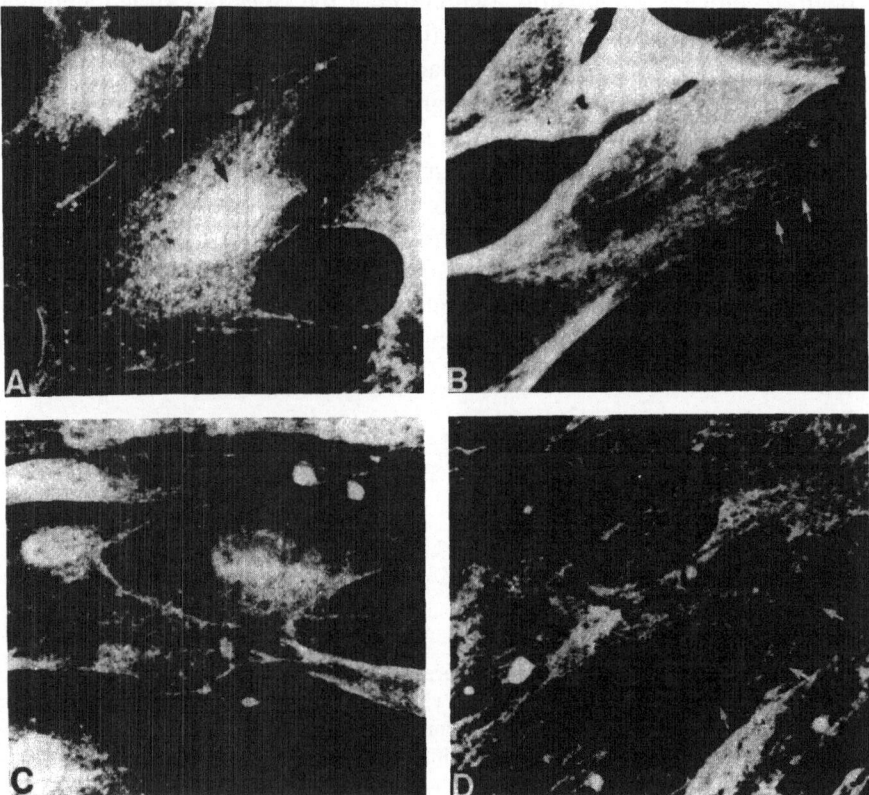

Figure 10. Immunofluorescence localization of fodrin and protein 4.1 in fixed and permeabilized capillary endothelial cells (RFC) grown on various matrix components. RFC were grown to confluency on bacteriologic plastic petri dishes coated with either fibronectin (Fn), type I/III collagen (I/III), type IV collagen (IV), or laminin (Ln). The cultures were fixed with 3.5% fresh paraformaldehyde, permeabilized with 0.2% Triton X-100, then stained with polyclonal antibodies to protein 4.1 or fodrin. Representative fields of RFC stained for 4.1 after culture on matrices of (A) Fn, (C) I/III, (e) IV, or (G) Ln. Protein 4.1 localizes in diffuse though slightly granular patterns and staining extends throughout

is allowed to gel. In the absence of angiogenic factor, the endothelial cells form luminal structures rapidly in the collagen gel but do not enter the densely packed and highly ordered corneal stroma. If angiogenic factor is placed in the center of the corneal plug, endothelial cells migrate into the corneal stroma and form tube-like structures. This system has several advantages over the other systems mentioned, namely, a defined spatial starting point (the corneal stroma/collagen gel interface) and directed angiogenesis into a cell-generated isotropic matrix.

2. Matrix Components as Solid-Phase Autocrine/Paracrine Factors

While informative, these tissue culture systems have a complex end point, angiogenesis, which presents difficulties in analysis and interpretation. One difficulty encountered, relevance to the *in vivo* state, can be tested by the use of *in*

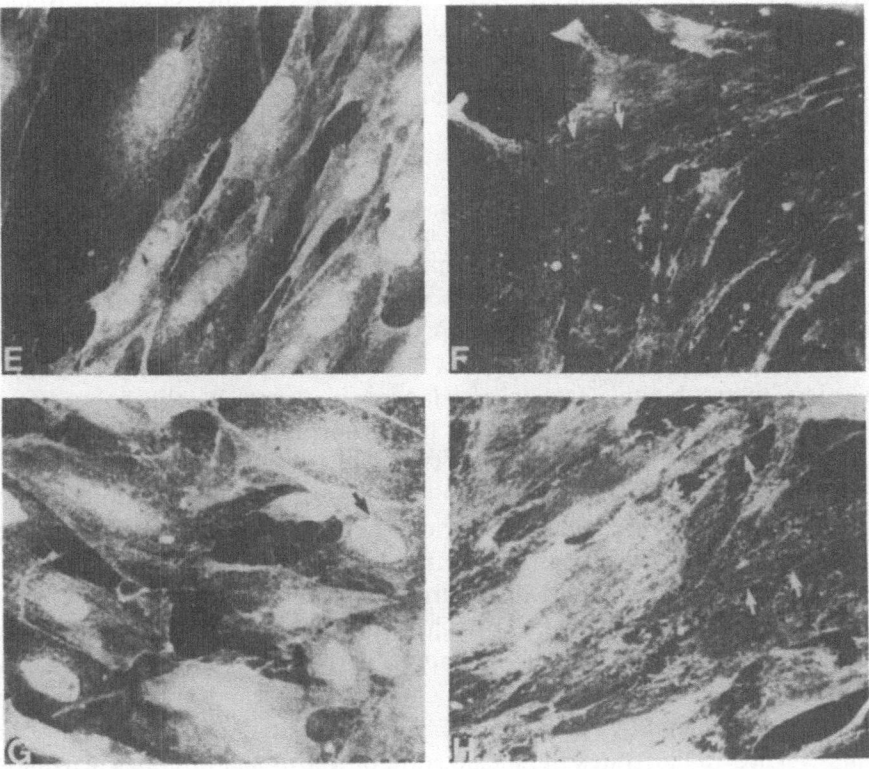

Figure 10. (continued)   the cell, with prominent nuclear staining (solid arrows in A, C, E, and G). This fluorescence pattern is not significantly altered by changing the substratum. RFC grown on (B) Fn, (D) I/III, (F) IV, or (H) Ln and stained with a polyclonal antibody to fodrin. Fodrin localization differs markedly from that of protein 4.1, most notably in the lack of nuclear staining. Fodrin is distributed throughout the cell body in a dense, granular pattern that tends to be organized in linear arrays. Solid arrows highlight selected prominent arrays in B, D, F, and H. Original magnification, × 400.

vivo models of angiogenesis such as corneal neovascularization and the chick chorioallantoic membrane models. A second problem, the understanding of the roles of existing and newly synthesized matrix components in the process of angiogenesis, is more difficult to evaluate. In this instance, detailed analyses of cell–matrix interactions in the various phases of the phenomenon of angiogenesis (activation, migration, proliferation, differentiation, stabilization, and regression) must be performed.[27] We have begun to examine the effects of solid-phase extracellular matrix components on proliferation and the roles of matrix binding proteins in this process.[6,7] Specifically, we have found that certain basement membrane components (laminin) elicit/permit high proliferative rates in cultures of rat epididymal fat pad and bovine adrenal cortical capillary endothelial cells, while others (type IV collagen) elicit/permit lower proliferative rates in otherwise identical cultures. In addition, we have found that either of these solid-phase components can modulate the proliferative rate of capillary endothelial cells grown in the presence of the other component. Furthermore, these *in vitro* observations have been correlated with the temporally ordered appearance of these two matrix components in an *in vivo* model of angiogenesis (silver nitrate-induced corneal injury), suggesting a potential role for these components as solid-phase modulators of cell proliferation and differentiation during the angiogenic response (Fig. 9).[7]

3. Cell Surface Matrix Binding Proteins

A large body of data gathered in our laboratory and in many other laboratories have demonstrated unequivocally that matrix influences endothelial cell behavior *in vitro*.[27] However, the mechanism by which the information present in the composition and organization of the matrix is transferred to the cellular machinery is as yet unknown. Therefore, another area that we have been actively investigating is the role of matrix binding proteins (for laminin and type IV collagen) in affecting microvascular endothelial cell behavior by organization of cortical and filamentous cytoskeletal elements. In preliminary studies, in collaboration with Dr. Lance Liotta of the NIH, using the monoclonal antibody LR2, we have found laminin receptor to be present on capillary endothelial cells.[44] In microvascular endothelial cells cultured at low density, the laminin receptor is localized in a diffuse perinuclear pattern and continuous linear arrays that reorganize over time to linear punctate arrays as the cells achieve higher cell density. The linear arrays colocalize with actin microfilaments as visualized by NBD-phallacidin. We have also examined the distribution of the cortical cytoskeletal components fodrin and protein 4.1 in cultured microvascular endothelial cells plated on various matrix components. Figure 10 illustrates the localization of fodrin and protein 4.1 using specific polyclonal antibodies.[19,36] Intracellular localization of protein 4.1 is intense, revealing a diffuse but slightly granular immunofluorescence pattern and a prominent nuclear staining (note the solid arrows in Fig. 10A, C, E, and G). In contrast, staining for fodrin reveals no nuclear localization and a dense granular pattern that can be discerned as linear arrays (note the arrows in Fig. 10B, D, F, and H). These linear patterns appear to colocalize with actin microfilaments as detected using double-label immunofluorescence (data not shown). Unlike the large vessel

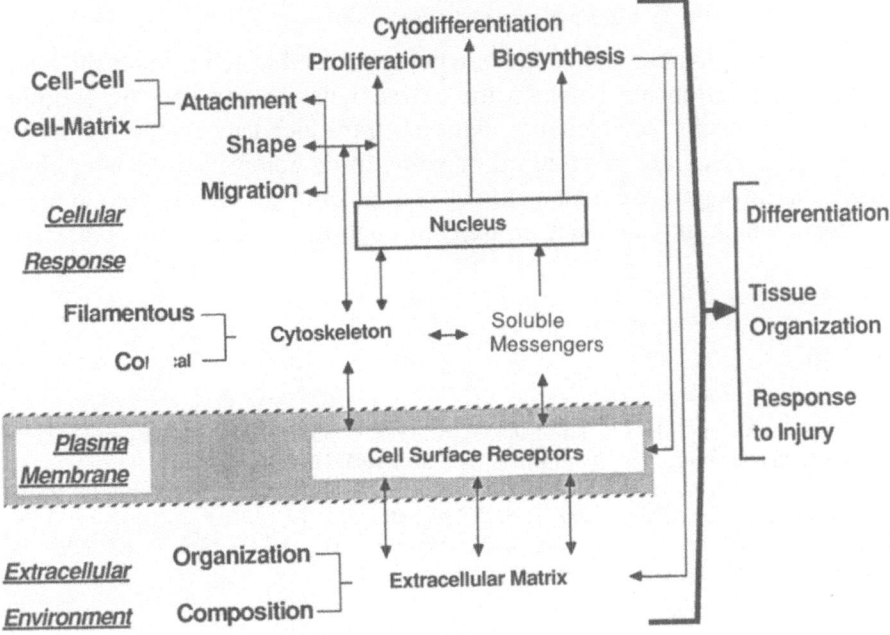

Figure 11. Schematic representation of cell–matrix interactions illustrating how information resident in the composition and/or organization of the extracellular matrix (ECM) could be transduced into the interior of the cell. In this scheme, cell surface matrix "receptors" would be organized by the ECM and they themselves, or in concert with subcortical proteins, e.g., fodrin, ankyrin, protein 4.1, vinculin, would: (1) organize the cortical and filamentous cytoskeleton, and (2) initiate the production of soluble second messengers. These changes would in turn permit/direct a cellular response which could include: (1) cytoskeletal rearrangements which would directly cause changes in cell–cell and cell–matrix attachment, shape, and migration, (2) cytoskeleton-mediated and soluble second messenger-mediated transduction of information to the nucleus resulting in cell proliferation, cytodifferentiation, and the biosynthesis and expression of new matrix "receptors" and ECM components. The ability of the cell to continually respond to and modify the ECM would allow for a flexible, coordinated cellular behavior resulting in differentiation, complex tissue organization, and response to injury.

cells, microvascular endothelial cells do not appear to modulate the intracellular distribution of protein 4.1 and fodrin in response to changes in matrix composition as detected at the light microscopic level. This difference may reflect: (1) utilization of different cytoskeletal proteins by large vessel and microvascular endothelial cells in interaction with cell surface matrix receptors, (2) subtle redistribution of these proteins which cannot be detected on a light microscopic level, or (3) an inappropriate tissue culture system which does not allow for "normal" cell surface receptor–cytoskeleton interactions. Immunoelectron microscopic localization of fodrin and protein 4.1 *in situ* and in various culture systems should provide more reliable information and better insight into the organization and functions of these proteins.

IV. CONCLUDING REMARKS

Thus, although incomplete, data are being accrued which support the concept of a dynamic relationship between the extracellular matrix and the endothelial cell. The mechanisms by which information is transduced across the plasma membrane to the nucleus are most certainly going to be complex, involving physical transduction through the cytoskeleton as well as through soluble second messenger systems which may also act, in part, through the cytoskeleton (Fig. 11).

REFERENCES

1. Brown, S. S., Malinoff, H. L., and Wicha, M. S., 1983, Connectin: Cell surface protein that binds both laminin and actin, *Proc. Natl. Acad. Sci. USA* **80**:5927–5930.
2. Carley, W., Milici, A. J., and Madri, J. A., 1986, Extracellular matrix specificity for the differentiation of capillary endothelial cells, *Fed. Proc.* **45**:1152.
3. Chen, W.-T., Hasegawa, E., Hasegawa, T., Weinstock, C., and Yamada, K. M., 1985, Development of cell surface linkage complexes in cultured fibroblasts, *J. Cell Biol.* **100**:1103–1114.
4. Fitzgerald, L. A., Charo, I. F., and Phillips, D. R., 1985, Human and bovine endothelial cells synthesize membranes similar to human platelet glycoproteins IIb and IIIa, *J. Biol. Chem.* **260**:10893–10896.
5. Folkman, J., and Haudenschild, C., 1980, Angiogenesis in vitro, *Nature* **32**:551–556.
6. Form, D. M., and Madri, J. A., 1985, Proliferation of microvascular endothelial cells in vitro: Modulation by extracellular matrix, *Fed. Proc.* **44**:1660.
7. Form, D. M., Pratt, B. M., and Madri, J. A., 1986, Endothelial cell proliferation during angiogenesis: In vitro modulation by basement membrane components, *Lab. Invest.* **55**:521–530.
8. Gotlieb, A. I., Spector, W., Wong, M. K. K., and Lacey, C., 1984, In vitro re-endothelialization: Microfilament bundle reorganization in migrating porcine endothelial cells, *Arteriosclerosis* **4**:91–96.
9. Gotlieb, A. I., Subrahmanyan, L., and Kalnins, V. I., 1983, Microtubule-organizing centers and cell migration: Effect of inhibition of migration and microtubule disruption in endothelial cells, *J. Cell Biol.* **96**:1266–1272.
10. Hay, E. D., 1982, *Cell Biology of Extracellular Matrix*, Plenum Press, New York.
11. Hayman, E. G., Pierschbacher, M. D., and Rouslahti, E., 1985, Detachment of cells from culture substrate by soluble fibronectin peptides, *J. Cell Biol.* **100**:1948–1954.
12. Hinter, H., Fritsch, P. O., Foidart, J.-M., Stingl, G., Schuler, G., and Katz, S. I., 1980, Expression of basement membrane zone antigens at the dermo-epibolic junction in organ cultures of human skin, *J. Invest. Dermatol.* **74**:200–204.
13. Horwitz, A., Duggan, K., Greggs, R., Decker, C., and Buck, C., 1985, The cell substrate attachment (CSAT) antigen has properties of a receptor for laminin and fibronectin, *J. Cell Biol.* **101**:2134–2144.
14. Ingber, D. E., Madri, J. A., and Jamieson, J. D., 1981, Role of basal lamina in neoplastic disorganization of tissue architecture, *Proc. Natl. Acad. Sci. USA* **78**:3901–3905.
15. Ingber, D. E., Madri, J. A., and Jamieson, J. D., 1985, Basement membrane as a spatial organizer of polarized epithelia. I. Neoplastic disorganization of pancreatic epithelium correlates directly with loss of basement membrane, *Am. J. Pathol.* **121**:248–260.
16. Ingber, D. E., Madri, J. A., and Jamieson, J. D., 1986, Basement membrane as a spatial organizer of polarized epithelia. II. Exogenous basement membrane regulates cell shape and reorients pancreatic epithelial tumor cells in vitro, *Am. J. Pathol.* **122**:129–139.
17. Kirschner, M., and Mitchison, T., 1986, Beyond self-assembly: From microtubules to morphogenesis, *Cell* **45**:329–342.

18. Kurkinen, M., Taylor, A., Garrels, J. I., and Hogan, B. L. M., 1984, Cell surface associated proteins which bind native type IV collagen or gelatin, *J. Biol. Chem.* **259:**5915–5922.

19. Leto, T. L., Pratt, B. M., and Madri, J. A., 1986, Mechanisms of cytoskeletal regulation: Modulation of aortic endothelial cell protein band 4.1 by extracellular matrix, *J. Cell. Physiol.* **127:**423–431.

20. Liotta, L., Rao, C. N., and Wewer, U., 1986, Biochemical interactions of tumor cells with the basement membrane, *Annu. Rev. Biochem.* **55:**1037–1057.

21. Lwebuga-Mukasa, J., Thulin, G., Madri, J. A., Barrett, C., and Warsaw, J., 1984, An acellular human amnionic membrane model for in vitro culture of type II pneumocytes: The role of the basement membrane on cell morphology and function, *J. Cell. Physiol.* **121:**215–225.

22. Lwebuga-Mukasa, J., Ingbar, D., and Madri, J. A., 1986, Repopulation of a human alveolar matrix by adult rat type II pneumocytes in vitro: A novel system for type II pneumocyte culture, *Exp. Cell Res.* **162:**423–435.

23. Maciag, T., Kadish, J., Wilkins, L., Stemerman, M. P., and Weinstein, R., 1982, Organizational behavior of human umbilical vein endothelial cells, *J. Cell Biol.* **94:**511–520.

24. Madri, J. A., 1982, Endothelial cell–matrix interactions in hemostasis, in: *Progress in hemostasis and Thrombosis* (T. H. Spaet, ed.), Grune & Stratton, New York, pp. 1–24.

25. Madri, J. A., Dreyer, B., Pitlick, F. A., and Furthmayr, H., 1980, The collagenous components of the subendothelium: Correlation of structure and function, *Lab. Invest.* **43:**303–315.

26. Madri, J. A., and Pratt, B. M., 1986, Endothelial cell–matrix interactions: In vitro models of angiogenesis, *J. Histochem. Cytochem.* **34:**85–91.

27. Madri, J. A., and Pratt, B. M., 1988, Angiogenesis, in: *The Molecular and Cellular Biology of Wound Repair* (R. Clark and P. Henson, eds.), Plenum Press, New York (in press).

28. Madri, J. A., and Stenn, K. S., 1982, Aortic endothelial cell migration. I. Matrix requirements and composition, *Am. J. Pathol.* **106:**180–186.

29. Madri, J. A., and Williams, S. K., 1983, Capillary endothelial cell cultures: Phenotypic modulation by matrix components, *J. Cell Biol.* **97:**153–165.

30. Milici, A. J., Furie, M. B., and Carley, W. W., 1985, The formation of fenestrations and channels by capillary endothelium in vitro, *Proc. Natl. Acad. Sci. USA* **82:**6181–6185.

31. Mollenhauer, J., and von der Mark, K., 1983, Isolation and characterization of a collagen-binding glycoprotein from chondrocyte membranes, *EMBO J.* **2:**45–50.

32. Montesano, R., Orci, L., and Vassakkum, P., 1983, In vitro rapid organization of endothelial cells into capillary-like networks is promoted by collagen matrices, *J. Cell Biol.* **97:**1648–1652.

33. Palotie, A., Tryggvason, K., Peltonen, L., and Seppa, H., 1983, Components of subendothelial aorta basement membrane: Immunohistochemical localization and role in cell attachment, *Lab. Invest.* **49:**362–370.

34. Porter, R., and Whelan, J., 1984, *Basement Membranes and Cell Movement*, Pitman, London.

35. Pratt, B. M., Form, D., and Madri, J. A., 1985, Endothelial cell–extracellular matrix interactions, in: *Biology, Chemistry and Pathology of Collagen* (R. Fleischmajer, B. R. Olsen, and K. Kuhn, eds.), New York Academy of Sciences, New York, pp. 274–288.

36. Pratt, B. M., Harris, A. S., Morrow, J. S., and Madri, J. A., 1984, Mechanisms of cytoskeletal regulation: Modulation of aortic endothelial cell spectrin by extracellular matrix, *Am. J. Pathol.* **117:**349–354.

37. Pytela, R., Pierschbacher, M. D., and Ruoslahti, E., 1985, Identification and isolation of a 140kD cell surface glycoprotein with properties expected of a fibronectin receptor, *Cell* **40:**191–198.

38. Rao, N. C., Barsky, S. H., Terranova, V. P., and Liotta, L. A., 1983, Isolation of a tumor cell laminin receptor, *Biochem. Biophys. Res. Commun.* **111:**804–808.

39. Reidy, M. A., and Silver, M., 1985, Endothelial regeneration. VII. Lack of intimal proliferation after defined injury to rat aorta, *Am. J. Pathol.* **118:**173–177.

40. Ruoslahti, E., and Pierschbacher, M. D., 1986, Arg-gly-asp: A versatile cell recognition signal, *Cell* **44:**517–518.

41. Wong, A. J., Pollard, T. D., and Herman, I. M., 1983, Actin filament stress fibers in vascular endothelial cells in vivo, *Science* **219:**867–869.

42. Woods, A., Hook, M., Kjellen, L., Smith, C. G., and Rees, D. A., 1984, Relationship of heparan

sulfate proteoglycans to the cytoskeleton and extracellular matrix of cultured fibroblasts, *J. Cell Biol.* **99**:1743–1753.

43. Yannariello-Brown, J., and Madri, J. A., 1985, Aortic endothelial cells synthesize specific binding proteins for laminin and type IV collagen, *J. Cell Biol.* **101**:333a.

44. Yannariello-Brown, J., Tchao, N. K., Liotta, L., and Madri, J. A., 1985, Co-distribution of the laminin receptor with actin microfilaments in permeabilized aortic and microvascular endothelial cells, *J. Cell Biol.* **101**:333a.

V

Endothelial Cell Anticoagulant and Fibrinolytic Activities

9

Assembly and Function of the Protein C Anticoagulant Pathway on Endothelium

Charles T. Esmon

I. INTRODUCTION

Many aspects of the protein C anticoagulant pathway are very unusual. Although the system functions as an anticoagulant, two of the necessary components, proteins C and S, depend on vitamin K for their biosynthesis.[15,20,22,24,67,76,80] Thus, oral anticoagulants not only inhibit the coagulation pathways, but also inhibit the anticoagulant pathway. This has been associated with a rare complication of oral anticoagulant therapy, coumarin-induced skin necrosis, in which the necrotic area is characterized by fibrin deposition in the microvasculature. In several patients examined retrospectively, low levels of protein C have been observed which probably contributed to the coagulopathy.[8,61,73]

A second unusual property is that protein S, although vitamin K dependent, does not function as a precursor to a serine protease,[19,84] but serves instead as a binding protein which complexes with activated protein C to enhance the association of both proteins with membrane surfaces.[42,78] In addition, protein S circulates both free and bound to the regulatory protein of the complement system, C4b-binding protein.[17,18] Although the function of this complex in the complement system is unknown,[17] the complex does not function in the anticoagulant pathway.[14]

A third unusual feature of the pathway is that thrombin, the terminal enzyme of the coagulation pathway, triggers the protein C anticoagulant pathway. To accomplish this, complex formation between thrombin and thrombomodulin (TM) alters the macromolecular specificity of thrombin. The central features of the alteration are (1) protein C activation is enhanced,[27] (2) TM blocks fibrinogen,[28,43,44,56] factor V,[28] and platelet[34,56] activation and inhibits protein S inactivation,[29] and (3) TM does not alter significantly the rate of antithrombin III inhibition of thrombin,[44] although in some settings thrombin inhibition may be enhanced.[43]

The overall features of the system suggest that it could constitute a focal

9

9

9

Charles T. Esmon • Thrombosis/Hematology Research Program, Oklahoma Medical Research Foundation, Oklahoma City, Oklahoma 73104.

point where modulation of the expression of any one of the components, including the cell surface receptors, could lead to a transient imbalance predisposing toward a thrombotic state. The major focus of this review is to develop the hypothesis that both humoral and cellular components of this system are altered in disease states leading to an increased propensity for thrombosis. To develop this concept, it is useful to briefly review the development of our current understanding of the system. It is not the intent of this brief chapter to provide a comprehensive review of the literature. For those interested in more comprehensive reviews, these have been published recently (Seegers,[75] Esmon,[25] Esmon and Esmon,[26] Dahlback[17] and other chapters therein, Laemmli and Griffin,[50] Clouse and Comp[11]).

II. PROTEIN C—STRUCTURAL DOMAINS

In human plasma, protein C circulates at approximately 3–5 μg/ml. Both our own unpublished work and that of Miletich et al.[58] have shown that protein C exists in multiple forms: single chain ($\approx$ 10% of total) and two chains with two subforms, α ($\approx$ 70%) and β ($\approx$ 20%). In the single-chain form, a Lys-Arg dipeptide bridges between the chains of the two-chain form. This Lys-Arg dipeptide can be released upon activation. The chemistry of protein C activation is relatively simple as shown in Fig. 1. With two-chain protein C, activation involves cleavage of protein C between an Arg at residue 12 and an Ile at residue 13.[2,46] The active site of this serine protease is located on the heavy chain and both the Gla residues and the β-hydroxyaspartic acid are located on the light chain. The β-hydroxyaspartic acid is at residue 71 which is part of the region sharing homology with epidermal growth factor.[2,23,36,77] The Gla residues are essential for optimal calcium binding and although the function of the β-hydroxyaspartic acid is uncertain, it has been implicated in binding either Ca^{2+} [59,79] or Fe^{2+}.[37]

One extremely convenient and useful structural feature of protein C is that there is a hinge region composed of hydrophobic residues that links the Gla domain to the EGF region of the light chain. This region is extremely sensitive to proteolytic cleavage with chymotrypsin.[33] This provides a method to rapidly and selectively remove the Gla domain. The resultant form of protein C, referred to as "Gla domainless protein C" or "GD-PC," can still be converted to the

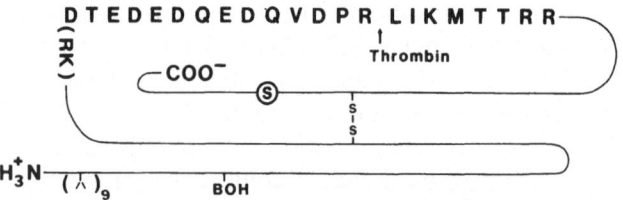

Figure 1. Human protein C structure. λ, γ-carboxyglutamic acid; BOH, β-hydroxyaspartic acid; s, the active site serine. The activation site is indicated by the arrow. The two amino acids in parentheses correspond to the Lys-Arg residues that join the two chains in single-chain protein C. The sources used to derive this figure are Kisiel[46] and Beckmann et al.[2]

active serine protease as predicted from the structural information given in Fig. 1. This modification renders the protein incapable of interacting with membranes with high affinity and thus provides a useful reagent for probing the function of the endothelial cell membrane in protein C activation as will be discussed below.

III. EXPRESSION OF PROTEIN C ANTICOAGULANT ACTIVITY

Activating protein C alone is insufficient to elicit anticoagulant activity. Although activated protein C is the protease involved in inactivating factors Va and VIIIa, the two target proteins whose inactivation is responsible for the anticoagulant effect,[38,48,55,66,81,86] the activated protein C functions too slowly by itself to be effective. *In vitro*, additional requirements include protein S and a membrane surface either in the form of liposomes,[84] the platelet,[42] or the endothelial cell[78] surface. Thus, protein C is not only activated on the endothelial cell surface, but one of the sites of expression is also on this cell surface. It is interesting to note that the endothelial cell appears to bind both protein S and activated protein C with at least 10 times higher affinity than the platelet,[42,78] suggesting that this surface may be very important in expression of anticoagulant activity. Very recently, Walker[85] reported in an extremely provocative paper that an additional factor (protein S binding protein) is essential for optimal expression of activated protein C anticoagulant activity. Since many of the properties of the endothelial cell protein S receptor are suggestive of an endothelial cell protein involved in protein S binding (see below), it is tempting to speculate that the newly described protein S binding protein may be related to the endothelial cell surface protein S receptor. Proof of this will probably have to await the availability of monospecific antibodies to one or both proteins.

IV. DISCOVERY OF TM

The central problem with protein C as a potential anticoagulant was that little[71] or no protein C[47] is activated when blood clots *in vitro*. Furthermore, depleting protein C from plasma has no effect on standard coagulation assays.[13] These are both conditions where coagulation is essentially uncontrolled and thrombin generation is very high. Thus, protein C is activated too slowly in blood to function as a natural anticoagulant. Therefore, if protein C were to function as a natural anticoagulant, either an alternative enzyme had to be the activator, or thrombin activity toward protein C had to be augmented by some acceleratory factor. The possible existence of such an acceleratory factor was suggested by the observation that all of the proteases leading to thrombin formation involve both a regulatory protein (cofactor) and a surface on which to assemble. This concept initiated studies to find the regulatory protein. Since little protein C was activated *in vitro* when blood clotted, we assumed that the factor must reside on

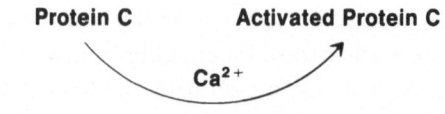

Figure 2. Assembly of the endothelial cell surface protein.

a vascular cell. Given the high concentration of endothelium *in vivo,* the endothelial cell was a reasonable candidate. The hypothesis is illustrated in Fig. 2.

The critical feature of this model hinges on the concept that thrombin must be bound for it to function as an effective protein C activator. If the TM concentration is very low, relatively little thrombin could bind and hence little protein C activation would occur. If one assumes the TM density is constant throughout the vasculature, then since the surface area of endothelium (and the number of cells exposed to the blood) increases approximately 1000-fold when blood moves from the macro- to the microcirculation,[10] the ability to form the thrombin–TM complex would also increase approximately 1000-fold as blood reaches the microvasculature. Thus, unlike the reaction catalyzed by the platelet surface, the efficiency of this system would be strongly dependent on the location in the vasculature. To optimally test the original hypothesis, it was useful to design a system in which thrombin and protein C would be exposed to the maximum endothelial cell surface possible. One system to achieve this is the Langendorff heart preparation in which perfusate is pumped through the coronary microcirculation. When thrombin was perfused through the microcirculation in the presence of protein C, anticoagulant activity emerged (Table 1). Neither thrombin nor protein C alone would elicit formation of this anticoagulant. Comparing the rate of protein C activation in the presence and absence of the coronary microcirculation revealed that perfusion enhanced protein C activation approximately 20,000 fold.[27]

The studies in the perfused heart were subsequently confirmed by kinetic analysis of protein C activation over cultured endothelial cells.[69] From analysis with active site-blocked thrombin derivatives,[69] it was clear that interaction of thrombin with this receptor was not dependent on the active site. With this information, it was possible to design an isolation procedure for TM.[32] The major

Table 1. Results of the Langendorff Heart Perfusion[a]

Perfusate	CT (sec)	Approximate concentration of activated protein C
Buffer ± perfusion	23	0
Protein C alone	23	0
Thrombin alone	22	0
Thrombin + protein C (4 sec)	60	0.5 μg/ml
Thrombin + protein C (30 min; no perfusion)	23	0

[a] Data derived from Esmon and Owen.[27]

Table 2. Properties of Isolated Rabbit Thrombomodulin[a]

Molecular weight	74,000
Detergent requirement for solubility	Yes
Stoichiometry with thrombin	1:1
Dependent on Ca^{2+} for protein C activation	Yes
K_d for thrombin	≤ 0.5 mN
K_m for protein C	8 μM
Inhibited by DIP-thrombin	Yes ($K_i = 0.5$ nM)

[a] Data from Owen and Esmon[69] and Esmon et al.[32,33]

step in the isolation involved immobilizing thrombin. The initial isolation was from detergent extracts of rabbit lung,[32] but since that time TM has been isolated from rat lung,[28a] bovine lung,[44] and human placenta[72] utilizing essentially similar methodologies. The properties of isolated rabbit TM are summarized in Table 2.

The isolation of TM allowed preparation of antibodies which could be used in studying the cellular specificity of TM. Both assays of the cellular specificity[69] and immunofluorescence studies in human tissue[57] and rabbits[21] demonstrate that TM is an endothelial cell-specific protein. The only exception noted was that it also is present on the syncytiotrophoblast of human placenta.[57]

V. COMPARISON OF PURIFIED TM WITH THE ENDOTHELIAL CELL SURFACE

The major difference between the activity on the endothelial cell surface and that observed with the purified TM is the affinity for the substrate. On the cell surface, this affinity is increased about 10-fold, suggesting that incorporation of TM into the endothelial cell surface expresses a substrate binding site. Recent studies[40] with purified rabbit TM have demonstrated that incorporation of TM into phosphatidylcholine vesicles revealed that even these synthetic vesicles which lacked surface charge allowed expression of the high-affinity substrate binding site. These results differ from those of Freyssinet et al.[39] who observed with human TM only a change in V_{max} (3.2-fold) and this was dependent on negatively charged lipids. These differences could either be species differences or, more likely, differences in TM structure and reconstitution methodology. Our recent results led us to modify our earlier model[26] to indicate that this site is tightly associated (probably covalently) with TM (Fig. 3). Further evidence that this site is associated with TM directly has been obtained recently. Kurosawa et al.[49] have shown that this substrate binding site can also be exposed by limited proteolysis of TM and hence the binding site is not a unique property of the endothelial cell surface, but rather constitutes a domain of the TM molecule. The central features of the mechanisms of TM action are illustrated in Fig. 3. Thrombin (T) interacts to form a 1:1 complex with TM. Interaction with thrombin is not dependent on Ca^{2+}. The latter has at least two distinct effects on the activation: (1) to induce or stabilize a protein C conformation that can be recognized by the T–TM

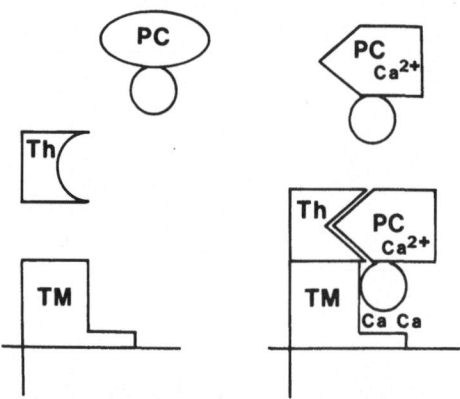

Figure 3. A model of protein C activation. The model represents our hypothesis that complex formation with TM alters the substrate recognition site and that TM has a direct binding site for the Gla domain of protein C.

complex[45] and (2) to enhance protein C binding to the surface. The first effect is mediated through a Gla-independent Ca^{2+} binding site.[33,45] Upon interaction with TM, the specificity of thrombin is altered. Whether this is mediated by steric effects[43,44] or a conformational change in the active site of thrombin[29,45] is unknown. We favor the conformational model because of (1) the specificity of the complex for the Ca^{2+}-stabilized conformer of protein C and (2) recent studies demonstrating that the environment near the active site serine is altered by complex formation with TM.[29]

VI. INFLUENCE OF TM ON THROMBIN SPECIFICITY

Regardless of the mechanism by which TM changes the specificity of thrombin, the change in specificity has interesting physiological implications. Upon complex formation with TM, fibrinogen clotting activity, factor V activation and platelet activation, and protein S inactivation are all decreased significantly, if not completely.[32,43,44,56] Recently, it has been shown that fibrinogen and TM are competitive with respect to thrombin binding.[43,44] The alteration in substrate specificity is summarized in Table 3.

It is clear from the summary that the overall influence of TM is to function as a vascular anticoagulant both by neutralizing thrombin and by accelerating protein C activation resulting in the generation of anticoagulant activity. The role

Table 3. Summary of Effects of Thrombomodulin on Thrombin Reactions[a]

Thrombin-catalyzed reactions inhibited by TM	Thrombin-catalyzed reactions altered relatively little by TM	Thrombin-catalyzed reactions accelerated by TM
Fibrinogen	ATIII	Protein C activation
Factor V	Protein C activation (in the absence of Ca^{2+})	

[a] References are cited in the text.

with respect to thrombin neutralization by antithrombin III is beginning to become clear. In the absence of alternative substrates, thrombin inhibition by antithrombin III has been reported to be either accelerated about 4-fold[43] or unaffected.[26,44] In either case in the plasma setting, thrombin inhibition would actually be enhanced at least 3- to 4-fold due to masking the binding site for alternative substrates, primarily fibrinogen.[44] Thus, TM functions to enhance thrombin inactivation by two distinct mechanisms: (1) instantaneous neutralization of thrombin's procoagulant functions and (2) acceleration of the reaction with antithrombin III either by directly accelerating inactivation[6,43] or by masking the protective effects of alternative substrates.[68]

VII. VARIATION OF TM PROPERTIES BETWEEN CELL LINES

At least *in vitro*, cultured cells of either bovine or rabbit origin can lose their capacity to interact in a high-affinity fashion with protein C.[30] The TM on many of the cell lines with reduced affinity for protein C was also found to have altered antigenic properties and reacted differently with both polyclonal and monoclonal antibodies. These findings are consistent with, but do not prove, the hypothesis that the substrate binding site is added to TM as a posttranslational modification. The chemical nature of this putative site is the subject of current investigation. This is also supported by the recent finding that heparinase can alter TM function.[6] Whether the differences are due to addition of heparinlike substances remains to be determined.

VIII. EXPRESSION OF ACTIVATED PROTEIN C ANTICOAGULANT ACTIVITY

Since its discovery,[60] membrane surfaces have been recognized as playing an essential role in the anticoagulant function of activated protein C. Most *in vitro* studies utilize negatively charged phospholipid membranes which also function as surfaces for the assembly of coagulation complexes.[54,67] Whether significant negatively charged phospholipid surface is accessible to plasma proteins under physiological conditions is speculative. Therefore, studies were initiated to determine if cells could support activated protein C anticoagulant activity. Both platelets[42] and the endothelial cell surface[78] were found to possess a limited number of specific binding sites for activated protein C. In both cell types, the binding required the presence of protein S. On bovine endothelium, the binding of activated protein C was approximately 200 times tighter than on the platelet surface (0.04 versus 11 nM). This observation combined with the trypsin sensitivity of the endothelial cell site suggests the possible existence of a binding protein. These observations are summarized in Fig. 4. Walker[85] has recently described a plasma protein S binding protein. The relationship between this protein and the cell surface binding site is unclear.

Formation and Function of Activated Protein C

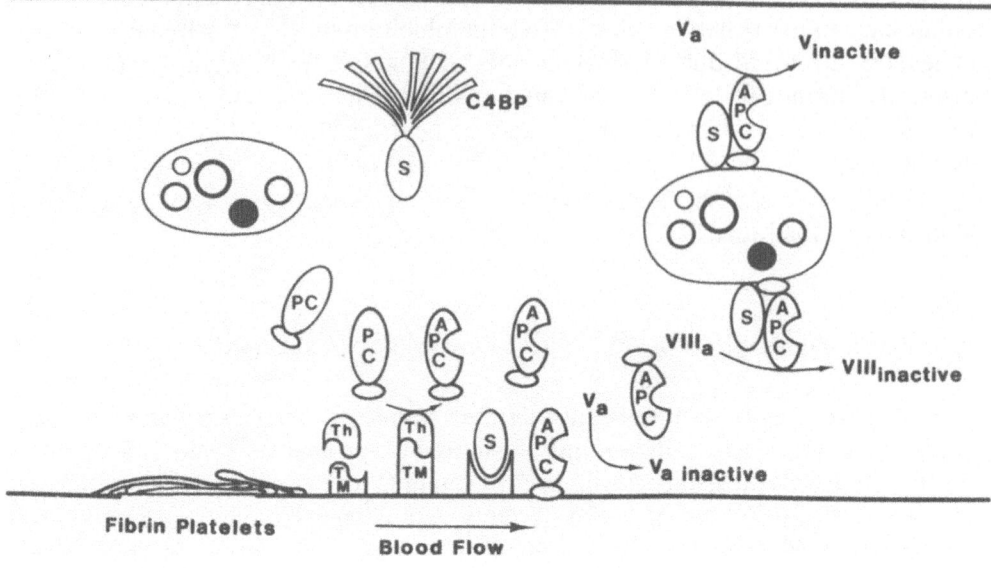

Figure 4. A model of the location and assembly of the complexes regulating coagulation. Reprinted from Esmon[24a] with permission of American Association for the Advancement of Science.

IX. REGULATION OF ENDOTHELIAL CELL TM ACTIVITY DURING INFLAMMATORY RESPONSES

One event which clearly alters the normal hemostatic balance is the inflammatory response. In septic shock, for instance, coagulation is no longer effectively controlled and disseminated intravascular coagulation results. Although tissue factor has long been recognized to be expressed during inflammation, the levels observed are rather low on both the monocyte[51,70] and endothelial cell[9,12,53] surface. We felt that one mechanism to enhance the coagulant effect of the tissue factor would be to depress the anticoagulant influence of the endothelial cell surface. Potential candidates for down-regulation were TM and the protein S binding sites. Three separate inflammatory stimuli have been examined at present: interleukin 1 (IL-1) tumor necrosis factor (TNF), and endotoxin. IL-1 had been shown by Gimbrone's group to initiate both tissue factor synthesis[4] and leukocyte binding site formation[5] and was therefore chosen as an initial inflammatory agent to investigate. Infusion of IL-1 into rabbits resulted, as expected, in formation of tissue factor activity on the aortic endothelium. Both the capacity of the endothelium to activate protein C and to catalyze factor Va inactivation were significantly reduced.[64] Interestingly, whereas tissue factor activity was maximal at 4–5 hr after injection and began to decline after 15 hr, both TM and the factor Va inactivation site were optimally depressed by 5–8 hr and remained low for at least 24 hr. This suggests that after a major inflammatory stimulus, the protein C an-

Table 4.   Changes in Endothelial Cell Coagulant Properties Following Exposure to
Endotoxin, IL-1, or TNF[a]

	Effect on tissue factor	Effect on TM	Effect on protein S–protein C function
Endotoxin	Increase (optimum at 4 hr)	Decrease to ~ 50% (maximum at 6–12 hr)	ND[b]
IL-1	Increase (optimum at 4–6 hr)	Decrease to 33% (maximum at 6 hr)	Decrease to < 10% (maximum at 6–10 hr)
TNF	Increase (optimum at ~ 12 hr)	Decrease to 20–30% (maximum at 8–12 hr)	Decrease to < 10% (maximum at 8–12 hr)

[a] Data taken from Moore et al.[62] for endotoxin; Nawroth et al.[64] for IL-1; and Nawroth and Stern[63] for TNF.
[b] ND, not determined.

ticoagulant pathway remains impaired for extended periods of time. This may be partially responsible for the response in the Schwartzman reaction.

To determine if the IL-1 effect was unique, TNF and endotoxin were examined. TNF had a similar response *in vitro* to that observed with IL-1 *in vivo*. Both the time course and extent of inhibition were similar.[63] This latter finding may be significant since Beutler et al.[3] have presented evidence suggesting that TNF may be one of the primary mediators of the lethal effects of endotoxin. This point will be developed in the section on clinical significance.

Although the changes in the hemostatic properties of the endothelium can be modulated by the monokines, endotoxin can have similar effects directly.[62] As with IL-1 and TNF, TM activity on human endothelial cells is depressed when they are exposed to endotoxin. With endotoxin, however, tissue factor expression precedes the decrease in TM activity. Because human endothelium expresses the protein S binding site to variable (and usually low) levels, it was not practical to study regulation of this site in culture. While the mechanism of these endotoxin-mediated changes is uncertain, the recent observation that endothelium can synthesize IL-1 when stimulated[65] suggests that the response may be mediated by endothelial synthesis of IL-1. If this proves correct, then many of the responses seen in septic shock may not necessarily be mediated only by macrophage-elaborated monokines.

The response of the endothelium to all three agents is remarkably similar (Table 4). From this similarity, a unified hypothesis can be drawn: Injury induces a unidirectional shift from the normal situation where anticoagulant activity predominates to a situation where coagulation is initiated and anticoagulation is compromised.

One question to arise from these studies is how reductions in TM and protein S binding sites would influence the hemostatic balance. While no one knows with certainty, my prejudice is that these changes would significantly alter the hemostatic balance. Reductions to the levels observed during inflammation probably result in decreased thrombin binding to the endothelium, decreased protein C activation, increased circulation of thrombin before inactivation, and impaired effectiveness of activated protein C. Taken together, this change in the protein

C anticoagulant pathway would likely be sufficient to initiate a severe coagulation imbalance that in turn could lead to either DIC or thrombosis, especially when tissue factor is present and inflammatory cells are adhering to the perturbed surface.

X. REVIEW OF THE PROTEIN C SYSTEM IN THROMBOTIC DISEASE

During the last 5 years, many reports of protein C and/or protein S deficiency in patients with recurrent thrombosis have appeared (for reviews see Refs. 11, 25, 41, 50). Although it is not clear that the low levels cause the thrombotic tendency, the high frequency of reduced protein C or protein S levels in familial thrombotic disease suggests that these deficiencies are a significant risk factor. At the present time, however, the extent of risk associated with the deficiencies is uncertain, especially since some family members with reduced levels are unaffected. Most of the patients described with homozygous protein C deficiency[7,74,87] have had severe thrombotic tendencies ultimately leading to death in infancy. Probably the best evidence that the protein C deficiency was a primary contributor to the observed thrombotic tendency comes from the studies[74] demonstrating that infusion of concentrates containing protein C into an infant with the deficiency prevented further thrombosis. It is interesting that neither heparin nor coumarin were equally effective. These observations suggest that protein C may trigger a unique response. Evidence for this hypothesis has recently been obtained in the baboon and will be reviewed below.

In addition to the hereditary deficiencies of these proteins which may constitute a significant risk of thrombosis, some very unusual acquired conditions may also contribute. One intriguing observation is that protein C levels decline very rapidly after the onset of anticoagulant administration.[20,24] Protein C levels decrease earlier than all other vitamin K-dependent proteins, except factor VII. This leads to the possibility of a transient imbalance in the coagulation system which in theory could lead to a thrombotic tendency. Although rare, a complication of coumarin therapy is skin necrosis. Several patients developing the microvascular coagulation have subsequently been shown to be protein C deficient.[8,61,73]

An additional mechanism of inhibition of the pathway involves neutralization of protein S activity. In normal individuals, protein S circulates both free and in complex with C4b-binding protein (C4b-BP). Only the free form functions in the anticoagulant pathway.[14] Patients have now been described with both hereditary[16] and acquired[15] abnormalities in the distribution of free and bound protein S. Among the situations where protein S is reduced is that of lupus erythematosus where chronic inflammatory insult is a common characteristic of the disease.[15] Recent studies suggest that C4b-BP is an acute-phase reactant[19,20] which may contribute to transient imbalance during a variety of traumatic challenges.

All of the above suggest that protein C, its activation and expression, are subject to intense regulation. From the limited available clinical studies, it is likely that protein C plays a critical role in regulation of coagulation. This prompted us

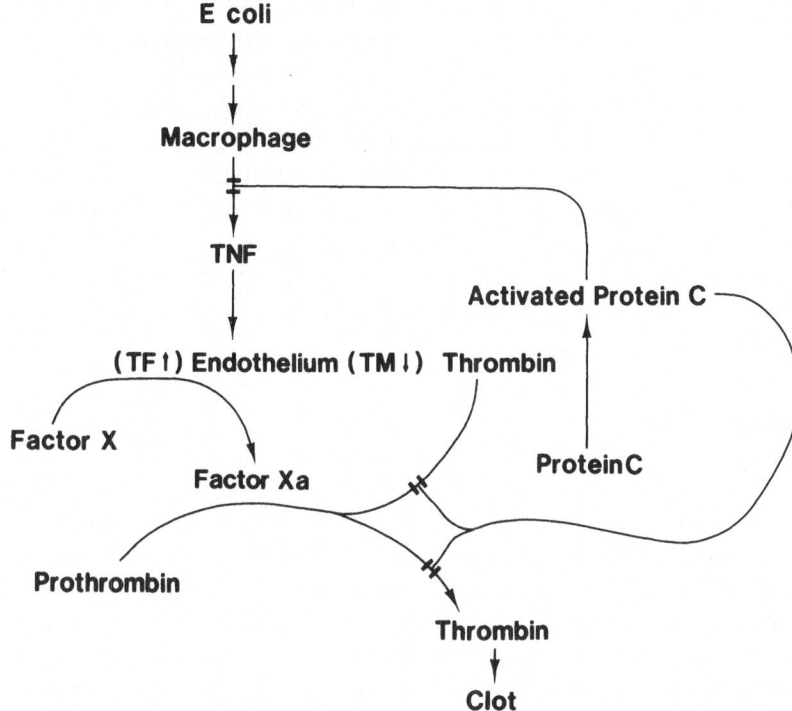

Figure 5. A potential pathway for the coordinate control of coagulation and inflammation by activated protein C. // indicate that the pathways of formation of the respective products are blocked by activated protein C.

to study the influence of protein C and activated protein C in the prevention of thrombosis and DIC. Preliminary experiments in the dog[1] demonstrate that activated protein C prevents thrombus extension without causing bleeding. In a different setting, activated protein C has now been shown to prevent both the coagulopathy and the lethal effects of *E. coli* in the baboon.[82] In addition to preventing coagulation in the baboon, activated protein C prevents elaboration of TNF.[82a] Thus, protein C activation and expression is not only controlled by TNF, but generation of activated protein C may control TNF formation. Potential control by these mechanisms is depicted in Fig. 5.

XI. PERSPECTIVES

As is apparent from this review, studies on protein C activation, expression, and the role of these processes in normal and pathological events are only recently beginning to emerge. Available data suggest that this pathway may be intensely regulated by the inflammatory process and that it may in turn regulate the inflammatory response. Clearly the details of the mechanism of involvement are yet to be established.

Given that protein C has been cloned[2] and functional expression has been achieved,[52] it should soon be possible to test the pharmacological effects of protein C in patients. When considering the multiple points in the pathway that are altered in disease states known to predispose toward thrombotic tendencies, I feel the promise of protein C is significant for managing severely affected patients.

The system presented here represents my best current understanding of how protein C and TM contribute to the regulation of coagulation. It is clear, however, that many new functions for the system are likely to emerge in the near future. As an example, the recent very provocative observation by Thompson and Salem[83] demonstrating that human TM also blocks prothrombin activation suggests a new function for the endothelium that transcends anything covered by this review.

XII. CONCLUDING REMARKS

The protein C anticoagulant pathway serves as a natural regulatory system to control coagulation. The critical steps in the pathway are (1) protein C activation and (2) assembly of protein S-activated protein C (APC) complexes on cell surfaces. Both the activation step and the expression of the anticoagulant activity occur on the endothelial cell surface. The activation of protein C involves complex formation between thrombin and thrombomodulin. Under physiological conditions, the complex catalyzes protein C activation at least 1,000 times faster than free thrombin. Complex formation not only enhances protein C activation, but also inhibits thrombin's procoagulant functions.

Modulation of these activities may constitute one mechanism for impaired regulation of coagulation. It is known that inflammatory stimuli can induce coagulation. Endotoxin and the monokines, tumor necrosis factor (TNF) and interleukin 1, can both lead to down regulation of thrombomodulin and the protein S binding sites. At similar times and concentrations, these vascular perturbants induce endothelial cell tissue factor activity. These findings may account in part for how endotoxin induces disseminated intravascular coagulation.

ACKNOWLEDGMENTS. The assistance of Dick Irish in the preparation of illustrations and of Karen Deatherage in the typing of the manuscript is greatly appreciated. The author's research discussed herein was supported by grants awarded by the National Institutes of Health (R01-30340 and R01-29807), and by an Established Investigator Award from the American Heart Association (80-167) with funds contributed in part by the Oklahoma Affiliate.

REFERENCES

1. Bang, N., Emerick, S., Yan, S., Long, G., Harms, C., Huss, C., Marks, C., Mattler, L., Comp, P., Esmon, N., and Esmon, C., 1986, Antithrombotic properties of activated human protein C, in: XXI Congress of the International Society of Haematology, Sydney, Australia, *Book of Abstracts*, p. 535 (abstract P-TH-132A-2).

2. Beckmann, R. J., Schmidt, R. J., Santerre, R. F., Plutzky, J., Crabtree, G. R., and Long, G. L., 1985, The structure and evolution of a 461 amino acid human protein C precursor and its messenger RNA, based upon the DNA sequence of cloned human liver cDNAs, *Nucleic Acids Res.* 13:5233–5247.

3. Beutler, B., Milsark, I. W., and Cerami, A. C., 1985, Passive immunization against cachectin/tumor necrosis factor protects mice from the lethal effect of endotoxin, *Science* 229:869–871.

4. Bevilacqua, M. P., Pober, J. S., Majeau, G. R., Cotran, R. S., and Gimbrone, M. A., Jr., 1984, Interleukin-1 (IL-1) induces biosynthesis and cell surface expression of procoagulant activity in human vascular endothelial cells, *J. Exp. Med.* 160:618–623.

5. Bevilacqua, M. P., Pober, J. S., Wheeler, M. E., Cotran, R. S., and Gimbrone, M. A., Jr., 1985, Interleukin 1 acts on cultured human vascular endothelium to increase the adhesion of polymorphonuclear leukocytes, monocytes, and related leukocyte cell lines, *J. Clin. Invest.* 76:2003–2011.

6. Bourin, M., Boffa, M., Bjork, I., and Lindahl, U., 1986, Functional domains of rabbit thrombomodulin, *Proc. Natl. Acad. Sci. USA* 83:5924–5928.

7. Branson, H., Katz, J., Marble, R., and Griffin, J. H., 1983, Inherited protein C deficiency and a coumarin-responsive chronic relapsing purpura fulminans syndrome in a neonate, *Lancet* 2:1165–1168.

8. Broekmans, A. W., Bertina, R. M., Loeliger, E. A., Hofman, V., and Klingeman, H. G., 1983, Protein C and the development of skin necrosis during anticoagulant therapy, *Thromb. Haemostasis* 49:251.

9. Brox, J. H., Osterud, B., Bjorklid, E., and Fenton, J. W., II, 1984, Production and availability of thromboplastin in endothelial cells: The effects of thrombin, endotoxin and platelets, *Br. J. Haematol.* 57:239–246.

10. Busch, C., Cancilla, P., DeBault, L., Goldsmith, J., and Owen, W., 1982, Use of endothelium cultured on microcarriers as a model for the microcirculation, *Lab. Invest.* 47:498–504.

11. Clouse, L. H., and Comp, P. C., 1986, The regulation of hemostasis: The protein C system, *N. Engl. J. Med.* 314:1298–1304.

12. Colucci, M., Balconi, R., Lorenzet, R., Pietra, A., Locati, D., Donati, M. B., and Semararo, N., 1983, Cultured human endothelial cells generate tissue factor in response to endotoxin, *J. Clin. Invest.* 71:1893–1896.

13. Comp, P. C., Nixon, R. R., and Esmon, C. T., 1984, Determination of protein C, an antithrombotic protein, using thrombin–thrombomodulin complex, *Blood* 63:15–21.

14. Comp, P. C., Nixon, R. R., Cooper, M. R., and Esmon, C. T., 1984, Familial protein S deficiency is associated with recurrent thrombosis, *J. Clin. Invest.* 74:2082–2088.

15. Comp, P. C., Vigano, S., D'Angelo, A., Thurnau, G., Kaufman, C., and Esmon, C. T., 1985, Acquired protein S deficiency occurs in pregnancy, the nephrotic syndrome and acute systemic lupus erythematosus, *Blood* 66:348a (abstract 1279).

16. Comp, P. C., Doray, D., Patton, D., and Esmon, C. T., 1986, An abnormal plasma distribution of protein S occurs in functional protein S deficiency, *Blood* 67:504–508.

17. Dahlback, B., 1984, Interaction between vitamin K-dependent protein S and the complement protein, C4b-binding protein: A link between coagulation and the complement system, *Semin. Thromb. Haemost.* 10:139–148.

18. Dahlback, B., and Stenflo, J., 1981, High molecular weight complex in human plasma between vitamin K-dependent protein S and complement component C4b-binding protein, *Proc. Natl. Acad. Sci. USA* 78:2512–2516.

19. Dahlback, B., Lundwall, A., and Stenflo, J., 1986, Primary structure of bovine protein S, *Proc. Natl. Acad. Sci. USA* 82:4199–4203.

20. D'Angelo, S. V., Comp, P. C., Esmon, C. T., and D'Angelo, A., 1986, Relationship between protein C antigen and anticoagulant activity during oral anticoagulation and in selected disease states. *J. Clin. Invest.* 77:416–425.

21. DeBault, L. E., Esmon, N. L., Olson, J. R., and Esmon, C. T., 1986, Distribution of the thrombomodulin antigen in the rabbit vasculature, *Lab. Invest.* 54:172–178.

22. DiScipio, R. G., and Davie, E. W., 1979, Characterization of protein S, a gamma-carboxyglutamic acid containing protein from bovine and human plasma, *Biochemistry* 18:899–904.

23. Drakenberg, T., Fernlund, P., Roepstorff, P., and Stenflo, J., 1983, β-Hydroxyaspartic acid in vitamin K-dependent protein C, *Proc. Natl. Acad. Sci. USA* **80**:1802–1806.
24. Epstein, D. J., Begum, P. W., Bajaj, S. P., and Rapaport, S. I., 1984, Radioimmunoassays for protein C and factor X: Plasma antigen levels in abnormal hemostatic states, *Am. J. Clin. Pathol.* **82**:573–581.
24a. Esmon, C. T., 1987, The regulation of natural anticoagulant pathways, *Science* **235**:1348–1352.
25. Esmon, C. T., 1984, Protein C, in: *Progress in Hemostasis and Thrombosis* (T. Spaet, ed.), Vol. 7, Grune & Stratton, New York, pp. 25–54.
26. Esmon, C. T., and Esmon, N. L., 1984, Protein C activation, *Semin. Thromb. Haemost.* **10**:122–130.
27. Esmon, C. T., and Owen, W. G., 1981, Identification of an endothelial cell cofactor for thrombin-catalyzed activation of protein C, *Proc. Natl. Acad. Sci. USA* **78**:2249–2252.
28. Esmon, C. T., Esmon, N. L., and Harris, K. W., 1982, Complex formation between thrombin and thrombomodulin inhibits both thrombin-catalyzed fibrin formation and factor V activation, *J. Biol. Chem.* **257**:7944–7947.
28a. Esmon, C. T., Esmon N. L., Saugstad, J., and Owen, W. G., 1982 Activation of protein C by a complex between thrombin and an endothelial cell surface protein, in: *Pathobiology of the Endothelial Cell* (H. Nossel, ed.), Academic Press, New York, pp 121–136.
29. Esmon, C. T., Esmon, N. L., Kurosawa, S., and Johnson, A. E., 1987, Interaction of thrombin with thrombomodulin, *Ann. N.Y. Acad. Sci.* **485**:221–227.
30. Esmon, C. T., Harris, K. W., Comp, P. C., Esmon, N. L., Nawroth, P. P., and Stern, D. M., 1987, Regulation of natural anticoagulant pathways of blood coagulation, Proceedings Volume of 1986 UCLA Symposium on Proteases in Biological Control and Biotechnology, in: *Proteases in Biological Control and Biotechnology* (D. D. Cunningham and G. L. Long, eds. Vol. 57, Alan R. Liss, New York, pp. 229–234.
31. Esmon, C. T., Taylor, F. B., Jr., Hinshaw, L. B., Chang, A., Comp, P. C., Ferrell, G., and Esmon, N. L., 1987, Protein C, isolation and potential use in prevention of thrombosis, Proceedings Volume of International Association of Biological Standardization Symposium on Standardization in Blood Fractionation Including Coagulation Factors, Melbourne, Australia, *Dev. Biol. Stand.* **67**:51–57.
32. Esmon, N. L., Owen, W. G., and Esmon, C. T., 1982, Isolation of a membrane-bound cofactor for thrombin-catalyzed activation of protein C, *J. Biol. Chem.* **257**:859–864.
33. Esmon, N. L., DeBault, L. E., and Esmon, C. T., 1983, Proteolytic formation and properties of γ-carboxyglutamic acid-domainless protein C, *J. Biol. Chem.* **258**:5548–5553.
34. Esmon, N. L., Carroll, R. C., and Esmon, C. T., 1983, Thrombomodulin blocks the ability of thrombin to activate platelets, *J. Biol. Chem.* **258**:12238–12242.
35. Esmon, N. L., D'Angelo, A., Vigano-D'Angelo, S., Esmon, C. T., and Comp, P. C., 1987, Analysis of protein C and protein S in disease states, Proceedings Volume of International Association of Biological Standardization Symposium on Standardization in Blood Fractionation Including Coagulation Factors, Melbourne, Australia, *Dev. Biol. Stand.* **67**:75–82.
36. Fernlund, P., and Stenflo, J., 1982, Amino acid sequence of the light chain of bovine protein C, *J. Biol. Chem.* **257**:12170–12179.
37. Fowler, S. A., Paulson, D., Owen, B. A., and Owen, W. G., 1986, Binding of iron by factor IX, possible role for β hydroxyaspartic acid, *J. Biol. Chem.* **261**:4371–4372.
38. Fulcher, C. A., Gardiner, J. E., Griffin, J. H., and Zimmerman, T. S., 1984, Proteolytic inactivation of human factor VIII procoagulant protein by activated protein C and its analogy with factor V, *Blood* **63**:486–489.
39. Freyssinet, J., Gauchy, J., and Cazenave, J., 1986, The effect of phospholipids on the activation of protein C by the human thrombin–thrombodulin complex, *Biochem. J.* **238**:151–157.
40. Galvin, J. B., Kurosawa, S., Moore, K., Esmon, C. T., and Esmon, N. L., 1987, Reconstitution of rabbit thrombomodulin into phospholipid vesicles, *J. Biol. Chem.* **262**:2199–2205.
41. Griffin, J. H., 1984, Clinical studies of protein C, *Semin. Thromb. Haemost.* **10**:162–166.
42. Harris, K. W., and Esmon, C. T., 1985, Protein S is required for bovine platelets to support activated protein C binding and activity, *J. Biol. Chem.* **260**:2007–2010.

43. Hofsteenge, J., Taguchi, H., and Stone, S. R., 1986, Effect of thrombomodulin on the kinetics of the interaction of thrombin with substrates and inhibitors, *Biochem. J.* **237**:243–251.
44. Jakubowski, H. V., Kline, M. D., and Owen, W. G., 1986, The effect of bovine thrombomodulin on the specificity of bovine thrombin, *J. Biol. Chem.* **261**:3876–3882.
45. Johnson, A. E., Esmon, N. L., Laue, T. M., and Esmon, C. T., 1983, Structural changes required for activation of protein C are induced by Ca^{2+} binding to a high affinity site that does not contain γ-carboxyglutamic acid, *J. Biol. Chem.* **258**:5554–5560.
46. Kisiel, W., 1979, Human plasma protein C: Isolation, characterization and mechanism of activation by α-thrombin, *J. Clin. Invest.* **64**:761–769.
47. Kisiel, W., Ericsson, L. L., and Davie, E. W., 1976, Proteolytic activation of protein C from bovine plasma, *Biochemistry* **15**:4893–4900.
48. Kisiel, W., Canfield, W. M., Ericsson, E. H., and Davie, E. W., 1977, Anticoagulant properties of bovine plasma protein C following activation by thrombin, *Biochemistry* **16**:5824–5831.
49. Kurosawa, S., Galvin, J. B., Esmon, N. L., and Esmon, C. T., 1987, Proteolytic formation and properties of functional domains of thrombomodulin, *J. Biol. Chem.* **262**:2206–2212.
50. Laemmli, B., and Griffin, H. J., 1985, Formation of the fibrin clot: The balance of procoagulant and inhibitory factors, *Clin. Hematol.* **14**:281–342.
51. Levy, G. A., Schwartz, B. S., Curtiss, L. K., and Edgington, T. S., 1981, Plasma lipoprotein induction and suppression of the generation of cellular procoagulant activity in vitro: The requirements for cellular collaboration, *J. Clin. Invest.* **67**:1614–1622.
52. Little, S. P., Bang, N. U., Long, G. L., Jaskunas, S. R., Marks, C. A., Mattler, L. E., and Daugherty, L. L., 1985, Amplification of recombinant human protein C (PC) expression in cultured Chinese hamster ovary cells (CHO), *Thromb. Haemost.* **54**:0990 (abstract).
53. Lyberg, T., Galdal, K. S., Evensen, S. A., and Prydz, H., 1983, Cellular cooperation in endothelial cell thromboplastin synthesis, *Br. J. Haematol.* **53**:85–95.
54. Mann, K. G., 1984, Membrane bound enzyme complexes, in: *Progress in Thrombosis and Haemostasis* (T. Spaet, ed.), Vol. 17, Grune & Stratton, New York, pp. 1–24.
55. Marlar, R. A., Kleiss, A. J., and Griffin, J. H., 1982, Mechanism of action of human activated protein C, a thrombin-dependent anticoagulant enzyme, *Blood,* **59**:1067–1072.
56. Maruyama, I., Salem, H. H., Ishii, H., and Majerus, P. W., 1985a, Human thrombomodulin is not an efficient inhibitor of procoagulant activity of thrombin, *J. Clin. Invest.* **75**:987–991.
57. Maruyama, I., Bell, C. E., and Majerus, P. W., 1985b, Thrombomodulin is found on endothelium of arteries, veins, capillaries, lymphatics, and on syncytiotrophoblast of human placenta, *J. Cell Biol.* **101**:363–371.
58. Miletich, J. P., Leykam, J. F., and Broze, G. J., Jr., 1983, Detection of single chain protein C in human plasma, *Blood* **62**(Suppl. 11):1127 (abstract).
59. Morita, T., Isaacs, B. S., Esmon, C. T., and Johnson, A. E., 1984, Derivatives of blood coagulation factor IX contain a high affinity Ca^{2+} binding site that lacks γ-carboxyglutamic acid, *J. Biol. Chem.* **259**:5698–5704.
60. Mammen, E. F., Thomas, W. R., Seegers, W. H., 1960, Activation of purified prothrombin to autoprothrombin I or autoprothrombin II (platelet cofactor II) or autoprothrombin II-A, *Thromb. Diath. Haemorrh.* **5**:218–250.
61. McGehee, W. G., Klotz, T. A., Epstein, D. J., and Rapaport, S. I., 1984, Coumarin necrosis associated with hereditary protein C deficiency, *Ann. Intern. Med.* **100**:59–60.
62. Moore, K. L., Andreoli, S. P., Esmon, N. L., Esmon, C. T., and Bang, N. U., 1987, Endotoxin enhances tissue factor and suppresses thrombomodulin expression in vascular endothelium *in vitro*, *J. Clin. Invest.* **79**:124–130.
63. Nawroth, P. P., and Stern, D. M., 1986, Modulation of endothelial cell hemostatic properties by tumor necrosis factor, *J. Exp. Med.* **163**:740–745.
64. Nawroth, P. P., Handley, D. A., Esmon, C. T., and Stern, D. M., 1986a, Interleukin-1 induces endothelial cell procoagulant while suppressing cell surface anticoagulant activity, *Proc. Natl. Acad. Sci. USA* **83**:3460–3464.
65. Nawroth, P. P., Bank, I., Handley, D., Cassimeris, J., Chess, L., and Stern, D., 1986b, Tumor necrosis factor/cachectin interacts with endothelial cell receptors to induce release of interleukin 1, *J. Exp. Med.* **163**:1363–1375.

66. Nesheim, M. E., Canfield, W. M., Kisiel, W., and Mann, K. G., 1982, Studies on the capacity of factor Xa to protect factor Va from inactivation by activated protein C, *J. Biol. Chem.* **257:**1433–1447.

67. Nelsestuen, G. L., Kisiel, W., and DiScipio, R. G., 1978, Interaction of vitamin K-dependent proteins with membranes, *Biochemistry* **17:**2134–2138.

68. Owen, W. G., 1985, Regulation of expression and inhibition of thrombin, *Thromb. Haemost.* **54:**57 (abstract).

69. Owen, W. G., and Esmon, C. T., 1981, Functional properties of an endothelial cell cofactor for thrombin-catalyzed activation of protein C, *J. Biol. Chem.* **256:**5532–5535.

70. Rickles, F. R., Levin, J., Hardin, J. A., Barr, C. F., and Conrad, M. E., 1977, Tissue factor generation by human mononuclear cells: Effects of endotoxin and dissociation of tissue factor generation from mitogenic response, *J. Lab. Clin. Med.* **89:**792–803.

71. Salem, H., Broze, G., Miletich, J., and Majerus, P., 1983, Human coagulation factor Va is a cofactor for the activation of protein C, *Proc. Natl. Acad. Sci. USA* **80:**1584–1588.

72. Salem, H. H., Maruyama, I., Ishii, H., and Majerus, P. W., 1984, Isolation and characterization of thrombomodulin from human placenta, *J. Biol. Chem.* **259:**12246–12251.

73. Samama, M., Horellou, M. H., Soria, J., Conard, J., and Nicolas, G., 1984, Successful progressive anticoagulation in a severe protein C deficiency and previous skin necrosis at the initiation of oral anticoagulant therapy, *Thromb. Haemost.* **51:**132–133.

74. Sells, R. H., Marlar, R., Montgomery, R. R., Desphande, G. N., and Humbert, J. R., 1984, Severe homozygous protein C deficiency, *J. Pediatr.* **105:**409–413.

75. Seegers, W. H., 1981, Protein C and autoprothrombin II-A, *Semin. Thromb. Haemost.* **7:**257–262.

76. Stenflo, J., 1976, A new vitamin K-dependent protein: Purification from bovine plasma and preliminary characterization, *J. Biol. Chem.* **251:**355–363.

77. Stenflo, J., and Fernlund, P., 1982, Amino acid sequence of the heavy chain of bovine protein C, *J. Biol. Chem.* **257:**12180–12190.

78. Stern, D. M., Nawroth, P. P., Harris, K., and Esmon, C. T., 1986, Cultured bovine aortic endothelial cells promote activated protein C–protein S-mediated inactivation of factor Va, *J. Biol. Chem.* **261:**713–718.

79. Sugo, T., Bjork, I., Holmgren, A., and Stenflo, J., 1984, Calcium-binding properties of bovine factor X lacking the γ-carboxyglutamic acid-containing region, *J. Biol. Chem.* **259:**5705–5710.

80. Sugo, T., Dahlback, B., Holmgren, A., and Stenflo, J., 1986, Calcium binding of bovine protein S: Effect of thrombin cleavage and removal of the γ-carboxyglutamic acid-containing region, *J. Biol. Chem.* **261:**5116–5120.

81. Suzuki, K., Stenflo, J., Dahlback, B., and Teodorsson, B., 1983, Inactivation of human coagulation factor V by activated protein C, *J. Biol. Chem.* **258:**11914–1920.

82. Taylor, F. B., Jr., Chang, A., Esmon, C. T., D'Angelo, A., Vigano-D'Angelo, S., and Blick, K. E., 1986, Protein C prevents the coagulopathic and lethal effects of *E. coli* infusion in the baboon, *J. Clin. Invest.* **79:**918–925.

82a. Taylor, F. B., Jr., Stern, D. M., Nawroth, P. P., Esmon, C. T., Hinshaw, L. B., and Blick, K. E., 1986, Activated protein C prevents E coli induced coagulopathy and shock in the primate, *Circulation* **74** (Suppl II) :64 (abstract 259).

83. Thompson, E. A., and Salem, H. H., 1986, Inhibition by human thrombomodulin of factor Xa-mediated cleavage of prothrombin, *J. Clin. Invest.* **78:**13–17.

84. Walker, F. J., 1981, Regulation of activated protein C by protein S: The role of phospholipid in factor Va inactivation, *J. Biol. Chem.* **256:**11128–11131.

85. Walker, F. J., 1986, Identification of a new protein involved in the regulation of the anticoagulant activity of activated protein C: Protein S-binding protein, *J. Biol. Chem.* **261:**10941–10944.

86. Walker, F. J., Sexton, P. W., and Esmon, C. T., 1979, Inhibition of blood coagulation by activated protein C through selective inactivation of activated factor V, *Biochim. Biophys. Acta* **571:**333–342.

87. Seligsohn, U., Berger, A., Abend, M., Rubin, L., Attias, D., Zivelin, A., and Rapaport, S., 1984, Homozygous protein C deficiency manifested by massive venous thrombosis in the newborn, *N. Engl. J. Med.* **310:**559–562.

The Biochemistry and Physiology of Anticoagulantly Active Heparin-like Molecules

James A. Marcum and Robert D. Rosenberg

I. INTRODUCTION

By the late 19th century, several investigators had noted loss of thrombin activity upon addition of thrombin to plasma and concluded that a specific inactivator of this enzyme, antithrombin, was present within the blood.[13,42] In the early 20th century, McLean[39] isolated a substance from the liver and demonstrated its potent anticoagulant properties. Howell and Holt[20] subsequently termed this substance heparin. The anticoagulant effect of this material on purified procoagulants was examined by Brinkhous *et al.*[8] who showed that heparin was effective as an anticoagulant only in the presence of a plasma component that they termed heparin cofactor. The laboratories of Waugh and Fitzgerald[59] and Monkhouse *et al.*[41] reported data which suggested that the plasma antithrombin activity and plasma heparin cofactor activity are intimately related. In addition, Abildgaard[1] isolated small quantities of an α_2-globulin from human plasma that functioned in both capacities. Rosenberg and Damus[49] provided the first direct evidence that plasma antithrombin activity and plasma heparin cofactor activity reside in the same molecule. In this review, we summarize progress in elucidating the biochemistry and cell biology of the heparin (heparan sulfate)–antithrombin interactions and outline the important physiologic role of this mechanism within the cardiovascular system.

II. HEPARIN–ANTITHROMBIN INTERACTION

Human antithrombin has a molecular weight of about 58,000 as well as an isoelectric point of 5.11.[49] The approximate shape of the protease inhibitor re-

James A. Marcum • Department of Pathology, Beth Israel Hospital and Harvard Medical School, and Department of Biology, Massachusetts Institute of Technology, Cambridge, Massachusetts 02139. *Robert D. Rosenberg* • Department of Medicine, Beth Israel Hospital and Harvard Medical School, and Department of Biology, Massachusetts Institute of Technology, Cambridge, Massachusetts 02139.

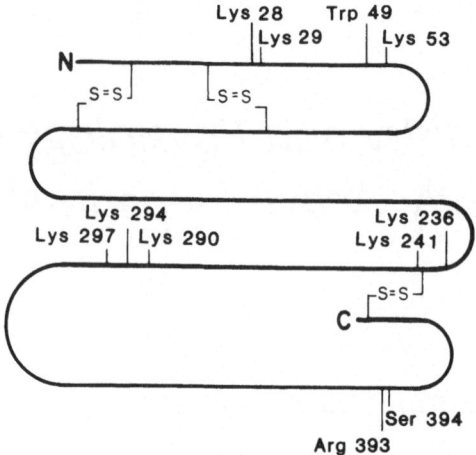

Figure 1. Critical sites within antithrombin.

sembles that of a prolate ellipsoid with an axial ratio of $\sim 5.$[44] The concentration of antithrombin within human plasma is about 150 μg/ml.[43] The complete primary structure of this protease inhibitor has been reported by Petersen *et al.*[47] The human gene that codes for antithrombin has been identified as well, and the amino acid sequence derived from the cDNA for the protease inhibitor agrees with the primary structure of the molecule.[48] In Fig. 1 are depicted the locations of the functionally important domains of antithrombin such as the enzyme-binding region, the potential heparin-binding sites, the conformation-sensitive tryptophan residue, and the S–S cross-bridges.

Rosenberg and Damus[49] uncovered the mechanism by which thrombin is inactivated by antithrombin. These investigators demonstrated that the protease inhibitor neutralizes the enzyme by forming a 1:1 stoichiometric complex between the two components via a reactive site (arginine)–active center (serine) interaction. Complex formation occurs at a relatively slow rate in the absence of heparin. However, in the presence of mucopolysaccharide which binds to lysine residues on antithrombin, enzyme–inhibitor interaction are accelerated dramatically.[49] The heparin-induced acceleratory phenomenon appears to be due to an allosteric alteration in the position of a critical arginine residue of the protease inhibitor enabling this amino acid moiety to interact more readily with the active center site of thrombin (Fig. 2).

This model of antithrombin action has been extended by a number of observations. Bjork *et al.*[6] have provided evidence that the arginine reactive site of antithrombin is located at Arg_{393}–Ser_{394} within the carboxy-terminus of this protein. Ferguson and Finlay[16] have demonstrated that the reduction of a single Cys_{239}–Cys_{430} bridge within antithrombin produces a dramatic decrease in its ability to bind heparin, but does not alter the capacity of the protease inhibitor to inactivate factor Xa or thrombin in the absence of the mucopolysaccharide. The reactive site arginine residue lies within this S–S loop, and in addition there appears to be a cluster of lysine residues in this region which have been implicated as a binding site for heparin (see below).

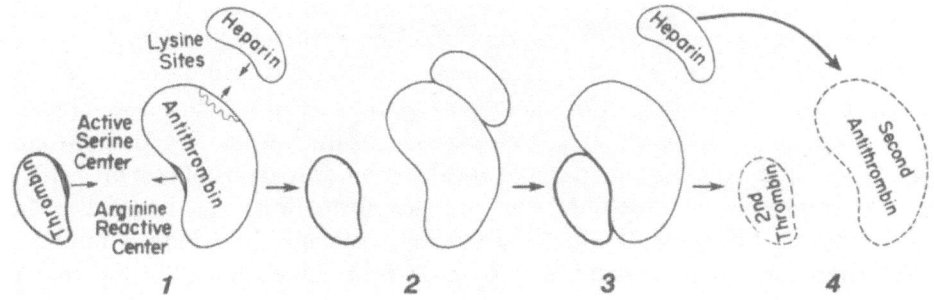

Figure 2. The overall mechanism of action of heparin and antithrombin.

Several groups have also claimed that a critical tryptophan residue is located within the heparin-binding site of antithrombin.[7,57] Their conclusion is based on the observations that modification of Trp$_{49}$ prevents antithrombin from binding heparin, produces a dramatic reduction in the ability of the mucopolysaccharide to accelerate thrombin–protease inhibitor interactions, but does not alter the capacity of the protein to inactivate the enzyme in the absence of the complex carbohydrate. Karp et al.[26] utilizing Trp$_{49}$-modified antithrombin have shown with quantitative binding techniques that this chemical alteration results in only a modest reduction ($\sim$ 10-fold) in the avidity of the protein for heparin. However, when the chemically altered protease inhibitor was saturated with the glycosaminoglycan, quantitative kinetic methods revealed a much larger decrease ($\sim$ 500-fold) in the heparin-dependent acceleration of thrombin–antithrombin and factor Xa–antithrombin interactions. These experimental data suggest that Trp$_{49}$ lies close to, but not within, a major binding region of antithrombin. It is of interest to point out that Lys$_{28}$, Lys$_{29}$, and Lys$_{53}$ are in proximity to Trp$_{49}$. It would appear likely that additional positively charged residues in conjunction with Lys$_{236}$, Lys$_{241}$, Lys$_{290}$, Lys$_{294}$, and Lys$_{297}$ may constitute the essential sites for binding of the mucopolysaccharides (see below).

The interactions of heparin and antithrombin with the various serine proteases of the hemostatic mechanism have been described in extensive detail.[9,14,15,18,50,53] In addition, the physicochemical mechanism by which the mucopolysaccharide accelerates the neutralization of hemostatic enzymes by antithrombin has also been examined. Jordan et al.,[23,24] utilizing fluorescence spectroscopy and polarization fluorescence spectroscopy, demonstrated that low-molecular-weight heparin (M_r 6500) binds antithrombin with a stoichiometry of 1:1 and a dissociation constant of 5.74×10^{-8} M. These investigators also determined the binding of the glycosaminoglycan to thrombin as well as factors IXa and Xa. Their data indicated that the stoichiometry of the heparin–thrombin interaction is 2:1 with two equivalent dissociation constants of 8×10^{-7} M, that the stoichiometry of heparin–factor IXa interactions is 1:1 with a dissociation constant of 2.58×10^{-7} M, and that the stoichiometry of the heparin–factor Xa interaction is 1:1 with a dissociation constant of 8.73×10^{-6} M.

The above investigators also estimated the initial velocities of factors IXa and Xa as well as thrombin neutralization by antithrombin as a function of low-molecular-weight heparin concentration.[23,24] Their results indicate that heparin-dependent enhancement in the rates of neutralization of these proteases by antithrombin requires binding of the mucopolysaccharide to the protease inhibitor but not necessarily to the enzyme. Comparisons of the various kinetic constants suggest that direct binding of heparin to antithrombin is responsible for an ~ 1000-fold acceleration of enzyme–inhibitor complex formation. The interaction between thrombin or factor IXa and the mucopolysaccharide, which is bound to antithrombin, provides an additional 4- to 15-fold enhancement in the rate of enzyme neutralization. The binding of factor Xa to heparin, which is bound to antithrombin, does not augment the velocity of factor Xa inactivation. Therefore, the acceleration of antithrombin action is due mainly to the binding of heparin to the protease inhibitor, whereas the binding of heparin to the hemostatic enzymes accounts for either none (factor Xa) or 1–2% of the total rate enhancement (factor IXa and thrombin).

Recently, several groups[21,25] have shown that high-molecular-weight heparin accelerates specific hemostatic enzyme–protease inhibitor interactions to a greater degree than those outlined above. These results may be ascribed to multiple antithrombin-binding sites of heparin as well as more potent approximation phenomena. The intricate kinetic events outlined above may be of greater importance with regard to heparin-like molecules found on the endothelium of blood vessels (see below). It was also observed throughout the above studies that neutralization of the hemostatic enzymes by the heparin–antithrombin complex results in release of the mucopolysaccharide from the protease inhibitor on a one-to-one molar basis and that the binding of heparin to the hemostatic enzyme–antithrombin complex is about 100 to 1000 times weaker than the interaction of the glycosaminoglycan with the free protease inhibitor.[25] These data indicate that heparin catalyzes the above interactions by initiating multiple rounds of protease–protease inhibitor complex formation (Fig. 2).

III. STRUCTURE–FUNCTION RELATIONSHIP BETWEEN HEPARIN AND ANTITHROMBIN

Heparin contains several critical domains which allow the mucopolysaccharide to bind antithrombin, to activate the protease inhibitor, and to accelerate neutralization of hemostatic enzymes by catalyzing the action of antithrombin. In the mid-1970s Lam et al.[27] showed that only a small fraction of the glycosaminoglycan binds the protease inhibitor and is responsible for virtually all of the anticoagulant activity of the complex carbohydrate. These findings were rapidly confirmed by two other groups[2,19] and allowed subsequent resolution of the regions on heparin which are involved in its unique anticoagulant function.

Oosta et al.[45] randomly cleaved highly active heparin with chemical techniques and isolated mucopolysaccharide fragments of varying size which bind tightly to the protease inhibitor. Oligosaccharides of 8 to 16 sugar residues accelerate factor Xa–antithrombin interactions but do not catalyze the neutralization of other hemostatic enzymes by the protease inhibitor (Domain 1). Heparin fragments of 16 residues or greater accelerate thrombin–antithrombin as well as factor Xa–antithrombin interactions by directly activating antithrombin (Domain 2). Mucopolysaccharide chains of 22 residues or greater contain additional structural element(s) required for approximating free thrombin with protease inhibitor (Domain 3). These latter segments of the mucopolysaccharide are also needed to accelerate factor IXa–antithrombin and factor XIa–antithrombin interactions.[45] Thus, heparin contains several discrete structural regions which are responsible for its interactions with antithrombin and its modulation of the biologic activities of the protease inhibitor.

Rosenberg et al.[51,52] have utilized highly active heparin as well as fragments of the mucopolysaccharide to define Domain 1. This region contains a critical tetrasaccharide sequence composed of a nonsulfated iduronic acid residue on the nonreducing end followed by an N-acetylglucosamine 6-O-sulfate, a glucuronic acid moiety, and an N-sulfated/O-sulfated glucosamine residue on the reducing end of the oligosaccharide. This unique sequence is not possessed by the anticoagulantly inactive heparin. Leder[28] isolated a sulfatase which removed an ester sulfate from the third carbon position of glucosamine and postulated that this substituent could represent a critical functional group within the anticoagulantly active glycosaminoglycan. Lindahl et al.[29] in collaboration with Leder, confirmed this supposition by showing with the above enzyme that a 3-O-sulfate substituent exists on the reducing end glucosamine residue of the above tetrasaccharide. These investigations were substantiated by chemical synthesis of the above structures by Choay et al.[12] These various investigators[3,11,30] have also demonstrated that an octasaccharide which contains the unique tetrasaccharide region constitutes Domain 1 of the heparin molecule. The chemical structure of this domain is provided in Fig. 3.

Direct studies of the interaction of the above octasaccharide with the protease inhibitor revealed that it is responsible for about 8.7–10.2 kcal/mole of binding energy.[5] The contributions of individual residues of Domain 1 have been evaluated by comparing the avidity of synthetic oligosaccharides as well as deaminative- and enzyme-cleavage fragments of the natural octasaccharide for antithrombin as outlined in Fig. 3. Based on the above experiments[5] the relative importance of individual monosaccharides within Domain 1 have been estimated and are provided at the bottom of Fig. 3. The contribution of the nonreducing end IdA or GlcA (residue 1) and GlcA (residue 3) to the total binding energy of the octasaccharide is minimal, while the contributions of the 6-O-SO$_3$ of GlcNAc (residue 2) and the 3-O-SO$_3$ of GlcNSO$_3$-6-O-SO$_3$ (residue 4) to the binding energy of the octasaccharide are each about 4–5 kcal/mole. The 3-O-SO$_3$ group of residue 4 is functionally linked to the 6-O-SO$_3$ group of residue 2 such that both of these moieties are required for the binding of the octasaccharide to the protease inhibitor. Indeed, the absence of either of the above sulfate groups leads to the same 4–5 kcal/mole loss in binding energy. The apparent lack of contribution by residues 1 and 3 to the interaction of the octasaccharide with antithrombin does not indicate necessarily that these moieties are without importance. It is possible that these

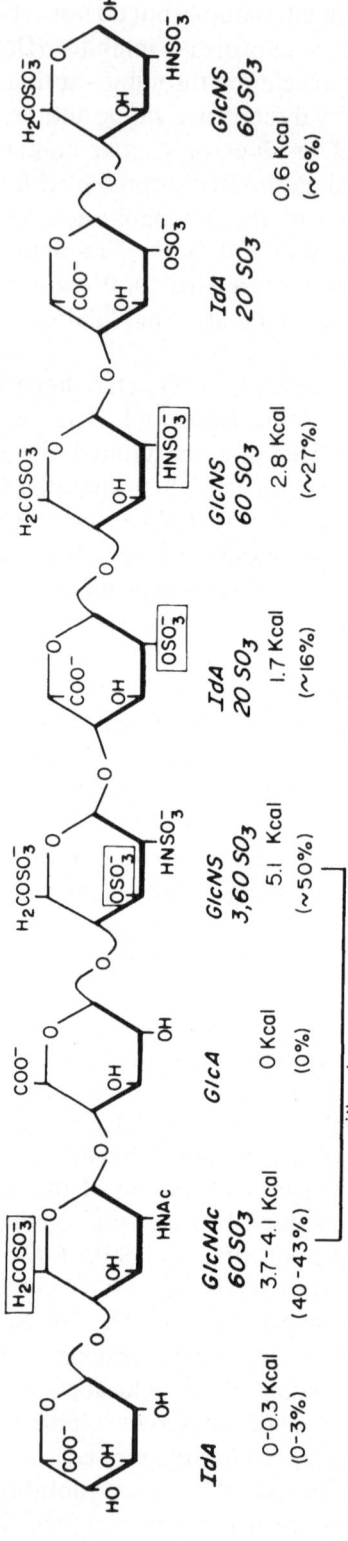

Figure 3. The structure of Domain 1 of anticoagulantly active heparin. The relative importance of the various residues with regard to antithrombin binding, is provided below each monosaccharide unit.

two nonsulfated uronic acid units function as critical spacers for the 6-O-SO$_3$ of residue 2 and the 3-O-SO$_3$ of residue 4. The contributions of the 2-O-SO$_3$ of IdA (residue 5) and the N-SO$_3$ of Glc-6-O-SO$_3$ (residue 6) to the binding of the octasaccharide to antithrombin are 1.7 and 2.8 kcal/mole, respectively, whereas those of residues 7 and 8 are about 0.6 kcal/mole.

Considerable attention has been devoted to examining the mechanism by which Domain 1 induced conformational changes within antithrombin as well as accelerates factor Xa–protease inhibitor interactions. Rosenberg and Damus[49] initially showed that heparin binds to antithrombin via a small number of lysine residues. Based on the NMR and X-ray data of the repeating units of heparin, Villaneuva[58] suggested that Lys$_{290}$, Lys$_{294}$, and Lys$_{297}$ within the α-helical region of the protease inhibitor could be matched for maximal interaction with the 6-O-SO$_3$ of residue 2, the N-SO$_3$ of residue 4, and the N-SO$_3$ of residue 6 of the octasaccharide. Recent evidence provided by Atha et al.[4,5] indicates that the sulfate groups of residues 2 and 6 are important for binding of heparin to antithrombin and that the 3-O-SO$_3$ group of residue 4 is also critical for complex formation.[5] Although the latter moiety appears to be out of range of Lys$_{294}$, it might interact with a different helical segment in the N-terminal region of antithrombin, such as lysine residues near Trp$_{49}$.

The above model is supported by the following data: modification of Trp$_{49}$ near the N-terminus of antithrombin blocks the heparin-enhanced inhibition of factor Xa,[7,57] and fluorescence-transfer studies suggest that one or two lysines such as Lys$_{28}$, Lys$_{29}$, and/or Lys$_{53}$ located near Trp$_{49}$ are essential for the binding of the polysaccharide to antithrombin.[46] Recently, Rosenberg and co-workers (unpublished data) have examined the contribution of the 3-O sulfated glucosamine (residue 4), as well as the 6-O sulfated glucosamine (residue 2), to the conformational changes of antithrombin, as measured by the intrinsic fluorescence of the protease inhibitor, and to the acceleration of factor Xa–antithrombin interactions. Utilizing synthetic tetrasaccharides and pentasaccharides of well-defined compositions, they demonstrated that the 3-O sulfate group of residue 4 accounted for the majority of the enhancement in the intrinsic fluorescence of antithrombin, whereas both ester sulfates of residues 2 and 4 are equally important in facilitating complex formation between antithrombin and factor Xa. The above data establish that the 3-O sulfated and 6-O sulfated glucosamine moieties of Domain 1 contribute, specifically, to the conformational changes of antithrombin which are functionally coordinated with the anticoagulant activity of the mucopolysaccharide. This transition of antithrombin could lead to the repositioning of the Arg$_{393}$–Ser$_{394}$ locale or to the formation of a secondary enzyme complexing site on the protease inhibitor, which is ultimately responsible for the acceleration of factor Xa neutralization.

Kinetic examination and ultraviolet circular dichroism spectroscopy of the above system suggest that Domain 2 of heparin interacts with a separate area of antithrombin and triggers a conformational transition of the protease inhibitor distinct from that induced by Domain 1. On the one hand, Oosta et al.[45] have studied the physical interactions of heparin oligosaccharides, which possess Domain 2, with thrombin and antithrombin. These researchers determined the rate

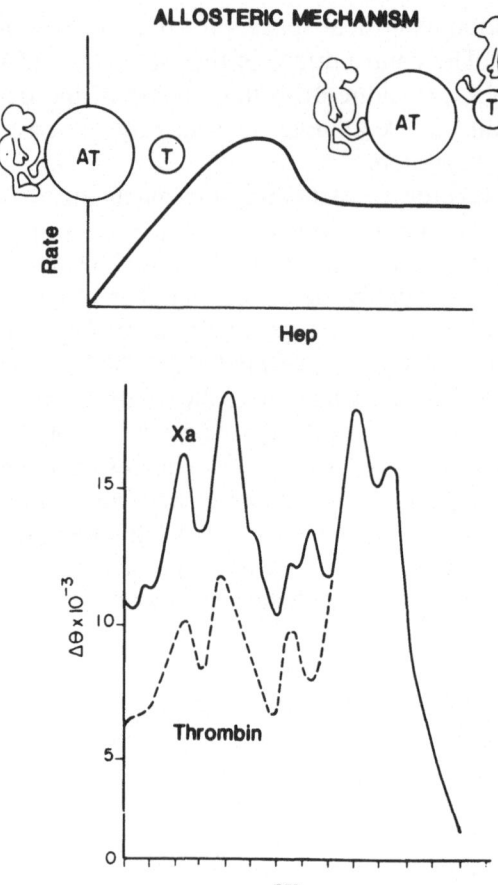

Figure 4. The kinetic and spectroscopic evidence for the multiple functional domains of heparin.

of neutralization of the enzyme by the protease inhibitor in the presence of varying concentrations of these mucopolysaccharide fragments. Their investigations show that the rate of neutralization of thrombin by antithrombin in the presence of heparin oligosaccharide concentrations which saturate binding sites on both proteins is only slightly decreased when compared to the rate of neutralization of thrombin by antithrombin in the presence of oligosaccharide fragment levels which saturate binding sites on the protease inhibitor but not the enzyme (Fig. 4). The above data exclude the possibility that acceleration of thrombin–antithrombin interactions by heparin oligosaccharides which possess Domain 2 is due to binding of the mucopolysaccharide, which is complexed with the protease inhibitor, to the enzyme. Therefore, it is conceivable that Domain 2 interacts with an additional site on antithrombin and thereby induces a second conformational transition in the protease inhibitor, which is required for rapid neutralization of thrombin.

On the other hand, Stone et al.[55] investigated the interaction of antithrombin with oligosaccharides (~ 8 to 18 monosaccharide units) as well as with heparin (M_r 6500 and 22,000) and observed two major chiral absorption spectra. The first spectral pattern is seen when oligosaccharides of ~ 8 to 14 residues interact with the protease inhibitor, and is typical of the Domain 1–antithrombin interactions

(Fig. 4). The second spectral pattern is noted when oligosaccharides of ~ 18 or more monosaccharide units bind to the protease inhibitor, and is produced by the sum of Domain 1– and Domain 2–antithrombin interactions (Fig. 4). The latter circular dichroism spectra are similar to those produced by Domain 1–protease inhibitor complexes, except for alterations within the ranges 292–282 and 275–255 nm. The subtraction of the first spectra from the second revealed a shallow negative band between 300 and 275 nm with potential negative minima at 290 and 283 nm as well as a deep negative band between 275 and 255 nm with possible negative minima at 268 and 262 nm. Conformational changes about a disulfide bridge(s) could account for this chiral absorption profile.

The above additional transitions are strongly correlated with the binding of Domain 2 of the mucopolysaccharide to antithrombin. Indeed, the extension of the 14-monomer oligosaccharide by 4 residues permits this species to accelerate thrombin–antithrombin interactions as well as to induce the second type of spectral transition. These observations provide evidence that Domain 2 of heparin exerts its effect by binding to an area of antithrombin distinct from that of Domain 1 and inducing an additional set of conformational changes within the protease inhibitor that are critical for the acceleration of thrombin–antithrombin interactions. For example, Domain 2 of the heparin molecule may bind to additional lysine residues such as Lys_{236}, Lys_{241}, or other amino acid groups within the Cys_{239}–Cys_{430} loop. These interactions could induce a torsion of the S–S bridge which might either reposition the Arg_{393}–Ser_{394} locale or trigger the formation of a secondary binding site such that thrombin, factor IXa, and factor XIa are able to complex more quickly with the protease inhibitor.

IV. VASCULAR TISSUE HEPARAN SULFATE

The above sections outlined in detail the structure–function relationship of the heparin–antithrombin system. The data suggested that these natural components might be important in the normal regulation of the hemostatic events. Indeed, mast cells have long been known to be the site of anticoagulantly active heparin synthesis.[17] Unfortunately, these cells are located beneath the endothelium, and would force one to hypothesize that biologically active mucopolysaccharides are discharged into the blood by mast cells which are located outside the vascular lumen. This scenario does not appear likely since there is no convincing evidence for significant blood levels of endogenous anticoagulantly active heparin.

Damus et al.[14] postulated that heparin or related polysaccharides present on the vascular endothelium endowed these surfaces with thrombo-resistant properties, but did not speculate as to the site of synthesis of these macromolecules. The early studies of Teien et al.[56] showed that heparan sulfate (a heparin-like species with increased amounts of GlcA and GlcNAc-6-O-SO$_3$) is found within the aorta and exhibits trace amounts of anticoagulant activity. However, the manner by which the above component expresses this biologic property, and its precise

Table 1. Affinity Fractionation of Heparan Sulfates[a]

| Component | Heparin-like activity (USP units/mg) | |
	Anti-factor IIa activity	Anti-factor Xa activity
	Bovine aorta	
Prior to affinity fractionation	0.14 ± 0.01	0.46 ± 0.03
	(n = 3)	(n = 3)
After affinity fractionation	11.3 ± 1.36	55.4 ± 4.81
	(n = 3)	(n = 5)
	Cerebral microvessels	
Prior to affinity fractionation	2.81 ± 0.04	2.79 ± 0.10
	(n = 3)	(n = 3)
After affinity fractionation	36.8 ± 0.86	40.7 ± 0.61
	(n = 6)	(n = 3)

[a] Values are expressed as mean ± S.D. No biologic activity was observed after pretreatment of samples with purified heparinase or when buffer was substituted for antithrombin.

location within vascular tissue were not determined. In more recent investigations, Marcum and Rosenberg[35] have isolated heparan sulfate from enriched endothelial cell preparations derived from calf cerebral microvessels as well as aortas and affinity-fractionated these glycosaminoglycans into two separate populations by employing immobilized antithrombin. The aortic and cerebral microvascular heparan sulfates which complexed with the protease inhibitor exhibited a significant augmentation in the specific thrombin-inhibitory and factor Xa-inhibitory activities (Table 1). These two fractions from aortas and cerebral vessels constituted 0.3 and 4.2%, respectively, of the chemical mass of the complex carbohydrate and accounted for the great bulk of anticoagulant activity of the starting material.

Histologic examination of the above vascular preparations revealed the existence of a small population of mast cells (< 1% of the total nucleated cells). These metachromatic-staining cells could be responsible for the anticoagulantly active mucopolysaccharides within the above vascular tissues. In order to define the cellular source of anticoagulantly active heparan sulfate within the blood vessels, Marcum et al.[33] isolated retinal microvessels by modifications of previously reported techniques[40] and demonstrated that these vascular products were free of mast cells as suggested by earlier studies.[22] The anticoagulant activity of these preparations was quantitated with a double-antibody radioimmunoassay for the thrombin–antithrombin complex (Table 2) and was similar to that observed for the aortic and cerebral vascular tissue.[33,35] The biologic potency of the retinal microvascular product appears to be due to heparan sulfate proteoglycans. On the one hand, ~ 60% of the anticoagulant activity of these components bind to immobilized antithrombin. Furthermore, the anticoagulant activity of these components was completely destroyed with purified *Flavobacterium* heparinase which specifically cleaves heparin and heparan sulfate to fragments which do not possess biologic activity. On the other hand, anticoagulant activity could not be measured in retinal microvascular preparations when the products were precipitated with

Table 2. Isolation and Characterization of Retinal Microvasculature

	ACE activity[a] (units/mg)	Heparin-like activity[b] (USP units/g)	Mast cell number[c] (percent cells)
Homogenate	921		
Microvessel preparation	2624	0.060 ± 0.015	N.D.
Percoll gradient			
Band A (1.033 g/ml)	2683	0.060 ± 0.018	N.D.
Band B (~1.075 g/ml)	<15.0	<0.001	N.D.

[a] Angiotensin-converting enzyme (ACE) activity is defined as nmoles of L-histidyl-L-leucine liberated per minute per mg protein. Values represent the mean of two separate experiments.

[b] Heparin-like activity is expressed as USP units per gram of wet weight microvascular tissue. Values represent the mean ± S.D. for seven separate fractionations.

[c] The number of mast cells as a percent of total nucleated cells. N.D., none detected.

10% trichloroacetic acid prior to proteolytic treatment. Proteoglycans are insoluble under these conditions. Thus, the above data suggest that heparan sulfate proteoglycans with anticoagulant activity are synthesized by cells which compose the vascular tissue, particularly endothelial cells.

V. HEPARAN SULFATE PROTEOGLYCANS FROM ENDOTHELIAL CELLS

To demonstrate that endothelial cells synthesize anticoagulantly active heparan sulfate, Marcum et al.[36–38] isolated capillary endothelial cells from rat and mouse epididymal fat pads, and macrovascular endothelial cells from bovine aorta. Rat and bovine cellular elements were cloned from single endothelial cells to exclude the possibility that a minor contamination of the cultures with additional cell types could be responsible for the production of anticoagulantly active heparan sulfate. The mucopolysaccharides were extracted from the above cells, and the anticoagulant activity of these samples was determined by quantitating the acceleration of thrombin–antithrombin complex formation with a specific radioimmunoassay for the enzyme–inhibitor complex (Table 3). This biologic property was completely eliminated by incubating the samples with purified Flavobacterium heparinase. The anticoagulant activity of endothelial cells obtained from rat and mouse epididymal fat pads was not expressed when Trp$_{49}$-modified antithrombin was substituted for the native protease inhibitor, indicating that the endothelial cell-derived heparan sulfate activates antithrombin in a manner similar to commercial heparin. Furthermore, all of these anticoagulant active macromolecules could be harvested after a brief exposure of cells to dilute trypsin, implying that these components are located on the surface of these cells.

Cloned bovine aortic and rat epididymal fat pad endothelial cells were labeled metabolically with $Na_2^{35}SO_4$, and mucopolysaccharides were obtained after extensive digestion of these cells with a variety of proteolytic enzymes. Radiolabeled glycosaminoglycans were then isolated by DEAE–Sephacel chromatography. The anticoagulantly active mucopolysaccharide emerged from the ion-exchange ma-

Table 3. Heparin-like Activity from Cultured Endothelial Cells

Cell type	Anticoagulant activity[a] ($\times 10^{-3}$ unit/10^6 cells)
Bovine aortic endothelial cells	
Cloned	
A	0.82 ± 0.15 ($n = 3$)
B	1.58 ± 0.37 ($n = 3$)
C	0.85 ± 0.29 ($n = 3$)
D	1.79 ± 0.57 ($n = 3$)
E	1.05 ± 0.06 ($n = 2$)
Noncloned	0.94 ± 0.44 ($n = 3$)
Mouse epididymal endothelial cells	
+/+	2.38 ± 0.51 ($n = 3$)
W/W^v	2.25 ± 0.28 ($n = 3$)
Rat epididymal endothelial cells	
Cloned	5.93 ± 0.29 ($n = 3$)
Noncloned	5.63 ± 1.33 ($n = 4$)
Bovine aortic smooth muscle cells	N.D.[b] ($n = 4$)

[a] Mean $\pm$ S.E. The data were corrected for losses of biologic activity incurred during isolation of crude preparations of heparan sulfate from cloned and noncloned cells.

[b] N.D., none detected.

trix prior to chrondroitin sulfate and could be completely degraded by purified *Flavobacterium* heparinase. These two molecular properties are typical of heparan sulfate.[36] This mucopolysaccharide was then affinity-fractionated into two separate populations with immobilized antithrombin. The heparan sulfate derived from bovine aortic and rat fat pad endothelial cells, which bound tightly to the protease inhibitor, represented ~ 1 and ~ 10%, respectively, of the starting mass and exhibited specific activities of 1.16 USP anticoagulant units/10^6 ^{35}S-cpm and 0.72 USP anticoagulant unit/10^6 ^{35}S-cpm, respectively. Heparan sulfate obtained from bovine aortic and rat fat pad endothelial cells, which interacted minimally with the protease inhibitor, constituted ~ 99 and ~ 90%, respectively, of the mucopolysaccharide mass and possessed specific anticoagulant potencies of < 0.0002 USP anticoagulant unit/10^6 ^{35}S-cpm and < 0.002 USP anticoagulant unit/10^6 ^{35}S-cpm, respectively.

The structures of affinity-fractionated heparan sulfates derived from cloned bovine aortic and rat epididymal fat pad endothelial cells were compared to mucopolysaccharide obtained from the same sources but with minimal affinity for the protease inhibitor.[36,37] The data showed that the affinity-fractionated heparan sulfates exhibited a significant increase in the amount of GlcA → AMN-3,6-*O*-SO$_3$ and/or GlcA → GlcN-3-*O*-SO$_3$ when contrasted with the corresponding mucopolysaccharides which possessed minimal affinity for the protease inhibitor (Table 4). These disaccharides represent unique markers for the presence of Domain 1 of heparin-like molecules which are able to bind and activate antithrombin (see above). Furthermore, the above biochemical studies of affinity-fractionated mucopolysaccharides obtained from the cloned macrovascular and microvascular

Table 4. Disaccharide Composition of Bovine Aortic Endothelial
Cell Heparan Sulfate

Disaccharide[b]	Percentage[a]	
	Affinity-fractionated	Depleted
GlcA → AMN 2-O-SO-3	5.5 ± 0.6	7.0 ± 1.5
GlcA → AMN 6-O-SO$_3$	11.8 ± 0.8	10.4 ± 1.0
IdA → AMN 6-O-SO$_3$	10.9 ± 1.5	8.7 ± 1.5
IdA → AMN 2-O-SO$_3$	56.0 ± 3.6	60.8 ± 3.1
GlcA → AMN 3-O-SO$_3$	5.5 ± 0.8	1.5 ± 0.7
IdA → AMN 2-O-SO$_3$ 6-O-SO$_3$	10.3 ± 1.3	11.6 ± 1.8
GlcA → AMN 3,6-O-(SO$_3$)$_2$	N.D.[c]	N.D.

[a] Values are mean ± S.E. and represent the averaged data of two separate experiments.
[b] Abbreviations: GlcA, glucuronic acid; IdA, iduronic acid; AMN, anhydromannitol.
[c] N.D., none detected; < 1%.

endothelial cells confirmed the heparan sulfate-like nature of these macromolecules. First, the low pH nitrous acid treatment of these components revealed that 40% of the amino groups are N-acetylated. Heparan sulfate is endowed with significant numbers of these substituents, whereas heparin possesses few of these moieties. Second, the relative distributions of the various disaccharides of these glycosaminoglycans indicate that they are characterized by an average of ∼ 1.5 N- and O-sulfate groups/disaccharide. Heparan sulfate contains ∼ 1.0 sulfate group/disaccharide, whereas heparin possesses ∼ 2.5 sulfate groups/disaccharide.

Cloned bovine aortic endothelial cells were also incubated with $Na_2^{35}SO_4$ as well as various tritiated amino acids and then solubilized with a guanidine–detergent solution (Fig. 5). Double-labeled macromolecules were initially isolated by DEAE–Sephacel chromatography (Fig. 5). The radiolabeled components (peaks DEAE-I and DEAE-II) were then gel-filtered on Sepharose CL4B, and an initial peak (CL4B-1) of ^{3}H- and ^{35}S-labeled species eluted with a K_{av} of ∼ 0.2, while a second peak (CL4B-II) of radiolabeled macromolecules eluted with a K_{av} ranging from 0.45 to 0.7. Treatment of peak CL4B-I with papain resulted in a shift of the K_{av} of radioactivity from ∼ 0.2 to ∼ 0.5 indicating that these components are proteoglycans, whereas analysis of peak CL4B-II on Sepharose 4B revealed that these substances coeluted with mucopolysaccharide chains derived by proteolytic digestion of the cellular elements ($K_{av} \simeq 0.5$). The incubation of the radiolabeled material from either fraction with mucopolysaccharidases of

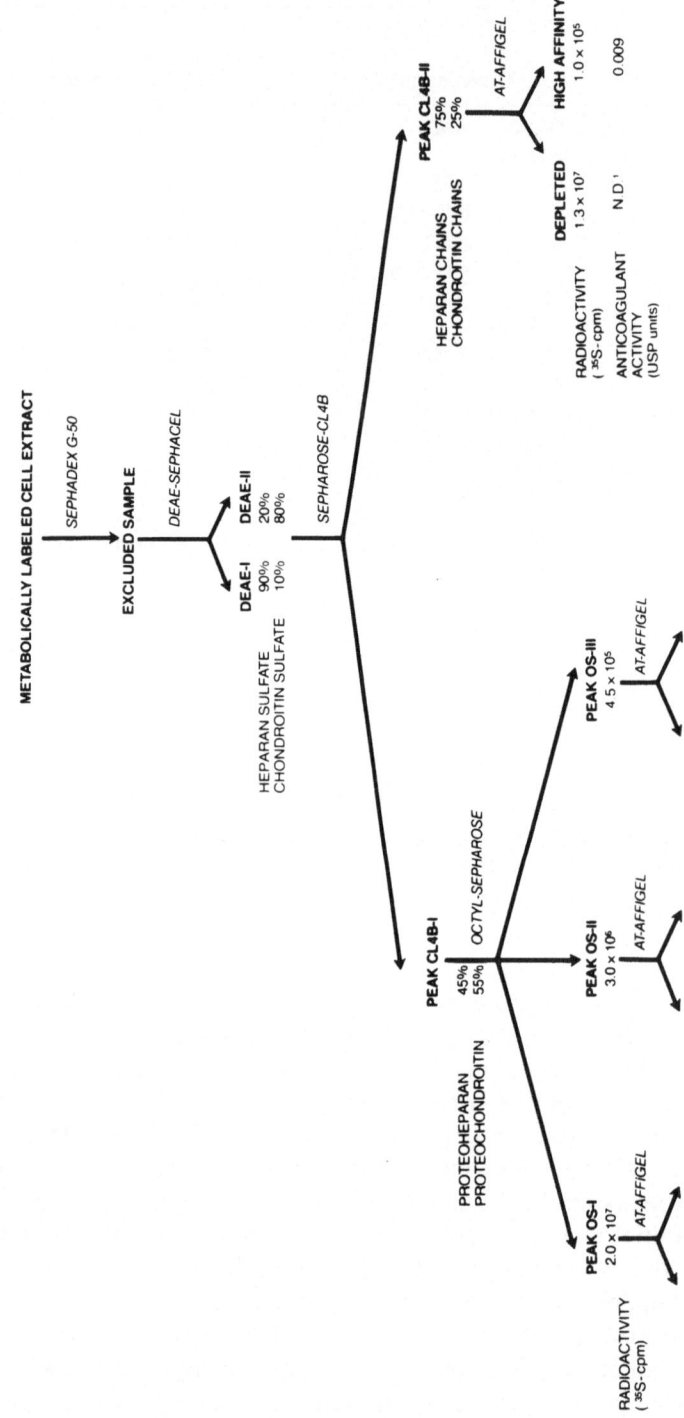

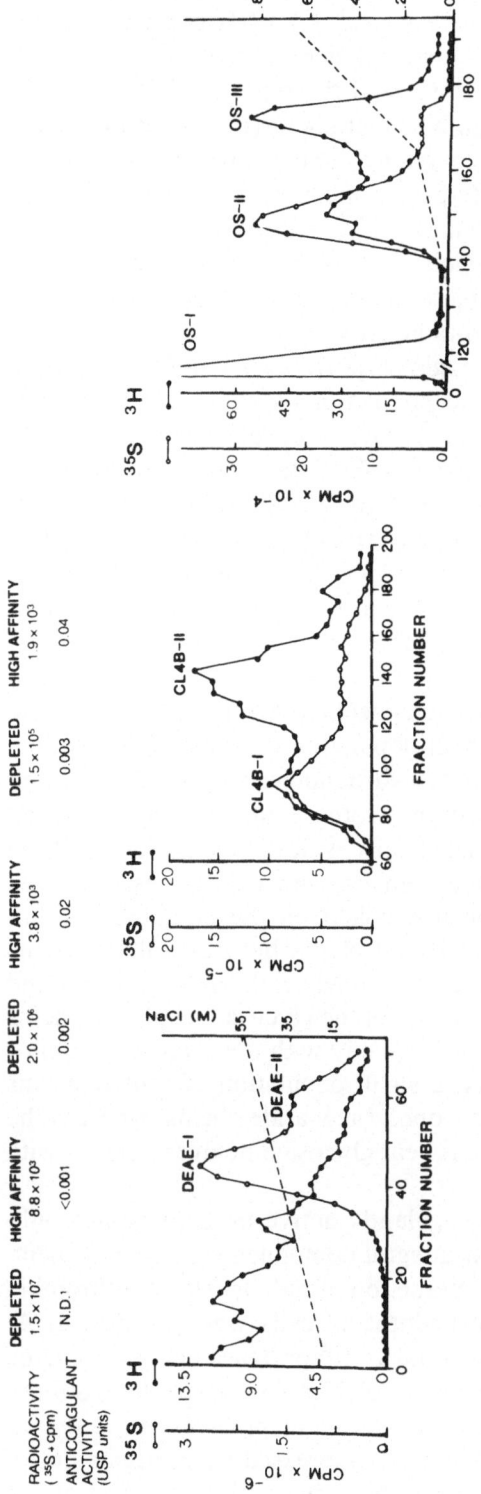

Figure 5. Isolation of radiolabeled proteoglycans from cloned endothelial cells. Radiolabeled material obtained from metabolically labeled cloned endothelial cells and desalted on Sephadex G-50, was applied to DEAE–Sephacel, and bound material was eluted with a linear salt gradient. The above radiolabeled macromolecules (peaks DEAE-I and DEAE-II) were gel-filtered on Sepharose CL4B. Radiolabeled proteoglycans obtained from gel filtration (peak CL4B-I) were applied to octyl-Sepharose, and bound material was eluted with a Triton X-100 gradient. Proteoglycans (peaks OS-I, OS-II, and OS-III) as well as heparan sulfate chains (peak CL4B-II) were affinity-fractionated employing immobilized antithrombin (AT-Affigel). N.D., none detected.

known substrate specificities showed that peak CL4B-I contained a mixture of proteoheparan and proteochondroitin, whereas peak CL4B-II consisted of heparan and chondroitin chains (Fig. 5). Radiolabeled proteoglycans from peak CL4B-I were next applied to octyl–Sepharose, and three peaks of ^{3}H- and ^{35}S-labeled material were eluted with a Trition X-100 gradient (Fig. 5). Peaks OS-II and OS-III, which constitute partially degraded and native proteoglycans, emerged from the column at added detergent concentrations of $\sim$ 0.06 and $\sim$ 0.26%, respectively, and possessed $\sim$ 13 and $\sim$ 2%, respectively, of the starting radioactive sulfate (Fig. 5). Peak OS-I, which most likely represents a badly degraded proteoglycan without a hydrophobic domain, did not bind to the matrix and accounted for $\sim$ 85% of the initial ^{35}S-radioactivity (Fig. 5).

The proteoglycans with hydrophobic regions (peaks OS-II and OS-III) were then affinity-fractionated into two separate populations with immobilized antithrombin (Fig. 5). The hydrophobic heparan sulfate proteoglycans which bound tightly to the protease inhibitor represented $<$ 1% of the starting material and exhibited a specific anticoagulant activity of $\sim$ 5 to 21 USP anticoagulant units/ 10^6 ^{35}S-cpm. The hydrophobic heparan sulfate proteoglycans which interacted weakly with the protease inhibitor constituted $>$ 99% of the starting material and possessed an anticoagulant potency of $\sim$ 0.001 to 0.02 USP unit/10^6 ^{35}S-cpm. Similar results have been obtained from studies conducted with cloned rat epididymal rat pad endothelial cells.

The above data indicate that all of the anticoagulantly active heparan sulfate chains may be covalently coupled to a small population of unique hydrophobic core proteins. To test this model, high-affinity hydrophobic proteoglycans as well as those with minimal ability to interact with the protease inhibitor were obtained from cloned rat epididymal fat pad endothelial cells as described in Fig. 5. These two subsets of proteoglycans were treated with papain, and the resultant free mucopolysaccharide chains were rechromatographed on the immobilized antithrombin. The results indicated that virtually all of the heparan sulfate chains derived from proteoglycans which could interact strongly with antithrombin bound tightly to the affinity matrix, whereas few, if any, of the glycosaminoglycan chains of the proteoglycan which could interact only weakly with the protease inhibitor complexed with the affinity matrix. Thus, a small population of proteoglycans contain all of the anticoagulantly active mucopolysaccharide chains, whereas the remaining larger population of proteoglycans bear glycosaminoglycan chains with minimal biologic potency.

The hydrophobic nature of the anticoagulantly active proteoheparans indicates that these substances may represent integral components of the cell membrane.[60] To examine this hypothesis, the interaction of radiolabeled antithrombin with the surface of cloned bovine aortic endothelial cells was quantitated. At $\sim$ 2 days postconfluence, these cells possess 5.8×10^4 protease inhibitor binding sites/cell with an apparent dissociation constant of 12.4 nM. These high-affinity antithrombin receptors are heparan sulfate membrane components since the interaction of the protease inhibitor is completely suppressed by pretreatment of endothelial cells with purified *Flavobacterium* heparinase (Fig. 6). The above

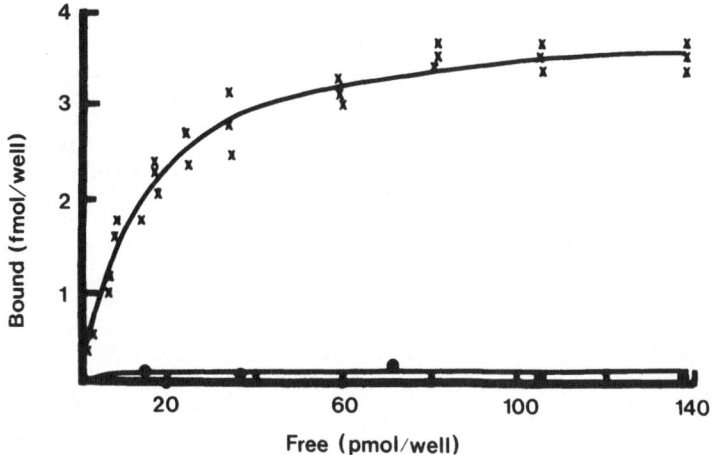

Figure 6. The binding of [^{125}I]antithrombin to bovine aortic endothelial cells. X, untreated endothelial cells; ●, endothelial cells treated with *Flavobacterium* heparinase.

results are in excellent agreement with previous studies utilizing radiolabeled antithrombin and bovine aortic segments.[54]

VI. PHYSIOLOGICAL ROLE OF PROTEOHEPARAN SULFATE

Several investigators have provided evidence both to support and to refute the hypothesis that anticoagulantly active heparan sulfate proteoglycans are present on the luminal surface of the endothelium and could endow blood vessels with nonthrombogenic properties.[10,31,32] Marcum *et al.*[34,38] have attempted to resolve this issue by demonstrating that intact blood vessels possess heparan sulfate proteoglycans which accelerate antithrombin action.

To this end, rat and mouse hindlimb preparations were perfused *in situ* with purified thrombin until a constant level of the enzyme was present in the effluent, at which time purified antithrombin was infused into the perfusion stream. The amount of thrombin–antithrombin complex generated within the vasculature was estimated by a specific radioimmunoassay for this interaction product. The rate of enzyme–inhibitor complex formation in the rat and mouse hindlimb preparations was augmented by as much as 19-fold and 15-fold, respectively, when compared to the amount of interaction product generated in the absence of the mucopolysaccharide (Table 5). To demonstrate that the antithrombin accelerating activity present within the hindlimb vasculature is due to heparan sulfate, a purified preparation of *Flavobacterium* heparinase was recirculated through the hindlimb vasculature prior to perfusion of the hemostatic components. Pretreatment of the luminal surface resulted in a reduction in the amount of enzyme–inhibitor complex generated to uncatalyzed levels (Table 5). The anatomical location of these mucopolysaccharides within the hemicorpus preparation has also been examined. Buffer was recirculated through the hindlimb preparation for extended

Table 5.   In Situ Generation of Thrombin–Antithrombin Complex in the Hindlimb
Preparation

Proteins perfused through the hindlimb preparation[b]	Thrombin–antithrombin complex[a]		
		Mouse	
	Rat	+/+	W/Wᵛ
Thrombin followed by native antithrombin	0.94 ± 0.17 (n = 13)	0.48 ± 0.04 (n = 13)	0.48 ± 0.05 (n = 14)
Thrombin followed by modified antithrombin	0.22 ± 0.04 (n = 4)	0.05 ± 0.03 (n = 6)	0.05 ± 0.03 (n = 6)
Thrombin after treatment with purified heparinase followed by antithrombin	0.08 ± 0.01 (n = 4)	0.05 ± 0.02 (n = 5)	0.05 ± 0.01 (n = 7)
Uncatalyzed amount	0.09	0.03	0.03

[a] Values are mean ± S.E. Enzyme–inhibitor complex formation within the cannula was subtracted from the original data to obtain the amount of complex generated within the vasculature of the rat (pmoles/20 sec) and mouse (pmoles/17 sec).

[b] Thrombin was perfused through hindlimb vasculature at an initial concentration of 5.4 nM, and antithrombin was perfused through rat and mouse vasculature at initial concentrations of 0.18 and 0.09 μM, respectively.

periods of time, and the presence of mucopolysaccharide within the recirculated buffer was ascertained by determining the acceleration of thrombin–antithrombin complex formation. No anticoagulant activity was detected in the buffer, implying that the anticoagulantly active molecules are tightly complexed with the endothelial cell surface.[34]

To show that heparan sulfate present on the endothelium functions in a manner similar to commercial heparin, antithrombin modified at Trp_{49} was substituted for native protease inhibitor. This change in components produced a dramatic decrease in the amounts of thrombin–antithrombin complex generated within the hindlimb preparation to uncatalyzed levels (Table 5). These observations indicate that the heparan sulfate present on the endothelium potentiates thrombin–antithrombin interactions via a mechanism dependent upon the same critical tryptophan residue which is required for heparin-induced acceleration of protease inhibitor action (see above).

To evaluate whether platelet factor 4 (PF_4), a small polypeptide released from the α-granules of platelets, can modulate the acceleratory phenomena, purified human PF_4 was added to an antithrombin bolus prior to injection into the rat hindlimb perfusion stream. Addition of PF_4 to antithrombin decreased enzyme–inhibitor complex formation to the uncatalyzed rate.[34] This platelet peptide has been shown to interfere with binding of heparin to antithrombin, thereby inhibiting the acceleration of enzyme–inhibitor complex formation. These data suggest that PF_4, after secretion from activated platelets, may play a role in thrombogenesis by neutralizing anticoagulantly active mucopolysaccharides present on the vascular endothelium.

To assess the contribution of mast cells to the maintenance of blood fluidity,

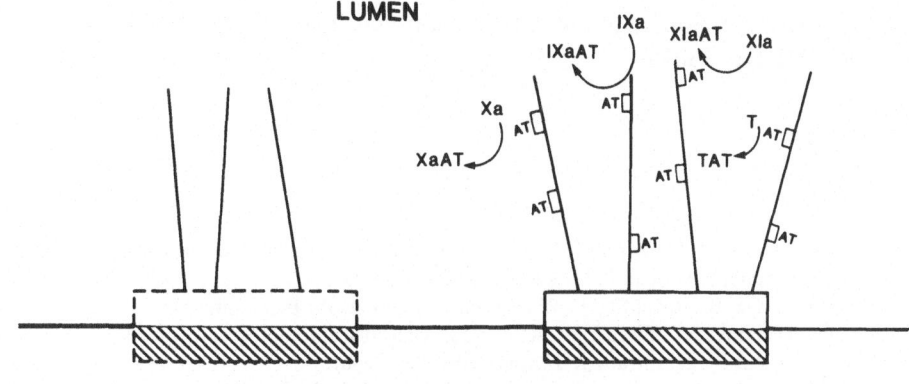

Figure 7. Model of anticoagulantly active (solid-line figure) and inactive (dashed-line figure) proteoheparans positioned in cell membrane (hatched lines).

the hindlimb vasculature of mast cell-deficient mice (W/W^v) and littermates containing normal levels of mast cells ($+/+$), was perfused with purified human thrombin and antithrombin.[38] The generation of thrombin–antithrombin complex by the vascular system of W/W^v and $+/+$ mice was enhanced to a comparable extent (Table 5). These observations indicate that heparan sulfate proteoglycans synthesized by the endothelium but not mast cells are responsible for maintaining the nonthrombogenic properties of blood vessels.

In the light of the above observations, we suggest that a small percentage ($< 1\%$) of circulating plasma antithrombin is normally bound to a specific subset of heparan sulfate proteoglycans synthesized by macrovascular or microvascular endothelial cells. This behavior allows the protease inhibitor to be selectively activated at blood surface interfaces where serine proteases of the intrinsic coagulation cascade are generated. Antithrombin is thereby critically positioned to neutralize these hemostatic enzymes and to protect natural surfaces against thrombus formation (Fig. 7). The catalytic nature of these interactions would ensure that this nonthrombogenic barrier is continually regenerated. Indeed, one can estimate from the above data that the amount of anticoagulant activity generated by the total number of endothelial cells ($\sim 1.4 \times 10^{10}$) within the cardiovascular system of a human is about 10 USP units of biologic activity. Given that the volume of the microvasculature is roughly 5% of the total plasma volume (~ 150 ml for an 80-kg person), we can calculate that the average level of heparin-like activity within the capillary bed of humans is ~ 0.1 USP unit/ml. This relatively high concentration of anticoagulantly active heparan sulfate would ensure that coagulation enzymes would be rapidly inhibited by antithrombin within the microcirculation. Alterations in the synthesis and/or placement of the anticoagulantly active heparan sulfate proteoglycans on the surface of microvascular and macrovascular endothelial cells could be responsible for arterial or venous thrombotic disease in humans.

ACKNOWLEDGMENT. Supported in part by National Institutes of Health Grants HL 34800 and PO1-HL 33014.

REFERENCES

1. Abildgaard, U., 1968, Highly purified antithrombin III with heparin cofactor activity prepared by disc gel electrophoresis, *Scand. J. Clin. Lab. Invest.* **21:**89–91.
2. Andersson, L. O., Barrowcliffe, T. W., Holmer, E., Johnson, E. A., and Sims, G. E. C., 1976, Anticoagulant properties of heparin fractionated by affinity chromatography on matrix-bound antithrombin III and by gel filtration, *Thromb. Res.* **9:**575–581.
3. Atha, D. H., Stephens, A. W., Rimon, A., and Rosenberg, R. D., 1984, Sequence variation in heparin octasaccharides with high affinity for antithrombin III, *Biochemistry* **23:**5801–5812.
4. Atha, D. H., Stephens, A. W., and Rosenberg, R. D., 1984, Evaluation of critical groups required for binding of heparin to antithrombin, *Proc. Natl. Acad. Sci. USA* **81:**1030–1034.
5. Atha, D. H., Lormeau, J.-C., Petitou, M., Choay, J., and Rosenberg, R. D., 1985, Contribution of monosaccharide residues in heparing binding to antithrombin III, *Biochemistry* **24:**6723–6729.
6. Bjork, I., Jackson, C. M., Jornvall, H., Lavine, K. K., Nordling, K., and Salsgiver, W. J., 1982, The active site of antithrombin, *J. Biol. Chem.* **257:**2406–2411.
7. Blackburn, M. N., and Sibley, C. C., 1980, The heparin binding site of antithrombin III, *J. Biol. Chem.* **255:**824–826.
8. Brinkhous, K. M., Smith, H. P., Warner, E. D., and Seegers, W. H., 1939, The inhibition of blood clotting: An unidentified substance which acts in conjunction with heparin to prevent the conversion of prothrombin to thrombin, *Am. J. Physiol.* **125:**683–687.
9. Broze, G. J., Jr., and Majerus, P. W., 1980, Purification and properties of human coagulation factor VII, *J. Biol. Chem.,* **255:**1242–1250.
10. Busch, C., and Owen, W. G., 1982, Identification *in vitro* of an endothelial cell surface cofactor for antithrombin III: Parallel studies with isolated perfused rat hearts and microcarrier cultures of bovine endothelium, *J. Clin. Invest.* **69:**726–729.
11. Choay, J., Lormeau, J.-C., Petitou, M., Sinay, P., Casu, B., Oreste, P., Torri, G., and Gatti, G., 1980, Anti-Xa active heparin oligosaccharides, *Thromb. Res.* **18:**573–578.
12. Choay, J., Petitou, M., Lormeau, J.-C., Sinay, P., Casu, B., and Gatti, G., 1983, Structural–activity relationship in heparin: A synthetic pentasaccharide with high-affinity for antithrombin III and eliciting high anti-factor Xa activity, *Biochem. Biophys. Res. Commun.* **116:**492–499.
13. Contejean, C., 1895, Recherches sur les injections intraveineuses de peptone et leur influence sur la coagulabilité du sang chez le chien, *Arch. Physiol. Norm. Pathol.* **7:**45–53.
14. Damus, P. S., Hicks, M., and Rosenberg, R. D., 1973, Anticoagulant action of heparin, *Nature* **246:**355–357.
15. Davie, E. W., and Ratnoff, O. D., 1964, Waterfall sequence of intrinsic blood clotting, *Science* **145:**1310–1312.
16. Ferguson, W. S., and Finlay, T. H., 1983, Localization of disulfide bond in human antithrombin III required for heparin-accelerated thrombin activation, *Arch. Biochem. Biophys.* **221:**304–307.
17. Galli, S. J., Dvorak, A. M., and Dvorak, H. F., 1984, Mast cells and basophils: Morphologic insights into their biology, secretory patterns and function, *Prog. Allergy* **34:**1–141.
18. Godal, H. C., Rygh, M., and Laake, K., 1974, Progressive inactivation of purified factor VII by heparin and antithrombin III, *Thromb. Res.* **5:**773–775.
19. Hook, M., Bjork, I., Hopwood, J., and Lindahl, U., 1976, Anticoagulant activity of heparin: Separation of high-activity and low-activity heparin species by affinity chromatography on immobilized antithrombin, *Fed. Eur. Biochem. Soc. Lett.* **66:**90–93.
20. Howell, W. H., and Holt, E., 1918, Two new factors in blood coagulation: Heparin and pro-antithrombin, *Am. J. Physiol.* **47:**328–341.
21. Hoylaerts, M., Owen, W. G., and Collen, D., 1984, Involvement of heparin chain length in the heparin-catalyzed inhibition of thrombin by antithrombin III, *J. Biol. Chem.* **259:**5670–5677.

22. Jarrott, B., Hjelle, J. T., and Spector, S., 1979, Association of histamine with cerebral microvessels in regions of bovine brain, *Brain Res.* **168:**323–330.
23. Jordan, R. E., Oosta, G. M., Gardner, W. T., and Rosenberg, R. D., 1980, The binding of low-molecular-weight heparin to hemostatic enzymes, *J. Biol. Chem.* **255:**10073–10080.
24. Jordan, R. E., Oosta, G. M., Gardner, W. T., and Rosenberg, R. D., 1980, The kinetics of hemostatic enzyme–antithrombin interactions in the presence of low-molecular-weight heparin, *J. Biol. Chem.* **255:**10081–10090.
25. Jordan, R. E., Favreau, L. V., Braswell, E. H., and Rosenberg, R. D., 1982, Heparin with two binding sites for antithrombin or platelet factor 4, *J. Biol. Chem.* **257:**400–406.
26. Karp, G. I., Marcum, J. A., and Rosenberg, R. D., 1984, The role of tryptophan residues in heparin–antithrombin interactions, *Arch. Biochem. Biophys.* **233:**712–720.
27. Lam, L. H., Silbert, J. E., and Rosenberg, R. D., 1976, The separation of active and inactive forms of heparin, *Biochem. Biophys. Res. Commun.* **69:**570–577.
28. Leder, I. G., 1980, A novel 3-*O*-sulfatase from human urine acting on methyl-2-deoxy-2-sulfamino-alpha-d-glucopyranoside 3-sulfate, *Biochem. Biophys. Res. Commun.* **94:**1183–1189.
29. Lindahl, U., Backstrom, G., Thunberg, L., and Leder, I. G., 1980, Evidence for a 3-Osulfated d-glucosamine residue in the antithrombin-binding sequence of heparin, *Proc. Natl. Acad. Sci. USA* **77:**6551–6555.
30. Lindahl, U., Backstrom, G., and Thunberg, L., 1983, The antithrombin-binding sequence in heparin: Identification of an essential 6-*O* sulfate group, *J. Biol. Chem.* **258:**9826–9830.
31. Lollar, P., and Owen, W. G., 1980, Clearance of thrombin from the circulation in rabbits by high-affinity binding sites on the endothelium: Possible role in the inactivation of thrombin by antithrombin III, *J. Clin. Invest.* **66:**1222–1330.
32. Lollar, P., MacIntosh, S. C., and Owen, W. G., 1984, Reaction of antithrombin III with thrombin bound to the vascular endothelium: Analysis in a recirculating perfused rabbit heart preparation, *J. Biol. Chem.* **259:**4335–4338.
33. Marcum, J. A., Fritze, L., Galli, S. J., Karp, G., and Rosenberg, R. D., 1983, Microvascular heparin-like species with anticoagulant activity, *Am. J. Physiol.* **245:**H725–H733.
34. Marcum, J. A., McKenney, J. B., and Rosenberg, R. D., 1984, The acceleration of thrombin–antithrombin complex formation in rat hindquarters via naturally occurring heparin-like molecules bound to the endothelium, *J. Clin. Invest.* **74:**341–350.
35. Marcum, J. A., and Rosenberg, R. D., 1984, Anticoagulantly active heparin-like molecules from vascular tissue, *Biochemistry* **23:**1730–1737.
36. Marcum, J. A., and Rosenberg, R. D., 1985, Heparinlike molecules with anticoagulant activity are synthesized by cultured endothelial cell, *Biochem. Biophys. Res. Commun.* **126:**365–372.
37. Marcum, J. A., Atha, D. H., Fritze, L. M. S., Nawroth, P., Stern, D., and Rosenberg, R. D., 1986, Cloned bovine aortic endothelial cells synthesize anticoagulantly active heparan sulfate proteoglycan, *J. Biol. Chem.* **261:**7507–7517..
38. Marcum, J. A., McKenney, J. B., Galli, S. J., Jackman, R. W., and Rosenberg, R. D., 1986, Anticoagulantly active heparinlike molecules from mast cell-deficient mice, *Am. J. Physiol.* **250:**H879–H888.
39. McLean, J., 1916, The thromboplastic action of cephalin, *Am. J. Physiol.* **41:**250–257.
40. Meezan, E., Brendel, K., and Carlson, E. C., 1974, Isolation of a purified preparation of metabolically active retinal blood vessels, *Nature* **251:**65–67.
41. Monkhouse, F. C., France, E. S., and Seegers, W. H., 1955, Studies on the antithrombin and heparin cofactor activities of a fraction absorbed from plasma by aluminum hydroxide, *Circ. Res.* **3:**397–402.
42. Morowitz, P., 1968, *The Chemistry of Blood Coagulation,* Thomas, Springfield, Ill.
43. Murano, G., Williams, L., Miller-Andersson, M., Aronson, D., and King, C., 1980, Some properties of antithrombin III and its concentration in human plasma, *Thromb. Res.* **18:**259–262.
44. Nordenman, B., Nystrom, C., and Bjork, I., 1977, The size and shape of human and bovine antithrombin III, *Eur. J. Biochem.* **78:**195–203.
45. Oosta, G. M., Gardner, W. T., Beeler, D. L., and Rosenberg, R. D., 1981, Multiple functional domains of the heparin molecule, *Proc. Natl. Acad. Sci. USA* **78:**829–833.

46. Pecon, J. M., and Blackburn, M. N., 1984, Pyridoxylation of essential lysines in the heparin-binding site of antithrombin III, *J. Biol. Chem.* **259**:935–938.
47. Petersen, E. E., Dudek-Wojciechowska, G., Sottrup-Jensen, L., and Magnusson, S., 1979, The primary structure of antithrombin III (heparin-cofactor): Partial homology between alpha₁-antitrypsin and antithrombin III, in: *The Physiological Inhibitors of Coagulation and Fibrinolysis* (D. Collen, B. Wiman, and M. Verstraete, eds.), Elsevier/North-Holland, Amsterdam, pp. 43–54.
48. Prochownik, E. V., and Orkin, S. H., 1984, In vivo transcription of a human antithrombin III "minigene," *J. Biol. Chem.* **259**:15386–15392.
49. Rosenberg, R. D., and Damus, P. S., 1973, The purification and mechanism of action of human antithrombin–heparin cofactor, *J. Biol. Chem.* **248**:6490–6505.
50. Rosenberg, J. S., McKenna, P., and Rosenberg, R. D., 1975, Inhibition of human factor IXa by human antithrombin–heparin cofactor, *J. Biol. Chem.* **250**:8883–8888.
51. Rosenberg, R. D., Armand, G., and Lam, L., 1978, Structure–function relationship of heparin species, *Proc. Natl. Acad. Sci. USA* **75**:3065–3069.
52. Rosenberg, R. D., and Lam, L. H., 1979, Correlation between structure and function of heparin, *Proc. Natl. Acad. Sci. USA* **76**:1218–1222.
53. Stead, N., Kaplan, A. P., and Rosenberg, R. D., 1976, Inhibition of activated factor XII by antithrombin–heparin cofactor, *J. Biol. Chem.* **251**:6481–6488.
54. Stern, D., Nawroth, P., Marcum, J. A., Handley, D., Rosenberg, R. D., Kisiel, W., and Stern, K., 1985, Interaction of antithrombin III with bovine aortic segments: Role of heparin in binding and enhanced anticoagulant activity, *J. Clin. Invest.* **75**:272–279.
55. Stone, A. L., Beeler, D., Oosta, G., and Rosenberg, R. D., 1982, Circular dichroism spectroscopy of heparin–antithrombin interactions, *Proc. Natl. Acad. Sci. USA* **79**:7190–7194.
56. Teien, A. N., Abildgaard, U., and Hook, M., 1976, The anticoagulant effect of heparan sulfate and dermatan sulfate, *Thromb. Res.* **8**:859–867.
57. Villanueva, G. B., Perret, V., and Danishefsky, I., 1980, Tryptophan residue at the heparin binding site in antithrombin III, *Arch. Biochem. Biophys.* **203**:453–457.
58. Villanueva, G. B., 1984, Predictions of the secondary structure of antithrombin III and the location of the heparin-binding site, *J. Biol. Chem.* **259**:2531–2536.
59. Waugh, D. F., and Fitzgerald, M. A., 1956, Quantitative aspects of antithrombin aad heparin in plasma, *Am. J. Physiol.* **184**:627–639.
60. Yanagishita, M., and Hascall, V. C., 1984, Proteoglycans synthesized by rat ovarian granulosa cells in culture: Isolation, fractionation, and characterization of proteoglycans associated with cell layer, *J. Biol. Chem.* **259**:10260–10269.

The Fibrinolytic System of Cultured Endothelial Cells

Scott A. Curriden, Thomas J. Podor, and David J. Loskutoff

I. INTRODUCTION

A. Overview

Vascular homeostasis is achieved through a complex set of reactions and interactions that occur largely at the intimal surface of the blood vessel and involve platelets, leukocytes, coagulation factors, and the vessel wall itself.[1,2] The coordinated action of a vast array of highly specific serine proteinases and associated inhibitors is the hallmark of this ongoing process,[3] and guarantees not only the formation of a stable thrombus at specific sites of vascular injury, but also its timely removal in order to restore blood vessel patency.[4] The thrombus itself consists primarily of platelets. It is "solidified" through the action of thrombin, an enzyme which proteolytically converts fibrinogen into fibrin and in the process binds the platelets to each other and to the vessel wall. The removal of the thrombus (thrombolysis) requires the degradation of this fibrin by plasmin formed specifically on the clot surface.[5] The action of thrombin is regulated by antithrombin III[6] while that of plasmin is limited by α_2-antiplasmin.[7,8] Although the identity and biochemical properties of the components comprising the coagulation system are well established (see Refs. 3, 9 for review), the nature of the molecules that initiate and control the fibrinolytic system of plasma is only now beginning to be understood. Fibrinolytic components such as plasminogen activators (PAs) and plasminogen activator inhibitors (PAIs) exist in plasma in trace amounts, making their isolation from and analysis in blood difficult.[10] Much of the recent progress in this field can be attributed to the decision of a number of investigators to isolate and study PAs and PAIs in cell culture systems where their concentrations are considerably higher, and their rates of synthesis can be experimentally manipulated.[10-12] These studies have made it abundantly clear that the fibrinolytic system is quite complex, consisting not only of multiple forms of PA, but also of

Scott A. Curriden, Thomas J. Podor, and David J. Loskutoff • Scripps Clinic and Research Foundation, La Jolla, California 92037.

multiple PAIs that may regulate their activity. Many of these studies suggest that the endothelium may be the hub of the vascular fibrinolytic system. Not only are most of the plasma fibrinolytic components synthesized by endothelial cells, but the endothelial cell plasma membrane and extracellular matrix may also provide surfaces upon which many of the fibrinolytic components are assembled. The immediate proximity of endothelium to blood-borne regulatory elements further emphasizes the potential role of endothelial cells in initiating and regulating vascular fibrinolysis.

We have been studying the fibrinolytic system of cultured bovine aortic endothelial cells (BAEs) in order to better understand this complex system and its regulation. The purpose of this review is to provide a brief summary of this system as we now understand it. For background, we will begin with a brief discussion of the fibrinolytic system of plasma. The fibrinolytic components produced by BAEs will then be discussed and compared to those present in plasma. Finally, we will identify molecules which regulate the synthesis of fibrinolytic components and the overall fibrinolytic activity of endothelial cells, and discuss their mechanism of action.

B. The Fibrinolytic System of Plasma

In its simplest form, the fibrinolytic system consists of plasminogen and those molecules that convert this inactive proenzyme into the active, trypsin-like enzyme, plasmin.[10] Although it has been viewed primarily as the system responsible for the dissolution of blood clots, it is now clear that its components also function in a variety of other biological processes, including ovulation, embryo implantation, macrophage activation, neoplasia, breast involution, tissue repair, and neovascularization.[5,10,12] These observations have led to the suggestion that plasminogen activation may be "one of the broadest of the fundamental processes of physiology."[11]

1. Plasminogen Activators

The primary fibrinolytic enzyme is plasmin although other enzymes have been implicated in vascular fibrinolysis as well.[13] Plasmin is not normally present in blood. It is derived from the circulating zymogen plasminogen through limited proteolytic cleavage by PAs.[5,11,12] PAs are highly specific serine proteinases which cleave an Arg–Val bond in plasminogen, converting it from a single polypeptide chain of approximately 90,000 daltons to a two-chain, disulfide-linked molecule of the same size.[14] Because plasminogen is present in most body fluids, changes in the availability and activity of PAs appear to contribute most significantly to the dynamic regulation of this enzyme system. PAs can be classified into two distinct groups, the urokinase-type (uPAs) and the tissue-type (tPAs) molecules. Although these PAs originate from separate genes[15,16] and are immunologically distinct molecules,[17] they both activate plasminogen through cleavage of the same peptide bond. However, the mechanisms of activation differ considerably. Unlike uPA, tPA requires a cofactor before its activity can be expressed, which, interestingly, seems to be fibrin itself.[18]

tPA has a high affinity for fibrin, but not for fibrinogen. It is in fact a poor PA in the absence of fibrin since its K_m for plasminogen is 60 μM and the concentration of plasminogen in blood is approximately 2 μM.[19] However, when fibrin is formed, both tPA and plasminogen bind to it, lowering the K_m of tPA for plasminogen to approximately 0.1 μM. As a consequence, plasmin formation occurs on the fibrin surface and not free in the circulation. tPA has also been shown to bind to or be stimulated by denatured proteins,[20] extracellular matrix components,[21] and cell surfaces,[22,23] interactions which may also promote plasminogen activation. Native tPA consists of 527 amino acids and has a molecular weight of approximately 72,000,[12,15,24] although the size may vary depending on its source and the extent of glycosylation. Two distinct portions of tPA can be distinguished after cleavage with plasmin. The "A" or heavy chain contains 275 amino acids (approximate M_r 36,000) and includes the NH_2-terminal portion of the molecule as well as two structures which are similar to the "kringles" found in plasminogen, single-chain uPA (scuPA), and prothrombin.[15] It also contains a region which shows homology with the "finger-like" structures in fibronectin[25] and a region termed the growth factor domain which shows homology with epidermal growth factor.[26] The "B" or light chain of tPA consists of 253 amino acids (approximate M_r 32,000), originates from the COOH-terminal end of the molecule, and contains the active-site region of the enzyme.[12] This portion of tPA shows considerable homology with the light chain of other serine proteinases.[15,27] Ny et al.[28] have shown that the domains of tPA are encoded by at least 14 exons, which are separated from each other by introns. The domains of tPA appear to function independently of each other,[29] supporting the hypothesis that the tPA gene evolved by exon shuffling.[27,28] Both the finger domain and the kringle-2 domain of tPA appear to be involved in fibrin binding.[30] tPA has been localized both immunologically[31,32] and enzymatically[33] to the vascular endothelium, suggesting that the endothelial cell is its site of origin.

uPA has been isolated from plasma in both a 33,000- and a 54,000-approximate-M_r form,[5,34,35] the smaller derived from the larger by proteolytic processing. Both forms are fully active and are rapidly inactivated by treatment with diisopropylfluorophosphate (DFP). Evidence is, however, accumulating from a number of studies to indicate that uPA is actually synthesized in the single-chain (scuPA) form.[35-40] scuPA (approximate M_r 54,000) is resistant to DFP and must therefore be considered a proenzyme. Sequence data suggest that the A chain (residues 1-157) forms the NH_2-terminus of scuPA while the B chain (residues 159-411) forms the COOH-terminal portion. The scuPA gene is also organized into distinct functional regions or domains, each sharing homology with other proteins.[16] These include a growth factor domain which shows homology with epidermal growth factor, a single kringle domain which shows strong homology to one of the kringle structures in plasminogen and tPA, and the active site domain. The growth factor domain appears to be involved in the binding of uPA to cells.[41]

As mentioned, urokinase has no affinity for fibrin and was long thought to lack fibrin specificity.[34] Recent evidence, however, suggests that the scuPA form of urokinase may exhibit clot specificity, although by an entirely different mechanism.[42,43] This specificity does not appear to result from the binding of scuPA to fibrin, but rather from the fibrin-mediated dissociation of scuPA from its competitive inhibitor.[43]

2. Fibrinolytic Inhibitors

Inhibitors of fibrinolysis are an integral part of the fibrinolytic system, functioning to regulate the formation and activity of plasmin.[44,45] There are two types of fibrinolytic inhibitors: those that inhibit plasmin (antiplasmins) and those that inhibit PA (PAIs). Inhibitors of plasmin are readily detected in platelets,[46] the vascular wall,[47] and a number of other nonvascular tissues and cells, but most remain poorly characterized. Plasma itself contains at least five antiplasmins although it is now clear that α_2-antiplasmin is the physiologically relevant one.[7,8,45]

Although it was historically difficult to demonstrate the presence of a biologically significant PAI in normal plasma,[5,48] a number of recent reports suggest the existence of such a molecule in cells and in the plasma of some individuals at risk to develop thrombotic problems.[10,49,50] In fact, it is now clear that there are multiple PAIs, although only two of these have been detected in plasma.[50-56] In an effort to clarify the confusion within the published nomenclature, these PAIs have now been classified into four distinct groups based on immunologic criteria.* The PAI-1 group of inhibitors includes those previously termed the "endothelial cell-type" PAI,[55-64] the "rapid" or "fast-acting" inhibitor,[10,49,50,52-54,65,66] the "dexamethasone-induced inhibitor,"[67,68] and the "β-migrating" PAI.[69] PAI-1 has been detected in plasma,[10,49,50,52-56,70] serum, and platelets,[55,64,65,69,71] and in a variety of normal and transformed cells[67,68,72-74] including endothelial cells.[55,57-63,75] It appears to be the physiological inhibitor of tPA.[10,50,66,76] The second type of inhibitor (PAI-2) was originally called the "placental-type" inhibitor.[77,78] Although it is not normally detected in plasma, its concentration increases dramatically during pregnancy.[51] PAI-2 has also been reported in a variety of normal and transformed cells and appears to be primarily a urokinase inhibitor.[39,78,79] PAI-3,[80] although present in plasma may be primarily an inhibitor of activated protein C,[81] and protease hexin,[82,83] the fourth group of PAIs, has not been detected in plasma in significant amounts. Considerable evidence now exists to suggest that neutralization of tPA in plasma by PAI-1 may play an important role in regulating the initiation of vascular fibrinolysis.[10,50] In addition, in various clinical and experimental cases of hypofibrinolysis, depressed levels of plasma fibrinolytic activity have been attributed to elevated PAI-1 activity.[10,49,50,70,84-89] Whether the elevated inhibitory activity of plasma originates from endothelial cells, platelets, or elsewhere is not yet clear.

---

* This nomenclature is based on the recommendation of the Subcommittee on Fibrinolysis, International Committee on Thrombosis and Hemostasis, Jerusalem, June 1986 (Collen, D., 1986, Report of the meeting on fibrinolysis, Jerusalem, Israel, June 2, 1986, *Thromb. Haemost.* **56**:415–416).

3. Other Factors

A number of other factors appear to play significant roles in the regulation of the fibrinolytic system. In general, these are endothelial cell-derived products and include thrombospondin,[90] fibronectin,[21] thrombomodulin,[91] protein S,[92-94] and protein S-binding proteins.[95,96] Protein C, a vitamin K-dependent plasma protein, may also influence plasma fibrinolysis.[97-100] Finally, specific binding sites and receptors for fibrinolytic components appear to exist on the surface of endothelial cells[22,23,90,101,102] and on the subendothelial matrix,[21] adding yet additional levels of complexity to the vascular fibrinolytic system.

II. FIBRINOLYTIC SYSTEM OF CULTURED ENDOTHELIAL CELLS

Many of the fibrinolytic components present in plasma are actually synthesized by endothelial cells. In addition, the endothelial cell surface and its extracellular matrix may contribute to the regulation of vascular fibrinolysis. In this section we will summarize some of the approaches we have used and results we have obtained using BAEs in culture.

A. Materials and Methods

1. Cells

The bovine aortic endothelial cell has proven to be a useful model system for studying fibrinolytic components. Bovine aorta are readily available and BAEs are easily removed in nearly pure form upon collagenase treatment.[103] The BAEs grow well in minimum essential medium (MEM) plus 10% calf serum (serum should be screened for ability to support long-term cultures and for optimal production of fibrinolytic components) and do not require additional growth factors such as endothelial cell growth factor (ECGF). The BAEs grow for many generations, enabling them to be cloned in order to ensure homogeneity.[17,57] These cells do not require fibronectin coating of plasticware and will remain as monolayers for extended periods with only periodic feeding. The above properties make BAEs ideal for large-scale production of conditioned medium (CM). To do so, the monolayers are washed thoroughly 1 day after feeding and then placed in serum-free medium for 24 hr. This CM is collected and used for assays or as a source for the purification of PAI-1. After collecting CM, the cells are then fed with serum-containing medium and after 24 hr the production cycle is repeated.[57] When metabolic labeling of PAI-1 is required, CM is prepared using leucine-deficient medium in the presence of L-[3,4,5-³H]leucine.[57] To prepare cellular extracts, monolayers are washed thoroughly in phosphate-buffered saline (PBS) and then extracted with Triton X-100.[104]

Human umbilical vein endothelial cells (HUVECs) are also used in several studies. These cells are isolated from fresh umbilical veins, again using collagenase treatment.[105] The cells are plated on fibronectin-coated surfaces and grown in

Medium 199 containing 20% fetal calf serum and supplemented with ECGF and heparin.[106] Cells cultured in this way generally grow well for four or five passages. Other cell lines used in our studies are obtained from the American Type Culture Collection (ATCC) and cultured as instructed.

2. Proteins

a. PAI-1. PAI-1 is purified from BAE serum-free CM by fractionation on concanavalin A–Sepharose followed by preparative SDS–polyacrylamide gel electrophoresis (SDS-PAGE).[57] For experiments in which kinetic constants will be determined, a second SDS-free procedure has been developed.[107] Serum-free BAE CM is again passed over a concanavalin A–Sepharose column. After elution with α-methyl-D-mannoside, inhibitor-containing fractions (as determined by tPA binding assay and reverse fibrin autography, described below) are pooled, treated with DFP, and loaded onto a Sephacryl S-200 column. Inhibitor-containing fractions are once again pooled and then are passed over a column of Blue B Agarose. This column is eluted with a salt gradient and inhibitor-containing fractions are pooled and concentrated using ultrafiltration. The concentrated material is loaded onto a Bio-Gel P-60 column and after elution the fractions are subjected to SDS–PAGE in order to assess their purity.

b. Plasminogen. Plasminogen is isolated from human plasma by affinity chromatography on lysine–Sepharose.[108]

c. tPA. Originally, a four-step procedure was used for the purification of tPA.[24] A simpler two-step procedure has been developed based on the binding of tPA to and elution from a column containing anti-tPA monoclonal antibodies linked to Sepharose 4B, followed by affinity chromatography on lysine–Sepharose.[109]

3. Assays

a. [^{125}I]Fibrin Plate Assay. The [^{125}I]fibrin plate assay is a two-step assay which measures net fibrinolytic activity of samples (CM, cell extract, live growing cells) added to 24-well tissue culture plates.[75,110] First, the PA source (e.g., cell extract, CM) is diluted in a buffer and added to the [^{125}I]fibrin-coated wells in the presence of plasminogen. The PA converts the plasminogen into plasmin. The plasmin which is generated then interacts with insoluble ^{125}I-labeled fibrin which previously had been dried onto the bottom of the wells and the resulting soluble [^{125}I]fibrin degradation products are released and detected with a gamma counter. The released counts per minute are compared to a uPA or tPA standard curve and expressed relative to the total counts per minute released upon treatment with trypsin. The results indicate the amount of active PA which is present (providing antiplasmin is not present in the sample). The total antifibrinolytic activity of a sample (i.e., anti-PA plus antiplasmin) can be established by measuring the capacity of the sample to depress the rate of uPA (or tPA)-mediated fibrinolytic activity in the standard assay.

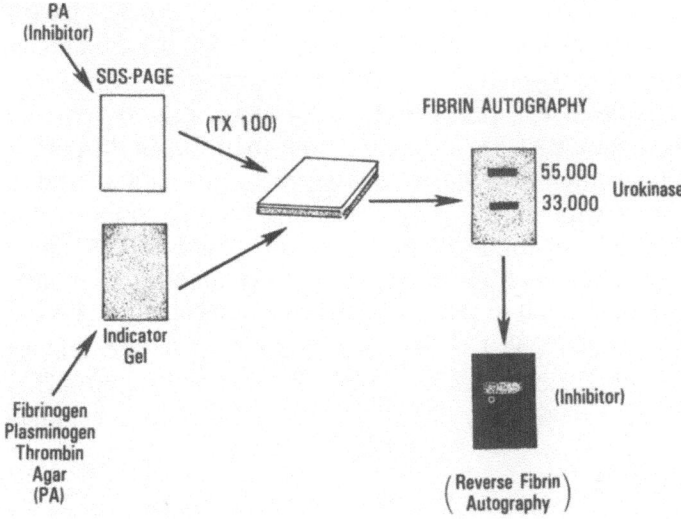

Figure 1. Schematic comparison of fibrin autography and reverse fibrin autography. In fibrin autography, a sample containing PA [in this example, a mixture of high (55,000)- and low (33,000)-molecular-weight urokinase] is subjected to SDS–PAGE. The acrylamide gel is soaked in 2.5% Triton X-100 to neutralize the SDS and then placed onto a fibrin–agar indicator film containing plasminogen. Lytic zones that appear in the fibrin result from the localized formation of plasmin and thus correspond to the positions of PA in the acrylamide gel. In reverse fibrin autoography (shown in parentheses), a sample containing inhibitor is processed in a similar manner but is then assayed on a fibrin–agar indicator film containing not only plasminogen but also PA. Upon incubation, the entire fibrin film gradually lyses because of the plasmin formed throughout. Regions of the film containing inhibitor are relatively resistant to lysis. Thus, inhibitor activity in the acrylamide gel is revealed by the formation of opaque, lysis-resistant areas in the cleared indicator film. (Reproduced from Analytical Biochemistry.[112]).

b. Fibrin Autography and Reverse Fibrin Autography. Fibrin autography and reverse fibrin autography are methods which allow visualization of PA activity and PAI activity, respectively (Fig. 1). To prepare fibrin agar indicator films, agarose is mixed with warm (37°C) PBS containing plasminogen and thrombin. Plasminogen-free fibrinogen in warm PBS is added and the solution is mixed rapidly and poured onto a warm glass plate. After SDS–PAGE, the polyacrylamide gels are soaked in a Triton X-100 solution to remove the SDS, blotted dry with a paper towel, and placed on the surface of the fibrin–agar indicator film. The indicator film is allowed to develop in a humid incubator, and is then photographed at various times. The dark areas of the indicator film correspond to plasmin-mediated lytic zones initiated by the interaction of PA from the polyacrylamide gel with plasminogen in the indicator (fibron autography).[17,111] For reverse fibrin autography, the indicator films are prepared with fibrin containing a PA. Upon incubation, the PA converts the plasminogen into plasmin, and the plasmin formed throughout the film hydrolyzes the opaque fibrin and results in a general clearing (lysis) of the indicator. The white areas that develop in these gels reflect areas relatively resistant to such lysis, and result from the presence of inhibitors in the polyacrylamide gel which diffuse into the indicator.[104,112]

c. *tPA-Binding Assay.* The tPA-binding assay is used to measure the amount of functionally active PAI-1 in a sample.[113] Purified tPA in PBS is allowed to bind to U-bottom microtiter wells during an overnight incubation (cold) and the wells are then washed with buffer. Any remaining binding sites on the plastic are "blocked" during incubation with bovine serum albumin (BSA) and washed again (there is a wash after each of the steps). SDS-treated test samples and purified PAI-1 standard are added to the wells and incubated to allow PAI-1/PA complex formation. The bound PAI-1 is detected by incubation with rabbit antiserum to PAI-1 followed by incubation with ^{125}I-labeled goat anti-rabbit IgG. The wells are removed and the radioactivity in each is determined with a gamma counter.

B. Results

1. PAs

Cultured BAEs are fibrinolytically active and in general secrete 2–8 IU of total PA activity per 10^6 cells per 16 hr (the specific activity of uPA is 100,000 IU/mg).[5,114] Experiments were performed to determine whether this fibrinolytic activity resulted from the presence of single or multiple forms of PA. CM and cell extracts were first fractionated by SDS–PAGE. PAs are stable to SDS and can be assayed following electrophoresis in the presence of SDS by fibrin autography (see Section II.A). These cells were found to produce a complex pattern of PAs corresponding to M_r of approximately 52,000, 74,000, and 116,000 (Fig. 2, lane 1). Thus, the fibrinolytic activity of these cells results from the presence of at least three distinct PA forms.

Further experiments showed that the high-M_r forms (74,000 and 116,000) specifically bound to fibrin, suggesting that they were tPAs. The low-M_r form behaved like uPA since it showed no association with fibrin under any conditions.[115] These conclusions were further supported by experiments which demonstrated that antiserum prepared against tPA only neutralized the activity of the high-M_r forms while antiserum prepared against uPA only neutralized the activity of the low-M_r form.[17] It should be noted that while uPA has been detected in and even purified from endothelial cells,[116–118] uPA may not be produced by endothelium *in vivo*. Primary and early passage BAEs (unpublished observations) and HUVECs[58,59,119,120,121] do not appear to consistently produce uPA. In addition, uPA was not detected by histologic techniques in endothelium *in situ*.[32]

2. PAI-1

As already mentioned, four distinct classes of PAI have been detected to date. The first class, PAI-1, has been shown to be produced by endothelial cells, and it has been detected in endothelial cells from several species (bovine,[104] human,[58–61] porcine[61]). Although originally thought to be an endothelial cell-spe-

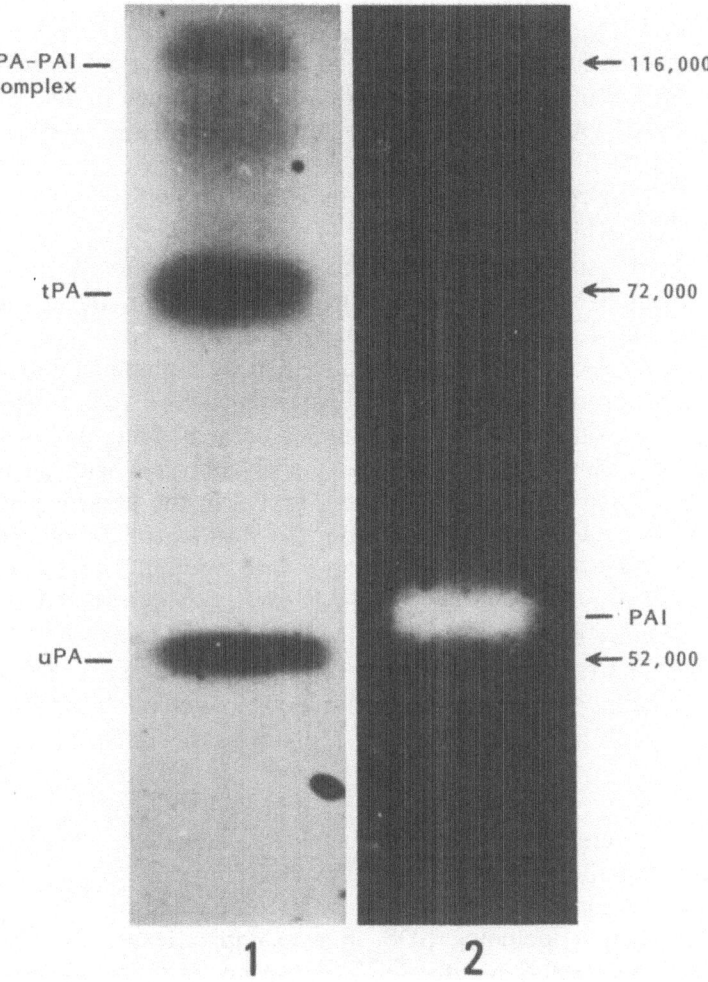

Figure 2. Analysis of BAE CM by fibrin and reverse fibrin autography. BAE CM (100 μl) was fractionated by SDS–PAGE and analyzed for the presence of PAs by fibrin autography (lane 1) and for inhibitors by reverse fibrin autography (lane 2). (Reproduced from Journal of Cellular Biochemistry.[114])

cific protein, it has now been shown to be produced by a variety of cells including human hepatocytes and rat hepatoma cells,[67,68,72] human granulosa cells,[74] human melanoma cells, and transformed fibroblasts.[73,113] It is also present in human platelets, plasma, and serum.[52–56,62,64,65,70,71] PAI-1 differs from PAI-2 and protease nexin in that it exhibits β-mobility when analyzed by agarose zone electrophoresis.[69] Moreover, it inhibits tPA as well as uPA, while PAI-2, PAI-3, and protease nexin inhibit uPA but not tPA. [79,80,82,107] In this section, we will summarize some of the biochemical properties of PAI-1 isolated from BAEs. We will also summarize what we have been able to learn about PAI-1 from our studies of the human PAI-1 gene.

The presence of PAI-1 in CM collected from confluent BAEs was first revealed when a fibrin indicator film similar to the one shown in lane 1 of Fig. 2 for the detection of PAs, spontaneously lysed because of the presence of contaminating tPA in the fibrin preparation.[104] However, one distinct area of the film did not lyse. This lysis-resistant zone in the indicator film was subsequently shown to result from PAI-1 which had diffused out of the SDS gel into the fibrin indicator film and prevented lysis. Refinement of this convenient system for detection of PAI activity resulted in the development of the reverse fibrin autography technique[104,112] (Fig. 1, and Fig. 2, lane 2).

Purification of BAE PAI-1 was accomplished by a combination of concanavalin A affinity chromatography and preparative SDS–PAGE.[57] The purified PAI-1 had an approximate M_r of 50,000 under both reducing and nonreducing conditions. It was shown to be a single-chain glycoprotein with an isoelectric point between 4.5 and 5. When BAEs were grown in the presence of L-[3,4,5-^{3}H]leucine and the PAI-1 was purified from the resultant metabolically labeled CM, an ^{3}H-labeled protein which comigrated with purified PAI-1 during SDS–PAGE was isolated and shown to have PAI activity. This result demonstrated that PAI-1 is actually synthesized by BAEs and not simply an inhibitor present in the growth media (serum) which has been adsorbed to and subsequently released from the cells.[122] This PAI-1 was a major biosynthetic product of BAEs, representing 2.5–12% of the total labeled protein synthesized and secreted in a 24-hr period.[57]

It was fortuitous (with regard to detection of PAI activity) that SDS–PAGE was used both in reverse fibrin autography analysis and in the purification scheme. It is now clear that the vast majority of the PAI-1 produced by BAEs is in an inactive ("latent") form. Activation can be accomplished by treatment with a variety of denaturants including SDS. For example, less than 0.01 U/μl of PAI-1 activity was detected in untreated CM, but medium treated with SDS (1.7 mM), guanidine HCl (4 M), urea (12 M), or KSCN (6 M) contained 0.9, 1.9, 0.8, and 0.5 U/μl, respectively.[123] This effect was dose-dependent with respect to the particular agent used. Activation did not appear to result from the removal of either a small dialyzable component from the medium, or of a high-M_r component that was bound to the latent PAI-1. Rather, it seemed to result from a conformational change in the molecule induced by each of the activating agents.[123] The identity of the biologic equivalent of these denaturants remains to be elucidated. The recognized role of surfaces in coagulation (e.g., phospholipids, platelets)[124] and fibrinolysis (e.g., fibrin,[19] cells[125]) raises the possibility that cellular, lipoprotein, and protein surfaces may be involved in the activation of latent PAI-1 in vivo.

The BAE PAI-1 in CM is an unusually stable molecule in that full activity can be recovered from samples exposed to 1 M acetic acid, 5% 2-mercaptoethanol, or denaturants like SDS, guanidine, and urea.[104] These properties are in contrast to the other PAIs and to plasma protease inhibitors which in general are rapidly inactivated by such treatments.[10,50,126]

Although BAE PAI-1 was stable under conditions which inactivate most other inhibitors, it was extremely sensitive to oxidation.[127] Oxidants such as chloramine-T, N-chlorosuccinimide, and hydrogen peroxide caused a dose-dependent decrease in PAI-1 activity. Studies with α-1-protease inhibitor (α-1-PI) had previously shown that α-1-PI is also extremely sensitive to oxidation and that this sensitivity involves conversion of its P_1 reactive site methionine into methionine sulfoxide by oxidants.[128] The biologic activity of α-1-PI could be restored by treatment of the oxidized inhibitor with methionine sulfoxide peptide reductase, an enzyme which specifically reduces methionine sulfoxide residues in proteins.[129] PAI-1 was found to be even more sensitive to oxidation than α-1-PI and oxidatively inactivated PAI-1 was also reactivated by treatment with the reductase.[127] These results suggested that a methionine residue(s) is also critical for PAI-1 activity. DNA sequence analysis of human PAI-1 (described below) revealed that a methionine residue is present in the reactive center of the inhibitor.

PAI-1 is an antiactivator and can neutralize the activity of both tPA and uPA. Inhibition is associated with the formation of enzyme–inhibitor complexes which are not dissociated by SDS.[57] The tPA/PAI-1 complex has an approximate M_r of 116,000 and upon analysis by Western blotting was recognized by antibodies both to tPA and to PAI-1.[130] Unexpectedly, most of the tPA antigen in CM collected from both BAEs and human endothelial cells is in complex with PAI-1.[58,130,131] Whether these complexes are assembled by the cell prior to secretion, or are formed after secretion, remains to be determined.

To perform accurate kinetic analysis of the interaction between BAE PAI-1 and PAs, it was necessary to purify PAI-1 in the absence of SDS since SDS is difficult to remove from proteins and might interfere with the interaction.[132] Therefore, a procedure was developed to purify PAI-1 in the absence of denaturants.[107] This purified molecule could then be activated by guanidine HCl, a denaturant which is easily removed by dialysis. The dissociation constants for the interaction of guanidine-activated PAI-1 with uPA and with tPA (single- and two-chain forms) were determined.[107] The K_d for the uPA reaction was 2.3×10^{-13} M while the K_ds for single-chain and two-chain tPA were 8×10^{-13} and 1.1×10^{-13} M, respectively. Complex formation with PAI-1 involved a biphasic reaction for all three reactants, consisting of a fast, second-order, initial reaction followed by a much slower, first-order reaction. The initial, second-order, rate constant (k_1) was 1.6×10^8 M^{-1} sec^{-1} for uPA, 3.5×10^7 M^{-1} sec^{-1} for single-chain TPA, and 1.5×10^8 M^{-1} sec^{-1} for two-chain tPA. The first-order rate constants (k_2) were all between 2×10^{-4} and 9×10^{-4} sec^{-1}. Taken together, these kinetic studies indicate that PAI-1 interacts more rapidly and more tightly with both uPA and single- and two-chain tPA than PAI-2,[79,133] PAI-3,[80] or protease nexin,[82] and suggest that PAI-1 is the primary physiological inhibitor of single-chain tPA.[107]

The entire coding sequence for human PAI-1 has been obtained from a λgt$_{11}$ expression library containing human placental cDNA inserts.[134] This cDNA library was expressed in E. coli and screened with an affinity-purified IgG specific for BAE PAI-1. Positive colonies were found to synthesize a high-M_r fusion protein which could be detected by Western blotting with the antibody and which demonstrated PAI activity when analyzed by reverse fibrin autography. Sequence

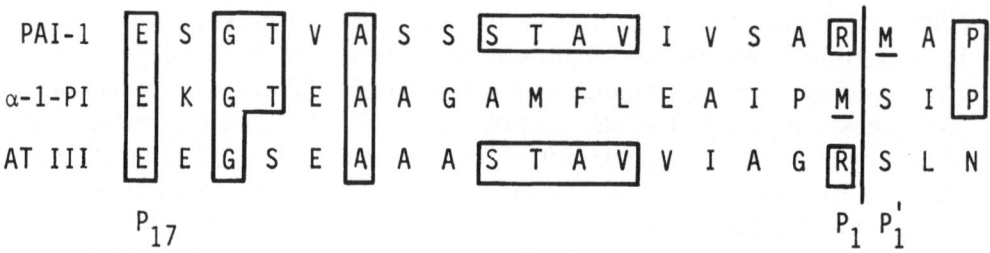

Figure 3. Comparison of PAI-1, α-1-PI, and AT III. Alignment of the sequences of PAI-1, α-1-PI, and AT III around their reactive centers. The reactive site peptide bonds are indicated by the vertical line, and the terminology of the reactive site residues is adapted from Travis and Salvesen.[44] The reactive site methionines are underlined, and amino acids in α-1-PI and AT III homologous to those in PAI-1 are boxed. (Modified from a figure in Proceedings of the National Academy of Sciences, USA.[134])

analysis of positive clones revealed that human PAI-1 consists of 379 amino acids and has approximately 30% homology over its entire length with both α-1-PI and antithrombin III (ATIII). The homology between PAI-1 and the other inhibitors in the region around the presumptive reactive center is shown in Fig. 3. Thus, PAI-1 is clearly a member of the serine protease inhibitor family (serpins[135]). This family of inhibitors evolved from a common ancestral protein over 500 million years ago, and includes molecules which inhibit all the major serine protease cascades of the blood (coagulation, complement, and fibrinolytic). Two features of the PAI-1 deduced amino acid sequence may clarify some of the unusual properties of PAI-1. First, the mature protein lacks cysteine residues, a finding which may account for this inhibitor's high resistance to loss of activity upon reduction. The presence of a methionine residue near the presumptive reactive center may also explain the sensitivity of PAI-1 to oxidative inactivation.

3. Regulation of Fibrinolytic Activity

The net fibrinolytic activity of BAEs changes with the growth state of the cells,[121,136] and in response to a number of agents. For example, simply feeding confluent monolayers with media containing fresh serum caused an 80% decrease in intracellular fibrinolytic activity as compared to unfed controls.[121] Similar decreases were observed when the cells were incubated in the presence of thrombin,[136,137] dexamethasone,[136] or calcium ionophore.[136] Endotoxin and interleukin 1 (IL-1) have also been demonstrated to decrease the fibrinolytic activity of endothelial cells.[89,138,139] The presence of both fibrinolytic activators (uPA, tPA) and inhibitors (PAI-1) in the same sample (i.e., cell extracts and CM) not only makes quantitation of total fibrinolytic activity difficult, but also makes it difficult to establish the mechanism of these altered fibrinolytic states. Clearly, these decreases cannot be assumed to be due simply to a decrease in PA levels, just as the increase in fibrinolytic activity observed when BAEs are treated with activated protein C (APC)[98,99] cannot be assumed to be due simply to an increase in PA levels. Thus, it is the ratio of active PAs to active PAI-1 which determines net fibrinolytic activity. Any change in the "fibrinolytic balance" results in altered fibrinolytic activity. A number of examples exist in the literature in which altered fibrinolytic activity reflects altered PA.[5,10,12] We will now give four examples which demonstrate that net fibrinolytic activity may also be modulated at the PAI-1 level, either in conjunction with changes in PA (e.g., thrombin) or independent of any large changes in PA levels (e.g., endotoxin, IL-1, and APC).

a. *Endotoxin.* When screening culture sera for growth-promoting activity, we noticed a large variation in PAI-1 levels in the resulting CM.[140] These PAI-1 levels varied dramatically between vendors and even between serum lots from the same vendor, implying that factors in the calf serum were capable of regulating PAI-1 production. Characterized sera which had low endotoxin levels were associated with low PAI-1 production (unpublished observation) and low levels of PAI-1 mRNA,[140] making endotoxin a possible regulatory agent. Studies by Colucci *et al.*[89] and others have demonstrated that endotoxin indeed stimulates PAI activity both *in vitro* using HUVECs[89,141,142] or bovine pulmonary artery endothelial cells[138] and *in vivo* using rabbits[89] or rats.[142] Additionally, plasma from patients with septicemia showed markedly increased levels of PAI,[89] raising the possibility that suppression of fibrinolytic activity by endotoxin (through stimulation of PAI) may contribute to the pathogenesis of disseminated intravascular coagulation (DIC) associated with bacterial infection.

b. *IL-1.* IL-1 is a soluble mediator of the immune response, and is produced by a variety of cells including mononuclear phagocytes. Many biologic activities have been attributed to this molecule including that of endogenous pyrogen and inducer of acute-phase protein production.[143] Interestingly, the most potent inducers of IL-1 production are microbial organisms and their products, including endotoxins.[143] When HUVECs were treated with IL-1 there was a large increase in PAI-1 antigen (400–800%) and PAI activity (up to tenfold).[139,142,144] The tPA antigen level also decreased (approximately 50%) upon IL-1 treatment.[139] Interestingly, a small *in vivo* increase in plasma PAI activity has been reported after infusion of IL-1 into rats.[142] The mechanism by which IL-1 stimulates PAI production and the link, if any, between stimulation by endotoxin (exogenous pyrogen) and IL-1 (endogenous pyrogen) remain to be demonstrated although it has been shown that endothelial cells are capable of secreting IL-1 and that this secretion is stimulated by endotoxin.[145-147] IL-1 stimulation of PAI-1 is consistent with its role as an inducer of acute-phase proteins since studies have shown that PAI-1 is an acute-phase protein.[85,86]

c. *Thrombin.* Thrombin was shown to decrease fibrinolytic activity in cultured BAEs.[137] This decrease in activity may reflect an increase in PAI-1 activity since incubation of confluent HUVECs with thrombin for 24 hr caused a sixfold increase in PAI-1 activity.[148] However, the level of tPA antigen also increased upon thrombin treatment (approximately fourfold). Thus, even though both PAI-1 activity and tPA antigen were increased in the medium, comparison of the

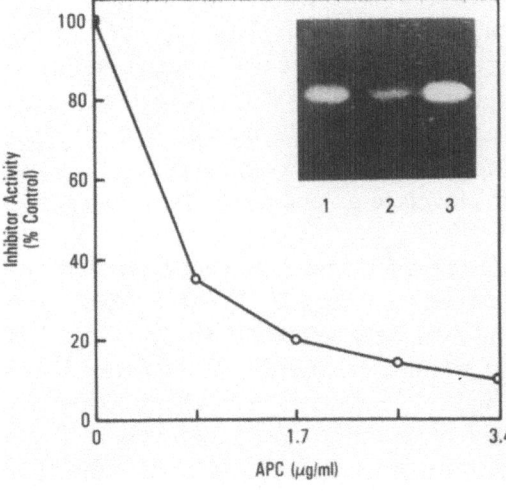

Figure 4. Effect of APC on purified PAI-1. Increasing amounts of APC were mixed with purified PAI-1 (100 μl, 50 ng/ml) and the mixtures were immediately added to [^{125}I]fibrin plates containing plasminogen and uPA. Changes in the inhibitory activity of the PAI-1 were then monitored by the [^{125}I]fibrin plate method as described in Section II.A. The results were normalized to the inhibitory activity of PAI-1 incubated in the absence of APC. Inset: analysis of samples by reverse fibrin autography. Purified PAI-1 (100 μl, 50 ng/ml) was incubated for 30 min at 37°C in the absence of APC (lane 1), or in the presence of either 3.4 μg/ml APC (lane 2) or 3.4 μg/ml DFP-APC (lane 3). After incubation, each of these samples was treated with 0.1% 2-mercaptoethanol to inactivate the APC and then fractionated by SDS–PAGE and analyzed by reverse fibrin autography. (Modified from a figure in Fibrinolysis.[152])

relative increases suggested that the net effect of thrombin would be to decrease fibrinolytic activity and thereby possibly protect clots from premature lysis. This suggestion is supported by the finding that endothelial cells produce IL-1 after exposure to thrombin.[146] The role of thrombin *in vivo* may be even more complex because thrombin activates protein C in the presence of thrombomodulin.[91]

d. APC. Protein C is a vitamin K-dependent plasma protease zymogen that upon activation exhibits both anticoagulant and profibrinolytic activities.[149–151] Recent studies have demonstrated that endothelial cells (HUVECs and BAEs) produce protein S, a cofactor which stimulates APC activity.[92,93] Thus, endothelial cells may be able to regulate the level of APC activity in their vicinity.

Bovine APC was shown by Comp and Esmon[97] to have a pronounced profibrinolytic effect in blood when infused into dogs. These authors suggested that this effect was due to the release of tPA from the endothelium. To further investigate this hypothesis, BAEs were cultured in the presence of bovine APC and shown to have elevated fibrinolytic activity, both in CM and in cell extracts.[98] Similar results were obtained using human APC and human endothelial cells,[99] even though human APC does not appear to be profibrinolytic *in vivo*.[100] The bovine APC caused a dose-dependent increase in the activity of all PA forms, but did not affect the activity of purified tPA, uPA, or tPA/PAI-1 complexes.[152] Preincubation of APC with PAI-1 prior to the addition of tPA, prevented complex formation, suggesting that APC interacts directly with the inhibitor itself. In direct experiments to test this possibility, we showed that APC caused a dose-dependent decrease in PAI-1 activity (Fig. 4), and that complexes between purified bovine APC and PAI-1 readily form and can still be detected after SDS–PAGE.[153] APC itself appears to be inhibited as a consequence of this interaction (unpublished observation). Thus, it seems that the APC-mediated increase in the fibrinolytic activity of these cells is due primarily to a decrease in PAI-1 activity, and not an increase in PA.

III. DISCUSSION

It should be apparent from these studies that the fibrinolytic system of endothelial cells is complex, and that its regulation involves considerably more than altering the rate of PA production. In fact, it seems likely that endothelium may influence vascular fibrinolytic activity through several direct and indirect mechanisms. In most instances, the details of these mechanisms remain to be defined. In this last section, we will briefly summarize the results discussed in previous sections, organizing them into a general discussion of the potential role of endothelium in vascular fibrinolysis. At the same time, we will attempt to point out those observations we consider to be important but whose significance remains unclear in terms of current models of fibrinolysis.

Endothelium can directly influence vascular fibrinolysis through the synthesis of fibrinolytic components. Cultured bovine and human endothelial cells have the potential to produce not only uPAs and tPAs, but also PAI-1, a molecule clearly involved in regulating the fibrinolytic system of these cells and of plasma. We must also consider the possibility that endothelium may directly influence vascular fibrinolysis through the production of other PAIs in addition to PAI-1. This possibility has yet to be addressed, and may add to the complexity of this system. These observations imply that endothelium *in situ* may have similar potential. However, a limited number of observations seem to indicate that uPA is not produced by endothelium.[32,119,120] In addition, although PAI-1 is a major biosynthetic product of these cells, it is a trace component of plasma.[52,55,56] Thus, uPA and PAI-1 production by cultured endothelial cells may not reflect the real situation *in vivo*, but rather may represent changes due to the adaptation of these cells to cell culture (e.g., prolonged growth when they are normally quiescent, growth in serum instead of plasma, and so on). The low concentration of these fibrinolytic components in plasma relative to cell culture medium may result from the presence of components in plasma that suppress the synthesis of these molecules by endothelium *in vivo*. Alternatively, cell culture serum may contain factors that stimulate the synthesis of these molecules by cultured endothelial cells. These factors may be released from cells as whole blood is converted into serum, an idea supported by the finding that the level of PAI-1 synthesis varies dramatically with the type of serum employed in the culture medium.[140] Another possibility to account for the apparent discrepancy between the cell culture and *in situ* observations is that endothelium *in situ* may make these components (i.e., uPA, PAI-1) but only at specific times (e.g., during the growth stage of wound repair; during angiogenesis), and thus not be detected during routine histologic screening of tissue. These latter suggestions are supported by the finding that the fibrinolytic activity of cultured BAEs changes dramatically with their growth state.[38,154,155]

The above considerations emphasize the inherent difficulty in attempting to extrapolate directly from cell culture to the *in vivo* situation. It is clear that a number of questions must still be resolved before the precise role of endothelial fibrinolytic components in vascular fibrinolysis can be established. The first concerns the observation that the majority of the tPA elaborated by endothelial cells is in fact in an inactive complex with PAI-1. It is unclear whether tPA/PAI-1 complexes serve a unique role in this system or are simply an artifact of cell culture. For example, can the tPA activity of such complexes be recovered under some conditions? Do complexes have a half-life in the circulation that is different from that of each of the individual components? Do complexes still bind to fibrin and cell surfaces, and if so, what are the consequences of these interactions? The significance of the finding that the majority of PAI-1 present in CM is present in an inactive, latent form is also unknown. In this regard, the intracellular form of PAI-1 appears to be active,[76,156] suggesting that PAI-1 becomes latent upon secretion. The secreted form has a short half-life, rapidly decaying into the latent form after secretion. At least two different mechanisms can be postulated to account for these observations. The first is that PAI-1 is inherently unstable and that components exist inside the cell to stabilize it. The second potential mechanism is that PAI-1 is inherently stable but becomes inactivated by extracellular processes once secreted. The sensitivity of the active form to inactivation by thrombin, APC, and oxidants is consistent with the latter possibility. Finally, and as mentioned earlier, uPA binds to a receptor on cell surfaces, and endothelial cells appear to have such a receptor.[101] The role of the uPA receptor, and the importance of this interaction, remain unclear. Does the interaction of uPA with this receptor[41,125,157] protect it from inhibitors or localize it at cell surfaces to perform specific functions? The absence of uPA at the surface of endothelium *in situ*[32] would seem to argue against a physiological role for the receptor. However, the observation that receptor number can be modulated[158] raises the possibility again that the expression of this receptor may be important, but only during specific physiological changes in endothelium (e.g., angiogenesis).

Not only do cultured endothelial cells produce a variety of components which may directly alter vascular fibrinolysis, but the rate of synthesis of each of these components is subject to modulation by external factors. Clearly, subtle stimulation or suppression of one of these components may have dramatic effects on net fibrinolytic activity, primarily by altering fibrinolytic balance. Although changes in tPA and uPA production are well documented in other cells,[5,10,12] there is little information on the regulation of these components in endothelial cells. However, the recent demonstration that PAI-1 synthesis by endothelial cells is sensitive to a wide variety of agents (e.g., IL-1, APC, thrombin, endotoxin) suggests that the modulation of this molecule may be especially important in regulating the net fibrinolytic activity of endothelium.

It should be noted that a variety of agents appear to alter tPA release *in vivo*,[5,10,12] suggesting that endothelium may also regulate the vascular fibrinolytic system through the selective release of stored components. Cell culture models to study storage and release of fibrinolytic components are in general lacking, and little information is available about whether tPA, uPA, PAI-1, and tPA/PAI-

1 complexes are also stored, and specifically released. Thus, it is difficult to discuss the exact contribution of stored fibrinolytic components to vascular fibrinolysis.

Finally, the circulating fibrinolytic components, whether released from endothelium or not, clearly may bind to the endothelial cell surface, and to the subendothelial matrix.[21–23,41,90,101,125,157,159–161] To date, plasminogen, tPA, and uPA have been shown to bind in a manner which is rapid, specific, and saturable. It is unclear whether tPA/PAI-1 complexes or PAI-1 itself can also bind to these surfaces, although PAI-1 appears to be incorporated into the matrix of fibroblasts and endothelial cells.[162,163] In any case, it seems likely that endothelium may also influence fibrinolysis by providing surfaces upon which many of the fibrinolytic components are assembled. The kinetics of plasminogen activation by bound components are often considerably improved over those of the fluid-phase components.[22,90] Thus, these surfaces may contribute to vascular fibrinolysis in a manner which is analogous to that of fibrin.

ACKNOWLEDGMENTS. The authors wish to thank C. Hekman, D. Lawrence, J. Mimuro, K. Roegner, M. Sawdey, R. Schleef, N. Wagner, L. Wing, and A. von Zonneveld for their suggestions and ongoing contributions to the work presented in sections of this chapter. We also thank P. Tayman for expert secretarial assistance. This work was supported in part by NIH Grants HL 22289, HL 16411, and HL 33985, and in part by Eli Lilly and Company.

REFERENCES

1. Williams, W. J., 1983, Disorders of hemostasis–thrombosis, in: *Hematology* (W. J. Williams, E. Beutler, A. J. Erslev, and M. A. Lichtman, eds.), McGraw–Hill, New York, pp. 1474–1488.
2. Thorgeirsson, G., and Robertson, A., 1978, The vascular endothelium—Pathobiologic significance, *Am. J. Pathol.* **93**:803–848.
3. Jackson, C. M., and Nemerson, Y., 1980, Blood coagulation, *Annu. Rev. Biochem.* **49**:765–811.
4. Astrup, T., 1978, Fibrinolysis: An overview, in *Progress in Chemical Fibrinolysis and Thrombolysis* (J. F. Davidson, R. M. Rowan, M. M. Samama, and P. C. Desnoyers, eds.), Vol. 3, Raven Press, New York, pp. 1–89.
5. Collen, D., 1980, On the regulation and control of fibrinolysis, *Thromb. Haemost.* **43**:77–89.
6. Rosenberg, R. D., 1977, Chemistry of the hemostatic mechanism and its relationship to the action of heparin, *Fed. Proc.* **36**:10–18.
7. Collen, D., 1976, Identification and some properties of a new fast-reacting plasmin inhibitor in human plasma, *Eur. J. Biochem.* **69**:209–216.
8. Moroi, M., and Aoki, N., 1976, Isolation and characterization of α_2-plasmin inhibitor from human plasma, *J. Biol. Chem.* **251**:5956–5965.
9. Davie, E. W., Fujikawa, K., Kurachi, K., and Kisiel, W., 1979, The role of serine proteases in the blood coagulation cascade, *Adv. Enzymol.* **48**:277–318.
10. Erickson, L. A., Schleef, R. R., Ny, T., and Loskutoff, D. J., 1985, The fibrinolytic system of the vascular wall, in: *Clinics in Haematology* (Z. M. Ruggeri, ed.), Vol. 14, Saunders, Philadelphia, pp. 513–529.
11. Astrup, T., 1966, Tissue activators of plasminogen, *Fed. Proc.* **25**:42–51.
12. Bachman, F., and Kruithof, E. K. O., 1984, Tissue plasminogen activator: Chemical and physiological aspects, *Semin. Thromb. Haemost.* **10**:6–17.

13. Plow, E., and Edgington, T. S., 1975, An alternative pathway for fibrinolysis. I. The cleavage of fibrinogen by leukocyte proteases at physiologic pH, *J. Clin. Invest.* **56**:30–38.

14. Robbins, K. C., Summaria, L., Hsieh, B., and Shah, R. J., 1967, The peptide chains of human plasmin: Mechanism of activation of human plasminogen to plasmin, *J. Biol. Chem.* **242**:2333–2342.

15. Pennica, D., Holmes, W. E., Kohr, W. J., Harkins, R. N., Vehar, G. A., Ward, C. A., Bennett, W. F., Yelverton, E., Seeburg, P. H., Heyneker, H. L., Goeddel, D. V., and Collen, D., 1983, Cloning and expression of human tissue-type plasminogen activator cDNA in E. Coli, *Nature* **301**:214–221.

16. Verde, P., Stoppelli, M. P., Galeffi, P., Di Nocera, P., and Blasi, F., 1984, Identification and primary sequence of an unspliced human urokinase poly (A)$^+$ RNA, *Proc. Natl. Acad. Sci. USA* **81**:4727–4731.

17. Levin, E. G., and Loskutoff, D. J., 1982, Cultured bovine endothelial cells produce both urokinase and tissue-type plasminogen activators, *J. Cell Biol.* **94**:631–636.

18. Thorsen, S., Glas-Greenwalt, P., and Astrup, T., 1972, Differences in the binding to fibrin of urokinase and tissue plasminogen activator, *Thromb. Diath. Haemorrh.* **28**:65–74.

19. Hoylaerts, M., Rijken, D. C., Lijnen, H. R., and Collen, D., 1982, Kinetics of the activation of plasminogen by human tissue plasminogen activator: Role of fibrin, *J. Biol. Chem.* **257**:2912–2919.

20. Radcliffe, R., and Heinze, T., 1981, Stimulation of tissue plasminogen activator by denatured proteins and fibrin clots: A possible additional role for plasminogen activator? *Arch. Biochem. Biophys.* **211**:750–761.

21. Salonen, E., Saksela, O., Vartio, T., Vaheri, A., Nielsen, L., and Zeuthen, J., 1985, Plasminogen and tissue-type plasminogen activator bind to immobilized fibronectin, *J. Biol. Chem.* **260**:12302–12307.

22. Hajjar, K., Harpel, P., Jaffe, E., and Nachman, R., 1986, Binding of plasminogen to cultured human endothelial cells, *J. Biol. Chem.* **261**:11656–11662.

23. Liu, C. Y., Wallen, P., and Handley, D. A., 1985, Preparation of active iodinated and gold-labelled tissue plasminogen activator and their binding to fibrin and endothelial cells, *Thromb. Haemost.* **54**:60a.

24. Rijken, D. C., and Collen, D., 1981, Purification and characterization of the plasminogen activator secreted by human melanoma cells in culture, *J. Biol. Chem.* **256**:7035–7041.

25. Patthy, L., Trexler, M., Vali, Z., Banyai, L., and Varadi, A., 1984, Kringles: Modules specialized for protein binding. Homology of the gelatin-binding region of fibronectin with the kringle structures of proteases, *FEBS Lett.* **171**:131–136.

26. Banyai, L., Varadi, A., and Patthy, L., 1983, Common evolutionary origin of the fibrin binding structures of fibronectin and tissue-type plasminogen activator, *FEBS Lett.* **163**:37–41.

27. Patthy, L., 1985, Evolution of the proteases of blood coagulation and fibrinolysis by assembly from modules, *Cell* **41**:657–663.

28. Ny, T., Elgh, F., and Lund, B., 1984, The structure of human tissue-type plasminogen activator gene: Correlation of intron and exon structures to functional and structural domains, *Proc. Natl. Acad. Sci. USA* **81**:5355–5359.

29. van Zonneveld, A. J., Veerman, H., and Pannekoek, H., 1986, Autonomous functions of structural domains on human tissue-type plasminogen activator, *Proc. Natl. Acad. Sci. USA* **83**:4670–4674.

30. van Zonneveld, A. J., Veerman, H., and Pannekoek, H., 1986, On the interaction of the finger and the kringle-2 domain of tissue-type plasminogen activator with fibrin, *J. Biol. Chem.* **261**:14214–14218.

31. Todd, A. S., 1959, The histological localization of fibrinolysin activator, *J. Pathol. Bacteriol.* **78**:281–283.

32. Rijken, D. C., Wijngaards, G., and Welbergen, J., 1980, Relationship between plasminogen activator and the activators in blood and the vascular wall, *Thromb. Res.* **18**:815–830.

33. Nilsson, I. M., and Pandolfi, M., 1976, Assay of fibrinolytic activity of the vessel wall, in: *Progress in Chemical Fibrinolysis and Thrombosis* (J. F. Davidson, M. M. Samama, and P. C. Desnoyers, eds.), Vol. 2, Raven Press, New York, pp. 1–13.

34. Wallen, P., 1977, Activation of plasminogen with urokinase and tissue activator, in: *Thrombosis and Urokinase* (R. Paoletti and S. Sherry, eds.), Academic Press, New York, pp. 91–102.

35. Wun, T. C., Schleuning, W. D., and Reich, E., 1982, Isolation and characterization of urokinase from human plasma, *J. Biol. Chem.* **257:**3276–3283.

36. Nielsen, L. S., Hansen, J. G., Skriver, L., Wilson, E. L., Kaltoft, K., Zeuthen, J., and Dano, K., 1982, Purification of a zymogen to plasminogen activator from human glioblastoma cells by affinity chromatography with monoclonal antibody, *Biochemistry* **21:**6410–6415.

37. Stump, D. C., Thienpont, M., and Collen, D., 1986, Urokinase-related proteins in human urine: Isolation and characterization of single-chain urokinase (pro-urokinase) and urokinase–inhibitor complex, *J. Biol. Chem.* **261:**1267–1273.

38. Loskutoff, D. J., 1987, The fibrinolytic system of cultured endothelial cells: Insights into the role of endothelium in thrombolysis, in: *Vascular Endothelium in Hemostasis and Thrombosis* (M. A. Gimbrone, ed.), Churchill Livingstone, Edinburgh, pp. 120–141.

39. Vassali, J. D., Dayer, J. M., Wohlwend, A., and Belin, D., 1984, Concomitant secretion of prourokinase and of a plasminogen activator-specific inhibitor by cultured human monocytes-macrophages, *J. Exp. Med.* **159:**1653–1668.

40. Stump, D., Lijnen, H., and Collen, D., 1986, Purification and characterization of single-chain urokinase-type plasminogen activator from human cell cultures, *J. Biol. Chem.* **261:**1274–1278.

41. Stoppelli, M. P., Tacchetti, C., Cubellis, M. V., Corti, A., Hearing, V. J., Cassani, G., Appella, E., and Blasi, F., 1986, Autocrine saturation of pro-urokinase receptors on human A431 cells, *Cell* **45:**675–684.

42. Gurewich, V., Pannell, R., Louie, S., Kelley, P., Suddith, R. L., and Greenlee, R., 1984, Effective and fibrin-specific clot lysis by a zymogen precursor form of urokinase (pro-urokinase): A study *in vitro* and in two animal species, *J. Clin. Invest.* **73:**1731–1739.

43. Lijnen, H. R., Zamarron, C., Blaber, M., Winkler, M. E., and Collen, D., 1986, Activation of plasminogen by pro-urokinase. I. Mechanism, *J. Biol. Chem.* **261:**1253–1258.

44. Travis, J., and Salvesen, G. S., 1983, Human plasma proteinase inhibitors, *Annu. Rev. Biochem.* **52:**655–709.

45. Aoki, N., and Harpel, P. C., 1984, Inhibitors of the fibrinolytic enzyme system, *Semin. Thromb. Haemost.* **10:**24–41.

46. Moore, S., Pepper, D. S., and Cash, J. D., 1975, The isolation and characterization of a platelet-specific β-globulin (β-thromboglobulin) and the detection of anti-urokinase and antiplasmin released from thrombin-aggregated washed human platelets, *Biochim. Biophys. Acta* **379:**360–369.

47. Hegt, V. N., 1977, Localization and distribution of fibrinolysis inhibition in the walls of human arteries and veins, *Thromb. Res.* **10:**121–133.

48. Korninger, C., and Collen, D., 1981, Neutralization of human extrinsic (tissue-type) plasminogen activator in human plasma: No evidence for a specific inhibitor, *Thromb. Haemost.* **46:**662–665.

49. Wiman, B., Csemiczky, G., Marsk, L., and Robbe, H., 1984, The fast inhibitor of tissue plasminogen activator in plasma during pregnancy, *Thromb. Haemost.* **52:**124–126.

50. Sprengers, E. D., and Kluft, C., 1987, Plasminogen activator inhibitors, *Blood* **69:**381–387.

51. Lecander, I., and Astedt, B., 1986, Isolation of a new specific plasminogen activator inhibitor from pregnancy plasma, *J. Haematol.* **62:**221–228.

52. Chmielewska, J., Ranby, M., and Wiman, B., 1983, Evidence for a rapid inhibitor to tissue plasminogen activator in plasma, *Thromb. Res.* **31:**427–436.

53. Verheijen, J. H., Chang, G. T. G., and Kluft, C., 1984, Evidence for the occurrence of a fast-acting inhibitor for tissue-type plasminogen activator in human plasma, *Thromb. Haemost.* **51:**392–395.

54. Kruithof, E. K. O., Ransijn, A., and Bachmann, F., 1983, Inhibition of tissue plasminogen activator by human plasma, in: *Progress in Fibrinolysis* (J. F. Davidson, R. Bachman, C. A. Bouvier, and E. K. O. Kruithof, eds.), Vol. 6, Churchill Livingstone, Edinburgh, pp. 365–369.

55. Erickson, L. A., Hekman, C. M., and Loskutoff, D. J., 1985, The primary plasminogen-activator inhibitors in endothelial cells, platelets, serum, and plasma are immunologically related, *Proc. Natl. Acad. Sci. USA* **82:**8710–8714.

56. Korninger, C., Wagner, O., and Binder, B. R., 1985, Tissue plasminogen activator inhibitor in

human plasma: Development of a functional assay system and demonstration of a correlating M_r = 50,000 antiactivator, *J. Lab. Clin. Med.* **105**:718-724.

57. van Mourik, J. A., Lawrence, D. A., and Loskutoff, D. J., 1984, Purification of an inhibitor of plasminogen activator (antiactivator) synthesized by endothelial cells, *J. Biol. Chem.* **259**:14914–14921.

58. Levin, E. G., 1983, Latent tissue plasminogen activator produced by human endothelial cells in culture: Evidence for an enzyme–inhibitor complex, *Proc. Natl. Acad. Sci. USA* **80**:6804-6808.

59. Philips, M., Juul, A. G., and Thorsen, S., 1984, Human endothelial cells produce a plasminogen activator inhibitor and a tissue-type plasminogen activator–inhibitor complex, *Biochim. Biophys. Acta* **802**:99-110.

60. Dosne, A. M., Dupuy, E., and Bodevin, E., 1978, Production of a fibrinolytic inhibitor by cultured endothelial cells derived from human umbilical vein, *Thromb. Res.* **12**:377-387.

61. Emeis, J. J., van Hinsbergh, V. W. M., Verheijen, J. H., and Wijngaards, G., 1983, Inhibition of tissue-type plasminogen activator by conditioned medium from cultured human and porcine vascular endothelial cells, *Biochem. Biophys. Res. Commun.* **110**:392-398.

62. Thorsen, S., and Philips, M., 1984, Isolation of tissue-type plasminogen activator-inhibitor: Evidence for a rapid plasminogen activator inhibitor, *Biochim. Biophys. Acta* **802**:111-118.

63. Sprengers, E. D., Verheijen, J. H., van Hinsbergh, V. W. M., and Emeis, J. J., 1984, Evidence for the presence of two different fibrinolytic inhibitors in human endothelial cell conditioned medium, *Biochim. Biophys. Acta* **801**:163-170.

64. Erickson, L. A., Ginsberg, M. H., and Loskutoff, D. J., 1984, Detection and partial characterization of an inhibitor of plasminogen activator in human platelets, *J. Clin. Invest.* **74**:1465-1472.

65. Sprengers, E. D., Akkerman, J. W. N., and Jansen, B. G., 1986, Blood platelet plasminogen activator inhibitor: Two different pools of endothelial cell type plasminogen activator inhibitor in human blood, *Thromb. Haemost.* **55**:325-329.

66. Colucci, M., Paramo, J. A., and Collen, D., 1986, Inhibition of one-chain and two-chain forms of human tissue-type plasminogen activator by the fast-acting inhibitor of plasminogen activator in vitro and in vivo, *J. Lab. Clin. Med.* **108**:53-59.

67. Coleman, P. L., Barouski, P. A., and Gelehrter, T. D., 1982, The dexamethasone-induced inhibitor of fibrinolytic activity in hepatoma cells: A cellular product which specifically inhibits plasminogen activation, *J. Biol. Chem.* **257**:4260-4264.

68. Loskutoff, D. J., Roegner, K., Erickson, L. A., Schleef, R. R., Huttenlocher, A., Coleman, P. L., and Gelehrter, T. D., 1986, The dexamethasone-induced inhibitor of plasminogen activator in hepatoma cells is antigenically related to an inhibitor produced by bovine aortic endothelial cells, *Thromb. Haemost.* **55**:8-11.

69. Erickson, L. A., Hekman, C. M., and Loskutoff, D. J., 1986, Denaturant induced stimulation of the β-migrating plasminogen activator inhibitor in endothelial cells and serum, *Blood* **68**:1298-1305.

70. Haggroth, L., Mattsson, C., Felding, P., and Nilsson, I. M., 1986, Plasminogen activator inhibitors in plasma and platelets from patients with recurrent venous thrombosis and pregnant women, *Thromb. Res.* **42**:585-594.

71. Booth, N. A., Anderson, J. A., and Bennett, B., 1985, Platelet release protein which inhibits plasminogen activators, *J. Clin. Pathol.* **38**:825-830.

72. Sprengers, E. D., Princen, H. M. G., Kooistra, T., and van Hinsbergh, V. W. M., 1985, Inhibition of plasminogen activators by conditioned medium of human hepatocytes and hepatoma cell line Hep G2, *J. Lab. Clin. Med.* **105**:751-758.

73. Wagner, O. F., and Binder, B., 1986, Purification of an active plasminogen activator inhibitor immunologically related to the endothelial type plasminogen activator inhibitor from the conditioned media of a human melanoma cell line, *J. Biol. Chem.* **261**:14474-14481.

74. Ny, T., Bjersing, L., Hsueh, A. J. W., and Loskutoff, D. J., 1985, Cultured granulosa cells produce two plasminogen activators and an antiactivator, each regulated differently by gonadotropins, *Endocrinology* **116**:1666-1668.

75. Loskutoff, D. J., and Edgington, T. S., 1977, Synthesis of a fibrinolytic activator and inhibitor by endothelial cells, *Proc. Natl. Acad. Sci. USA* **74**:3903-3907.

76. Hekman, C. M., and Loskutoff, D. J., 1986, Kinetic analysis of the β-migrating plasminogen

activator inhibitor (β-PAI) purified from bovine aortic endothelial cells (BAEs), *Fibrinolysis* **1**(Suppl.):46.

77. Astedt, B., Lecander, I., Brodin, T., Lundblad, A., and Low, K., 1986, Purification of a specific placental plasminogen activator inhibitor by monoclonal antibody and its complex formation with plasminogen activator, *Thromb. Haemost.* **53**:122–125.

78. Holmberg, L., Lecander, I., Persson, B., and Astedt, B., 1978, An inhibitor from placenta specifically released binds urokinase and inhibits plasminogen activator released from ovarian carcinoma in tissue culture, *Biochim. Biophys. Acta* **544**:128–137.

79. Kruithof, E. K. O., Vassalli, J. D., Schleuning, W. D., Mattaliano, R. J., and Bachmann, F., 1986, Purification and characterization of a plasminogen activator inhibitor from the histiocytic lymphoma cell line U-937, *J. Biol. Chem.* **261**:11207–11213.

80. Stump, D. C., Thienpont, M., and Collen, D., 1986, Purification and characterization of a novel inhibitor of urokinase from human urine, *J. Biol. Chem.* **261**:12759–12766.

81. Heeb, M. J., Espana, F., Geiger, M., Collen, D., Stump, D. C., and Griffin, J. H., 1987, Immunological identity of heparin-dependent plasma and urinary protein C inhibitor and plasminogen activator inhibitor-3, *J. Biol. Chem.* **262**: 15813–15816.

82. Scott, R. W., Bergman, B. L., Bajpai, A., Hersh, R. T., Rodriguez, H., Jones, B. N., Barreda, C., Watts, S., and Baker, J. B., 1985, Protease nexin: Properties and a modified purification procedure, *J. Biol. Chem.* **260**:7029–7034.

83. Howard, E. W., and Knaver, D. J., 1986, Human protease nexin-I: Further characterization using a highly specific polyclonal antibody, *J. Biol. Chem.* **261**:684–689.

84. Pizzo, S., Fuchs, H., Doman, K., Petruska, D., and Berger, H., 1986, Release of tissue plasminogen activator and its fast-acting inhibitor in defective fibrinolysis, *Arch. Intern. Med.* **146**:188–191.

85. Juhan-Vague, I., Aillaud, M. F., De Cock, F., Philip-Joet, C., Arnaud, C., Serradimigni, A., and Collen, D., 1985, The fast-acting inhibitor of tissue-type plasminogen activator is an acute phase reactant protein, in: *Progress in Fibrinolysis* (J. F. Davidson, M. B. Donati, and S. Coccheri, eds.), Vol. 7, Churchill Livingstone, Edinburgh, pp. 146–149.

86. Kluft, C., Verheijen, J. H., Jie, A. F. H., Rijken, D. C., Preston, F. E., Sue-Ling, H. M., Jespersen, J., and Aasen, A. D., 1985, The postoperative fibrinolytic shutdown: A rapidly reverting acute phase pattern for the fast-acting inhibitor of tissue-type plasminogen activator after trauma, *Scand. J. Clin. Lab. Invest.* **45**:605–610.

87. Brommer, E. J. P., Verheijen, J. H., Chang, G. T. G., and Rijken, D. C., 1984, Masking of fibrinolytic response to stimulation by an inhibitor of tissue-type plasminogen activator in plasma, *Thromb. Haemost.* **52**:154–156.

88. Nilsson, I. M., and Tengborn, L., 1984, A family with thrombosis associated with high level of tissue plasminogen activator inhibitor, *Haemostasis* **14**:24.

89. Colucci, M., Paramo, J. A., and Collen, D., 1985, Generation in plasma of a fast-acting inhibitor of plasminogen activator in response to endotoxin stimulation, *J. Clin. Invest.* **75**:818–824.

90. Silverstein, R., Harpel, P., and Nachman, R., 1986, Tissue plasminogen activator and urokinase enhance the binding of plasminogen to thrombospondin, *J. Biol. Chem.* **261**:9959–9965.

91. Esmon, C. T., and Esmon, N. L., 1984, Protein C activation, *Semin. Thromb. Hemost.* **10**:122–130.

92. Fair, D. S., Marlar, R. A., and Levin, E. G., 1986, Human endothelial cells synthesize protein S, *Blood* **67**:1168–1171.

93. Stern, D., Brett, J., Harris, K., and Naworth, P., 1986, Participation of endothelial cells in the protein C–protein S anticoagulant pathway: The synthesis and release of protein S, *J. Cell Biol.* **102**:1971–1978.

94. Walker, F., 1984, Protein S and the regulation of activated protein C, *Semin. Thromb. Hemost.* **10**:131–138.

95. Walker, F. J., 1986, Identification of a new protein involved in the regulation of the anticoagulant activity of activated protein C: Protein S-binding protein, *J. Biol. Chem.* **261**:10941–10944.

96. Stern, D. M., Naworth, P. D., Harris, K., and Esmon, C. T., 1986, Cultured bovine aortic endothelial cells promote activated protein C–protein S-mediated inactivation of factor Va, *J. Biol. Chem.* **261**:713–718.

97. Comp, P. C., and Esmon, C. T., 1981, Generation of fibrinolytic activity by infusion of activated protein C into dogs, *J. Clin. Invest.* **68**:1221–1228.
98. Sakata, Y., Curriden, S., Lawrence, D., Griffin, J. H., and Loskutoff, D. J., 1985, Activated protein C stimulates the fibrinolytic activity of cultured endothelial cells and decreases antiactivator activity, *Proc. Natl. Acad. Sci. USA* **82**:1121–1125.
99. van Hinsbergh, V. W. M., Bertina, R. M., van Wijngaarden, A., van Tilburg, N. H., Emeis, J. J., and Haverkate, F., 1985, Activated protein C decreases plasminogen activator inhibitor activity in endothelial cell-conditioned medium, *Blood* **65**:444–451.
100. Colucci, M., Stassen, J. M., and Collen, D., 1984, Influence of protein C activation on blood coagulation and fibrinolysis in squirrel monkeys, *J. Clin. Invest.* **74**:200–204.
101. Miles, L., Levin, E., and Plow, E., 1986, Localization of fibrinolytic proteins on peripheral blood cells and endothelial cells, *Circulation* **74**(Suppl. II):246.
102. Naworth, P. P., and Stern, D. M., 1985, A pathway of coagulation on endothelial cells, *J. Cell. Biochem.* **28**:253–264.
103. Booyse, F. M., Sedlak, B. J., and Rafelson, M. E., Jr., 1975, Culture of arterial endothelial cells, *Thromb. Diath. Haemorrh.* **34**:825–839.
104. Loskutoff, D. J., van Mourik, J. A., Erickson, L. A., and Lawrence, D., 1983, Detection of an unusually stable fibrinolytic inhibitor produced by bovine endothelial cells, *Proc. Natl. Acad. Sci. USA* **80**:2956–2960.
105. Gimbrone, M. A., Jr., 1976, Culture of vascular endothelium, in: *Progress in Hemostasis and Thrombosis* (T. H. Spaet, ed.), Vol. 3, Grune & Stratton, New York, pp. 1–28.
106. Thornton, S. C., Mueller, S. N., and Levine, E. M., 1983, Human endothelial cells: Use of heparin in cloning and long-term serial cultivation, *Science* **282**:623–625.
107. Hekman, C. M., and Loskutoff, D. J., 1987, Kinetic analysis of the interactions between plasminogen activator inhibitor 1 and both urokinase and tissue plasminogen activator, *Arch. Biochem. Biophys.* (in press).
108. Deutsch, D. G., and Mertz, E. G., 1970, Plasminogen: Purification from human plasma by affinity chromatography, *Science* **170**:1095–1096.
109. Schleef, R., Sinha, M., and Loskutoff, D. J., 1985, Characterization of two monoclonal antibodies against human tissue-type plasminogen activator, *Thromb. Haemost.* **53**:170–175.
110. Unkeless, J. C., Tobia, A., Ossowski, L., Quigley, J. P., Rifkin, D. B., and Reich, E., 1973, An enzymatic function associated with transformation of fibroblasts by oncogenic viruses, *J. Exp. Med.* **137**:85–111.
111. Granelli-Piperno, A., and Reich, E., 1978, A study of proteases and protease–inhibitor complexes in biological fluids, *J. Exp. Med.* **148**:223–234.
112. Erickson, L. A., Lawrence, D. A., and Loskutoff, D. J., 1984, Reverse fibrin autography: A method to detect and partially characterize protease inhibitors after sodium dodecyl sulfate–polyacrylamide gel electrophoresis, *Anal. Biochem.* **137**:454–463.
113. Schleef, R. R., Sinha, M., and Loskutoff, D. J., 1985, Immunoradiometric assay to measure the binding of a specific inhibitor to tissue-type plasminogen activator, *J. Lab. Clin. Med.* **106**:408–415.
114. Loskutoff, D. J., Ny, T., Sawdey, M., and Lawrence, D., 1986, The fibrinolytic system of cultured endothelial cells: Regulation by plasminogen activator inhibitor, *J. Cell. Biochem.* **32**:273–280.
115. Loskutoff, D. J., and Mussoni, L. M., 1983, Interactions between fibrin and the plasminogen activators produced by cultured endothelial cells, *Blood* **62**:62–68.
116. Booyse, F. M., Scheinbuks, J., Radek, J., Osikowicz, G., Feder, S., and Quarfoot, A. J., 1981, Immunological identification and comparison of plasminogen activator forms in cultured normal human endothelial cells and smooth muscle cells, *Thromb. Res.* **24**:495–504.
117. Booyse, F. M., Osikowicz, G., Feder, S., and Scheinbuks, J., 1984, Isolation and characterization of a urokinase-type plasminogen activator ($M_r = 54,000$) from cultured human endothelial cells indistinguishable from urinary urokinase, *J. Biol. Chem.* **259**:7198–7205.
118. Bykowska, K., Levin, E. G., Rijken, D. C., Loskutoff, D. J., and Collen, D., 1982, Characterization of a plasminogen activator secreted by cultured bovine endothelial cells, *Biochim. Biophys. Acta* **703**:113–115.
119. Kristensen, P., Larsson, L. I., Nielsen, L. S., Grondahl-Hansen, J., Andreasen, P. A., and

Dano, K., 1984, Human endothelial cells contain one type of plasminogen activator, *FEBS Lett.* **168**:33–36.

120. Goldsmith, G. M., Ziats, N. P., and Robertson, A. L., 1981, Studies on plasminogen activator and other proteases in subcultured human vascular cells, *Exp. Mol. Pathol.* **35**:257–264.

121. Levin, E. G., and Loskutoff, D. J., 1979, Comparative studies of the fibrinolytic activity of cultured vascular cells, *Thromb. Res.* **15**:869–878.

122. Rohrlich, S., and Rifkin, D. B., 1981, Isolation of the major serine protease inhibitor from the 5-day serum-free conditioned medium of human embryonic lung cells and demonstration that it is fetuin, *J. Cell. Physiol.* **109**:1–15.

123. Hekman, C. M., and Loskutoff, D. J., 1985, Endothelial cells produce a latent inhibitor of plasminogen activators that can be activated by denaturants, *J. Biol. Chem.* **260**:11581–11587.

124. Cochrane, C. G., and Griffin, J. H., 1982, The biochemistry and pathophysiology of the contact system of plasma, *Adv. Immunol.* **33**:241–306.

125. Vassalli, J. D., Baccino, D., and Belin, D., 1985, A cellular binding site for the M_r 55,000 form of the human plasminogen activator, urokinase, *J. Cell Biol.* **100**:86–92.

126. Mullertz, S., 1978, Natural inhibitors of fibrinolysis, in: *Progress in Chemical Fibrinolysis and Thrombolysis* (J. F. Davidson, R. M. Rowan, M. M. Samama, and P. C. Desnoyers, eds.), Vol. III, Raven Press, New York, pp. 213–237.

127. Lawrence, D. A., and Loskutoff, D. J., 1986, Inactivation of plasminogen activator inhibitor by oxidants, *Biochemistry* **25**:6351–6355.

128. Travis, J., Owen, M., George, P., Carrell, R., Rosenberg, S., Hallewell, R. A., and Barr, P. J., 1985, Isolation and properties of recombinant DNA produced variants of human α-1-proteinase inhibitor, *J. Biol. Chem.* **260**:4384–4389.

129. Brot, N., Weissbach, L., Werth, J., and Weissbach, H., 1981, Enzymatic reduction of protein-bound methionine sulfoxide, *Proc. Natl. Acad. Sci. USA* **78**:2155–2158.

130. Loskutoff, D. J., 1985, The fibrinolytic system of cultured endothelial cells: Deciphering the balance between plasminogen activation and inhibition, in: *Progress in Fibrinolysis* (J. F. Davidson, M. B. Donati, and S. Coccheri, eds.), Vol. 7, Churchill Livingstone, Edinburgh, pp. 15–22.

131. Rijken, D. C., van Hinsberg, V. W. M., and Sens, E. H. C., 1984, Quantitation of tissue-type plasminogen activator in human endothelial cell cultures by use of an enzyme immunoassay, *Thromb. Res.* **33**:145–153.

132. Laskowski, M., Jr., and Sealock, R. W., 1971, Protein proteinase inhibitors—Molecular aspects, in: *The Enzymes* (P. D. Boyer, ed.), Vol. III, Academic Press, New York, pp. 375–473.

133. Christensen, U., Holmberg, L., Bladh, B., and Astedt, B., 1982, Kinetics of the reaction between urokinase and an inhibitor of fibrinolysis from placental tissue, *Thromb. Haemost.* **48**:24–26.

134. Ny, T., Sawdey, M., Lawrence, D., Millan, J. L., and Loskutoff, D. J., 1986, Cloning and sequence of a cDNA coding for the human β-migrating endothelial-cell type plasminogen activator inhibitor, *Proc. Natl. Acad. Sci. USA* **83**:6776–6780.

135. Carrell, R., and Travis, J., 1985, α-1-Antitrypsin and the serpins: Variation and countervariation, *Trends Biochem. Sci.* **10**:20–24.

136. Levin, E. G., and Loskutoff, D. J., 1982, Regulation of plasminogen activator production by cultured endothelial cells, *Ann. N.Y. Acad. Sci.* **401**:184–194.

137. Loskutoff, D. J., 1979, Effect of thrombin on the fibrinolytic activity of cultured bovine endothelial cells, *J. Clin. Invest.* **64**:329–332.

138. Crutchley, D. J., and Conanan, L. B., 1986, Endotoxin induction of an inhibitor of plasminogen activator in bovine pulmonary artery endothelial cells, *J. Biol. Chem.* **261**:154–159.

139. Bevilacqua, M. P., Schleef, R. R., Gimbrone, M. A., and Loskutoff, D. J., 1986, Regulation of the fibrinolytic system of cultured human vascular endothelium by interleukin 1, *J. Clin. Invest.* **78**:587–591.

140. Sawdey, M., Ny. T., and Loskutoff, D. J., 1986, Messenger RNA for plasminogen activator inhibitor, *Thromb. Res.* **41**:151–160.

141. Dubor, F., Dosne, A. M., and Chedid, L. A., 1986, Effect of polymyxin B and colimycin on induction of plasminogen antiactivator by lipopolysaccharide in human endothelial cell culture, *Infect. Immun.* **52**:725–729.

142. Emeis, J. J., and Kooistra, T., 1986, Interleukin 1 and lipopolysaccharide induce an inhibitor of tissue-type plasminogen activator in vivo and in cultured endothelial cells, *J. Exp. Med.* **163**:1260–1266.

143. Krakauer, T., 1986, Human interleukin 1, in: *CRC Critical Reviews in Immunology* (M. Z. Atassi, ed.), Vol. 6, CRC Press, Boca Raton, Fla., pp. 213–244.

144. Nachman, R. L., Hajjar, K. A., Silverstein, R. L., and Dinarello, C. A., 1986, Interleukin 1 induces endothelial cell synthesis of plasminogen activator inhibitor, *J. Exp. Med.* **163**:1595–1600.

145. Wagner, C. R., Vetto, R. M., and Burger, D. R., 1985, Expression of I-region-associated antigen (Ia) and interleukin 1 by subcultured human endothelial cells, *Cell. Immunol.* **93**:91–104.

146. Stern, D. M., Bank, I., Naworth, P. P., Cassimeris, J., Kisiel, W., Fenton, J. W., II, Dinarello, C., Chess, L., and Jaffe, E. A., 1985, Self-regulation of procoagulant events on the endothelial cell surface, *J. Exp. Med.* **162**:1223–1235.

147. Miossec, P., Cavender, D., and Ziff, M., 1986, Production of interleukin 1 by human endothelial cells, *J. Immunol.* **136**:2486–2491.

148. Gelehrter, T. D., and Sznycer-Laszuk, R., 1986, Thrombin induction of plasminogen activator-inhibitor in cultured human endothelial cells, *J. Clin. Invest.* **77**:165–169.

149. Kisiel, W., Canfield, W. M., Ericsson, L. H., and Davie, E. W., 1977, Anticoagulant properties of bovine plasma protein C following activation by thrombin, *Biochemistry* **16**:5824–5831.

150. Marlar, R. A., Kleiss, A. J., and Griffin, J. H., 1982, Mechanism of action of human activated protein C, a thrombin-dependent anticoagulant enzyme, *Blood* **59**:1067–1072.

151. Seegers, W. H., McCoy, L. E., Groben, H. D., Sakuragawa, N., and Agrawal, B. B. L., 1972, Purification and some properties of autoprothrombin II-A: An anticoagulant perhaps also related to fibrinolysis, *Thromb. Res.* **1**:443–460.

152. Sakata, Y., Griffin, J. H., and Loskutoff, D. J., 1988, Effect of activated protein C on the fibrinolytic components released by cultured bovine aortic endothelial cells, *Fibrinolysis* **2**:7–15.

153. Sakata, Y., Loskutoff, D. J., Gladstone, C., Hekman, C., and Griffin, J. H., 1986, Mechanisms of protein C-dependent clot lysis: Role of plasminogen activator inhibitor, *Blood* **68**:1218–1223.

154. Loskutoff, D. J., and Levin, E., 1984, Properties of plasminogen activators produced by endothelial cells, in: *Biology of Endothelial Cells* (E. A. Jaffe, ed.), Nijhoff, The Hague, pp. 200–208.

155. Levin, E. G., and Loskutoff, D. J., 1980, Serum-mediated suppression of cell-associated plasminogen activator activity in cultured endothelial cells, *Cell* **22**:701–707.

156. Levin, E., 1986, Quantitation and properties of the active and latent plasminogen activator inhibitors in cultures of human endothelial cells, *Blood* **67**:1309–1313.

157. Bajpai, A., and Baker, J., 1985, Cryptic urokinase binding sites on human foreskin fibroblasts, *Biochem. Biophys. Res. Commun.* **133**:475–482.

158. Grimaldi, G., DiFiore, P., Locatelli, E. K., Falco, J., and Blasi, F., 1986, Modulation of urokinase plasminogen activator gene expression during the transition from quiescent to proliferative state in normal mouse cells, *EMBO J.* **5**:855–861.

159. Knudsen, B., Silverstein, R., Leung, L., Harpel, P., and Nachman, R., 1986, Binding of plasminogen to extracellular matrix, *J. Biol. Chem.* **261**:10765–10771.

160. Harpel, P., Chang, T., and Verderber, E., 1985, Tissue plasminogen activator and urokinase mediate the binding of glu-plasminogen to plasma fibrin I: Evidence for new binding sites in plasmin-degraded fibrin I, *J. Biol. Chem.* **260**:4432–4440.

161. Lucas, M., Fretto, L., and McKee, P., 1983, The binding of human plasminogen to fibrin and fibrinogen, *J. Biol. Chem.* **258**:4249–4256.

162. Laiho, M., Saksela, O., Andreasen, P. A., and Keski-Oja, J., 1986, Enhanced production and extracellular deposition of the endothelial-type plasminogen activator inhibitor in cultured human lung fibroblasts by transforming growth factor-β, *J. Cell Biol.* **103**:2403–2410.

163. Mimuro, J., Schleef, R. R., and Loskutoff, D. J., 1987, Extracellular matrix of cultured bovine aortic endothelial cells contains functionally active type 1 plasminogen activator inhibitor, *Blood* **70**:721–728.

VI

Endothelial Cell Procoagulant Activity

Vascular Endothelium
Functional Modulation at the Blood Interface

Michael A. Gimbrone, Jr., and Michael P. Bevilacqua

I. INTRODUCTION

Vascular endothelium, in its simplest conception, is an anatomical barrier—a partition that separates the intravascular compartment from the rest of the cells and tissues of the body. From a functional viewpoint, however, there is ample evidence that the cells comprising this interface are dynamic partners in multiple, complex interactions involving macromolecular and cellular constituents of blood as well as vessel wall components, such as smooth muscle and extracellular matrix. As the contents of this volume illustrate, the list of vital functions of vascular endothelium is diverse and ever-increasing. Furthermore, it is apparent that the functional status of the vascular lining is as important as its anatomical integrity.[1,2] This is especially true since many of the relevant cell surface and metabolic properties of endothelium are inducible, rather than constitutive, and thus may be subject to physiologic regulation and potential pathologic derangement.

Studies in our laboratory have been directed toward defining various stimuli that can influence the functional properties of the endothelial–blood interface in ways that may be relevant to the pathogenesis of vascular diseases, such as thrombosis and atherosclerosis. These stimuli include viral transformation,[3–6] bacterial products,[7] hemodynamic forces,[8–11] and various humoral mediators.[12–22] In this chapter, we briefly review the results of recent studies which indicate that two vital properties of the endothelium—cell surface expression of procoagulant activity and adhesivity for blood leukocytes—can be regulated by certain inflammatory/immune mediators, in particular interleukin 1 (IL-1) (Fig. 1). Our findings strongly support the emerging concept of vascular endothelium as a functionally dynamic interface[23] and provide new insights into its active role in inflammation and thrombosis.

Michael A. Gimbrone, Jr., and Michael P. Bevilacqua • Vascular Research Division, Department of Pathology, Brigham and Women's Hospital and Harvard Medical School, Boston, Massachusetts 02115.

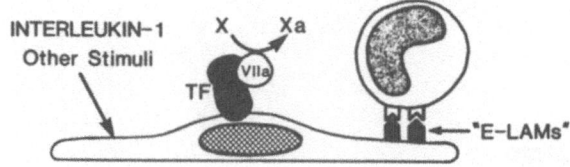

Figure 1. A schematic representation of the direct action of interleukin 1 and other stimuli on a vascular endothelial cell resulting in the expression of (1) a tissue factor (TF)-like procoagulant activity, which can trigger the coagulation cascade via a factor VII- and X-dependent mechanism, and (2) endothelial–leukocyte adhesion molecules (E–LAMs), putative surface structures involved in the adherence of blood leukocytes to the vascular endothelial lining *in vivo*.

II. EXPERIMENTAL PROCEDURES

A. Cell Culture System

Essentially all of these experiments were performed in homologous human *in vitro* systems, consisting of cultured human umbilical vein endothelial cells (HUVEC; usually in the form of confluent monolayers), purified natural or recombinant human polypeptide mediators (cytokines), human plasma (coagulation) factors, and normal leukocytes isolated from human peripheral blood or cultured human leukocyte cell lines.

HUVEC were isolated from two to five sterile cord segments, pooled, and grown in primary culture with Medium 199 (M199; M.A. Bioproducts, Bethesda) containing 20% fetal calf serum (FCS; GIBCO, Grand Island, NY) and antibiotics, as previously described.[24] HUVEC were serially passaged with M199–20% FCS supplemented with endothelial cell growth factor (50–100 μg/ml; Meloy Laboratories, Springfield, VA) and porcine intestinal heparin (50–100 μg/ml; Sigma Chemical Company, St. Louis, MO) in 75-cm^2 Costar tissue culture flasks (Costar, Cambridge, MA) coated with purified human plasma fibronectin (1 μg/cm^2; Meloy Laboratories) or 0.1% gelatin (Bactogelatin 0143-02; Difco Laboratories, Detroit, MI). Most experiments were performed with HUVEC at passage levels two to four. For certain comparative studies, a line of SV40-transformed HUVEC (SVHEC-F),[3] or strains of adult human endothelial cells isolated from iliac artery or thoracic aorta, saphenous vein, bovine aortic endothelial cells, and human dermal fibroblasts[18–20] were also utilized.

For experimental use, each cell type was plated (2–4 × 10^4 cells/well) and grown to confluence (3–7 days) in 16-mm tissue culture wells (Cluster 24, Costar) or on 15-mm plastic coverslips, which were precoated with fibronectin (1–5 μg/cm^2) for uniform endothelial growth.

B. Cytokine Preparations

Human IL-1, isolated as a mixture of two (or more) ∼ 17,000-dalton polypeptides from the supernatant of *Staphylococcus albus*-stimulated human monocytes by immunosorption and Sephadex chromatography, as described by Dinarello *et al.*,[25] was obtained from Genzyme, Inc. (Boston, MA). This material was provided in a sterile solution (0.15 M NaCl with 5% FCS) containing 100 U/ml thymocyte costimulation activity, < 1.0% T cell growth factor activity, < 1 U/ml interferon activity, and undetectable endotoxin (*Limulus* assay). Two distinct molecular species of recombinant human IL-1, rIL-1α and rIL-1β,[26,27] expressed from different cDNAs in *E. coli*, were obtained as purified proteins (10^8 U/mg protein in thymocyte costimulation assays) in sterile phosphate-buffered saline (1000 U/ml) with 0.1% bovine serum albumin (BSA), also from Genzyme. Human recombinant tumor necrosis factor (rTNF), expressed from a cDNA clone in *E. coli* and purified to homogeneity,[28] was provided by Dr. Walter Fiers, Laboratory of Molecular Biology, The State University of Ghent (Ghent, Belgium). This material had 1.9×10^7 U/ml activity by the L929 cytotoxicity assay. The IL-1 preparations were active, and rTNF was inactive, in a comitogenesis assay using the D10.G4.1 T cell line (courtesy of Dr. A. Abbas, Brigham and Women's Hospital, Boston).

Other cytokines, including human interleukin 2, α-, β-, and γ-interferon, bacterial endotoxin, and various antibodies were obtained as previously described.[13,19,20,22] Prior to treatment with cytokines, cultured monolayers were washed with RPMI-1640 (M.A. Bioproducts) containing either 10% FCS or 0.1% BSA, or Tyrode's buffer containing 0.1% BSA. Appropriate cytokine or control preparations then were added and incubations (37°C, 5% CO_2–air) carried out for up to 28 hr, whereupon each well was prepared for evaluation of procoagulant activity or leukocyte–endothelial adhesion.

C. Functional Assays

1. Endothelial Procoagulant Activity (PCA)

At the end of the pretreatment phase, each endothelial monolayer was washed three times with 0.5 ml of RPMI-1640. To determine total cellular PCA, a standard one-stage clotting (plasma recalcification) assay was performed, as previously described,[17] at 37°C, using glass tubes containing 100 μl of citrate-treated, pooled, normal donor, platelet-poor plasma, or coagulation factor VII-, IX-, or X-deficient plasma (George King Bio-Medical, Overland Park, KS) to which 100 μl of cell lysate (frozen–thawed three times, scrape harvested) and 100 μl of $CaCl_2$ (30 mM) were added. In certain experiments, cell surface-expressed PCA was assayed directly in the culture wells on intact viable monolayers, using a modified clotting assay.[17,18] Milliunits of PCA were defined by standard curves developed with rabbit brain thromboplastin (Sigma) and normal, platelet-poor, citrate-treated human plasma; 10^3 mU of PCA corresponded to a clotting time of 20 sec in the standard assay.

2. Leukocyte–Endothelial Adhesion Assay

Polymorphonuclear leukocytes (PMNL) and monocytes were isolated, as previously described in detail,[6,16,19] from anticoagulated whole blood collected by venipuncture from normal donors of both sexes. PMNL suspensions were > 95% pure by Wright–Giemsa staining and contained < 1% platelet contamination by phase-contrast microscopy. Monocyte-enriched populations (78–88% pure) were prepared by modified density gradient centrifugation.[19] For certain experiments, monocytes (> 90% pure) were also isolated by counterflow centrifugation elutriation.[19] The purity of the monocyte suspensions was determined using several criteria, including: (1) size (as assessed by the Coulter Cell Counter, model ZF),

(2) morphology, evaluated by phase-contrast microscopy, (3) histochemical staining for nonspecific esterase activity, and (4) indirect immunoperoxidase staining, using a monoclonal antimonocyte antibody (MO-2, Bethesda Research Laboratories) and antilymphocyte antibodies (T11 and B1, kindly provided by Dr. L. Nadler, Boston).

The human monocyte-like cell line U937[29] (kindly provided by Dr. C. Bianco, New York Blood Center, New York) and the human promyelocytic cell line HL-60[30] (kindly provided by Dr. V. Kelly, Brigham and Women's Hospital, Boston) were cultured in RPMI-1640 medium with 25 mM HEPES and 10% FCS. To facilitate the quantitation of their adherence, washed peripheral blood cells or the leukocyte cell lines were radiolabeled with [^{111}In]oxine (^{111}In, Amersham Corp., Arlington Heights, IL) as previously described.[16]

The adhesion of human leukocytes and leukocyte cell lines (and other comparable cells, such as platelets and erythrocytes) was measured using quantitative monolayer adhesion assays, as previously described by our laboratory.[6,16,19] Test wells (16-mm diameter) containing confluent endothelial monolayers on plastic coverslips were washed three times and then incubated with 0.5 ml of unlabeled or radiolabeled blood cell suspensions, for defined periods varying from 10 to 60 min at 37°C, under static conditions, in their respective assay media (Tyrode's–0.1% BSA for PMNL, RPMI–0.1% BSA for monocytes, RPMI–1% FCS for HL-60 cells and U937 cells). Typically, the blood cell concentrations were 5×10^6/ml for PMNL, 4×10^5/ml for monocytes, and 2×10^6/ml for the cell lines. At the end of the assay incubation, the endothelial monolayers were subjected to a standardized wash procedure to remove unbound cells. Each coverslip was washed by repeated passage (three times) through the air–fluid interface of a 250-ml beaker containing assay media or Hanks's balanced salt solution with calcium and magnesium (M.A. Bioproducts). The number of leukocytes bound per square millimeter was determined by direct microscopic counting, or calculated from the monolayer-bound radioactivity and the known specific activity (counts per minute/cell) of the ^{111}In-labeled blood cell preparations. Pilot experiments indicated that visual and radiometric methods gave comparable results, and that the ^{111}In-labeling procedure did not significantly alter the adhesive characteristics of the blood cells. Experiments using ^{111}In-labeled leukocytes routinely included morphologic monitoring of at least one well from each experimental group in order to assess the intactness of the endothelial monolayer and the nature of the endothelial cell–leukocyte interaction.

III. RESULTS AND DISCUSSION

A. Endothelial Procoagulant Activity

1. Background

The continuous endothelial lining of the circulatory system, in effect, constitutes a nonthrombogenic container for blood. This vital property was appreciated as early as 1856 by Rudolph Virchow and formed part of his classic triad of predisposing factors for pathologic thrombosis.[31] Subsequently, it was shown that this "nonthrombogenic behavior," in part, reflected the failure of the endothelial cell surface to activate the plasma clotting system or to induce platelet adhesion. Initially, this property was considered to be more passive than active, i.e., the endothelial lining was viewed as a sort of insulation, preventing blood from interacting with more reactive subendothelial tissues. With the discovery that endothelial cells can synthesize prostacyclin,[32] the most potent naturally occurring inhibitor of platelet aggregation, a more active "antithrombotic" role for endothelium became apparent.[33]

With the increasing awareness of the functional complexities of endothelium, our appreciation of its multiple potential roles in hemostasis and thrombosis has also evolved. The current working concept[2] is that the vascular endothelial lining is the locus of a multifactorial "hemostatic/thrombotic balance" involving both *anti*- and *pro*thrombotic influences. The latter include not only the synthesis and secretion of adhesive cofactors, such as von Willebrand factor, thrombospondin, and fibronectin, and coagulation proteins such as factor V, but also the surface assembly and activation of procoagulant macromolecular complexes.[23,35] It has been suggested that these endothelial functions represent a mechanism for the localization of clot-promoting activity near the surface of the injured vessel wall and thus efficient hemostatic plug formation.[34,35]

As summarized here, recent experiments in our laboratory have shown that multifunctional immune/inflammatory mediators, such as IL-1 and TNF, can induce the biosynthesis and cell-surface expression of a tissue factor-like PCA in cultured human endothelial cells, thus potentially making them actively thrombogenic.

2. Experimental Observations

a. Monocyte-Derived IL-1 Induces Procoagulant Expression in Cultured Human Vascular Endothelial Cells. Intact cultured HUVEC monolayers exhibit essentially no detectable PCA at their surfaces, as measured by a one-stage plasma clotting assay (Table 1), thus mimicking the nonthrombogenic behavior of the normal vascular endothelial lining *in vivo*. Physical disruption of unstimulated HUVEC liberates some PCA, but this is relatively low compared to that found in lysates of cultured human dermal fibroblasts. Previous studies had suggested that the expression of PCA in cultured endothelium was inducible by treatment with bacterial endotoxins,[36,37] and as a consequence of cocultivation with blood leukocytes.[37]

In our studies,[17,18,20,21] we found that treatment of passaged HUVEC monolayers with purified human monocyte-derived IL-1 results in a concentration- and time-dependent rise in PCA. This induction was observed with as little as 0.2 U/ml of IL-1 and was maximal at 5–10 U/ml. Continuous exposure of endothelial monolayers to IL-1 resulted in a rapid rise in PCA which peaked between 3 and 6 hr and subsequently declined toward basal levels (Fig. 2). After 24 hr of continuous IL-1 exposure, endothelial PCA had returned to basal levels, and appeared

Table 1. Cell Surface Expression of Procoagulant Activity in
Cultured Human Endothelial Cells and Fibroblasts[a]

	Procoagulant activity (mU/ 10^5 cells)	
	Control	IL-1 treated
Endothelium		
Intact monolayer	0.5	120
Cell lysate	19	230
(percent surface expression)	(3)	(52)
Dermal fibroblasts		
Intact monolayer	240	256
Cell lysate	$\geq$500	$\geq$500
(percent surface expression)	(48)	(50)

[a] Replicate cultures of confluent HUVEC or human dermal fibroblasts were pretreated with control or IL-1 (10 U/ml)-containing media for 4 hr at 37°C and then assayed as intact monolayers (surface-available PCA) or as cell lysates (total cellular PCA). Data represent mean values from duplicate culture plates in one of three similar experiments. Adapted from Bevilacqua et al.[18]

to be refractory to restimulation by fresh IL-1 (see below). IL-1 induction of endothelial PCA was blocked by cycloheximide (10 µg/ml) or actinomycin D (5 µg/ml), thus suggesting a requirement for *de novo* protein/RNA synthesis. In contrast, treatment with aspirin (100 µM), which is sufficient to block arachidonate metabolism via the cyclooxygenase pathway, did not affect IL-1 PCA induction.

As seen in Table 1, a significant proportion of the IL-1-induced PCA activity was expressed at the surface of intact monolayers. Little or no PCA was released into the culture supernatant during these incubations. Thus, in effect, the IL-1

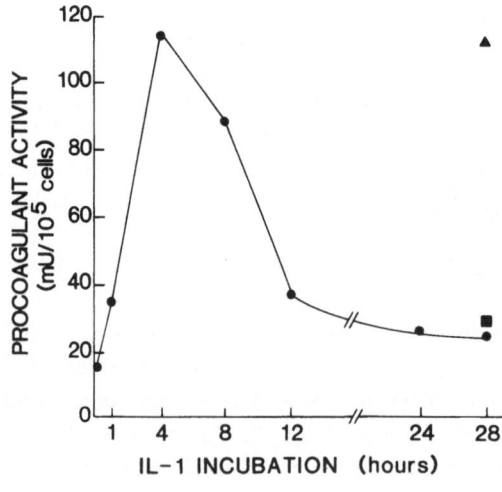

Figure 2. Time course of IL-1 induction of total cellular procoagulant activity in cultured HUVEC. Replicate cultures were incubated in RPMI–10% FCS medium containing 5 U/ml purified, human monocyte-derived IL-1 for intervals up to 28 hr. At 24 hr, certain control plates (▲) and IL-1-treated plates (■) were stimulated with fresh IL-1 (5 U/ml) for an additional 4 hr. Each point represents the mean of duplicate determinations from one of three similar experiments. Adapted from Bevilacqua et al.[17]

treatment caused a significant change in the normal nonthrombogenic nature of the intact endothelial cell surface.

A similar induction of endothelial PCA was observed when cultured adult human arterial and venous endothelial cells were treated with IL-1.[18,20] However, this effect appeared to be relatively selective for human endothelial cells since IL-1 did not stimulate PCA increases in several other cultured cell types. For example, human dermal fibroblasts (Table 1) and SV40-transformed HUVEC, both of which constitutively express large quantities of PCA, did not show increased total cellular or cell surface-available PCA upon IL-1 treatment.

Further experiments were carried out to characterize this IL-1-induced PCA.[17,18] IL-1-induced endothelial PCA was expressed in plasmas deficient in coagulation factor IX but not in plasmas deficient in factor VII or factor X. PCA activity was blocked by an antiserum to human apoprotein III (a kind gift of Professor H. Prydz, Oslo, Norway) and was also abolished by treatment of cell lysates with phospholipase. In addition, using a two-stage chromogenic assay with purified coagulation factors (kindly provided by Dr. D. Stern and W. Kisiel), we found that IL-1 induced PCA requires the presence of factor VIIa in order to activate factor X to Xa. Taken together, these observations strongly suggest that most, if not all, of the IL-1-induced PCA in our cultured human endothelial cells was tissue factor.

b. Induction of Endothelial PCA by Recombinant IL-1 and TNF. Recently, it has been established that human monocyte-derived IL-1 consists of several biochemically distinct species,[25] which appear to be encoded by at least two different genes.[26,27] In recent experiments,[20] we have established that *E. coli*-derived purified human recombinant IL-1 polypeptides, termed rIL-1α and rIL-1β, both are active in inducing endothelial PCA. The effects of these distinct molecular species of rIL-1 were also found to be concentration-dependent, time-dependent, and reversible. In related experiments, purified human rTNF (obtained from Dr. Walter Fiers, State University of Ghent, Belgium) also was found to induce endothelial PCA. rTNF represents another monocyte product which is genetically distinct from the two species of IL-1 previously tested. IL-1 and TNF preparations could be inactivated by heating (80°C, 15 min and 100°C, 5 min, respectively) and were unaffected by the presence of polymyxin B (50 μg/ml), thus distinguishing the active component of these preparations from endotoxin, which can also induce endothelial PCA.[18,20] Furthermore, a rabbit antiserum prepared to rTNF abolished the PCA-inducing activity of the rTNF preparations but did not inhibit the actions of natural human monocyte IL-1 or the two IL-1 species tested. Conversely, a second antiserum prepared to human monocyte-derived IL-1 blocked the induction of PCA by all of the IL-1 preparations tested, but did not significantly inhibit rTNF activity. These experiments helped to establish that the PCA-inducing activity of each of these preparations is in fact attributable to the specific monokine, rather than some other component, and, furthermore, that TNF is not a significant contaminant of the purified natural human monocyte-derived IL-1 preparations used in our studies.

In other experiments,[20] the effects of combined TNF and IL-1 treatments of endothelial monolayers were found to be additive, even at apparent maximal concentrations of the individual monokines. In addition, continuous preincubation of endothelial monolayers with human monocyte-derived IL-1 for 24 hr led to a state of hyporesponsiveness to rechallenge with fresh human monocyte IL-1 (see Fig. 2). However, challenge of IL-1-pretreated monolayers with TNF resulted in significant PCA expression.[20] These observations suggest that distinct stimulus recognition and transduction mechanisms for each of these monokines exist in endothelium, thus permitting their additive interplay in the expression of PCA.

3. Pathophysiologic Relevance

These studies have helped to define the activated monocyte-derived products IL-1 and TNF as potent inducers of tissue factor-like PCA in human endothelium. The fact that a significant proportion of this activity is expressed at the surface of intact living cells, indicates that these mediators can induce, in a reversible fashion, a profound change in the nonthrombogenic properties of the vascular endothelial surface. If similar effects are elicited by these monokines *in vivo*, this would strongly implicate these mediators as important stimuli of endothelial dysfunction. Indeed, a variety of immune and inflammatory processes are often associated with localized or disseminated intravascular coagulation. Fibrin generation in these settings, in part, may be a consequence of monokine induction of endothelial PCA. In addition to triggering blood clotting, tissue factor expression by endothelium may also be relevant to other local cellular effects, including stimulation of the secretion of various endothelial cell products by thrombin and other activated components of the coagulation cascade.[38–40] Thus, the induction of PCA at the endothelial surface may have implications beyond hemostasis and thrombosis.

In addition to the induction of tissue factor-like PCA, IL-1 and TNF also appear to influence other components of the coagulation and fibrinolytic mechanisms. Thus, recent studies from our laboratory and others[21,41,42] have indicated that IL-1 can interfere with the expression of endothelial fibrinolytic activity, in part through suppression of tissue plasminogen activator activity, as well as a profound augmentation in the secretion of plasminogen activator inhibitor. The net effect of these processes would be to foster the maintenance of fibrin once it had been generated via the procoagulant mechanism. Several recent studies also suggest that other components of the "endothelial hemostatic–thrombotic balance"[23] can be significantly influenced by monokines and other mediators[43–46] (see Chapter 13). Thus, a complex picture is emerging, in which the endothelial lining may be playing a pivotal role in the balance of pro- and anticoagulant influences under the modulating influence of mediators such as IL-1 and TNF.

B. Endothelial–Leukocyte Adhesion

1. Background

Enhanced margination and diapedesis of circulating blood leukocytes in the microvasculature are a hallmark of acute and chronic inflammation.[47–50] Focal intimal attachment of monocytes also is observed in large arteries as an early event in atherosclerotic lesion development.[51–53] In both arteries and veins, leukocyte interactions with the vessel wall may directly contribute to the development of vasculitis and thrombosis.[54,55] In each of these pathophysiologic settings, leukocyte adhesion to the luminal endothelial surface appears to be an essential early event.

Although it has long been suspected that changes in the vascular lining might contribute to the localization of leukocyte–vessel wall interactions, until recently, experimental evidence supporting an active role for the vascular endothelium was essentially lacking.[55] Recent studies using *in vitro* model systems[6,16,55–71] have helped to establish a more dynamic concept of leukocyte–endothelial adhesion and to better define the relative contributions of each cell in this interaction. It is now becoming apparent that the endothelial lining has the potential to actively contribute, via multiple mechanisms, to the topographical localization and temporal sequence of leukocyte–vessel wall interactions.

In particular, our laboratory has investigated the hypothesis that inflammatory/immune mediators can act *directly* on the endothelial lining to alter its adhesiveness for leukocytes. We have found that mediators such as IL-1 and TNF can stimulate cultured human endothelial monolayers to bind increased numbers of human PMNL, monocytes, and related cell lines.[18,19,72] This process appears to involve the inducible expression of endothelial cell structures, which we have called "endothelial–leukocyte adhesion molecules" or "E-LAMs."[18,19,72] Here, we will briefly review the experimental observations supporting our working concept of this *endothelial-dependent mechanism of leukocyte adhesion*, and some of its pathophysiologic implications.

2. Experimental Observations

a. IL-1 and Other Mediators Can Act Directly on Endothelial Cells to Increase Leukocyte Adhesion. In the absence of added stimuli (i.e., under basal conditions), 149 ± 13 human PMNL (1.2% of total added) adhered per mm^2 on confluent HUVEC monolayers (mean $\pm$ S.E., 16 experiments) in standardized 10-min adhesion assays.[19] After this relatively short incubation period, the majority ($> 90\%$) of the adherent PMNL were individually attached to the top surface of the monolayer with little evidence of transendothelial migration, upon microscopic examination (Fig. 3). *Selective* treatment of the HUVEC monolayers with human monocyte-derived IL-1,[18,19] or preparations of rIL-1,[18,72] resulted in a time-dependent (onset 30 min; peak 4 hr) and concentration-dependent (maximum ~ 5 U/ml) increase in leukocyte adhesion. In a large series of experiments, involving multiple blood donors and different HUVEC culture strains, selective pretreatment of confluent endothelial monolayers with 5 U/ml of IL-1 for 4 hr resulted in a 22.6 ± 1.9-fold stimulation of PMNL adhesion (mean $\pm$ S.E., 51 experiments) in a 10-min adhesion assay. Microscopic examination revealed that the adherent PMNL were uniformly attached across the surfaces of the endothelial monolayers and in many instances appeared spread. Neither extensive leukocyte aggregation nor morphologic evidence of endothelial injury (e.g., cell retraction or lysis) was

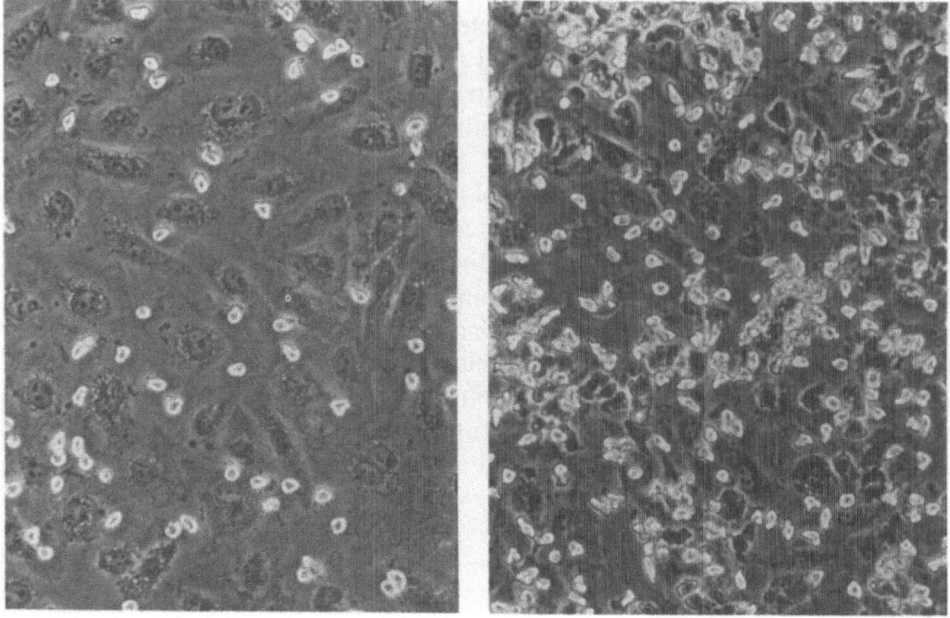

Figure 3. Phase-contrast pnotomicrographs ot the adhesion of human PMNL to control (A) and IL-1-treated (5 U/ml, 4 hr) (B) HUVEC monolayers in a standard 10-min adhesion assay. Adapted from Bevilacqua *et al.*[18]

observed. Introduction of IL-1 directly into the 10-min adhesion assay system, or selective pretreatment of PMNL suspensions for 10–20 min prior to their addition, did not promote PMNL adhesion to untreated endothelial monolayers. Similar results were obtained with cultured human endothelial monolayers isolated from saphenous veins, iliac arteries, or thoracic aortas.[18]

Human peripheral blood monocytes typically demonstrated higher basal levels of adhesion to untreated HUVEC monolayers ($11.2 \pm 0.8\%$ of total added; 10-min adhesion assay, 11 experiments) than did peripheral blood PMNL from the same donors ($\sim 1\%$ of total added).[18,19] Selective pretreatment of the endothelial monolayer with IL-1 (5 U/ml, 4 hr) resulted in a two- to fivefold increase in monocyte adhesion ($p < 0.005$), whereas the addition of IL-1 during the adhesion assay did not significantly alter adhesion. Endothelium-adherent monocytes were esterase-positive and were recognized by the antimonocyte monoclonal antibody MO-2. T lymphocytes (esterase-negative, T11 antigen-positive), which contaminated the monocyte preparations, also adhered in greater numbers to IL-1-treated endothelial monolayers than to control monolayers.

In order to characterize further the endothelium-dependent mechanisms involved in these adhesive interactions, we have made extensive use of two human leukocyte cell lines, the promyelocytic cell line HL-60[30] and the monocyte-like cell line U937.[29] Unlike normal blood leukocytes, in their undifferentiated state, these cell lines are "nonadherent," i.e., they do not bind to uncoated or serum-

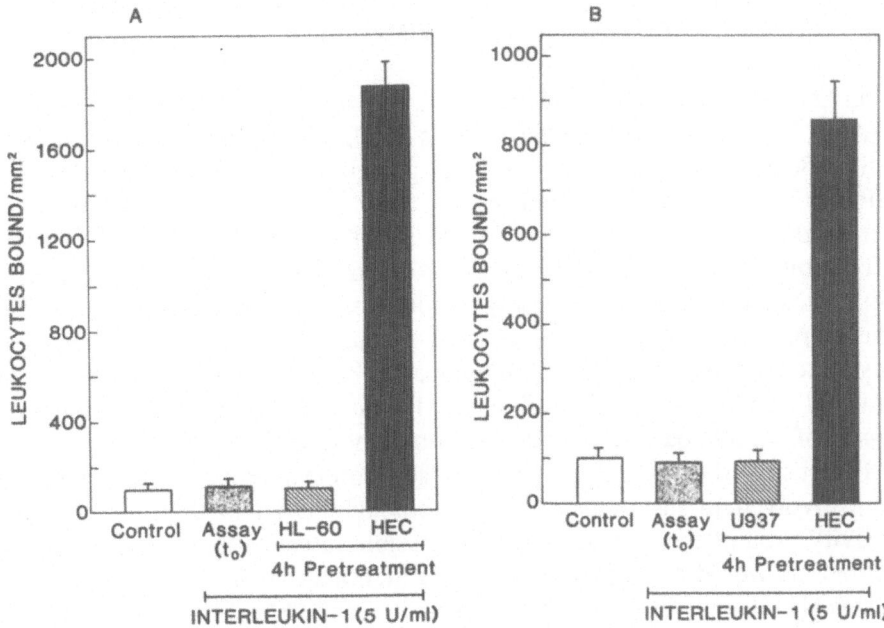

Figure 4. Effect of IL-1 on the adhesion of HL-60 cells (A) and U937 cells (B) to cultured human endothelial monolayers. Confluent HUVEC monolayers (passage 3) (solid bars), or leukocyte cell lines (hatched bars), were selectively pretreated with IL-1 (5 U/ml, 4 hr) in RPMI–10% FCS and then washed before use in a standardized 30-min leukocyte–monolayer adhesion assay. At the start of the adhesion assay, IL-1 (5 U/ml) was added to a parallel set of wells which had not been pretreated (Assay, t_0, stippled bar). Basal adhesion (no added mediator) is shown by white bar (Control). Each bar represents the mean ± S.D. of triplicate wells in one of four similar experiments.

coated tissue culture plastic or glass; furthermore, they show low levels of binding to unstimulated HUVEC monolayers.[18,19] However, the adhesion of both HL-60 and U937 to endothelium can be dramatically increased by selective pretreatment of HUVEC monolayers with IL-1 (Fig. 4). In contrast, direct addition of IL-1 to the adhesion assay (10–30 min) or selective pretreatment of either HL-60 or U937 cells (up to 4 hr) does not increase their adhesion to untreated HUVEC monolayers. These observations further emphasize the endothelium-directed action of the IL-1 stimulus and the endothelium-dependent nature of this inducible leukocyte adhesion mechanism.

We have also studied another monocyte-derived pleiotropic inflammatory mediator, TNF, and bacterial endotoxin, both of which can induce increased endothelial adhesivity for human blood leukocytes and leukocyte cell lines.[18,19,72] Again, through the use of selective pretreatment protocols and the HL-60 cell line (which unlike normal leukocytes appears to be unresponsive to direct stimulation by these mediators), a clear-cut endothelium-dependent effect can be demonstrated.[18,72] The ability of IL-1, TNF, and endotoxin to stimulate leukocyte adhesion to cultured endothelial cells has been demonstrated in several laboratories, in addition to ours, in experiments involving multiple types of leukocytes.[73–77]

b. Characterization of Endothelium-Dependent Mechanisms of Leukocyte Adhesion. IL-1 stimulation of adhesiveness of HUVEC monolayers for leukocytes is a time-dependent process that requires protein and RNA synthesis.[18,19,72] Inhibition of arachidonate metabolism, via the cyclooxygenase pathway, has no effect on this phenomenon.[18,19,72] Generation of soluble mediators by the IL-1-treated monolayer does not appear to be essential because the IL-1-stimulated endothelial supernatant was routinely removed prior to the adhesion assay,[19] and, furthermore, transfer of this conditioned medium does not promote leukocyte adhesion to a second endothelial monolayer.[19,72] Interestingly, we have recently found that conditioned medium from IL-1-treated HUVEC in fact contains an *inhibitor* of stimulated PMNL–endothelium adhesion.[78]

The observed similarities in time course, protein synthesis requirements, and stimuli involved in the induction of endothelial PCA and enhanced leukocyte adhesivity prompt the question of their possible interrelationship. The following observations would suggest that they are functionally distinct: (1) the leukocyte cell lines HL-60 and U937 show minimum adhesion to SV40-transformed HUVEC which constitutively express large amounts of surface PCA (unpublished observation); (2) a polyclonal antiserum to human tissue factor apoprotein III (a gift of Professor H. Prydz) blocks IL-1-induced PCA but not U937–HUVEC adhesion; (3) hirudin did not significantly influence leukocyte adhesion to IL-1-treated HUVEC, thus suggesting that thrombin is not essential for endothelial hyperadhesivity.[19]

Our current working hypothesis is that monokine-induced endothelial adhesiveness for leukocytes involves the surface expression of E–LAMs.[18,72,79] The following observations are consistent with this view: (1) Increased adhesivity is maintained following brief paraformaldehyde fixation of the IL-1-treated endothelial monolayer,[18,72] suggesting that ongoing cellular metabolism is not required; (2) in subconfluent monolayers, increased adhesion microscopically is observed over the surface of individual cells, not in the intervening noncellular spaces.[72] To directly test this hypothesis and to further define monokine-induced alterations in endothelial cell surface structures, we have developed monoclonal antibodies to IL-1- and TNF-treated HUVEC.[22,72,79] One antibody, H4/18, which was raised against IL-1-treated HUVEC, demonstrates substantial surface binding to IL-1- and TNF-treated monolayers but does not recognize control monolayers.[22] The induction of H4/18 antigen by IL-1 and TNF mimics the induction of increased endothelial adhesiveness for leukocytes: both occur in a concentration-, time- (peak ∼ 4 hr), and protein synthesis-dependent fashion. Flow cytometry studies demonstrated that, after treatment with monokines, essentially all of the endothelial cells within a culture expressed H4/18 antigen. Moreover, monoclonal H4/18 immunoprecipitates the same biosynthetically labeled polypeptides from IL-1- and TNF-treated HUVEC (unpublished observation). We have examined the effect of monoclonal H4/18 on endothelial cell–leukocyte adhesion. $F(ab')_2$ fragments of H4/18 have been found to cause partial inhibition of HL-60 cell binding to IL-1-treated HUVEC, whereas F(ab')$_2$ fragments of an irrelevant non-binding antibody (K16/16) and another antibody which binds to HUVEC (TS2/9) were without effect.[72] Further studies using other monoclonal antibodies and polyclonal antisera against IL-1- and TNF-treated HEC are in progress.

Additional candidates for E–LAMs are under investigation by our laboratory and others. For example, endothelial cell class II major histocompatibility (Ia) antigens can be induced by immune interferon,[13] and appear to be involved in endothelial cell–lymphocyte interactions.[14,15,79,80] However, it is unlikely that endothelial Ia antigens are primarily involved in the monokine-induced endothelial cell–leukocyte adhesion which we have studied since neither IL-1 nor TNF induces the expression of these antigens. Moreover, increased expression of endothelial Ia antigens by a 3-day immune interferon treatment did not result in increased PMNL or HL-60 cell adhesion (unpublished observation). Another cell surface protein, ICAM-1, which is found on a variety of cell types, including leukocytes and endothelial cells, appears to play a role in the adhesion of leukocytes to certain molecular targets.[81] Interestingly, ICAM-1 shows a constitutive expression on endothelial cells, and can be stimulated by IL-1, TNF, and also by immune interferon.[82,83] Another candidate, glycoprotein IIb/IIIa, is a cell surface macromolecular complex which recognizes the RGD peptide sequence found in a variety of adhesive proteins such as fibrinogen and fibronectin.[84] The recent observation that glycoprotein IIb/IIIa is synthesized by cultured endothelial cells[85] raises the interesting possibility that this complex alone or in conjunction with adhesive plasma proteins is also involved in endothelial cell–leukocyte adhesion.

Finally, a recent study,[77] which utilized monoclonal antibodies to the leukocyte membrane glycoprotein complex (CDw18), suggests that IL-1/TNF/LPS-induced "E–LAMs" may be interacting, at least in part, with these leukocyte surface structures. However, our observations with the HL-60 cell line (which appears to lack a functional CDw18 complex) indicate that one or more other leukocyte mechanisms may be involved.

In summary, there is increasing evidence that monokine-stimulated endothelial adhesivity for leukocytes involves the inducible expression at the endothelial surface of specialized structures which we have termed "endothelial–leukocyte adhesion molecules." The molecular nature of these putative E–LAMs and their mechanism(s) of interaction with leukocytes is an area of ongoing study.

3. Pathophysiologic Relevance

Adhesion of blood leukocytes to the endothelial lining *in vivo* is the essential first event in their emigration from the vascular space into the tissues of the body. The cellular and molecular mechanisms of this process are relevant to physiologic leukocyte trafficking, as well as the enhanced emigration characteristic of inflammatory states. Induction of E–LAMs by monokines and related mediators may represent an important mechanism for the localization of leukocyte adhesion at specific sites in the vascular lining. In contrast to rapidly acting, primarily leukocyte-directed inflammatory mediators, such as formyl-methionyl-leucyl-phenylalanine, the effect of these slower-acting, endothelium-directed stimuli may be to establish a sustained "portal of entry," favoring efficient leukocyte transmigration into an inflammatory focus.

The observation that monokine-treated human endothelial cells are hyperadhesive for several types of blood leukocytes, including neutrophils, monocytes, and lymphocytes,[18,19,74-76] suggests that this general mechanism may be operative in various pathophysiologic settings, including acute and chronic inflammation, delayed hypersensitivity, and atherogenesis. Multiple classes of E–LAMs may be involved, each acting as a "receptor" for a particular type of leukocyte; further, as recent observations suggest,[76] more than one class of leukocyte surface "acceptor molecule" may also be involved. The dynamic interplay of these inducible surface properties thus may contribute to the temporal and spatial orchestration of leukocyte–endothelial cell interactions *in vivo*.

IV. PERSPECTIVES

In the studies summarized here, we have emphasized the direct action of *exogenous* immune/inflammatory mediators on endothelial function. However, as several lines of investigation have recently revealed, the endothelial cell is potentially a source as well as a target for such mediators. In particular, cultured endothelial cells can be stimulated to release inflammatory lipids such as PAF,[45] polypeptide mitogens such as platelet-derived growth factor,[86] and IL-1 itself,[87-89] thus conceivably establishing a paracrine or autocrine mechanism locally within the vessel wall. Known stimuli of endothelial IL-1 production include bacterial endotoxin and TNF[87-89]; the former could derive from systemic (septicemia) or local (soft tissue infection) sources, while the latter might be secreted by infiltrating monocytes/macrophages.

Direct evidence for "activation" of human microvascular endothelium *in vivo*, in association with mononuclear leukocyte infiltration, has recently been obtained[90] using the monoclonal antibody H4/18. This antibody does not recognize normal endothelium *in vivo*, but does stain microvessels in the vicinity of mononuclear perivascular infiltrates in experimental delayed hypersensitivity reactions in skin, and in various other inflammatory conditions in which monokine/lymphokine generation might be expected to occur. These studies thus lend further credence to the concept of endothelial activation[90] and help bridge the gap between *in vitro* observations and *in vivo* pathophysiologic relevance.

In conclusion, it is clear that key functional properties of the vascular endothelium–blood interface, such as the expression of PCA and E–LAMs, are inducible and thus may be subject to topographic and temporal regulation. Further investigation of the interplay of relevant stimuli and cellular mechanisms of response undoubtedly will contribute to our understanding of the physiology of vascular endothelium and its potential for dysfunction.[91]

V. CONCLUDING REMARKS

With the realization that the endothelial cell is a dynamic, interactive component of the blood vessel wall, it has become apparent that the functional status of the vascular lining is as important as its anatomical integrity. Recent *in vitro* studies from our laboratory and others have identified several types of stimuli that can profoundly influence vascular endothelial functions, including viral infection/transformation, bacterial products, immune/inflammatory mediators, and fluid mechanical (hemodynamic) forces. In particular, two vital properties of the endothelial cell surface—expression of procoagulant (tissue factor) activity, and the display of E–LAMs—appear to be under the control of cytokines such as IL-1 and TNF. These actions may be relevant to the enhanced thrombogenicity and leukocyte reactivity associated with inflammation, atherosclerosis, and other vasculopathies *in vivo*. Further characterization of inducible endothelial surface properties and the mediators which regulate them may provide new insights into the pathogenesis of vascular disease and potential therapeutic strategies.

ACKNOWLEDGMENTS. We wish to acknowledge our colleagues and collaborators, especially Drs. R. S. Cotran, F. W. Luscinskas, D. L. Mendrick, J. S. Pober, and M. E. Wheeler who have actively contributed to various aspects of the studies summarized here. We also thank A. F. Brock, K. Case, L. A. Lapierre, G. Majeau, and D. Smith for their expert technical assistance, and C. Curtis for preparation of the manuscript. This research was supported primarily by grants from the National Institutes of Health.

REFERENCES

1. Gimbrone, M. A., Jr., 1981, Vascular endothelium and atherosclerosis, in: *Vascular Injury and Atherosclerosis* (S. Moore, ed.), Dekker, New York, pp. 25–52.
2. Gimbrone, M. A., Jr. (ed.), 1986, *Vascular Endothelium in Hemostasis and Thrombosis*, Churchill Livingstone, Edinburgh, pp. 1–250.
3. Gimbrone, M. A., Jr., and Fareed, G. C., 1976, Transformation of cultured human vascular endothelium by SV40 DNA, *Cell* 9:685–693.
4. Curwen, K., Gimbrone, M. A., Jr., and Handin, R. I., 1980, *In vitro* studies of thromboresistance: The role of prostacyclin (PGI_2) in platelet adhesion to cultured normal and virally transformed human vascular endothelial cells, *Lab. Invest.* 42:366–374.
5. Corkey, R. F., Corkey, B. E., and Gimbrone, M. A., Jr., 1981, Hexose transport in normal and SV-40 transformed human endothelial cells in culture, *J. Cell. Physiol.* 106:425–434.
6. Gimbrone, M. A., Jr., and Buchanan, M. R., 1982, Interactions of platelets and leukocytes with vascular endothelium: *In vitro* studies, in: *Symposium on Endothelium* (A. P. Fishman, ed.), Annals of the New York Academy of Sciences, New York, pp. 171–183.
7. Quesenberry, P. J., and Gimbrone, M. A., Jr., 1980, Vascular endothelium as a regulator of granulopoiesis: Production of colony-stimulating activity by cultured human endothelial cells, *Blood* 56:1060–1067.
8. Dewey, C. F., Jr., Gimbrone, M. A., Jr., Bussolari, S. R., and Davies, P. F., 1981, The dynamic response of vascular endothelial cells to fluid shear stress, *J. Biomech. Eng.* 103:177–185.
9. White, G. E., Gimbrone, M. A., Jr., and Fujiwara, K., 1983, Factors influencing the expression of stress fibers in vascular endothelial cells *in situ*, *J. Cell Biol.* 97:416–424.
10. Davies, P. F., Dewey, C. F., Jr., Bussolari, S. R., Gordon, E. J., and Gimbrone, M. A., Jr., 1984, Influence of hemodynamic forces on vascular endothelial function: *In vitro* studies of shear stress and pinocytosis in bovine aortic cells, *J. Clin. Invest.* 73:1121–1129.
11. Davies, P. F., Remuzzi, A., Gordon, E. J., Dewey, C. F., Jr., and Gimbrone, M. A., Jr., 1986,

Turbulent fluid shear stress induces vascular endothelial cell turnover *in vitro*, *Proc. Natl. Acad. Sci. USA* **83**:2114–2117.

12. Pober, J. S., and Gimbrone, M. A., Jr., 1982, Expression of Ia-like antigens by human vascular endothelial cells is inducible *in vitro*: Demonstration by monoclonal antibody binding and immunoprecipitation, *Proc. Natl. Acad. Sci. USA* **79**:6641–6645.

13. Pober, J. S., Gimbrone, M. A., Jr., Cotran, R. S., Reiss, C. S., Burakoff, S. J., and Ault, K. A., 1983, Ia expression by vascular endothelium is inducible by activated T-lymphocytes and human gamma interferon, *J. Exp. Med.* **157**:1339–1353.

14. Pober, J. S., Gimbrone, M. A., Jr., Collins, T., Cotran, R. S., Ault, K. A., Fiers, W., Krensky, A. M., Clayberger, C., Reiss, C. S., and Burakoff, S. J., 1984, Interactions of T lymphocytes with human vascular endothelial cells: Role of endothelial cell surface antigens, *Immunobiology* **168**:483–494.

15. Pober, J. S., Collins, T., Gimbrone, M. A., Jr., Libby, P., and Reiss, C. S., 1986, Inducible expression of class II major histocompatibility complex antigens and the immunogenicity of vascular endothelium, *Transplantation* **41**:141–146.

16. Gimbrone, M. A., Jr., Brock, A. F., and Schafer, A. I., 1984, Leukotriene B$_4$ stimulates polymorphonuclear leukocyte adhesion to cultured vascular endothelial cells, *J. Clin. Invest.* **74**:1552–1555.

17. Bevilacqua, M. P., Pober, J. S., Majeau, G. R., Cotran, R. S., and Gimbrone, M. A., Jr., 1984, Interleukin 1 (IL-1) induces biosynthesis and cell surface expression of procoagulant activity in human vascular endothelial cells, *J. Exp. Med.* **160**:618–623.

18. Bevilacqua, M. P., Pober, J. S., Wheeler, M. E., Cotran, R. S., and Gimbrone, M. A., Jr., 1985, Interleukin 1 (IL-1) activation of vascular endothelium: Effects on procoagulant activity and leukocyte adhesion, *Am. J. Pathol.* **121**:393–403.

19. Bevilacqua, M. P., Pober, J. S., Wheeler, M. E., Cotran, R. S., and Gimbrone, M. A., Jr., 1985, Interleukin 1 acts on cultured vascular endothelium to increase the adhesion of polymorphonuclear leukocytes, monocytes and related cell lines, *J. Clin. Invest.* **76**:2003–2011.

20. Bevilacqua, M. P., Pober, J. S., Majeau, G. R., Fiers, W., Cotran, R. S., and Gimbrone, M. A., Jr., 1986, Recombinant tumor necrosis factor induces procoagulant activity in cultured human vascular endothelium: Characterization and comparison with the actions of interleukin 1, *Proc. Natl. Acad. Sci. USA* **83**:4533–4537.

21. Bevilacqua, M. P., Schleef, R. R., Gimbrone, M. A., Jr., and Loskutoff, D. J., 1986, Regulation of the fibrinolytic system of cultured human vascular endothelium by interleukin 1, *J. Clin. Invest.* **78**:587–591.

22. Pober, J. S., Bevilacqua, M. P., Mendrick, D. L., Lapierre, L. A., Fiers, W., and Gimbrone, M. A., Jr., 1986, Two distinct monokines, interleukin 1 and tumor necrosis factor, each independently induce biosynthesis and transient expression of the same antigen on the surface of cultured human vascular endothelial cells, *J. Immunol.* **136**:1680–1687.

23. Gimbrone, M. A., Jr., 1986, Endothelium: Nature's blood container, in: *Vascular Endothelium in Hemostasis and Thrombosis* (M. A. Gimbrone, Jr., ed.), Churchill Livingstone, Edinburgh, pp. 1–13.

24. Gimbrone, M. A., Jr., 1976, Culture of vascular endothelium, in: *Progress in Hemostasis and Thrombosis*, Vol. 3 (T. Spaet, ed.), Grune & Stratton, New York, pp. 1–28.

25. Dinarello, C. A., Bernheim, H. A., Cannon, J. G., LoPreste, G., Warner, S. J. C., Webb, A. C., and Auron, P. E., 1985, Purified, ^{35}S-Met, ^{3}H-Leu-labelled human monocyte interleukin-1 (IL-1) with endogenous pyrogen activity, *Br. J. Rheumatol.* **24**(Suppl. 1):59–64.

26. Auron, P. E., Webb, A. C., Rosenwasser, L. J., Mucci, S. F., Rich, A., Wolfe, S. M., and Dinarello, C. A., 1984, Nucleotide sequence of human monocyte interleukin-1 precursor cDNA, *Proc. Natl. Acad. Sci. USA* **81**:7907–7911.

27. March, C. J., Mosley, B., Larsen, A., Cerretti, D. P., Braedt, G., Price, V., Gillis, S., Henney, C. S., Kronheim, S. R., Gradstein, K., Conlon, P. J., Hopp, T. P., and Cosman, D., 1985, Cloning, sequence and expression of two distinct human interleukin-1 complementary DNAs, *Nature* **315**:641–647.

28. Marmenout, A., Fransen, L., Tavernier, J., Van der Heyden, J., Tizard, J., Kawashima, E., Shaw, A., Johnson, M. J., Simon, D., Muller, R., Ruysschaert, M. R., Van Vliet, A., and Fiers,

W., 1985, Molecular cloning and expression of human tumor necrosis factor and comparison with mouse tumor necrosis factor, *Eur. J. Biochem.* **152**:512–519.

29. Sundstrom, C., and Nilsson, K., 1976, Establishment and characterization of a human histiocytic lymphoma cell line (U937), *Int. J. Cancer* **17**:565–577.

30. Collins, S. J., Gallo, R. C., and Gallager, R. E., 1977, Continuous growth and differentiation of human myeloid leukemia cells in suspension culture, *Nature* **270**:347–349.

31. Virchow, R., 1856, Phlogose und thrombose in gefessystem, Gesammelte Abhandlungen zur wissenschaftlichen medicine, Meidinger Sohn, Frankfurt-am-Main, pp. 458–463.

32. Weksler, B. B., Marcus, A. J., and Jaffe, E. A., 1977, Synthesis of prostaglandin I$_2$ (prostacyclin) by cultured human and bovine endothelial cells, *Proc. Natl. Acad. Sci. USA* **74**:3922–3926.

33. Moncada, S., Herman, A. G., Higgs, E. A., and Vane, J. R., 1977, Differential formation of prostacyclin (PGX or PGI$_2$) by layers of the arterial wall, *Thromb. Res.* **11**:323–344.

34. Stern. D. M., Drillings, M., Nossel, H. L., Hurlet-Jansen, A., LaGamma, K. S., and Owen, J., 1983, Binding of factors IX and IXa to cultured vascular endothelial cells, *Proc. Natl. Acad. Sci. USA* **80**:4119–4123.

35. Nawroth, P. P., and Stern, D. M., 1986, Endothelial cells as active participants in procoagulant reactions, in: *Vascular Endothelium in Hemostasis and Thrombosis* (M. A. Gimbrone, Jr., ed.), Churchill Livingstone, Edinburgh, pp. 14–39.

36. Colucci, M., Balconi, G., Lorenzet, R., Pietra, A., Locati, D., Donati, M. B., and Semeraro, N., 1983, Cutured human endothelial cells generate tissue factor in response to endotoxin, *J. Clin. Invest.* **71**:1893–1896.

37. Lydberg, T., Galdal, K. S., Evensen, S. A., and Prydz, H., 1983, Cellular cooperation in endothelial cell thromboplastin synthesis, *Br. J. Haematol.* **53**:85–95.

38. Daniel, T. O., Gibbs, V. C., Milfay, D. F., Garovoy, M. R., and Williams, L. T., 1986, Thrombin stimulates c-sis gene expression in microvascular endothelial cells, *J. Biol. Chem.* **261**:9579–9582.

39. Harlan, J. M., Thompson, P. J., Ross, R. R., and Bowen-Pope, D., F., 1986, α-Thrombin induces release of platelet-derived growth factor-like molecule(s) by cultured human endothelial cells, *J. Cell Biol.* **103**:1129–1133.

40. Gajdusek, C., Carbon, S., Ross, R., Nawroth, P., and Stern, D., 1986, Activation of coagulation releases endothelial cell mitogens, *J. Cell Biol.* **103**:419–428.

41. Emeis, J. J., and Kooistra, T., 1986, Interleukin 1 and lipopolysaccharide induce an inhibitor of tissue-type plasminogen activator in vivo and in cultured endothelial cells, *J. Exp. Med.* **163**:1260–1266.

42. Nachman, R. L., Hajjar, K. A., Silverstein, R. L., and Dinarello, C. A., 1986, Interleukin 1 induces endothelial cell synthesis of plasminogen activator inhibitor, *J. Exp. Med.* **163**:1595–1600.

43. Nawroth, P. P., and Stern, D. M., 1986, Modulation of endothelial cell hemostatic properties by tumor necrosis factor, *J. Exp. Med.* **163**:740–745.

44. Nawroth, P. P., Handley, D. A., Esmon, C. T., and Stern, D. M., 1986, Interleukin 1 induces endothelial cell procoagulant while suppressing cell-surface anticoagulant activity, *Proc. Natl. Acad. Sci. USA* **83**:3460–3464.

45. Bussolino, F., Brevario, F., Tetta, C., Aglietta, M., Mantovani, A., and Dejana, E., 1986, Interleukin 1 stimulates platelet-activating factor production in cultured human endothelial cells, *J. Clin. Invest.* **77**:2027–2033.

46. Rossi, V., Brevario, F., Ghezzi, P., Dejana, E., and Mantovani, A., 1985, Prostacyclin synthesis induced in vascular cells by interleukin 1, *Science* **229**:174–176.

47. Allison, F., Smith, M. R., and Wood, W. B., 1955, Studies on the pathogenesis of acute inflammation. I. The inflammatory reaction to thermal injury as observed in the rabbit ear chamber, *J. Exp. Med.* **102**:655–668.

48. Marchesi, V. T., and Florey, H. W., 1960, Electron micrographic observations on the emigration of leukocytes, *Q. J. Exp. Physiol.* **45**:343–348.

49. Grant, L., 1973, The sticking and emigration of white blood cells in inflammation, in: *The Inflammatory Process*, Vol. 2 (B. Zweifach, L. Grant, and L. McCluskey, eds.), Academic Press, New York, p. 205.

50. Wilkinson, P. C., and Lackie, J. M., 1979, The adhesion, migration, and chemotaxis of leukocytes in inflammation, *Curr. Top. Pathol.* **68**:48–88.

51. Gerrity, R. G., 1981, The role of the monocyte in atherogenesis. I. Transition of blood-borne monocytes into foam cells in the fatty lesions, *Am. J. Pathol.* **103**:181–190.
52. Joris, I., Zand, T., Nunnari, J. J., Krolikowski, F. J., and Majno, G., 1983, Studies on the pathogenesis of atherosclerosis. I. Adhesion and emigration of mononuclear cells in the aorta of hypercholesterolemic rats, *Am. J. Pathol.* **113**:341–358.
53. Faggiotto, A., Ross, R., and Harker, L., 1984, Studies of hypercholesterolemia in the nonhuman primate. I. Changes that lead to fatty streak formation, *Arteriosclerosis* **4**:323–340.
54. Schaub, R. G., Simmons, C. A., Koets, M. H., Romano, P. J., II, and Stewart, G. J., 1984, Early events in the formation of a venous thrombus following local trauma and stasis, *Lab. Invest.* **51**:218–224.
55. Harlan, J. M., 1985, Leukocyte–endothelial interactions, *Blood* **65**:513–525.
56. Lackie, J. M., and De Bono, D., 1977, Interactions of neutrophil granulocytes (PMNs) and endothelium *in vitro*, *Microvasc. Res.* **13**:107–112.
57. Beasley, J. E., Pearson, J. D., Carleton, J. S., Hutchings, A., and Gordon, J. L., 1978, Interaction of leukocytes with vascular cells in culture, *J. Cell Sci.* **33**:85–101.
58. Pearson, J. D., Carleton, J. S., Beasley, J. E., Hutchings, A., and Gordon, J. L., 1979, Granulocyte adhesion to endothelium in culture, *J. Cell Sci.* **38**:225–235.
59. Beasley, J. E., Pearson, J. D., Hutchings, A., Carleton, J. S., and Gordon, J. L., 1979, Granulocyte migration through endothelium in culture, *J. Cell Sci.* **38**:237–248.
60. Hoover, R. L., Briggs, R. T., and Karnovsky, M. J., 1978, The adhesive interaction between polymorphonuclear leukocytes and endothelial cells *in vitro*, *Cell* **14**:423–428.
61. Hoover, R. L., Folger, R., Haering, W. A., Ware, B. R., and Karnovsky, M. J., 1980, Adhesion of leukocytes to endothelium: Roles of divalent cations, surface charge, chemotactic agents and substrate, *J. Cell Sci.* **45**:73–86.
62. Macgregor, R. R., Macarak, E. J., and Kefalides, N. A., 1978, Comparative adherence of granulocytes to endothelial monolayers and nylon fiber, *J. Clin. Invest.* **61**:697–702.
63. Macgregor, R. R., Friedman, H. M., Macarak, E. J., and Kefalides, N. A., 1980, Virus infection of endothelial cells increases granulocyte adherence, *J. Clin. Invest.* **65**:1469–1477.
64. Boxer, L. A., Allen, J. M., and Baehner, R. L., 1980, Diminished polymorphonuclear leukocyte adherence, *J. Clin. Invest.* **66**:268–274.
65. Taylor, R. F., Price, T. H., Schwartz, S. M., and Dale, D. C., 1981, Neutrophil–endothelial cell interactions on endothelial monolayers grown on micropore filters, *J. Clin. Invest.* **67**:584–587.
66. Hoover, R. L., Karnovsky, M. J., Austen, K. F., Corey, E. J., and Lewis, R. A., 1984, LTB$_4$ modulates neutrophil endothelial interactions, *Proc. Natl. Acad. Sci. USA* **81**:2191–2193.
67. Zimmerman, G. A., and Hill, H. R., 1984, Inflammatory mediators stimulate granulocyte adherence to cultured human endothelial cells, *Thromb. Res.* **35**:203–217.
68. Tonnesen, M. G., Smedly, L. A., and Henson, P. M., 1984, Neutrophil–endothelial cell interactions: Modulation of neutrophil adhesiveness induced by complement fragments C5a and C5a des arg and formyl-methionyl-leucyl-phenylalanine *in vitro*, *J. Clin. Invest.* **74**:1581–1592.
69. Charo, C. F., Yuen, C., and Goldstein, I. M., 1985, Adherence of human polymorphonuclear leukocytes to endothelial monolayers: Effects of temperature, divalent cations, and chemotactic factors on the strength of adherence measured with a new centrifugation assay, *Blood* **65**:473–479.
70. Harlan, J. M., Schwartz, B. R., Reidy, M. A., Schwartz, S. M., Ochs, H. D., and Harker, L. A., 1985, Activated neutrophils disrupt endothelial monolayer integrity by an oxygen radical-independent mechanism, *Lab. Invest.* **52**:141–150.
71. DiCorletto, P. E., and de la Motte, C. A., 1985, Characterization of the adhesion of the human monocytic cell line U937 to cultured endothelial cells, *J. Clin. Invest.* **75**:1153–1161.
72. Bevilacqua, M. P., Wheeler, M. E., Pober, J. S., Fiers, W., Mendrick, D. L., Cotran, R. S., and Gimbrone, M. A., Jr., 1986, Endothelial-dependent mechanisms of leukocyte adhesion: Regulation by interleukin 1 and tumor necrosis factor, in: *Leukocyte Emigration and Its Sequelae* (H. Movat, ed.), Karger, Basel, pp. 79–93.
73. Dunn, C. J., and Fleming, W. E., 1985, The role of interleukin-1 in the inflammatory response with particular reference to endothelial cell–leukocyte adhesion, in: *The Physiologic, Metabolic,*

and Immunologic Actions of Interleukin-1 (M. J. Kluger, J. J. Oppenheim, and M. C. Powanda, eds.), Liss, New York, pp. 45–54.

74. Gamble, J. R., Harlan, J. M., Klebanoff, S. J., and Vadas, M. A., 1985, Stimulation of the adherence of neutrophils to umbilical vein endothelium by human recombinant tumor necrosis factor, *Proc. Natl. Acad. Sci. USA* **82:**8667–8671.

75. Cavender, D. E., Haskard, D. O., Joseph, B., and Ziff, M., 1986, Interleukin 1 increases the binding of human B and T lymphocytes to endothelial cell monolayers, *J. Immunol.* **136:**203–207.

76. Schleimer, R. P., and Rutledge, B. K., 1986, Cultured human vascular endothelial cells acquire adhesiveness for neutrophils after stimulation with interleukin 1, endotoxin, and tumor-promoting phorbol diesters, *J. Immunol.* **136:**649–654.

77. Pohlman, T. H., Stanness, K. A., Beatty, P. G., Ochs, H. D., and Harlan, J. M., 1986, An endothelial cell surface factor(s) induced in vitro by lipopolysaccharide, interleukin 1, and tumor necrosis factor increases neutrophil adherence (in part) by a CDw18-dependent mechanism, *J. Immunol.* **136:**4548–4553.

78. Wheeler, M. E., Bevilacqua, M. P., Luscinskas, F. W., Brock, A. F., and Gimbrone, M. A., Jr., 1986, Interleukin-1 treated endothelial cells produce an inhibitor of leukocyte–endothelial adhesion, *Fed. Proc.* **45:**450.

79. Bevilacqua, M. P., Pober, J. S., Wheeler, M. E., Mendrick, D. L., Fiers, W., and Gimbrone, M. A., Jr., 1986, Interleukin 1 (IL-1) and tumor necrosis factor (TNF) independently activate vascular endothelial cell functions, *Fed. Proc.* **45:**941.

80. Masuyama, J.-I., Minato, N., and Kano, S., 1986, Mechanisms of lymphocyte adhesion to human vascular endothelial cells in culture, *J. Clin. Invest.* **77:**1596–1605.

81. Rothlein, R., Dustin, M. L., Marlin, S. D., and Springer, T. A., 1986, A human intercellular adhesion molecule (ICAM-1) distinct from LFA-1, *J. Immunol.* **137:**1270–1274.

82. Dustin, M. L., Rothlein, R., Bhan, A. F., Dinarello, C. A., and Springer, T. A., 1986, A natural adherence molecule (ICAM-1): Induction by IL-1 and γ-IFN, tissue distribution, biochemistry and function, *J. Immunol.* **137:**1270–1274.

83. Pober, J. S., Gimbrone, M. A., Jr., Lapierre, L. A., Mendrick, D. L., Fiers, W., Rothlein, R., and Springer, T. A., 1986, Overlapping patterns of human endothelial cells by interleukin 1, tumor necrosis factor, and immune interferon, *J. Immunol.* **137:**1893–1896.

84. Pytela, R., Pierschbacher, M. D., Ginsberg, M. H., Plow, E. F., and Ruoslahti, E., 1986, Platelet membrane glycoprotein IIb/IIIa: Member of a family of Arg-Gly-Asp-specific adhesion receptors, *Science* **231:**1559–1562.

85. Fitzgerald, L. A., Charo, I. F., and Phillips, D. F., 1985, Human and bovine endothelial cells synthesize membrane proteins similar to human platelet glycoproteins IIb and IIIa, *J. Biol. Chem.* **260:**10893–10896.

86. Ross, R., Raines, E. W., and Bowen-Pope, D. F., 1986, The biology of platelet-derived growth factor, *Cell* **46:**155–169.

87. Stern, D. M., Bank, I., Nawroth, P. P., Cassimeris, J., Kisiel, W., Fenton, J. W., Dinarello, C., Chess, L., and Jaffe, E. A., 1985, Self-regulation of procoagulant events on the endothelial surface, *J. Exp. Med.* **162:**1223–1235.

88. Nawroth, P. P., Bank, I., Handley, D., Cassimeris, J., Chess, L., and Stern, D., 1986, Tumor necrosis factor/cachectin interacts with endothelial cell receptors to induce release of interleukin 1, *J. Exp. Med.* **163:**1363–1375.

89. Libby, P., Ordovas, J. M., Auger, K. R., Robbins, A. H., Birinyi, L. K., and Dinarello, C. A., 1986, Endotoxin and tumor necrosis factor induce interleukin-1 gene expression in adult human vascular endothelial cells, *Am. J. Pathol.* **124:**179–185.

90. Cotran, R. S., Gimbrone, M. A., Jr., Bevilacqua, M. P., Mendrick, D. L., and Pober, J. S., 1986, Induction and detection of a human endothelial activation antigen *in vivo, J. Exp. Med.* **164:**661–666.

91. Gimbrone, M. A., Jr., 1986, Endothelial dysfunction and the pathogenesis of atherosclerosis, in: *Atherosclerosis VII* (N. H. Fidge and P. J. Nestel, eds.), Elsevier, Amsterdam, pp. 367–369.

13

Endothelium and the Regulation of Coagulation

David M. Stern, Dean A. Handley, and Peter P. Nawroth

I. INTRODUCTION

The coagulation system was initially defined as a complex series of enzyme–substrate interactions occurring on phospholipid surfaces.[1,2] These studies have provided a kinetic basis for understanding the assembly of individual coagulation components into functionally active complexes in purified systems.[3-5] Addition of the platelet surface has enriched this model. The platelet is a natural surface which serves as an amplification mechanism for the activation of coagulation. The platelet is capable of promoting assembly of both the factor IXa–VIII–X complex and the prothrombinase complex, thus giving it a central role in procoagulant reactions.[6-10] Once the placelet is stimulated, however, there is no turning back: the platelet is an end-stage cell committed to the augmentation of thrombin formation. Thus, the platelet is not a sensitive regulator of anticoagulant and procoagulant events.

If one steps back from the coagulation cascade as it is thought to function *in vitro*, and places coagulation within its *in vivo* context, different considerations become evident. *In vivo*, the coagulation system plays an integral role in the maintenance of homeostasis, promoting the fluidity of blood within the vasculature and rapid clot formation in the extravascular space. Furthermore, the host response if pathophysiologic states often includes a contribution to the coagulation mechanism. The Shwartzmann reaction, a basic phenomenon in the biology of bacterial infections, for example, includes a prominent thrombotic component in addition to tissue infiltration by leukocytes. These considerations indicate that the coagulation system must be controlled by a tight balance of tonically-active opposing hemostatic mechanisms. The presence in the plasma of measurable levels of thrombin,[11] prothrombin fragment F_{1+2},[12] and fibrinopeptide A[13] under normal conditions, indicates that activation of the coagulation system is ongoing.

David M. Stern and Peter P. Nawroth • Department of Physiology and Cellular Biophysics, College of Physicians and Surgeons, Columbia University, New York, New York 10032. *Dean A. Handley* • Department of Physiology, Columbia University, New York, New York 10027.

In addition, evidence of the continuous function of anticoagulant mechanisms is ubiquitous. Levels of the plasmin-derived peptide Bβ 1–42 from fibrinogen/fibrin,[14] thrombin–antithrombin III complex,[15] and the protein C activation peptide,[16] are also measurable in homeostasis. From this *in vivo* perspective, the coagulation mechanism is clearly a sensitive effector system which must respond rapidly to the surrounding milieu.

These observations concerning *in vivo* coagulation indicate that a cell continuously exposed to the blood, which is capable of appropriately responding to environmental stimuli, could play a central role in the hemostatic system. We have turned our attention to the vessel wall since it is a ubiquitous vascular surface. In terms of intravascular coagulation, as the cells forming the luminal vascular surface, endothelium is strategically located to function in the regulation of procoagulant and anticoagulant mechanisms. This led to the hypothesis that important hemostatic events occur on the endothelial cell surface, with the cell playing a central regulatory role. In order for endothelium to function in this capacity, it would have to be capable of promoting clot formation, in addition to blocking activation of coagulation. Furthermore, endothelium would be required to maintain a balance of opposing hemostatic mechanisms by regulating the expression of receptors for both procoagulant and anticoagulant mechanisms. Finally, if this endothelial cell-based system is important in physiology, then abnormal interaction of coagulation proteins with their vessel wall receptors should lead to disorders characterized by excessive clotting or bleeding.

A consequence of this hypothesis, implicating a close relationship between the coagulation system and the vessel wall, is the potential for coagulation enzymes to influence endothelial cell physiology. Thus, on the one hand, modulation of receptors for coagulation factors on the cell surface can influence the course of coagulation reactions. On the other hand, coagulation proteins can become important factors governing vessel wall biology.

II. A PROCOAGULANT PATHWAY ON THE ENDOTHELIAL CELL SURFACE

In homeostasis, endothelium plays an important role in ensuring the fluidity of blood. Multiple anticoagulant mechanisms ranging from the protein C[17–19] and heparin–antithrombin III[20–23] system through to the fibrinolytic pathway[24,25] are operative in preventing activation of coagulation and fibrin deposition on the vessel wall. In view of the predominance of anticoagulant mechanisms in homeostasis, we set out to test if there was a balancing set of procoagulant reactions which could function on the endothelial cell surface. If procoagulant mechanisms were also operative, then endothelium could potentially play a role in localized fibrin formation. For these studies, endothelial cell monolayers were incubated with the procoagulant enzyme factor XIa, the appropriate zymogens, and fibrinogen. Activation of coagulation occurred resulting in thrombin formation as indicated by the time-dependent cleavage of fibrinopeptide A from fibrinogen.[26]

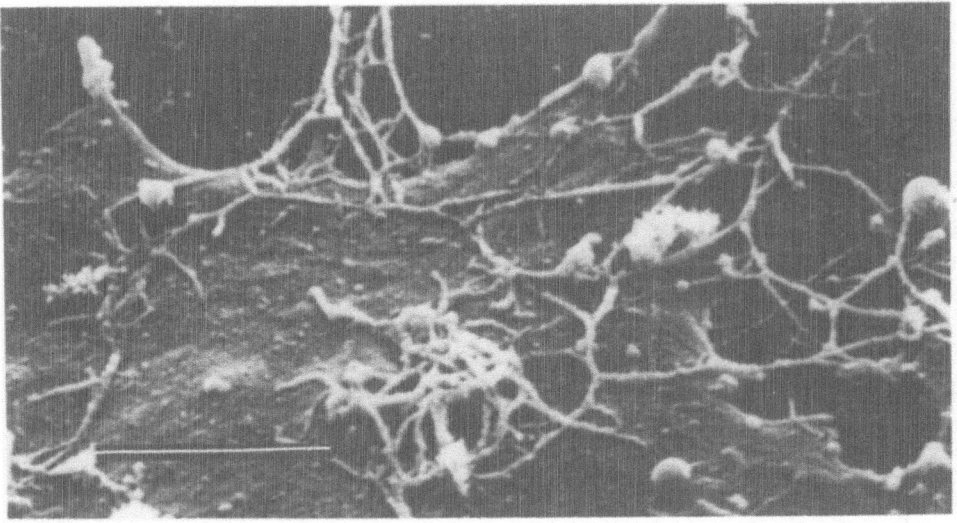

Figure 1. Fibrin clot on endothelial cells. Endothelial cells were incubated with factors XIa, IX, VIII, X, prothrombin, and fibrinogen. When the first definite fibrin strands were seen, monolayers were washed four times with albumin-free buffer and fixed for scanning electron microscopy in 3% glutaraldehyde in 0.15 M sodium cacodylate buffer.

Morphologically, formation of fibrin strands in close association with the endothelial cell surface was evident (Fig. 1). As these experiments were carried out, it was clear that fibrin strands, visible to the unaided eye, appeared randomly distributed over the monolayer. The fibrin strands seemed to form on the monolayers at the bottom of the tissue culture dishes and rise up to the surface of the culture fluid. At the electron microscopic level, some fibrin strands are observed to begin and end on the endothelial cell surface. Furthermore, as the clot forms, these endothelial cells, which were initially contiguous, retract from each other allowing for augmentation of the procoagulant response as hemostatic components contact the subendothelium. Similar changes in endothelial cell morphology in response to fibrin have been noted by Kadish *et al.*[27]

Thrombin formed on the endothelial cell surface could be quickly inactivated by anticoagulant pathways, or, if present in large enough quantities, could prime procoagulant mechanisms, such as activation of platelets. These considerations led us to examine the effect platelets have on the endothelial cell procoagulant pathway (Fig. 2). When endothelium was incubated with factors IXa, VIII, X, and prothrombin, thrombin formation was observed in the absence of platelets (Fig. 2). Though the rate of thrombin formation may appear somewhat slow, compared with that observed on synthetic phospholipids in the presence of all components of the prothrombinase complex,[6–8] activation of coagulation clearly occurs. Furthermore, in the absence of exogenous factor V, endothelial cell-derived factor V/Va[28,29] plays a central role as the cofactor for prothrombin activation (see Section II.B). When unactivated platelets are incubated with the same

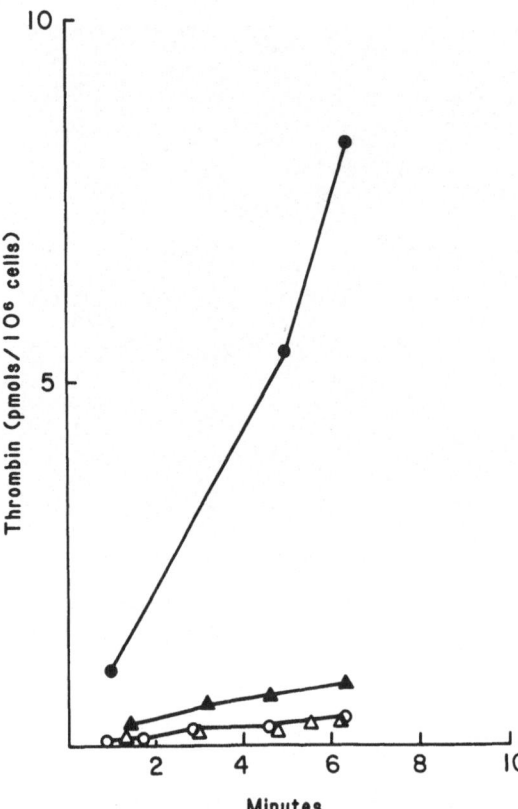

Figure 2. Effect of platelets on thrombin formation by endothelial cells. Monolayers of endothelial cells were incubated with factors IXa (0.9 nM), VIII (1.2 U/ml), X (175 nM), and prothrombin (1.5 μM) in the presence (●) or absence (○) of 1.1×10^8 platelets/ml. Another set of monolayers was incubated with the same coagulation proteins and anti-human factor V IgG (100 μg/ml) in the presence (▲) or absence (△) of platelets. Addition of normal human IgG (100 μg/ml) had no effect on prothrombin activation in the presence or absence of platelets. Addition of human anti-factor V IgG (100 μg/ml) did not effect factor X activation in the presence or absence of platelets in contrast to its effect on thrombin formation in the presence of platelets (▲). In each case, aliquots were removed at the indicated times and assayed in the chromogenic substrate assay.

coagulation proteins, in the absence of endothelium, thrombin formation is minimal. In contrast, addition of platelets to endothelium results in a dramatic increase in thrombin formation. The stimulatory effect of platelets is accounted for in large part on their release of factor V following activation. This is indicated by the abrogation of enhanced thrombin formation in the presence of an antibody to human factor V.[30] The latter antibody has been shown to block the coagulant activity of human factor V (platelets were of human origin in these experiments), but under these conditions did not appear to interact with the bovine factor V associated with the monolayers of bovine aortic endothelial cells. These data are consistent with the hypothesis that platelets can amplify the activation of coagulation initiated on the endothelial cell surface.

The observation that endothelium could participate in procoagulant reactions led us to undertake an examination of the mechanisms involved in activation of coagulation on the endothelial cell surface. Since platelets require stimulation in order to promote factor IXa–VIII-mediated activation of factor X (Fig. 2), this led us to examine whether endothelium could participate in this reaction.

A. Activation of Factors IX and X on the Endothelial Cell Surface

The first procoagulant factor we examined in relation to endothelium was factor IX and its activated form, factor IXa. This was chosen as a starting point

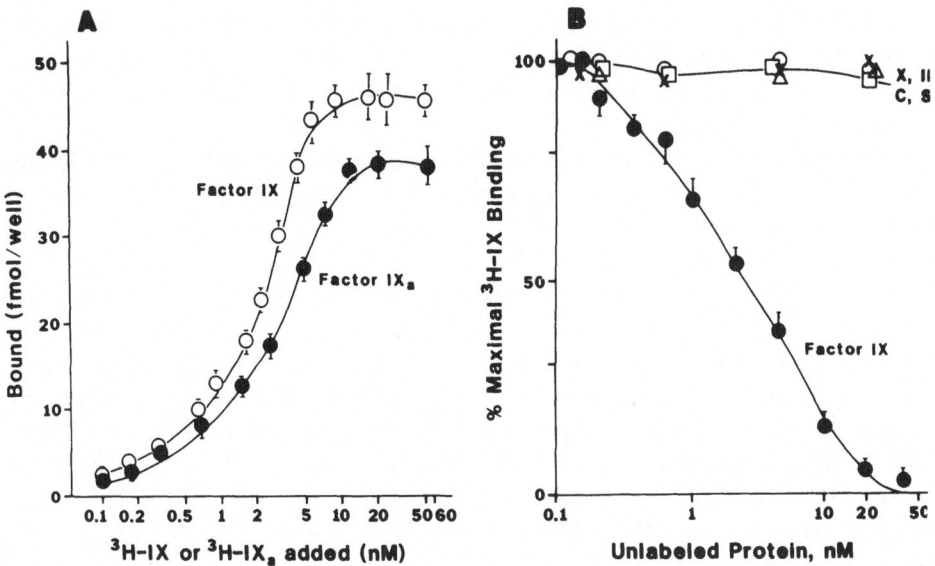

Figure 3. Interaction of factors IX and IXa with monolayers of cultured bovine aortic endothelial cells. (A) Saturability. Semilogarithmic plot of specifically bound [³H]factor IX (○) and [³H]factor IXa (●) versus the concentration of added radiolabeled protein. Error bars denote S.E.M. (B) Specificity. Competitive binding study in which inhibition of specific [³H]factor IX (2 nM) binding to endothelial cells is studied in the presence of unlabeled factors IX, X, prothrombin, proteins C and S.

since clearance studies of factor IX, carried out in patients and experimental animals, were consistent with the possibility of intravascular binding sites.[31–34] For example, when radiolabeled factor IX was infused into mice, Fuchs *et al.*[34] noted that an initial rapid phase of clearance could be blocked by the prior addition of a large excess of unlabeled factor IX. Furthermore, factor IXa is the most potent procoagulant enzyme in the induction of thrombus formation in the Wessler stasis model.[35] This thrombosis model employs vessels with an intact endothelium. The contrast between the potency of factor IXa, compared with factor Xa and thrombin, in an *in vivo* thrombosis model, and the considerably greater effectiveness of the latter enzymes to clot blood *in vitro*, suggested that factor IXa might function on the vessel wall.

These considerations stimulated us to examine in detail the interaction of factors IX/IXa with endothelium. Radioligand binding studies[36,37] were carried out with either the zymogen or enzyme and confluent monolayers of cultured endothelial cells (Fig. 3). Both factors IX and IXa bind to endothelium in a saturable manner with half-maximal occupancy of the sites at a ligand concentration of about 2–3 nM (Fig. 3A). The significance of this dissociation constant is evident when the K_d for factor IX–phospholipid interaction, 2 μM, is considered.[38] This was the first suggestion that the endothelial cell factor IX/IXa interaction site might involve more than phospholipid determinants alone. The potential physiological relevance of this coagulation factor–vessel wall interaction follows from the high probability that at the plasma concentration of factor IX, 70–100 nM, the endothelial cell sites will be saturated. Consistent with this hypothesis, experiments have demonstrated that endogenously bound factor IX could be eluted from the luminal surface of bovine aortic segments immediately after sacrifice of the animals.[39]

The specificity of the endothelial cell binding site for factor IX/IXa is clear from the results of experiments indicating that other vitamin K-dependent coagulation proteins, including factor X, prothrombin, protein C, and protein S, did not compete (Fig. 3B). As suggested by the results indicating that factors IX and IXa bind to endothelium in a similar fashion (Fig. 3A), they were equally effective competitive inhibitors. This is an unusual situation when viewed in the context of other coagulation factor–cell surface interactions. In most other cases, there is great specificity for the zymogen or enzyme form, especially the latter. For example, factor V does not compete effectively with factor Va in the association of the prothrombinase complex on the platelet surface.[40] Similarly, protease nexin does not bind or mediate the endocytosis of the zymogens, factor X and prothrombin, though it can interact with the enzyme forms.[41] In terms of factor IX/IXa–endothelial cell interaction, the similar binding of the enzyme and zymogen was an important issue to consider as we examined the functional implications of this ligand–receptor complex in the coagulation mechanism.

Since factor IXa could interact with the endothelial cell, the role of this cell surface–coagulation factor interaction in modulating the coagulant properties of factor IXa was examined. In coagulation, factor IXa activates factor X in the presence of the cofactor, factor VIII, and an appropriate cellular surface.[1,2] When endothelial cell monolayers[42] were incubated with increasing concentrations of factor IXa, in the presence of saturating levels of factors VIII and X, the rate of factor Xa formation rapidly increases at low concentrations of enzyme and reaches an apparent maximum (Fig. 4, top panel). Half-maximal rates of factor X activation are achieved at factor IXa concentrations of about 150–300 pM. This finding suggested that the interaction of factor IXa with a limited number of cellular binding sites was mediating factor X activation. Prompted by this hypothesis, the binding of factor IXa to endothelium was reexamined with special attention to the possible influence of other coagulation factors. When factor IXa binding was studied in the absence of other coagulation factors, or in the presence of only factor VIII, half-maximal occupancy of receptors occurred at 2–3 nM (Fig. 4, middle panel). This binding isotherm is clearly quite different from the factor IXa concentration-dependence observed in the kinetic studies (Fig. 4, top panel). Furthermore, the fact that factor IX was a competitor for these sites was an additional problem. Since the zymogen is always present in vast excess under physiologic conditions, it would appear to block any access of the enzyme to endothelium. The solution to these problems is seen from the experiment shown in the bottom panel of Fig. 4. Here, factor IXa binding was studied in the presence of both the cofactor, factor VIII, and the substrate, factor X. Under these conditions, factor IXa binding is half-maximal at about 200 pM, which is comparable to the results of the kinetic studies. This binding was relatively selective for the enzyme and represents a modulation of the shared factor IX/IXa site in the presence of the other components of the factor X activation mixture.[42] Consistent with this find-

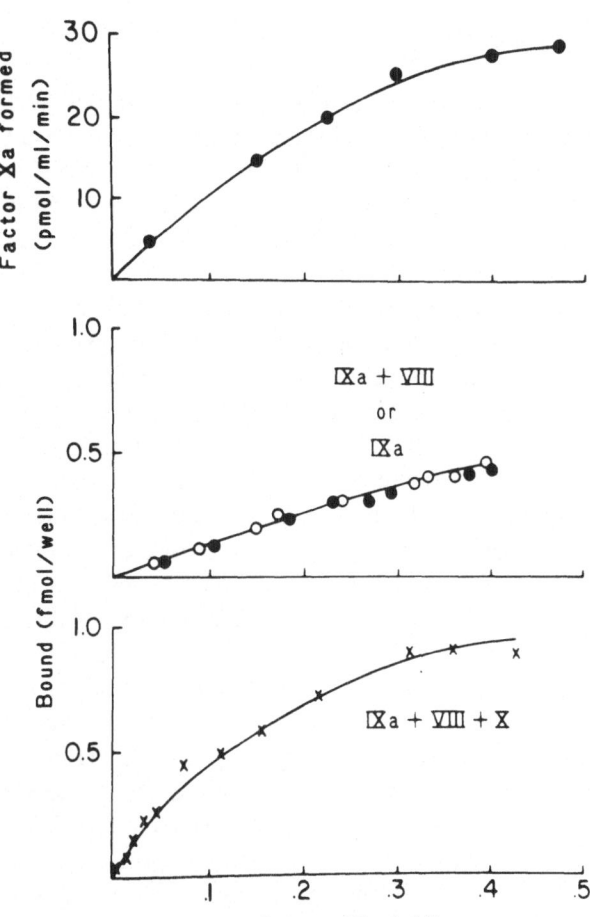

Figure 4. Interaction of factor IXa with bovine aortic endothelial cells. (Top) Factor IXa–VIII-mediated activation of factor X. Endothelial cell monolayers were incubated with increasing concentrations of factor IXa in the presence of factor VIII (2.4 U/ml) and factor X (300 pmole/ml) at 23°C. The initial rate of factor Xa formation is shown. (Middle) Binding of factor IXa to endothelium in the absence of other coagulation factors (IXa, ●) or in the presence of only factor VIII (IXa + VIII, ○). Endothelium was incubated with radiolabelled Factor IXa in the presence (○) or absence (●) of factor VIII (2.4 U/ml). Specific binding is shown. (Bottom) Binding of factor IXa to endothelium in the presence of factors VIII and X (IXa + VIII + X). Endothelium was incubated with radiolabeled factor IXa in the presence of factors VIII (2.4 U/ml) and X (300 pmole/ml).

ing, factor IX was a considerably less effective inhibitor than active site-blocked factor IXa (dansyl-Glu-Gly-Arg-factor IXa) in blocking factor IXa–VIII-mediated activation of factor X on the endothelial surface.[42] From this site, thus, factor IXa could function as a vessel-localized focal point for the activation of coagulation.

If factor IXa bound to its vessel wall site plays a role in the coagulation mechanism, it is important to understand the nature of the cellular receptor site. The specificity and high affinity of factor IX/IXa vessel wall interaction suggested that a unique receptor might be involved. Photoaffinity cross-linking studies and ligand blotting have provided data in support of this by indicating that the factor IX/IXa receptor involves, at least in part, a protein.[43] Currently, studies are under way to isolate and characterize this endothelial cell receptor.

Once the procoagulant enzyme factor IXa is formed, it can initiate a series of reactions culminating in thrombin formation on the endothelial cell surface. In homeostasis, however, factor IXa is not present in significant amounts. This led us to examine endothelial cell-dependent mechanisms of factor IX activation.

B. Induction of Tissue Factor in Endothelium

Tissue factor (TF) is a procoagulant cell surface cofactor which, in the presence of factor VII/VIIa, promotes the activation of factor X.[44] Characteristically, endothelium has very little TF activity.[45] This is in keeping with the predominance of anticoagulant mechanisms on quiescent endothelium. After stimulation, however, endothelium can synthesize and express TF. Induction of TF in endothelium was first shown following exposure of cultured endothelial cells to endotoxin[46-48] and has subsequently expanded to include many endothelial cell perturbants. Particularly notable in terms of the pathophysiology of disease states is induction of endothelial cell TF in response to mediators of inflammatory disease, tumor necrosis factor/cachectin (TNF) and interleukin 1 (IL-1).[49-53] When monolayers of cultured endothelium were incubated with TNF, in contrast to the cytotoxic effect of TNF on certain tumor cells *in vitro*,[54] toxic changes were not striking. Rather, TNF specifically modulated cellular hemostatic properties including the induction of TF (other alterations in endothelial cell hemostatic properties are described in Section IV). Induction of TF[51] occurred in a manner dependent on the dose of TNF added (Fig. 5A). The identification of this procoagulant activity as TF was confirmed by the factor VIIa-dependence of factor X activation, which could be prevented by anti-TF IgG (Table 1).[55] Although neoplastic tissue has been reported to have a direct factor X activating enzyme, TNF did not induce significant amounts of this activity in endothelium[96] (Fig. 5A, Table 1). TF activity induced by TNF was evident after a 2-hr lag, increased steadily up to 10 hr, and thereafter slowly declined (Fig. 5B). Procoagulant activity on the endothelial cell surface was not due to expression from a preformed pool, since lysates of control endothelial cells did not have significant TF activity. Furthermore, cycloheximide blocked the induction of TF by TNF, indicating a requirement for *de novo* protein synthesis (Fig. 5B). Thus, TNF induces endothelium to synthesize and express TF. Similar results have been observed with IL-1 and cultured endothelial cells.[49,52]

The TF-initiated pathway of coagulation may be a predominant route for activation of the clotting process. First, the alternative route for the activation of factor IX involves factor IXa, but severe deficiency of factor XI, in the absence of an acquired antibody to factor XI, is not invariably associated with a serious bleeding diathesis.[56] Second, subendothelial layers of the vessel wall,[45] as well as endothelial cells after perturbation, all express TF. Previous studies of the TF pathway have employed rather high concentrations of TF[44,57,58] to activate factor IX, and under these conditions factor X activation readily occurs in the absence of factors VIII and IX. Under physiologic conditions, however, these factors are essential for hemostasis since their deficiency leads to the hemophilias.[59] To examine this question further, factor X activation, mediated by factor VIIa and endothelial cell TF, was studied using intact monolayers of endothelium exposed to the perturbant endotoxin[26] (Fig. 6). In the presence of factors VIII and IX, considerably more factor X activation occurred than in their absence. Thus, on the perturbed endothelial cell surface, factors VIII and IX play an important role

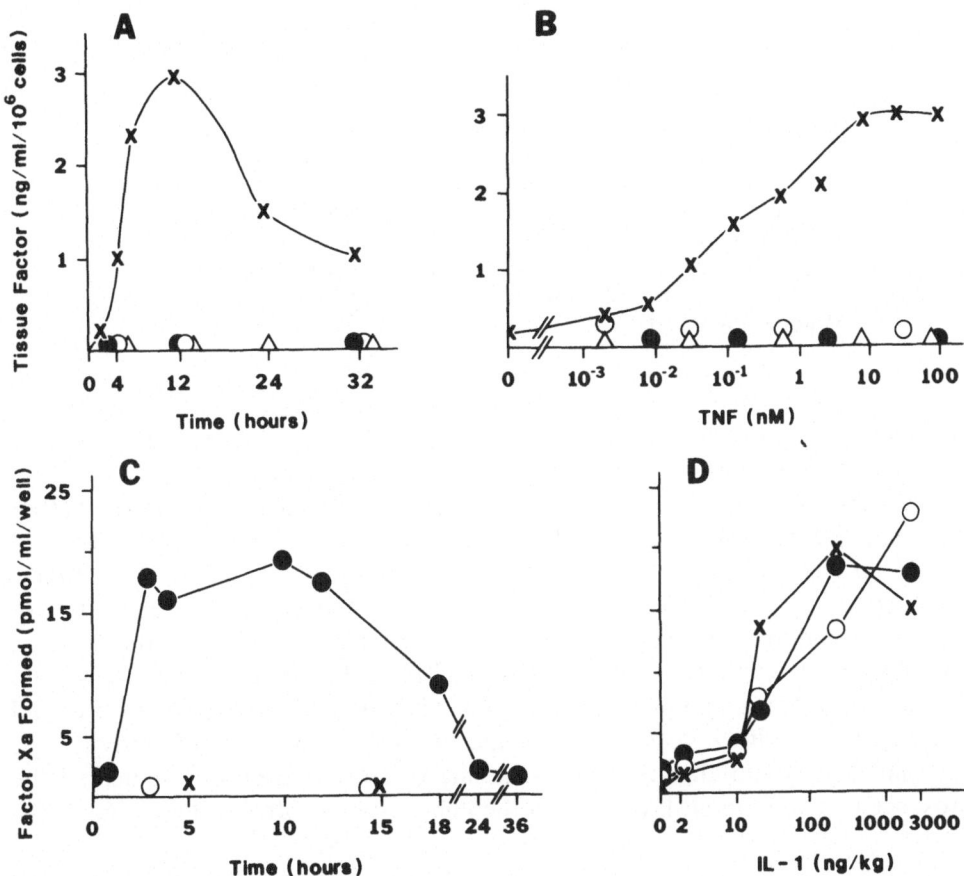

Figure 5. Induction of endothelial cell TF by TNF *in vitro* (A, B) and IL-1 in vivo (C, D). (B) Dependence of TF induction on the dose of TNF. Monolayers were incubated with the indicated concentration of TNF (X) for 12 hr. TF activity of the endothelium was assessed by coagulant assay. Where indicated, factor VII-deficient plasma replaced normal plasma (△). Heat-treated TNF (○) or cycloheximide (2 μg/ml) (●) was added to certain cultures. (A) Time course of TF induction by TNF. Monolayers were incubated in serum-free medium alone (▲) or in the presence of TNF (10 nM) (X). Cycloheximide (●) or heat-treated TNF (○) was added where indicated. TF was assayed by coagulant assay. Where indicated, factor VII-deficient plasma was used (△). (D) Dependence of TF induction on the dose of IL-1. Rabbits were infused with recombinant murine IL-1 at the indicated dose and 8 hr later sacrificed, aortas were isolated, and segments from proximal (○), mid (X), and distal (●) portions were assayed by monitoring factor VIIa-dependent factor X activation. In the absence of factor VIIa, no factor Xa was detectable. Factor Xa formed over 10 min is shown. (C) Time course of TF induction by IL-1. Rabbits were infused with recombinant murine IL-1 (0.5 μg/kg body wt, ●), heat-treated IL-1 (0.9 μg/kg body wt, ○), or control protein (0.8 μg/kg body wt, X). Animals were sacrificed at the indicated times and the TF activity of the native aortic endothelium was assayed by monitoring factor VIIa-dependent factor X activation.

Table 1.   Activation of Factor X by TNF-Treated Endothelial Cells[a]

Cell treatment	Assay reaction mixture	Factor Xa formed (pmole/10^6 cells)
None	VIIa, X	3 ± 1
TNF	VIIa, X	108 ± 17
TNF	X	4 ± 1
TNF	VIIa, X, anti-tissue factor IgG	11 ± 1

[a] Endothelial cell monolayers were incubated with or without TNF (10 nM) for 12 hr in serum-free medium and washed with incubation buffer. The tissue factor assay was then carried out using factors VIIa and X as described in the text. Values are expressed as factor Xa formed within 10 min. Anti-tissue factor IgG was generously provided by Dr. R. Bach, Mount Sinai School of Medicine, New York, N.Y.

in factor VIIa–TF-mediated activation of factor X. This is consistent with the physiologically important role of these factors in coagulation.

The induction of TF activity in endothelium *in vitro* led us to examine whether this could occur *in vivo*.[50] A physiologic mediator, IL-1, which induces endothelial cell TF *in vitro*, was selected. IL-1 was infused into rabbits, and at later times, the animals were sacrificed, aortas were excised, and hemostatic properties of the native endothelium were assessed. After the infusion of IL-1, there was a time-dependent increase of greater than tenfold in the TF activity expressed on the surface of aortic endothelium (Fig. 5D). TF activity was maximal by 5 hr and was reversible with a decline to baseline by 24 hr. Observed TF activity was associated with endothelium since scanning electron microscopy indicated the presence of a continuous layer of endothelium without evidence of adherent cells.

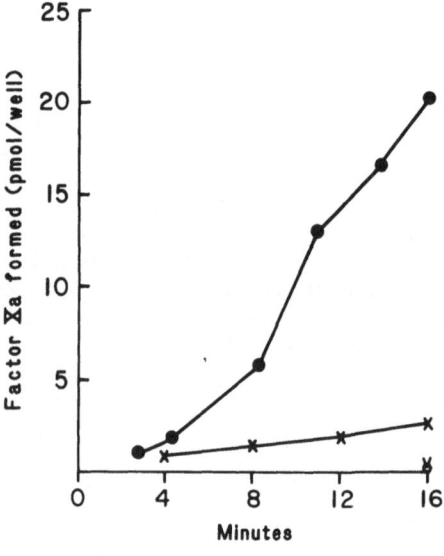

Figure 6. Endotoxin-treated endothelial cells and factor X activation. Monolayers of endotoxin-treated endothelial cells were incubated with factors VIIa (1.8 nM) and X (182 nM) in the presence (●) or absence (X) of factors IX (68 nM) and VIII (1.1 U/ml). Aliquots were removed at the indicated intervals and assayed in a chromogenic substrate assay. The mean of duplicates is plotted.

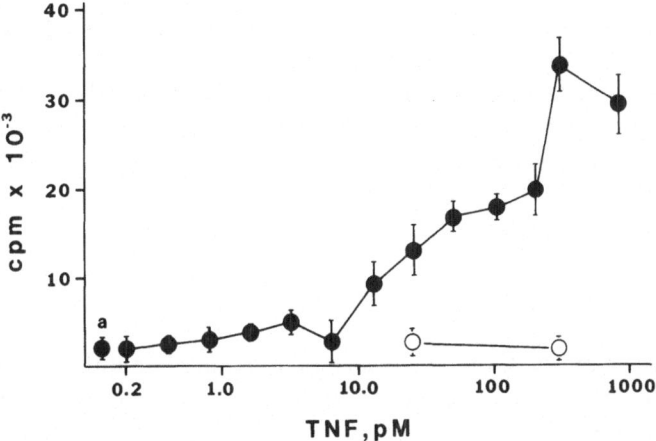

Figure 7. The effect of TNF concentration on the generation of IL-1 by endothelial cells. Endothelial cell monolayers were incubated for 18 hr at 37°C in serum-free medium in the presence of either the indicated concentration of TNF (●) or heat-treated TNF (○). Samples were then assayed in the thymocyte proliferation assay. The direct effect of TNF (0.8 nM) added to the thymocyte assay in the absence of endothelial cell supernatant is shown (a). The mean ± S.D. of triplicate determinations is shown.

TF activity was proportional to the dose of IL-1 infused and appeared to be uniformly distributed throughout different locations of the thoracic aorta (Fig. 5C). Thus, infusion of an inflammatory mediator, IL-1, can induce endothelial cell procoagulant activity *in vivo*. As will be discussed in Section IV, IL-1 suppresses endothelial cell anticoagulant mechanisms in addition to inducing procoagulant activity.

IL-1 is a particularly relevant endothelial cell stimulus since endothelium can synthesize and release this mediator in response to physiologic stimuli, such as monokines and thrombin.[60-66] When endothelium is incubated with endotoxin or TNF,[63,64] for example, IL-1 activity is elaborated into the medium. Using the thymocyte costimulation assay, IL-1 activity was clearly detectable in serially diluted supernatants of TNF-treated endothelial cells. Adsorption of this IL-1 activity in culture supernatant by an immobilized antibody to IL-1 suggests that IL-1 like molecules were responsible for the observed thymocyte proliferation. Consistent with this finding, Lomedico *et al.*,[66] using endothelial cell cultures from our laboratory, have shown that TNF-treated endothelium has increased amounts of mRNA hybridizable to IL-1 cDNA probes. Another laboratory has reported similar results as well.[65] Induction of IL-1 by TNF occurred in a time-dependent manner with increased IL-1 activity detectable in culture supernatants in 4–6 hr and a maximal level reached by 24–48 hr. Addition of cycloheximide blocked IL-1 elaboration, supporting the view that *de novo* protein synthesis plays an important role. Generation of IL-1 by endothelium was also dependent on the concentration of TNF incubated with the cultures (Fig. 7). IL-1 generation was half-maximal at a TNF concentration of 40–80 pM and reached a maximum at 300 pM. The significance of these levels is indicated by recent data from animal studies in which endotoxin/*Escherichia coli*-treated primates were demonstrated to have nanomolar levels of TNF in plasma. Thus, two consequences of TNF–endothelial cell interaction, TF and IL-1 production, could play a role in the pathophysiology of gram-negative septicemia. In addition, this suggests that the pathophysiology of disorders in which TNF plays an integral role involves not only direct action of TNF on cellular targets, but also an amplification mechanism involving IL-1 and its cellular targets.

C. Prothrombin Activation and Fibrin Formation

Activation of factor X on the surface of stimulated endothelium sets the stage for subsequent thrombin and fibrin formation. Effective thrombin formation requires the presence of factor Va, calcium, and an appropriate cellular surface in addition to the enzyme, factor Xa. When cultured or native[29,39] endothelium is incubated with factor Xa and prothrombin in calcium-containing buffer, thrombin formation is observed. Prothrombin activation could be blocked by adding antibody to factor V,[40] although no exogenous factor V/Va had been added to the reaction mixture.[29,39] This finding suggested that endothelium was providing the factor V-like activity for this reaction. Although the source of factor V could have been previous contact with serum-containing medium, experiments in serum-free medium, carried out by Rodgers and Shuman,[29] were consistent with the presence of endogenous factor V of endothelial cell origin. In this context, studies by Cerveny et al.[28] have directly indicated that endothelium synthesizes factor V/Va. The availability of endothelial cell factor V for prothrombin activation on confluent monolayers, however, is unclear. This mechanism appears to function more effectively on preconfluent endothelium and has been speculated to play a role in wound healing or other circumstances where cellular proliferation occurs.[29]

Assembly of the prothrombinase complex on the endothelial cell surface in the presence of added factor Va has been examined by Tracy et al.[67] Half-maximal rates of thrombin formation occurred at concentrations of factor Xa and Va similar to those when the same reaction was studied on platelets.[6–8] However, a considerable lag was observed on endothelium before thrombin formation occurred, compared with platelets.[67] Although the reason for this lag is unclear, the presence of inhibitory binding sites for factor Xa, such as the endothelial cell receptor thrombomodulin,[68] may provide an explanation. Another possibility includes a factor Xa binding site which internalizes surface-bound ligand and leads to its degradation.[69] Assembly of the prothrombinase complex on the endothelial cell surface, thus, may follow after the interaction of factor Xa with multiple binding sites. Although endothelium may not be as effective as platelets for rapidly forming thrombin, at least under the conditions studied so far, the vessel wall has been defined as an alternative surface for prothrombinase assembly.

Thrombin formed on the endothelial cell surface could be inactivated by anticoagulant mechanisms, such as antithrombin III bound to anticoagulantly-active vascular heparin-like molecules[20–24] or complexed with thrombomodulin,[17] initiating protein C activation (see Chapter 9). Alternatively, if enough thrombin formed, on perturbed endothelium for example, fibrin formation would result. As shown in Fig. 1, fibrin deposition can occur in close association with the endothelial cell surface. Although the experiment shown in Fig. 1 was carried out on cultured endothelium, fibrin deposition on endothelium has been seen in the *in vivo* setting as well, following infusion of IL-1[50] (see Section IV).

The experimental studies described in this section indicate that stimulated endothelium can initiate and propagate activation of the coagulation system. Although the physiologic significance of the endothelial procoagulant pathway is unclear at this time, the effectiveness of its function will probably be linked to the level of activity of anticoagulant mechanisms also operative on endothelium. A discussion of the interaction of anticoagulant and procoagulant mechanisms involving endothelium and their coordinate regulation by monokines is the subject of Section IV.

III. CONSEQUENCES OF THE GENERATION OF PROCOAGULANT ENZYMES IN CLOSE PROXIMITY TO THE ENDOTHELIAL CELL SURFACE

The endothelium is not a passive template for the generation of procoagulants, but can function as a cell capable of continuously responding to its environment. On the one hand, a coagulation enzyme produced on the cell surface can alter endothelial cell physiology. On the other hand, cellular events can modulate the outcome of coagulation reactions on the cell surface. This hypothesis will be examined by considering the consequences of factor Xa formation on the endothelial cell surface in terms of the coagulation mechanism and in relation to other aspects of endothelial cell physiology. The subjects which will be examined in detail in this section include factor Xa-induced release of endothelial cell mitogenic activity,[70] endocytosis of factors X and Xa,[69] and thrombin-induced release of endothelial cell mitogenic activity.[63] These examples, selected from work of ours carried out with several collaborators, are among many available examples of the dynamic nature of endothelial cell–coagulation factor interaction.

A. Factor Xa-Induced Release of Endothelial Cell Mitogenic Activity

A link between atherogenesis and coagulation has long been suspected from clinicopathological work. Fibrin deposition in atherosclerotic lesions is a well-known observation,[71–76] and a recent study indicates that an increased ratio of fibrin II to fibrinogen correlates with more advanced lesions.[76] Thrombin, the final coagulation enzyme in the pathway leading to fibrin formation, has been shown to release platelet-derived growth factor-like molecules ($PDGF_c$) from endothelium.[77] This enzyme also activates platelets, however, causing release of their intracellular stores of PDGF.[78] These considerations prompted us to examine the hypothesis that products of the activated coagulation system, before thrombin formation, could perturb endothelium resulting in release of growth factor activity.

When endothelium was incubated with factors IXa, VIII, and X, in addition to factor Xa formation, release of mitogenic activity occurred (Fig. 8)[70] (these studies were carried out in collaboration with Drs. Gadjusek and Ross). Release of mitogenic activity required factor Xa formation, since substitution of active site-blocked factor IXa (dansyl-Glu-Gly-Arg-factor IIXa) for native factor IXa prevented both factor Xa formation and mitogen release. Mitogenic activity was characterized as being due to $PDGF_c$ molecules based on competition studies using the PDGF radioreceptor assay and neutralization experiments with anti-PDGF antiserum. To examine which components of the reaction mixture were responsible for mitogen release, endothelial cell monolayers were incubated separately with factors IX, VIII, X, or their activated forms. Only factor Xa was effective in causing mitogen release. Factor Xa-induced release of mitogenic activity occurred in a thrombin-independent fashion. Mitogenic activity generated in response to factor Xa was elaborated in a dose-dependent manner and appeared to saturate at low concentrations of the enzyme, being half-maximal at about 0.5 nM (Fig. 9). These levels of factor Xa may be achieved under physiologic conditions and are reached when factor Xa is activated on the endothelial cell surface. The mechanism of factor Xa-induced mitogen release was further examined in studies employing actinomycin D and cycloheximide. Neither inhibitor diminished mitogen release in response to factor Xa, compared with untreated controls. Consistent with these results, cytoplasmic dot blot hybridization experiments showed no significant differences by densitometric scanning in endothelial cell RNA hybridizing with the v-sis probe whether cultures were incubated with factor Xa or not. The lack of a requirement for *de novo* transcription or translation suggests that factor Xa may trigger posttranslational events leading to elaboration of growth factor activity.

These data reflect the close relationship between the coagulation system and endothelium. Factor Xa, a product of the activated coagulation mechanism, functions both as a procoagulant, leading to thrombin formation, and as a modulator of endothelial cell physiology, resulting in enhanced release of mitogenic activity. Thrombin also releases mitogenic activity.[77] This suggests the existence of a series of interactions in which stimulation of endothelium initiates procoagulant reactions on the endothelial cell surface leading to factor Xa and thrombin formation. These enzymes could then potentiate the perturbed state by augmenting localized mitogen release, thus closing a circle of interaction by further promoting the generation of procoagulants.

B. Endocytosis of Factors X and Xa by Endothelium

Cellular processing of coagulation enzymes by endocytosis would be a direct mechanism for altering the course of clotting reactions on the endothelial cell surface. On the one hand, protease nexins[41] secreted by multiple cell types in tissue culture can bind to certain proteases and mediate their subsequent cell surface interaction and degradation. However, this is a relatively slow process since the protease nexin must be synthesized and released. On the other hand, direct binding of an enzyme to a preexisting binding site, followed by cellular processing, would be a more rapid mechanism. This led us to examine the pos-

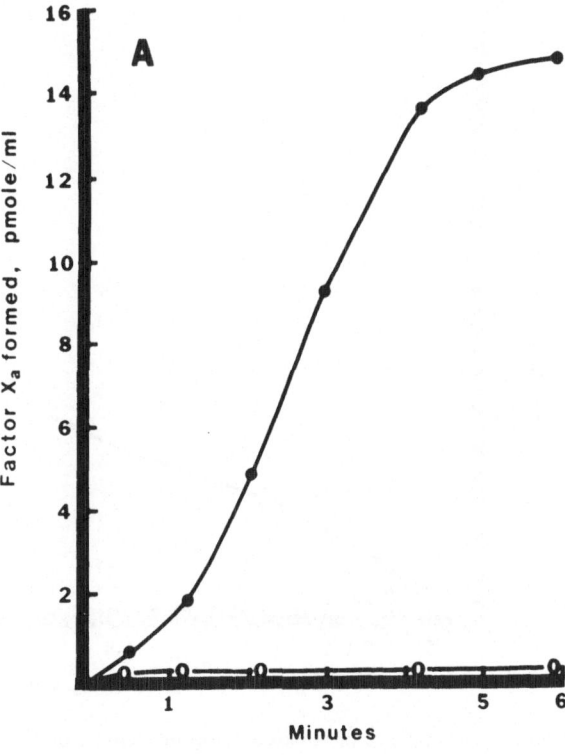

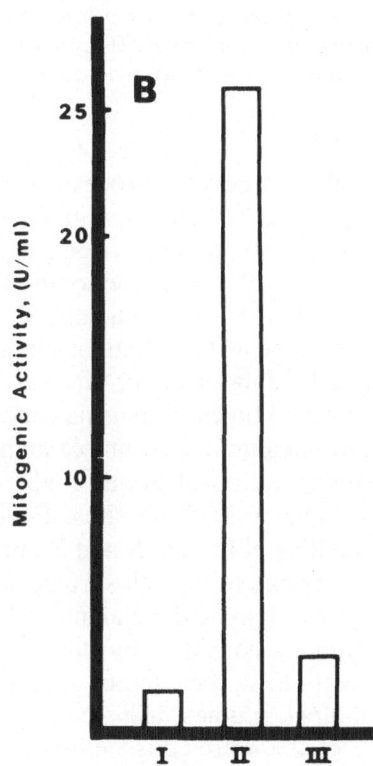

Figure 8. Activation of factor X on the endothelial cell surface and the elaboration of mitogenic activity. (A) Factor Xa formation. Endothelial cell monolayers were incubated with factor VIII (1.5 U/ml), factor X (300 nM), factor IXa (0.1 nM), or dansyl-Glu-Gly-Arg-factor IXa (0.2 nM). Aliquots of reaction mixture supernatant were withdrawn at the indicated times and assayed for factor Xa amidolytic activity. The mean of duplicates is shown. Where indicated, factor IXa (●) was replaced by dansyl-Glu-Gly-Arg-factor IXa (○) in the reaction mixture. (B) Elaboration of mitogenic activity. Endothelial cell monolayers were incubated for 3 hr without coagulation factors or with coagulation factors as described in A. Samples were assayed in the [³H]thymidine incorporation assay using 3T3 cells. The mean of triplicates is shown. Standard deviations were < 10%. I, no coagulation factors present; II, factors IXa, VIII, and X present; III, dansyl-Glu-Gly-Arg-factors IXa, VIII, and X present.

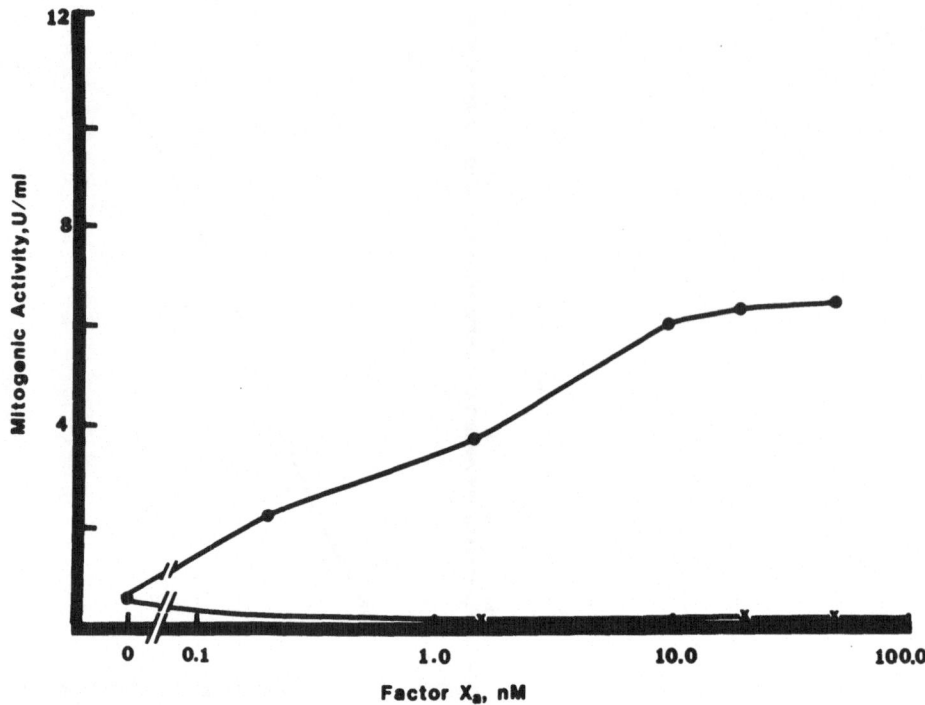

Figure 9. Dependence of released mitogenic activity on the concentration of factor Xa incubated with endothelial cells. Confluent endothelial cell cultures in serum-free medium were incubated 3 hr with increasing concentrations of factor Xa (●) and factor X (X). Conditioned medium was assayed for mitogenic activity in the [³H]thymidine incorporation assay using 3T3 cells. Data shown are the means of triplicate determinations. Standard deviations were < 15%.

sibility of cellular processing of the zymogen, factor X, and the enzyme form, factor Xa.[69] Clearance of factor Xa formed on the endothelial cell surface by endocytosis could provide a regulatory pathway involved in controlling assembly of the prothrombinase complex.

Studies with radiolabeled factors X and Xa indicated that each molecule interacted with a distinct class of binding sites on the endothelial cell surface. The binding of factor Xa was half-maximal at approximately 500 nM, which is about 2.5 times the plasma concentration. Factor Xa, in contrast, binds to cultured endothelium in a saturable manner with half-maximal occupancy of the sites at a concentration of about 1 nM. Factor X does not compete with factor Xa for occupancy of these sites. Following binding to their cell surface sites, cellular handling of factors X and Xa proceeds along different routes. Radiolabeled factor Xa, bound to the cell surface at 4°C, was subsequently internalized and degraded by a lysosome-dependent pathway. Similar experiments with factor X demonstrated a considerably slower rate of internalization and a negligible rate of degradation. Rather, factor X appeared to be released in an undegraded form. Morphologic studies, using colloidal gold conjugates prepared with factors X and Xa (Fig. 10), were consistent with the results of radioligand binding studies. At 4°C,

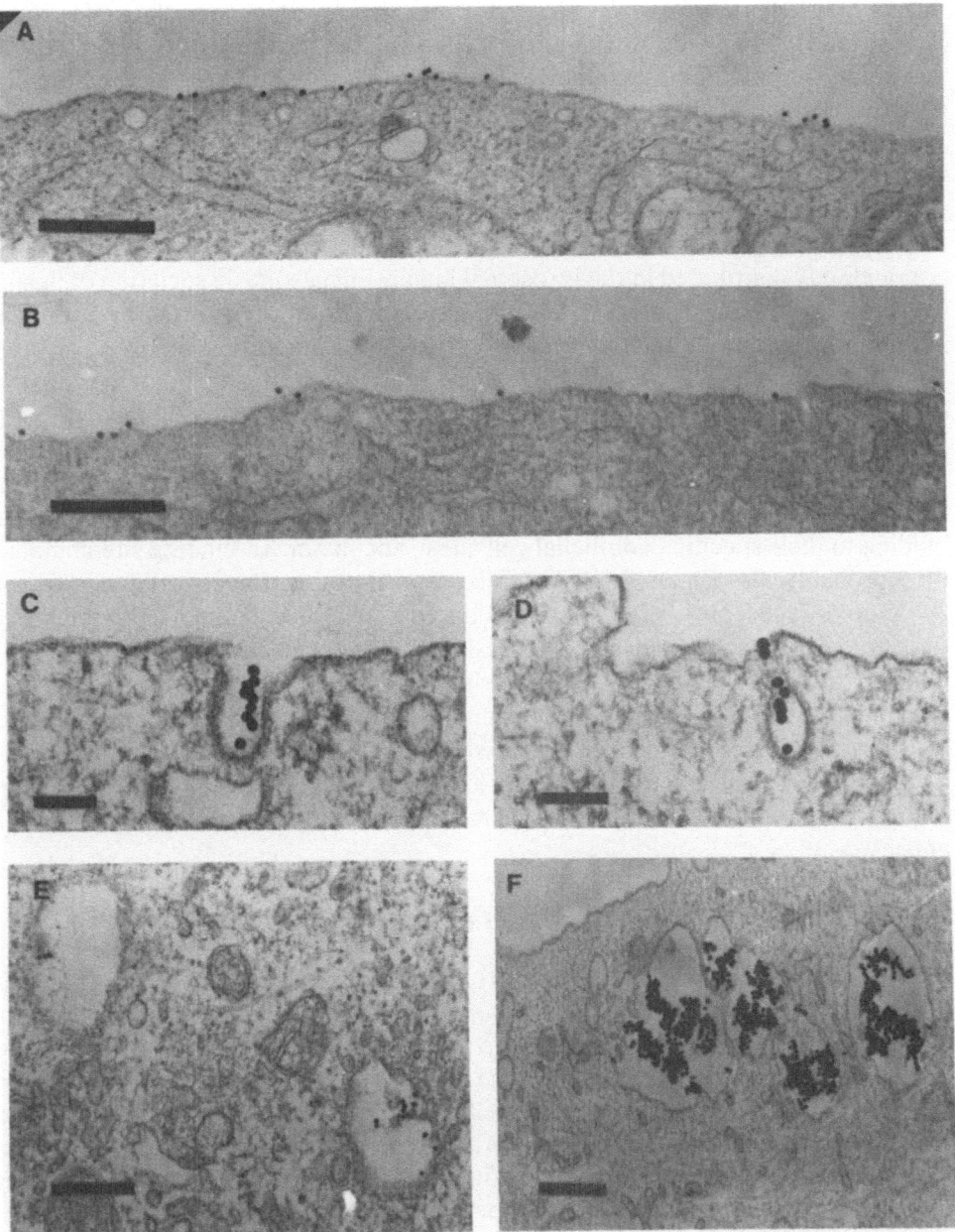

Figure 10. Binding and internalization of factors X and Xa conjugated to colloidal gold. Surface labeling of factor X–gold (A) and factor Xa–gold (B) after incubation with endothelial cells for 2 hr at 4°C, followed by rinsing and fixation at 4°C. In both cases, gold particles are randomly distributed at the outer membrane surface. Bars = 0.5 nm. After incubation at 1°C for 2 hr, cells were rapidly warmed to 37°C to initiate endocytosis of surface-bound factor X–gold (C) and factor Xa–gold (D). Endocytosis proceeds at coated pit regions involving clustering of the gold probes. Bars = 0.1 nm. After 2 hr at 37°C, there is limited accumulation of factor X–gold (E) in structures resembling lysosomes, whereas greater accumulation in these lysosome-like structures is seen in cells labeled with factor Xa–gold (F). Bars = 0.5 nm.

the gold particles were evenly distributed and closely attached to the cell surface (Fig. 10A,B). Following warming the cells to 37°C, clustering of gold particles and uptake at coated pit regions were seen (Fig. 10C,D). Finally factor Xa–gold particles accumulated in lysosome-like structures, whereas endocytosed factor X accumulated only to a limited extent intracellularly in lysosomes or elsewhere (Fig. 10E,F). The close association of factor X–gold probes with the inner lysosomal membrane suggests continued receptor interaction, and may account for the subsequent externalization of factor X. In contrast, factor Xa–gold probes are randomly distributed in the lysosomal lumen, suggesting dissociation of ligand from receptor. In each case, a 100-fold molar excess of ligand (factor X or Xa) not conjugated to gold inhibited binding and internalization by > 85%.

The role of endocytosis in regulating the availability of factor Xa on the endothelial cell surface was studied by examining factor IXa–VIII-mediated activation of factor X. When factor X activation was studied at 37°C, only 44% of the factor Xa formed was released into the supernatant. The remainder was subject to endocytosis and degradation. Thus, factors X and Xa are internalized after binding to their specific endothelial cell sites. The factor Xa-binding site complex then probably dissociates in acidic prelysosomal compartments, allowing factor Xa to enter lysosomes, where degradation would go on. Factor Xa processing is thus analogous to that observed for asialoglycoprotein or low-density lipoprotein.[79,80] Factor X, on the other hand, is slowly internalized, and principally shuttled back to the cell surface in a manner similar to the cellular processing of transferrin.[81] Extensive cellular processing of cell-bound coagulation factors indicates that the endothelial cell is not a passive template for assembly of coagulation factor complexes, but is a dynamic surface with its own regulatory mechanisms.

C. Thrombin-Induced Elaboration of Endothelial Cell Interleukin 1 (IL-1)

Inflammatory mediators such as IL-1[93-95] can induce many changes in endothelial cell coagulant properties. Induction of TF in endothelium provides one such example (Section II). This leads to the question of whether a product of the coagulation system could influence the participation of endothelial cells in the inflammatory response. Since endothelium is both responsive to IL-1[49,50] and synthesizes this inflammatory mediator,[60-64] the effect of coagulation factors on IL-1 production was investigated.[63] When factors IX, X, prothrombin and their activated forms were studied, only thrombin was found to be an effective inducer of endothelial cell IL-1 production at low concentrations. IL-1 elaboration in response to thrombin occurred in a time-dependent manner directly related to the concentration of enzyme added. Half-maximal release of IL-1 activity occurred about 6–8 hr after thrombin was added to cultures. Elaboration of IL-1 could be prevented by adding cycloheximide to cultures and IL-1 activity could be removed from culture supernatants with anti-IL-1 antibody. The thrombin concentration causing half-maximal release was about 0.075 U/ml with detectable levels of IL-1 produced down to 0.025 U/ml. Modified thrombins were examined to determine domains of the enzyme which could be mediating endothelial cell IL-1 release.

Native α-thrombin induced the elaboration of IL-1; however, blockade of the protease's active site, with antithrombin III or diisopropylfluorophosphate, rendered the molecule ineffective in mediating IL-1 release. Thus, the integrity of the active site of thrombin is required for thrombin to induce endothelial cell synthesis and release of IL-1.

Since IL-1 has been shown to induce TF activity in endothelial cells,[50,52,63] this suggested the possibility of self-regulation of procoagulant events on the endothelial cell surface. To examine this hypothesis, IL-1-containing supernatants from endothelial cells treated with thrombin were incubated with fresh cultures, after residual thrombin was inactivated.[63] TF activity was induced in the fresh endothelium by the conditioned medium, as long as IL-1 activity was present. Thrombin, by inducing endothelial cell IL-1 production, can augment the cellular procoagulant response: in addition to directly cleaving fibrinogen to form fibrin, IL-1 released in response to thrombin can interact with endothelium of other vascular beds, modulating hemostatic properties. These findings reflect the close relationship between the hemostatic and inflammatory systems.

IV. INTEGRATION OF THE ENDOTHELIAL CELL PROCOAGULANT PATHWAY AND ANTICOAGULANT MECHANISMS

Since blood does not form clots continuously as it passes over the endothelial cell surface, procoagulant mechanisms on the endothelial cell surface must be under tight control. This suggests the hypothesis that in homeostasis, or the quiescent state, endothelial cell anticoagulant mechanisms would predominate. In a stimulated, or perturbed state, this balance could be shifted to favor activation of coagulation on the vessel wall. To effectively shift the coagulant phenotype of endothelium would require both the induction of procoagulant activity and the suppression of anticoagulant mechanisms. Before describing the experiments carried out in our laboratory to test this hypothesis, the function of the vessel wall anticoagulant mechanism that we monitored during these studies will be outlined.

A. The Protein C Pathway

Multiple anticoagulant mechanisms are operative on the endothelial cell surface. Antithrombin III bound to anticoagulantly active heparin-like molecules on the luminal endothelial cell surface is one example of a vessel wall-based inhibitory mechanism.[20-23] Included among other mechanisms are the fibrinolytic activators/enzymes[24,25] and prostenoids.[82] In this section an anticoagulant mechanism which results in inactivation of the procoagulant cofactors, the protein C pathway, will be examined.

Recent studies have indicated that the protein C pathway is closely associated with the vessel wall.[17-19] Formation of the central enzyme in this pathway, activated protein C, results from the action of thrombin.[83] Following the interaction of thrombin with the endothelial cell receptor thrombomodulin,[17] the latter en-

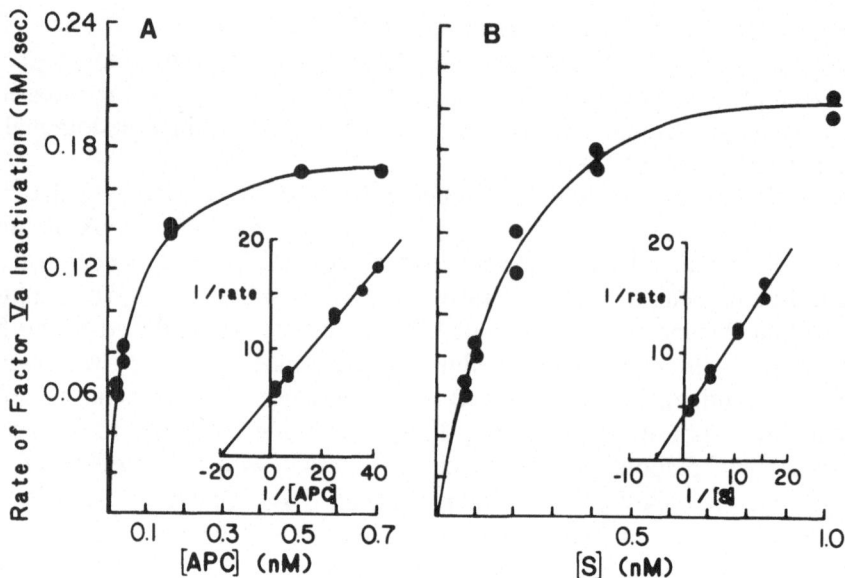

Figure 11. Factor Va inactivation on bovine endothelial cells: Effect of protein S and activated protein S and activated protein C (APC). (A) Endothelial cell monolayers were incubated with factor Va (80 nM), protein S (10 nM), and the indicated concentrations of APC. The initial rate of factor Va inactivation is plotted versus the concentration of APC. (Inset) Calculated using nonlinear regression analysis applied to the Michaelis–Menten equation. (B) Endothelial cell monolayers were incubated with factor Va (80 nM), APC (20 nM), and the indicated concentrations of protein S. The initial rate of factor Va inactivation is plotted versus the concentration of protein S. (Inset) Calculated using nonlinear regression analysis applied to the Michaelis–Menten equation.

zyme becomes an effective activator of protein C. Thus, a product of the pro-coagulant pathway, thrombin, can feed directly into the protein C mechanism by promoting formation of the anticoagulant enzyme activated protein C. One must consider, however, that activated protein C by itself is not a potent anticoagulant. Activated protein C requires the cofactor protein S and a surface to promote assembly of the activated protein C/protein S complex in order to optimally ex-press its anticoagulant activity.[84–86] This led to studies examining assembly of the activated protein C/protein S complex on the endothelial cell surface.[18] In addition, these experiments led to the recognition of endothelium as a site of synthesis of protein S.[19,90]

1. Assembly of Activated Protein C/Protein S Complex

When factor Va and activated protein C are incubated with cultured bovine aortic endothelial cells, rapid factor Va inactivation is dependent on the presence of protein S.[18] Kinetic studies (Fig. 11A) have indicated that endothelial cell-dependent factor Va inactivation was saturable with respect to activated protein C, with half-maximal rates occurring at an activated protein C concentration of about 0.05 nM. The enzyme system was also saturable with respect to protein S

(Fig. 11B), with half-maximal factor Va inactivation rates occurring at a protein S concentration of about 0.2 nM. This is considerably below the plasma level of protein S ($\sim$ 100 nM).

This finding suggested that a limited number of cellular binding sites might mediate the interaction of protein S and activated protein C with endothelium. Radioligand binding studies demonstrated specific, time-dependent, and saturable binding of [^{125}I]protein S to endothelium. Addition of activated protein C increased the affinity of protein S for its endothelial cell site. Binding of ^{125}I-labeled activated protein C to endothelial cell monolayers was absolutely dependent on the presence of protein S. Thus, the endothelial cell activated protein C/protein S complex appears to result from the interaction of activated protein C with the cell surface, which is facilitated by the presence of protein S. Other studies are in progress to further characterize the cell surface activated protein C/protein S complex in order to better define the stoichiometry and other parameters.

Recent studies by our and other laboratories have examined the endothelial cell activated protein C/protein S complex on human endothelium (previous studies on bovine aortic endothelium).[87] Although human endothelium supports activated protein C/protein S-mediated factor Va inactivation, the role of proteins is less clear, probably because protein S binding and function on the endothelial cell surface is subject to considerable variation. This may be due to the fact, as will be discussed later (Section IV.B), that assembly of the activated protein C/protein S complex can be blocked by subtle perturbation of endothelium. Thus, sensitivity of human endothelial cells to culture conditions could be the cause of such a perturbation.

2. Synthesis and Release of Protein S

Protein S, a regulatory vitamin K-dependent plasma protein, is an essential component of the protein C anticoagulant pathway.[83,84] As described in the previous section, protein S functions as a nonenzymatic cofactor which promotes binding of enzyme activated protein C to the endothelial cell surface and other membrane surfaces. The first suggestion that endothelium might elaborate functional protein S occurred during studies of assembly of activated protein C/protein S on the endothelial cell surface.[18] Following prolonged incubation of endothelial cells in serum-free medium, the rate of activated protein C-mediated factor Va inactivation steadily increased in the absence of exogenous protein S.[19] Subsequent experiments showed that this acceleration of the factor Va inactivation rate could be blocked by an antibody to protein S. Furthermore, clinical studies have shown that although the level of other vitamin K-dependent coagulation factors is decreased by 50% in patients with liver disease, protein S is decreased by only 25%.[88] This is consistent with the existence of extrahepatic sites of protein S synthesis. These findings led us[19] and Fair et al.[90] to examine in more detail the synthesis of protein S by endothelium. If endothelium did synthesize protein S, this would be another entry on the list of coagulation cofactors, in addition to the von Willebrand factor and factor V, made by this cell.[28,29]

Studies employing immunoprecipitation of ^{35}S-labeled endothelial cells and

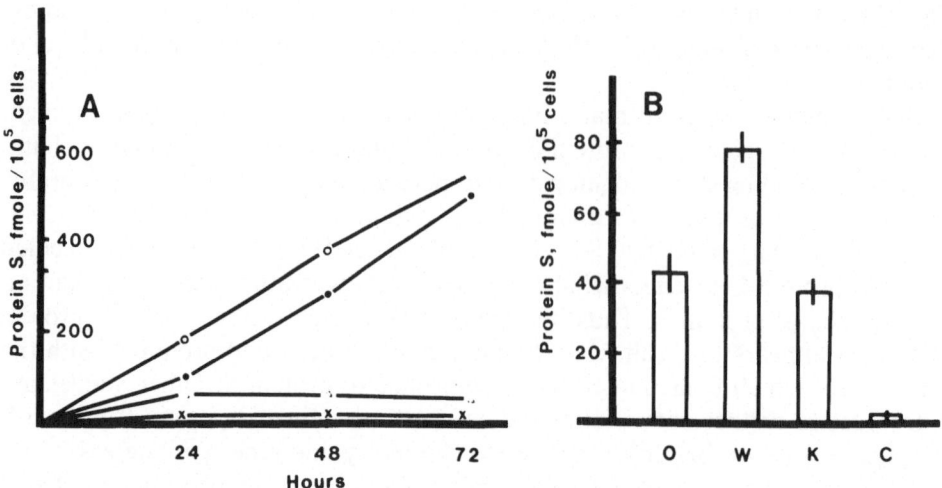

Figure 12. Protein S released from and remaining associated with endothelial cells. (A) Released protein S. Confluent endothelial cell monolayers maintained in serum-free medium were washed with dextran sulfate (10 mg/ml)-containing buffer and then incubated in serum-free medium alone (●), or supplemented with either cycloheximide (2 μg/ml) (△), vitamin K (25 μg/ml) (○), or the warfarin derivative 3(α-acetonylbenzyl)-4-hydroxycoumarin (1 μg/ml) (X). Aliquots of supernatant were taken at the indicated times and assayed for protein S antigen. The mean of duplicates is shown. (B) Intracellular protein S. After removal of the culture medium at 72 hr (see A above), monolayers were washed once in dextran sulfate (10 mg/ml)-containing buffer and solubilized with 1% Nonidet P-40. The radioimmunoassay for protein S antigen was then done. O, endothelial cells maintained in serum-free medium. W, endothelial cells maintained in serum-free medium that contained the warfarin derivative used in A (1 μg/ml). K, endothelial cells maintained in serum-free medium that contained vitamin K (25 μg/ml). C, endothelial cells maintained in serum-free medium that contained cycloheximide. The mean and S.E.M. are shown ($n = 7$).

Western blotting demonstrated that endothelium does synthesize protein S.[19] Functional studies of factor Va inactivation indicate that endothelial cell protein S can serve as a cofactor for activated protein C. These results were complemented by electron microscopy, which demonstrated intracellular localization of protein S antigen within cisternae of rough endoplasmic reticulum, the *trans* face of the Golgi, and a population of intracellular vesicles. Employing immunofluorescence, protein S was observed in endothelium but not vascular fibroblasts or smooth muscle cells.

Since endothelial cell protein S is physiologically significant in hemostasis only after its release from the intracellular pool, the secretion of protein S was important to examine (Fig. 12). In unsupplemented culture medium, release of protein S antigen occurred in a constitutive manner and was associated with an intracellular pool of about 40 fmole of protein S antigen per 10^5 cells. Addition of vitamin K to the medium had no significant effect on protein S release or intracellular accumulation, indicating that vitamin K is not limiting under these conditions. Supplementation of culture medium with the warfarin derivative 3(α-acetonylbenzyl)-4-hydroxycoumarin, in contrast, considerably decreased protein

S release and increased its intracellular accumulation. These results are consistent with studies of other cell types which have demonstrated that γ-carboxylation facilitates release of protein S. Cycloheximide blocked both the release of protein S and its intracellular accumulation, demonstrating the central role of *de novo* protein synthesis in protein S production by endothelium. The results of these studies on bovine endothelium have been complemented by similar observations on human endothelium from the laboratory of Fair *et al.*[90]

The presence of protein S within endothelium suggested the possibility of a mechanism for rapid release, in addition to ongoing constitutive production. Experiments with ionophores demonstrated that Ionomycin and the calcium ionophore A23187 were effective in releasing intracellular protein S. Recent studies have shown that a more physiologic mechanism resulting in rapid protein S release occurs following incubation of endothelium with low concentrations of norepinephrine.[91]

The synthesis of protein S by endothelium emphasizes the close relationship of the protein C/protein S pathway to the vessel wall. It also raises the question of which other γ-carboxyglutamic acid-containing proteins are made by endothelium. In this context, the function of a vitamin K-dependent carboxylase system in endothelium may explain the presence of γ-carboxyglutamic acid-containing proteins in calcified atherosclerotic plaques.[92]

B. The Role of Tumor Necrosis Factor/Cachectin (TNF) and IL-1 as Agents Which Can Modulate the Balance of Coagulant Activities on the Endothelial Cell Surface in Vitro and in Vivo

The coagulation mechanism is a sensitive effector system regulated by cellular receptors/cofactors. As a ubiquitous vascular surface, endothelium and its receptors contribute to the tone of the coagulation mechanism within the intravascular space. Studies of endothelial cell anticoagulant mechanisms clearly indicate that in homeostasis, pathways preventing activation of coagulation predominate. After stimulation of endothelium, however, one could speculate that anticoagulant mechanisms would be suppressed and procoagulant activities enhanced. If this were true, then prior to extensive vessel wall damage with denudation of endothelium, these cells could be capable of actively promoting localized coagulation. In the case of inflammation, for example, intravascular clot formation is a host defense potentially enhancing isolation of an inflammatory focus. Sepsis, which is associated with multiple abnormalities of intravascular coagulation, provided an opportunity to test this hypothesis. Our goal was to examine if a central mediator of the septic state, TNF,[97-99] could promote intravascular coagulation through the coordinated induction of endothelial cell procoagulant receptors/cofactors and suppression of binding sites involved in the function of anticoagulant mechanisms.

As a starting point, the hypothesis that endothelium is a target tissue for TNF was examined. Binding studies (Fig. 13) indicated that radiolabeled TNF bound in a saturable manner to endothelial cell monolayers.[64] Half-maximal occupancy of these sites occurred at a TNF concentration of 105 ± 40 pM and at saturation

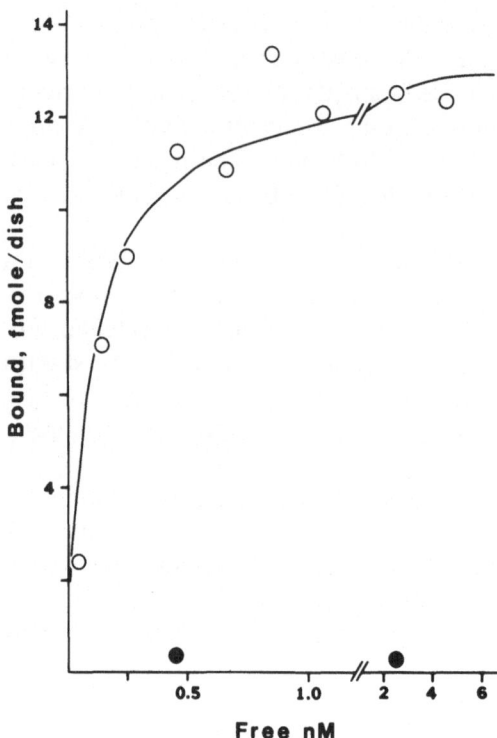

Figure 13. Saturability of [^{125}I]-TNF binding to endothelial cell monolayers. Endothelial cell monolayers were incubated for 2 hr at 4°C with the indicated concentrations of [^{125}I]-TNF alone (total binding) or [^{125}I]-TNF in the presence of a 300-fold molar excess of unlabeled TNF (nonspecific binding). Specific binding (total minus nonspecific binding) is plotted versus free [^{125}I]-TNF (O). Data were analyzed by a nonlinear least-squares program and the curve (——) indicates the best-fit line. Where indicated heat-inactivated [^{125}I]-TNF and unlabeled TNF were used (●).

there were $1.5 \pm 0.5 \times 10^3$ sites/cell. Heat-treated TNF, which lacks biologic activity,[100] does not bind to endothelium. The close relationship between the concentrations of TNF that induce endothelial cell TF (Fig. 5A), and induce the generation of IL-1 by endothelium (Fig. 7), and levels which result in occupancy of TNF binding sites on endothelium (Fig. 13) suggests that these sites may be involved in mediating the cellular effects of TNF. Recent data from Rubin *et al.*,[101] indicating that mouse L(M) cells sensitive to TNF bound the monokine specifically but resistant L(M) cells did not, support the concept of a relationship between the presence of TNF binding sites and a cellular response to TNF. The affinity of [I]^{125}-TNF for the endothelial cell surface sites is similar to that observed for other cells with specific binding sites.[101–103,109] Taken together, these data suggest that TNF bound to its endothelial cell binding sites exerts the cellular effects of this monokine.

In terms of the coagulation system, the consequences of TNF–receptor interaction result in a unidirectional shift in endothelial cell coagulant properties favoring activation of coagulation. Induction of procoagulant activity includes enhanced TF activity (Section II.B) and an increase in the number of factor IX/IXA binding sites.[104] Since endothelium has potent anticoagulant mechanisms, induction of procoagulant activity may not be sufficient to promote the development of a prethrombotic state. This led us to examine the effect of TNF on endothelial cell anticoagulant properties. For these studies, we monitored the protein C pathway since endothelium provides cofactor activity promoting the formation and function of activated protein C (Fig. 14). Endothelial cell-dependent activated protein C formation was in large part dependent on the presence of thrombomodulin, as indicated by the 75% inhibition of protein C activation in the presence of antithrombomodulin IgG (Fig. 14A). TNF resulted in a dose-dependent decrease in endothelial cell-dependent thrombin-mediated protein C activation (Fig. 14A). Decreased activated protein C formation was also dependent on the incubation time of endothelium with TNF, with an effect evident by 1 hr and the maximal effect after 6 hr (Fig. 14B). In addition to decreased protein C activation, the anticoagulant enzyme activated protein C functions less effectively on the endothelial cell surface after the cells have been exposed to TNF. In contrast to the rapid factor Va inactivation observed with control monolayers, after incubation with TNF, the rate of factor Va inactivation was reduced (Fig. 14C). The rate of factor Va inactivation declined in a time-dependent manner after the addition of TNF to cultures with negligible rates occurring after 8 hr (Fig. 14D). These changes in endothelial cell cofactor activity for the protein C pathway occur at concentrations similar to those which induce procoagulant activity (Fig. 5). Furthermore, studies by other investigators[105-107] have indicated that TNF decreases the net fibrinolytic activity of endothelial cells by altering the balance of plasminogen activator to plasminogen activator inhibitor. TNF-induced modulation of the fibrinolytic system also occurs for longer times (greater than 48 hr), similar to decreased endothelial cell function in the protein C pathway (Fig. 14). This contrasts with the transient rise and fall of endothelial cell TF induced by TNF (Fig. 5). These results indicate that TNF-induced modulation of endothelial cell hemostatic properties does include a coordinated increase in procoagulant activity and diminution of anticoagulant mechanisms. Alteration in coagulant phenotype of endothelium clearly favors activation of coagulation on the vessel surface. When similar studies were carried out *in vivo* in rabbits following infusion of IL-1,s;[50] similar modulation of hemostatic properties of the native aortic endothelium was seen. Furthermore, fibrin deposition on an apparently morphologically intact endothelium was observed *in vivo*. Thus, the net result of TNF– or IL-1–endothelial cell interaction is a unidirectional shift in the balance of anticoagulant and procoagulant mechanisms on the endothelial cell surface, from the quiescent state in which anticoagulant mechanisms predominate, to a stimulated state in which procoagulant activities are dominant.

In the case of TNF, one could speculate that modulation of hemostatic properties of the vessel wall, potentially resulting in occlusive thrombosis within the tumor vasculature, could play a role in TNF-mediated tumor necrosis. This would be consistent with pathologic findings indicating that hemorrhagic necrosis, with collapse of the vasculature around the tumor, is associated with TNF-induced tumor necrosis.[99] Preliminary studies, employing a murine model, support this notion and indicate that occlusive thrombosis limited to the tumor bed was seen by 2 hr after TNF infusion.[108] These thrombi originate as masses of fibrin closely associated with the endothelial cell surface, which are observed within 30 min of TNF administration.[108] This points to a potential beneficial effect of localized intravascular coagulation and emphasizes the importance of understanding mechanisms which underlie vessel wall hemostatic properties.

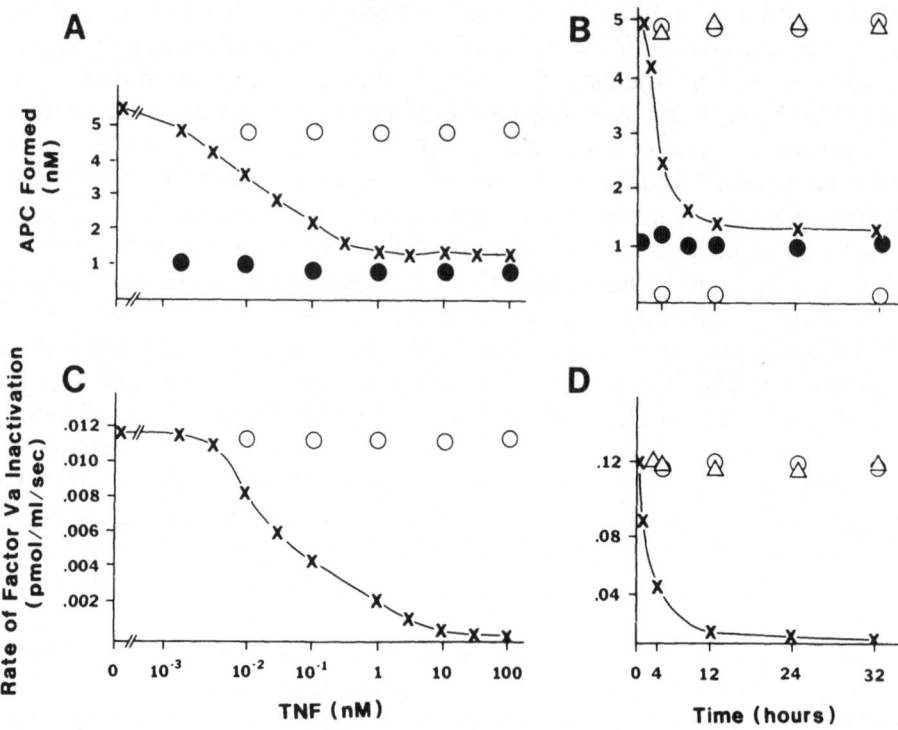

Figure 14. Effect of TNF on endothelial cell participation in the protein C/protein S pathway. (A, B) Activation of protein C on endothelial cell monolayers. (A) Dependence of decreased thrombin-mediated protein C activation on the dose of TNF. Monolayers were incubated with the indicated concentration of TNF (X) or heat-inactivated TNF (O). Endothelium was then incubated with thrombin and protein C to assess its ability to promote thrombin-mediated protein C activation. Where indicated, goat anti-rabbit thrombomodulin IgG (150 μg/ml) was preincubated with endothelium for 30 min (●). Control IgG from nonimmune animals had no effect on the assay. Results are expressed as activated protein C (APC) formed per 40 min per 10^5 cells. (B) Time course of decreased thrombin-mediated protein C activation. Monolayers were incubated with TNF (10 nM) (X), heat-inactivated TNF (O), or serum-free medium alone (△) for the indicated times, and washed. Where indicated, goat anti-rabbit thrombomodulin IgG (150 μg/ml) (generously provided by Dr. C. Esmon, Oklahoma Medical Research Foundation) was preincubated with endothelium for 30 min (●). Endothelium was then incubated with thrombin and protein C to assess its ability to promote thrombin-mediated protein C activation. Results are expressed as APC formed per 40 min per 10^5 cells. (C, D) Endothelial cell-dependent APC/protein S-mediated factor Va inactivation. (C) Dependence of decreased APC/protein S-mediated factor Va inactivation on the dose of TNF. Monolayers were incubated with the indicated concentration of TNF (X) or heat-treated TNF (O) for 12 hr. Endothelium was then assayed for the ability to promote APC/protein S-mediated factor Va inactivation. Results are expressed as the rate of factor Va inactivation per 10^5 cells. (D) Time course of decreased APC/protein S-mediated factor Va inactivation. Monolayers were incubated with TNF (10 nM) (X), heat-inactivated TNF (O), or serum-free medium alone (△) for the indicated times, and washed. Endothelium was then assayed for the ability to promote APC/protein S-mediated factor Va inactivation. Results are expressed as the rate of factor Va inactivation per 10^5 cells.

V. CONCLUDING REMARKS

As the cells forming the luminal vascular surface, endothelium is ideally situated to play an important role in the regulation of coagulation. The traditionally accepted inertness of endothelium in hemostatic reactions actually masks a delicate balance of opposing anticoagulant and procoagulant mechanisms operative on the cell surface. The common denominator of these endothelial cell-dependent hemostatic mechanisms is the expression of cell surface receptors for coagulation proteins. These receptors include constitutively expressed molecules, such as thrombomodulin, binding sites for factor IX and protein S, and inducible molecules, such as TF. In the setting of endothelial cell biology, the coagulant phenotype of the vessel wall can be modulated in response to environmental stimuli. Physiologic endothelial cell stimulators, such as TNF and IL-1, can induce a unidirectional shift in the balance of coagulant properties favoring activation of coagulation, through the coordinated induction of procoagulant and suppression of anticoagulant activities. These changes are mediated by altered expression of coagulation factor receptors on the endothelial cell surface and lead to a procoagulant pathway (Fig. 15).

These considerations lead to the hypothesis that early stages in the pathogenesis of thrombotic disease can occur on stimulated endothelium. Activation of coagulation on the perturbed endothelial cell can then recruit platelets and other components of the coagulation system, generating clotting enzymes in proximity to the vessel wall and ultimately leading to intravascular clot formation. On the one hand, a fibrin clot disrupts the architecture of the vessel wall, disturbing continuity of the endothelial cell monolayer and exposing subendothelium. The denuded vessel wall, with its clot buried in the subendothelium, is thus representative of an established lesion. On the other hand, coagulation proteases can interact directly with the endothelial cell causing release of molecules such as IL-1 and $PDGF_c$ mitogens. These second messengers then perpetuate the perturbed state and provide links between the coagulation mechanism and the host response in pathophysiologic states. Although activation of the coagulation mechanism may initially appear to be a nonspecific response, localization of this process to discrete areas of the vessel wall in response to locally produced mediators, such as IL-1,

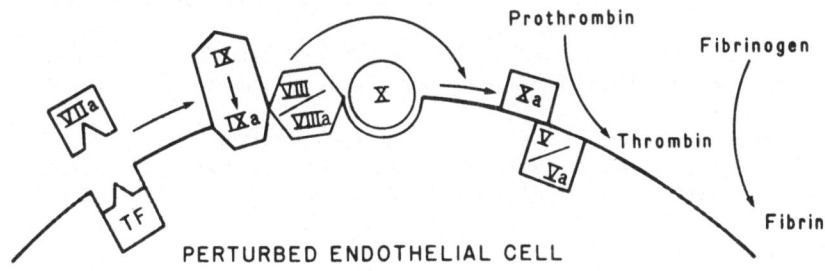

Figure 15. Schematic depiction of the endothelial cell procoagulant pathway on the surface of a perturbed endothelial cell. TF, tissue factor.

gives this system specificity and flexibility to respond to environmental stimuli. These findings place the coagulation physiology within the broader context of the biology of the vessel wall.

ACKNOWLEDGMENTS. The authors gratefully acknowledge the work of all of their collaborators in the many studies cited. This work was supported by NIH Grant HL-34624 (D.M.S.) and a Young Investigator Award from the Oklahoma Affiliate of the American Heart Association (D.M.S.). D.M.S. completed this work during the tenure of Young Investigator Award from the American Heart Association with funds contributed in part by the Oklahoma Affiliate.

REFERENCES

1. Davie, E., Fujikawa, K., Kurachi, K., and Kisiel, W., 1980, The role of serine proteases in the blood coagulation cascade, *Adv. Enzymol.* **48**:277–318.
2. Jackson, C., and Nemerson, Y., 1980, Blood coagulation, *Annu. Rev. Biochem.* **49**:765–811.
3. Nesheim, M. E., Pendergast, F. G., and Mann, K. G., 1979, Interactions of a fluorescent active site directed inhibitor of thrombin, *Biochemistry* **18**:996–1003.
4. Bach, R., Gentry, R., and Nemerson, Y., 1986, Factor VII-binding to tissue factor reconstituted phospholipid vesicles: Induction of cooperativity by phosphatidylserine, *Biochemistry* **25**:4007–4020.
5. van Dieijan, G., Tars, G., Rosing, J., and Hemker, H., 1981, The role of phospholipid and factor VIIIa in the activation of bovine factor X, *J. Biol. Chem.* **256**:3433–3442.
6. Miletich, J. C., Jackson, C., and Majerus, P., 1977, Interaction of coagulation factor Xa with human platelet, *Proc. Natl. Acad. Sci. USA* **74**:4033–4036.
7. Tracy, P. B., Nesheim, M. E., and Mann, K. G., 1981, Coordinate binding of factor Va and Xa to the unstimulated platelet, *J. Biol. Chem.* **256**:743–751.
8. Dahlback, B., and Stenflo, J., 1978, Binding of bovine coagulation factor Xa to platelets, *Biochem. J.* **23**:4938–4945.
9. Hultin, M. B., 1982, Role of human factor VIII in factor X activation, *J. Clin. Invest.* **69**:950–958.
10. Rosing, J., Van Rijn, J. L. M. L., Bevers, E. M., van Dieijen, G., Comfurius, P., and Zwaal, R. F. A., 1985, The role of activated human platelets in prothrombin and factor X activation, *Blood* **65**:319–332.
11. Shuman, M., and Majerus, P., 1976, The measurement of thrombin in clotting blood by radio-immunoassay, *J. Clin. Invest.* **58**:1249–1258.
12. Lau, H., Rosenberg, J., Beeler, D., and Rosenberg, R., 1979, The isolation and characterization of a specific antibody population directed against the prothrombin activation fragments F_1 and F_{1+2}, *J. Biol. Chem.* **254**:8751–8761.
13. Nossel, H. L., Yudelman, I., Canfield, R. E., Butler, V. P., Spanondis, K., Wilner, G. D., and Qureshi, G. D., 1974, Measurement of fibrinopeptide A in human blood, *J. Clin. Invest.* **54**:43–53.
14. Nossel, H. L., 1981, Clinical impact of hemostatic regulators, *Prog. Clin. Biol. Res.* **72**:163–177.
15. Lau, H., and Rosenberg, R., 1980, The isolation and characterization of a specific antibody population directed against thrombin–antithrombin complex, *J. Biol. Chem.* **255**:5885–5893.
16. Bauer, K., Kass, B., Beeler, D., and Rosenberg, R., 1984, Detection of protein C activation in humans, *J. Clin. Invest.* **74**:2033–2041.
17. Esmon, N. L., Owen, W. G., and Esmon, C. T., 1982, Isolation of a membrane bound cofactor for thrombin-catalyzed activation of protein C, *J. Biol. Chem.* **257**:859–864.
18. Stern, D., Nawroth, P., Harris, K., and Esmon, C., 1986, Cultured bovine aortic endothelial

cells promote activated protein C–protein S-mediated inactivation of factor Va, *J. Biol. Chem.* **261**:713–718.

19. Stern, D., Brett, J., Harris, K., and Nawroth, P., 1986, Participation of endothelial cells in the protein C–protein S anticoagulant pathway: The synthesis and release of protein S, *J. Cell Biol.* **102**:1971–1978.
20. Marcum, J. A., McKenney, J. B., and Rosenberg, R. D., 1984, Acceleration of thrombin–antithrombin complex formation in rat hindquarters via heparin-like molecules bound to the endothelium, *J. Clin. Invest.* **74**:341–350.
21. Marcum, J. A., and Rosenberg, R. D., 1984, Anticoagulantly active heparin-like molecules from vascular tissue, *Biochemistry* **23**:1730–1737.
22. Stern, D., Nawroth, P., Handley, D., Kisiel, W., Rosenberg, R., and Stern, K., 1985, Interaction of antithrombin III with bovine aortic segments, *J. Clin. Invest.* **75**:272–279.
23. Marcum, J., Ather, D., Fritze, L., Nawroth, P., Stern, D., and Rosenberg, R. D., 1986, Cloned bovine aortic endothelial cells synthesize anticoagulantly-active heparin sulfate proteoglycan, *J. Biol. Chem.* **261**:7507–7517.
24. Loskutoff, D. J., and Edgington, T. S., 1977, Synthesis of a fibrinolytic activator and inhibitor by endothelial cells, *Proc. Natl. Acad. Sci. USA* **74**:3903–3907.
25. Loskutoff, D. J., van Mourik, J. A., Erickson, L. A., and Lawrence, D., 1983, Detection of an unusually stable fibrinolytic inhibitor produced by bovine endothelial cells, *Proc. Natl. Acad. Sci. USA* **80**:2956–2960.
26. Stern, D., Nawroth, P., Handley, D., and Kisiel, W., 1985, An endothelial cell-dependent pathway of coagulation, *Proc. Natl. Acad. Sci. USA* **82**:2523–2527.
27. Kadish, J. L., Butterfield, C. E., and Folkman, J., 1979, The effect of fibrin on cultured vascular endothelial cells, *Tissue Cell* **11**:99–108.
28. Cerveny, T. J., Fass, D. N., and Mann, K. G., 1984, Synthesis of coagulation factor V by cultured aortic endothelium, *Blood* **63**:1467–1474.
29. Rodgers, G. M., and Shuman, M. A., 1983, Prothrombin is activated on vascular endothelial cells by factor X_a and calcium, *Proc. Natl. Acad. Sci. USA* **80**:7001–7005.
30. Hartubise, P., Coots, M., Jacob, D., Mahleman, A., and Glueck, H., 1979, Monoclonal $IgG_4(\lambda)$ with factor V inhibitory activity, *J. Immunol.* **122**:2119–2121.
31. Aggeler, P., 1961, Physiological basis for transfusion therapy in hemorrhagic disorders, *Transfusion* **1**:71–86.
32. Biggs, R., and Denson, K. W. E., 1963, The fate of prothrombin and factors VIII, IX and X transfused to patients deficient in these factors, *Br. J. Haematol.* **9**:532–541.
33. Thompson, A., Forney, A., Gentry, P., Smith, K., and Harker, L., 1980, Human factor IX in animals: Kinetics from isolated radiolabelled protein and platelet destruction following crude concentrate infusions, *Br. J. Haematol.* **45**:329–342.
34. Fuchs, H., Trap, H., Griffith, H., Roberts, H., and Pizzo, S., 1984, Regulation of factor IXa *in vitro* in human and mouse plasma and *in vivo* in the mouse, *J. Clin. Invest.* **73**:1696–1703.
35. Wessler, S., Reimer, S. M., and Sheps, M. C., 1959, Biologic assay of a thrombosis including activity in human serum, *J. Appl. Physiol.* **14**:943–946.
36. Heimark, R. L., and Schwartz, S., 1983, Binding of coagulation factors IX and X to the endothelial cell surface, *Biochem. Biophys. Res. Commun.* **111**:723–731.
37. Stern, D. M., Drillings, M., Nossel, H. L., Hurlet-Jensen, A., La Gamma, K., and Owen, J., 1983, Binding of factors IX and IXa to cultured vascular endothelial cells, *Proc. Natl. Acad. Sci. USA* **80**:4119–4123.
38. Nelsestuen, G. L., Kisiel, W., and DiScipio, R. G., 1978, Interaction of vitamin K-dependent proteins with membranes, *Biochemistry* **17**:2134–2141.
39. Stern, D. M., Nawroth, P. P., Kisiel, W., Handley, D., Drillings, M., and Bartos, J., 1984, A coagulation pathway on bovine aortic segments leading to generation of factor Xa and thrombin, *J. Clin. Invest.* **74**:1910–1921.
40. Tracy, P., Petersen, J., Nesheim, M., McDuffie, F., and Mann, K. G., 1979, Interaction of coagulation factor V and factor Va with platelets, *J. Biol. Chem.* **254**:10354–10361.
41. Scott, R., Bergman, B., Bajpai, A., Hersh, R., Rodriguez, H., Jones, B. N., Carreda, C., Watts, S., and Baker, J., 1985, Protease nexin, *J. Biol. Chem.* **260**:7029–7034.

42. Stern, D. M., Nawroth, P. P., Kisiel, W., Vehar, G., and Esmon, C. T., 1985, The binding of factor IXa to cultured bovine aortic endothelial cells, *J. Biol. Chem.* **260:**6717–6722.
43. Rimon, S., Savion, N., Nawroth, P., and Stern, D., 1985, The endothelial cell factor IX binding site involves a cell surface protein, *Blood* **66:**367 (abstract).
44. Nemerson, Y., and Bach, R., 1982, Tissue factor revisited, *Prog. Hemost. Thromb.* **6:**237–261.
45. Maynard, J. R., Dryer, B. E., Stemerman, M. D., and Pitlick, F. A., 1977, Tissue factor coagulant activity of cultured human endothelial and smooth muscle cells and fibroblasts, *Blood* **50:**387–396.
46. Lyberg, T., Galdal, K. S., Evensen, S. A., and Prydz, H., 1983, Cellular cooperation in endothelial thromboplastin synthesis, *Br. J. Haematol.* **53:**85–95.
47. Colucci, M., Balconi, G., Lorenzet, R., Pietra, A., Locati, D., Donati, M. D., and Semerano, M., 1983, Cultured human endothelial cells generate tissue factor in response to endotoxin, *J. Clin. Invest.* **71:**1893–1896.
48. Nawroth, P. P., Stern, D. M., Kisiel, W., and Bach, R., 1985, Cellular requirements for tissue factor generation by perturbed bovine aortic endothelial cells in culture, *Thromb. Res.* **40:**677–691.
49. Bevilacqua, M. P., Pober, J. S., Majeau, G. R., Cotran, R. S., and Gimbrone, M. A., 1984, Interleukin 1 induces biosynthesis and cell surface expression of procoagulant activity in human vascular endothelial cells, *J. Exp. Med.* **160:**618–621.
50. Nawroth, P. P., Handley, D., Esmon, C. T., and Stern, D. M., 1986, Interleukin 1 induces endothelial cell procoagulant while suppressing cell surface anticoagulant activity, *Proc. Natl. Acad. Sci. USA* **83:**3460–3464.
51. Nawroth, P. P., and Stern, D. M., 1986, Modulation of endothelial cell hemostatic properties by tumor necrosis factor, *J. Exp. Med.* **163:**740–745.
52. Bevilacque, M. P., Pober, J. S., Wheeler, M. E., Mendridx, D. L., Fiers, W., and Gimbrone, M. A., 1986, Interleukin 1 and tumor necrosis factor independently activate vascular endothelial cell functions, *Fed. Proc.* **45:**4576A.
53. Brox, J. H., Osterud, B., Bjorklid, E., and Fenton, J. W., II, 1984, Production and availability of thromboplastin in endothelial cells: The effects of thrombin, endotoxin and platelets, *Br. J. Haematol.* **57:**239–246.
54. Pennica, D., Nedwin, G., Palladino, M., Kohr, W., Aggarwal, B., and Goeddel, A., 1984, Tumor necrosis factor, *Nature* **312:**724–728.
55. Bach, R., Nemerson, Y., and Konigsberg, W., 1981, Purification and characterization of bovine tissue factor, *J. Biol. Chem.* **256:**8324–8331.
56. Stern, D., and Nossel, A., 1982, Acquired antibody to factor XI in a patient with congenital factor IX deficiency, *J. Clin. Invest.* **69:**1270–1276.
57. Jesty, J., and Silverberg, S. A., 1979, Kinetics of the tissue factor-dependent activation of coagulation factors IX and X in a bovine plasma system, *J. Biol. Chem.* **254:**12337–12345.
58. Zur, M., and Nemerson, Y., 1980, Kinetics of factor IX activation via the extrinsic pathway, *J. Biol. Chem.* **255:**5703–5707.
59. Rizza, C. R., 1972, The clinical features of clotting factor deficiencies, in: *Human Blood Coagulation, Haemostasis and Thrombosis* (R. Biggs, ed.), Blackwell, Oxford, pp. 210–224.
60. Windt, M. R., and Rosenwasser, L. J., 1984, Human vascular endothelial cells produce interleukin-1, *Lymphokine Res.* **3:**175A (abstract).
61. Shanahan, W., and Korn, J., 1984, Endothelial cells express IL-1-like activity as assessed by enhancement of fibroblast PGE synthesis, *Clin. Res.* **32:**666A (abstract).
62. Wagner, C., Standage, B., McCall, E., Vetto, R., and Burger, P., 1984, The role of endothelial cells as antigen-presenting cells in immunological responses, *Int. Symp. Biol. Vasc. Endothelial Cell,* 3rd, Boston, p. 18A.
63. Stern, D. M., Bank, I., Nawroth, P. P., Cassimeris, J., Kisiel, W., Fenton, J. W., II, Dinarello, C., Jaffe, E. A., and Chess, L., 1985, Self-regulation of procoagulant events on the endothelial surface, *J. Exp. Med.* **162:**1223–1235.
64. Nawroth, P., Bank, I., Handley, D., Cassimeris, J., Chess, L., and Stern, D., 1986, Tumor necrosis factor/cachectin interacts with endothelial cell receptors to induce release of interleukin 1, *J. Exp. Med.* **163:**1363–1375.

65. Libby, P., Ordoras, J. M., Auger, K. R., Robbins, A. H., Birinyi, L. K., and Dinarello, C. A., 1986, Endotoxin and tumor necrosis factor induce interleukin 1 gene expression in adult human vascular endothelial cells, *Am. J. Pathol.* **124:**179–185.
66. Lomedico, P., Kilian, P., Gubler, A., Stern, A., and Chizzonite, R., 1986, Molecular biology of interleukin-1, *Cold Spring Harbor Symp. Quant. Biol.* (in press).
67. Tracy, P., Boville, M., and Hoak, J., 1986, Regulation of prothrombinase activity of vascular cells, *J. Cell. Biochem.* **10A:**E100 (abstract).
68. Thompson, E. A., and Salem, H. H., 1986, Inhibition by human thrombomodulin of factor Xa mediated cleavage of prothrombin, *J. Clin. Invest.* **78:**13–17.
69. Nawroth, P., McCarthy, D., Kisiel, W., Handley, D., and Stern, D. M., 1985, Cellular processing of bovine factors X and Xa by cultured bovine aortic endothelial cells, *J. Exp. Med.* **162:**559–572.
70. Gadjusek, C., Carbon, S., Ross, R., Nawroth, P. P., and Stern, D. M., 1986, Activation of coagulation releases endothelial cell mitogens, *J. Cell Biol.* **103:**419–428.
71. Smith, E. B., Staples, E. G., and Dietz, M. S., 1979, Role of endothelium in sequestration of lipoprotein and fibrinogen aortic lesions, thrombi, and graft pseudo-intimas, *Lancet* **2:**812–816.
72. Smith, E. B., Alexander, K. G., and Massie, I. B., 1976, Insoluble fibrin in human aortic intima: Quantitative studies in the relationship between insoluble fibrin and soluble fibrinogen and LD-lipoprotein, *Atherosclerosis* **23:**19–39.
73. Smith, E. B., and Smith, R. H., 1976, Early changes in aortic intima, *Atherosclerosis Dev.* **1:**119–136.
74. Woolf, N., 1978, Thrombosis and atherosclerosis, *Br. Med. Bull.* **34:**137–142.
75. Pearson, T. A., Dillman, J., Solez, K., and Heptinstall, R. H., 1979, Monoclonal characteristics of organizing arterial thrombi: Significance in the origin and growth of human atherosclerotic plaques, *Lancet* **1:**7–11.
76. Bini, A., Fenoglio, J. F., Sobel, J., Owen, J., and Kaplan, K., 1985, Fibrinogen and fibrin related antigens in human thrombi and atherosclerotic lesions, *Thromb. Haemost.* **54:**165.
77. Harlan, J. G., Bowen-Pope, D., Thompson, P. J., and Ross, R., 1985, Alpha-thrombin induces secretion of platelet derived-growth factor-like molecules from cultured human endothelial cells, *Thromb. Haemost.* **54:**168 (abstract).
78. Witte, L., Kaplan, K., Nossel, H., Lages, B., Weiss, H., and Goodman, W. S., 1978, Studies on the release from human platelets of the growth factor for cultured human arterial smooth muscle cells, *Circ. Res.* **42:**402–421.
79. Ashwell, G., and Morell, A. G., 1974, The role of surface carbohydrates in the hepatic recognition of circulating glycoproteins, *Adv. Enzymol.* **41:**99–143.
80. Basu, S. K., Goldstein, J. L., and Brown, M. S., 1978, Characterization of the low density lipoprotein receptor in membranes prepared from human fibroblasts, *J. Biol. Chem.* **253:**3852–3861.
81. Ciechanover, A., Schwartz, A. L., and Lodish, H. I., 1985, Sorting and recycling of cell surface receptors and endocytosed ligands: The asialoglycoprotein and transferrin receptors, *J. Cell. Biochem.* **23:**107–122.
82. Weksler, B. B., Marcus, A. J., and Jaffe, E. A., 1977, Synthesis of prostaglandin I2 (prostacyclin) by cultured human and bovine endothelial cells, *Proc. Natl. Acad. Sci. USA* **74:**3922–3926.
83. Esmon, C. T., 1983, Protein C: Biochemistry, physiology and clinical implications, *Blood* **62:**455–558.
84. Walker, F. J., 1984, Protein S and the regulation of activated protein C, *Semin. Thromb. Hemost.* **10:**131–138.
85. Walker, F. J., 1980, Regulation of activated protein C by a new protein, *J. Biol. Chem.* **255:**5521–5524.
86. Walker, F. J., 1981, Regulation of protein C by protein S, the role of phospholipid in factor V_a inactivation, *J. Biol. Chem.* **256:**11128–11131.
87. Bovill, E. G., and Tracy, P. B., 1986, Human venous endothelial cells and arterial smooth muscle cells support activated protein C/protein S inactivation of factor Va, *Blood* **66:**354 (abstract).
88. Bertina, R., van Wijngaarden, V. W., Reinalda-Poot, J., Poort, S., and Bom, V., 1985, Deter-

mination of plasma protein S—the protein cofactor of activated protein C, *Thromb. Haemost.* **53**:268–272.

89. Jaffee, E. A., Hoyer, L. W., and Nachman, R. L., 1973, Synthesis of antihemophilic factor antigen by cultured human endothelial cells, *J. Clin. Invest.* **52**:2757.

90. Fair, D. S., Marlar, R. A., and Levin, G., 1986, Human endothelial cells synthesize protein S, *Blood* **67**:1168–1171.

91. Brett, J., Godman, G., Nawroth, P., and Stern, D., 1986, Norepinephrine stimulates release of endothelial cell protein S, *Fed. Proc.* **45**:1073 (abstract).

92. Levy, R., Lian, J., and Gallop, P., 1979, γ-Carboxyglutamic acid and atherosclerotic plaque, in: *Vitamin K Metabolism and Vitamin K-Dependent Proteins* (J. Suttie, ed.), University Park Press, Baltimore, pp. 269–273.

93. Oppenheim, J. J., and Gery, I., 1982, Interleukin is more than an interleukin, *Immunol. Today* **3**:113.

94. Mizel, S. B., 1982, The interleukins: Regulation of lymphocyte differentiation, proliferation, and functional activation, in: *Biological Response in Cancer* (E. Mihich, ed.), Plenum Press, New York, pp. 89–119.

95. Dinarello, C. A., 1984, Interleukin-1, *Rev. Infect. Dis.* **6**:51.

96. Gordon, S. G., and Cross, B. A., 1981, A factor x-activating cysteine protease from malignant tissue, *J. Clin. Invest.* **67**:1665.

97. Beutler, B., Milsark, I. W., and Cerami, A. C., 1985, Passive immunization against cachectin/tumor necrosis factor protects mice from lethal effects of endotoxin, *Science* **229**:869–871.

98. Beutler, B., and Cerami, A., 1986, Cachectin and tumor necrosis factor as two sides of the same biological coin, *Nature* **320**:584–588.

99. Old, L., 1985, Tumor necrosis factor, *Science* **230**:630–634.

100. Williamson, B., Carswell, E., Rubin, B., Prendergast, J., and Old, L. J., 1983, Human tumor necrosis factor produced by human B-cell lines: Synergistic cytotoxic interaction with human interferon, *Proc. Natl. Acad. Sci. USA* **80**:5397–5402.

101. Rubin, B. Y., Anderson, S. L., Sullivan, S. A., Williamson, B. D., Carswell, E. A., and Old. L. J., 1985, High affinity binding of ^{125}I-labelled human tumor necrosis factor to specific cell surface receptors, *J. Exp. Med.* **162**:1099–1104.

102. Baglioni, C., McCandless, S., Tavernier, J., and Fiers, W., 1985, Binding of tumor necrosis factor to high affinity receptors in HeLa and lymphoblastoid cells sensitive to growth inhibition, *J. Biol. Chem.* **260**:13395–13397.

103. Aggarwal, B. B., Essalu, T. E., and Hass, P. E., 1985, Characterization of receptors for human tumor necrosis factor and their regulation of γ-interferon, *Nature* **318**:665–667.

104. Nawroth, P. P., Cornelson, S., and Stern, D. M., 1986, Tumor necrosis factor upregulates factor IX/IXa binding sites and factor IXa–VIII-mediated factor Xa formation on endothelium, *Circulation* **74**(Suppl. II):233A.

105. Nachman, R. L., Hajjar, K., Silverstein, R. L., and Dinarello, C. A., 1986, Interleukin 1 induces endothelial cell synthesis of plasminogen activator inhibitor, *J. Exp. Med.* **163**:1595–1600.

106. Schleef, R. R., Bevilacqua, M. P., Gimbrone, M. A., and Loskutoff, D. J., 1986, Effects of interleukin 1 on fibrinolytic system of cultured human umbilical vein endothelial cells, *Fed. Proc.* **45**:5345A.

107. Emeis, J., and Kooistra, T., 1986, Interleukin 1 and lipopolysaccharide induce an inhibitor of tissue type plasminogen activator *in vivo* and in cultured endothelial cells, *J. Exp. Med.* **163**:1260–1266.

108. Nawroth, P. P., Handley, D. A., and Stern, D. M., 1986, Tumor necrosis factor-induced tumor killing involves coagulation, *Circulation* **74**(Suppl. II):233A.

109. Beutler, B., Mahoney, J., LeTrang, N., Pekala, P., and Cerami, A., 1985, Purification of cachectin, a lipoprotein lipase-suppressing hormone secreted by endotoxin-induced RAW264.7 cells, *J. Exp. Med.* **161**:984–996.

Endothelial Cell Response to Stress Factors

Heat-Shock Response as a Possible Model for (Patho)physiological Stress in Endothelial Cells

Nika V. Ketis and Morris J. Karnovsky

I. INTRODUCTION

The cells of most eukaryotes and prokaryotes respond to metabolic perturbations by the induction of a set of proteins termed the "stress" or "heat-shock proteins" (HSP).[3,78] These stress proteins appear in all organisms and have been most intensely studied during and after heat treatment.[3,78] Although functionally uncharacterized, a search for regulatory mechanisms controlling the expression of HSP at the cellular, biochemical, and molecular levels has begun to shed some light on their possible function in cells experiencing hyperthermia and physiological stress.

Cells exposed to hyperthermia exhibit altered morphology and reorganization at the levels of translation and transcription.[24,44,72,80,93] Several polypeptides are synthesized at elevated levels in mammalian cells incubated with a number of diverse agents. Due to the lack of defined function, the polypeptides are described in terms of their electrophoretic mobility on SDS–polyacrylamide gels: 20–30k, 70–73k, 80–90k, and 100–110k molecular weight proteins. The number and diversity of the agents that induce HSP are very high: they include anoxia,[72] amino acid analogues,[37] sulfhydryl-reacting reagents,[54] transition metal ions,[54] uncouplers of oxidative phosphorylation,[36] viral infections,[70] ethanol,[56] various antibiotics,[30] and certain ionophores[91] and chelators.

Most of the HSP can be detected in cells under normal, unstressed conditions. In *Drosophila*, the low-molecular-weight HSP are transcribed during early development.[97] Welch[91] has demonstrated increased phosphorylation of the 28k protein (HSP28) in rat embryo fibroblasts treated with tumor-promoting agents, calcium ionophore, or serum added to quiescent cells. In addition, it has been shown that HSP80, 90, and 100 in unstressed cells are extremely sensitive to extracellular

Nika V. Ketis and Morris J. Karnovsky • Department of Pathology, Harvard Medical School, Boston, Massachusetts 02115.

levels of glucose or calcium.[81,91,92] HSP90, which is highly phosphorylated, appears to play a role in a number of different hormone receptors[13,74,77] and is transiently associated with a number of different tyrosine kinases.[12] Recently, synthesis of HSP71, the most inducible form of the HSP family, has been observed during mouse embryogenesis[6] and in embryonal carcinoma cells,[7,65] and high levels of HSP71 synthesis have been reported in both embryonic and adult chicken erythrocytes.[66] The constitutive form of the HSP70 family, HSP73, appears to be very similar to Dr. Rothman's uncoating ATPase which is involved in the ATP-dependent uncoating of clathrin from coated vesicles.[87] The accumulated data suggest that HSP play essential roles not only in cells exposed to certain adverse conditions but also in unstressed cells.

Of the HSP thus far identified, the HSP70 family appears most conserved among different organisms. There are two major members of the HSP70 family in mammalian cells: the abundant constitutive form, HSP73, and the highly inducible form, HSP71. HSP71 has several distinct isoforms as determined by isoelectric focusing. The number of isoforms observed appears dependent on the cell type, the stress agent applied, and the amount and duration of stress. HSP71 appears cell cycle related,[95] and its expression is elevated in response to several oncogenes—the cellular oncogene C-myc, SV40, and adenovirus E1A.[40,41,70] It is clear that the HSP70 family plays an essential role in unstressed cells.

In view of the physiological and pathological "stress" to which endothelial cells are subjected, such as shear forces, hypoxia, hypertension, angiogenesis, acute inflammation, and neoplasia, the study of the expression and regulation of HSP in endothelial cells is of great interest. In an effort to understand the biology of a cell that may undergo physiological and pathological "stress," the *in vivo* synthesis and accumulation of "stress" proteins in endothelial cells of different origins were characterized. The implication of these findings is discussed.

II. MATERIALS AND METHODS

A. Cell Culture

Calf aortic endothelial cells were isolated as described by Booyse *et al.*[9] When passage cells were desired, primary endothelial cells were cloned, passaged, and grown in DMEM supplemented with 5% fetal calf serum (FCS) and 5% Nu-Serum (Collaborative Research, Waltham, MA), penicillin (100 μg/ml), streptomycin (100 μg/ml), and amphotericin (0.25 μg/ml). Calf brain capillary endothelium was isolated according to the method of Spatz *et al.*,[82] and rat epididymal endothelium was isolated as described by Wagner and Matthews.[89] Cells were identified as endothelium morphologically by their cobblestone appearance,[26] immunologically by staining with fluorescently labeled anti-factor VIII,[26] and enzymatically by assaying for angiotensin II-converting enzyme. Angiotensin II-converting enzyme activity was determined as described in the technical bulletin provided by Ventrex Laboratories (Portland, ME), the supplier of the radioactive substrate.

Radiolabeled proteins synthesized and secreted by confluent endothelial cells were prepared by incubating 10^5 to 10^6 cells with [^{35}S]methionine (80 μCi) or [^{3}H]proline (150 μCi) and 200–400 μl of fresh medium.

B. Gel Electrophoresis and Autoradiography

SDS–polyacrylamide gel electrophoresis (SDS–PAGE) was based on the discontinuous Tris–glycine system of Laemmeli.[46] Two-dimensional gel electrophoresis involved isoelectric focusing followed by SDS–PAGE as described by O'-Farrell.[71] Polyacrylamide gels containing radiolabeled proteins were processed for fluorography by treatment in Enlightning solution (New England Nuclear, Boston, MA) for 15 to 30 min. The gels were dried on filter paper using a slab gel dryer and exposed to X-ray film (Kodak, XAR-2).

Protein concentrations were determined by the method of Bradford.

C. Immunochemistry

Radiolabeled growth medium proteins were harvested in phosphate-buffered saline, pH 7.2, containing 0.1% SDS and 0.5% Triton X-100. The monoclonal antibody against platelet thrombospondin was obtained from Dr. D. F. Mosher. Its characterization has been provided elsewhere.[35] The immunoprecipitation of bovine aortic endothelial cell (BAEC)-derived thrombospondin was carried out as described by Mumby et al.[69]

D. Immunofluorescence

Immunofluorescence localization of thrombospondin in BAEC cultures was performed as per Raugi et al.,[73] using antithrombospondin immunoglobulin obtained from Dr. D. F. Mosher.

E. RNA Isolation

Confluent monolayers of endothelial cells were washed with buffered Hanks's balanced salt solution and detached with trypsin–EDTA. The cells were collected and resuspended while gently vortexing in lysis buffer containing 1% diethylpyrocarbonate (Sigma Chemical Co., St. Louis, MO), 100 mM NaCl, 10 mM $CuCl_2$, and 30 mM Tris-HCl, pH 7.4. Triton X-100 was added to a final concentration of 0.5%, and the suspension was immediately vortexed. After spinning out the nuclei (4000g for 5 min), the supernatants were transferred to Eppendorf tubes. Five volumes of cold 7.5 M guanidinium-HCl (Bethesda Research Laboratories, Bethesda, MD), 25 mM citrate, pH 7.0, and 0.5% N-laurylsarcosine were added while the suspension was mixed followed by the addition of 0.025 vol of 1 M acetic acid. Nucleic acids were precipitated overnight at −20°C in the presence of 0.5 vol of 95% ethanol. The pellet was collected at 4000g for 5 min and the supernatant carefully discarded. The precipitation of nucleic acids was repeated as described above using one-half the original volume of guanidinium-HCl solution without N-laurylsarcosine. NaCl (0.2 M) and 2 vol of 95% ethanol were added to the RNA pellets recovered by differential centrifugation. The nucleic acids were precipitated again overnight at −20°C. The final RNA pellet was collected by centrifugation, resuspended in a small volume of sterile distilled water, and stored at −70°C until used.

F. Cell-Free Protein Synthesis

Total RNA was translated in a cell-free rabbit reticulocyte lysate for 60 min as described by the supplier, New England Nuclear. Translation was monitored by the incorporation of [^{35}S]methionine into TCA-precipitable material. Translation samples were mixed 1:1 with loading buffer, and approximately equal numbers of counts were electrophoresed by SDS–PAGE. The electrophoretic profiles were analyzed by fluorography as previously described.

G. Northern Gel Hybridization and Slot Blot Analysis

Total RNA was then electrophorized on denaturing formaldehyde gels, blotted onto nitrocellulose, and hybridized to nick-translated probes. After hybridization at 60°C, the filters were washed with 0.1 × SSC (1 × SSC is 0.15 M NaCl plus 0.015 M sodium citrate) − 0.1% SDS at 55°C, dried, and exposed to X-ray film (Kodak, XAR-2). (The plasmid containing the human HSP71 probe was a gift of Dr. Lee Weber.)

The quantity of thrombospondin mRNA was determined by slot blotting by standard procedures suggested by the supplier (Schleicher and Schuell, Keene, NH); hybridization probe was prepared from the M1 insert by ECOR1 digestion and preparative agarose gel electrophoresis, followed by nick-translation to a specific activity of 10^8 cpm/μg. The hybridization and subsequent washes were done at high stringency.

H. Cell Survival

After the desired treatment, confluent monolayers of endothelium were trypsinized with 0.25% trypsin in Hanks's balanced salt solution buffered with Hepes, pH 7.0. Following the trypsinization, the cells were resuspended in 3 ml of growth medium containing 2 mg/ml of egg white albumin. Cell counts were determined with a particle cell counter (Coulter Counter Electronics). Cells were diluted and plated for cell survival determinations. Each determination was performed in triplicate. Plating efficiency was between 60 and 90%.

III. RESULTS AND DISCUSSION

A. The Effect of Hyperthermia on the Protein Profiles of Primary Cultures of Bovine Aortic Endothelial Monolayers

In this section, we will demonstrate that cellular proteins are differentially expressed in confluent primary cultures of BAEC cells under conditions of brief hyperthermic treatment or continuous heat treatment. Figure 1 illustrates the polypeptide profiles in one-dimensional polyacrylamide gels of cultures labeled with [³H]proline (Fig. 1A) or [³⁵S]methionine (Fig. 1B,C) for the times indicated after transferring from 37°C to 42.5°C (Fig. 1, A2), 43°C (Fig. 1, B2), or at 37°C without any heat treatment prior to labeling (Fig. 1, A1, B1, C1).

If BAEC are stressed at 42.5°C for 2 to 4 hr, as many as six major HSP (HSP28, 71, 73, 80, 90, and 100) are evident, with little decrease in cellular protein synthesis. In contrast, cells stressed at 43°C for extended periods of time show a marked decrease in normal cellular protein synthesis with selective increase in the synthesis of HSP71, 73, 80, 90, and 100. Note that HSP28 is not evident when cells are labeled with [³⁵S]methionine but becomes apparent when these cells are labelled with [³H]leucine (data not shown) or [³H]proline (Fig. 1, A2).

Brief hyperthermic treatment of BAEC (45°C, 10 min) followed by a return to normal culture conditions results in a rapid and dramatic change in the protein synthetic pattern of ³⁵S-labeled proteins analyzed by SDS–PAGE (Fig. 1C). The level of normal cellular protein synthesis is dramatically decreased with selective increase in the synthesis of HSP71 (Fig. 1, C2).

The HSP induced in BAEC appear similar to the stress proteins from a number of other animal cell types when analyzed by SDS–PAGE.[30,86] Differential expression of HSP has been reported for HSP70 during infection of human cells with adenovirus 5[70] and in chicken embryo fibroblasts exposed to amino acid analogues and arsenite.[36,37] This most likely reflects the underlying complexity in the regulation of gene expression in response to stress.

Surprisingly, HSP71 appears to be identical to a 71k polypeptide present in BAEC under unstressed conditions. This is clearly illustrated in Fig. 2, which shows autoradiographs of two-dimensional gels of ³⁵S-labeled total cellular proteins of cells exposed to normal temperatures (37°C) (Fig. 2A) and cells exposed to 42°C for 2 hr (Fig. 2B). Three of the six charge isomers (pI at 7.2 to 6.5) of HSP71 comigrate with the 71k polypeptide of control cells. Peptide mapping by limited proteolysis[17] of gel slices bearing HSP71 or the 71k polypeptide of control cells demonstrated that the patterns of protease V8 peptides from both sources were identical (data not shown). The pI for the endothelial HSP71 appears somewhat more basic than the major isoelectric charge isomer of HSP72 in HeLa cells whose pI is 5.6.[85] Note that the position of the 71k polypeptide from BAEC at the control temperature (37°C) is clearly distinguished from the larger, more acidic cognate (asterisk in Fig. 2A).

Not all cells in culture express basal levels of HSP71 even after prolonged passage. BAEC and bovine brain capillary cells (see Section III.D) are therefore somewhat unique in that they express low levels of what is typically only an inducible protein in other systems,[58,86] human cells being the exception. The presence of HSP71 in BAEC at 37°C is most likely a result of culture conditions. Others have shown that cultured endothelium has a low but definite rate of turnover unlike endothelium *in vivo* which may more accurately reflect a state of "injury" *in vivo*.[79] Confluent BAEC or human umbilical vein endothelium *in vivo* obtained by scraping the cells from intact arteries, contains low levels of mRNA for PDGF. However, cultured endothelium showed an 83-fold (bovine) or a 10-

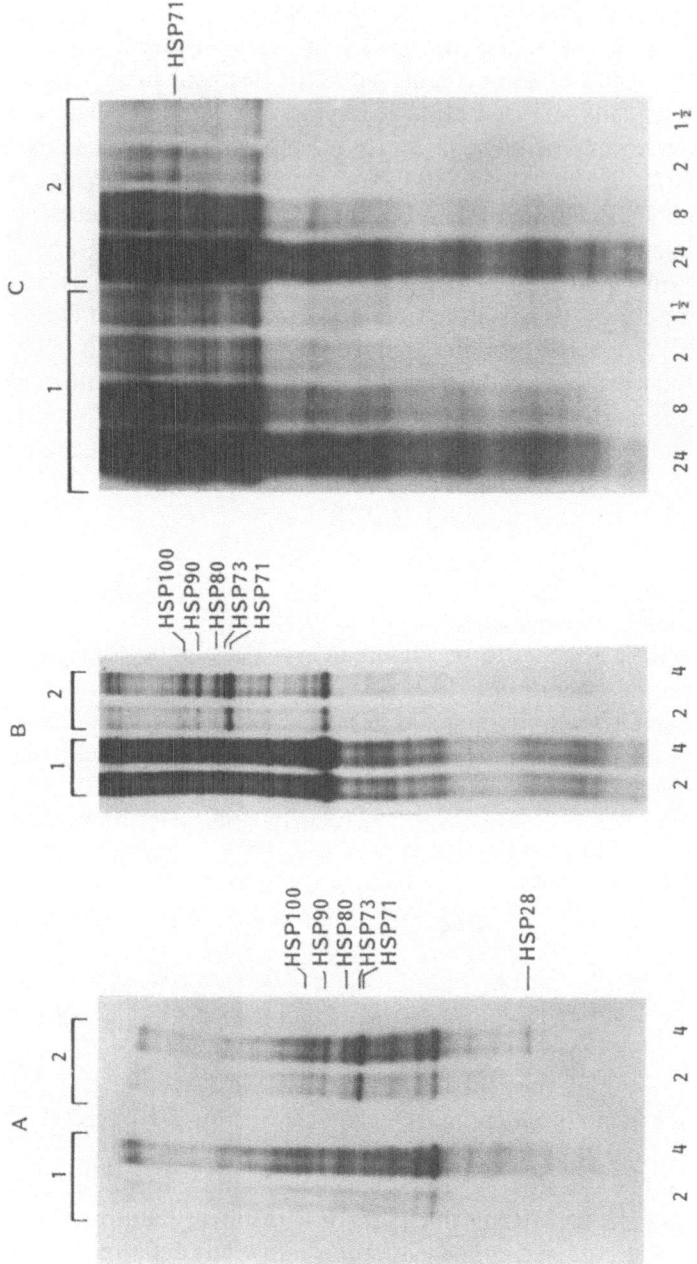

Figure 1. The effect of continuous heat treatment and exposure to brief hyperthermia on protein synthesis in BAEC. The BAEC were labeled with [³H]proline (A) or [³⁵S]methionine (B, C) for the times (hours) indicated. A1, B1, and C1 represent cells at 37°C with no heat-shock treatment; A2, cells at 42.5°C; B2, cells at 43°C; C2, cells transferred from 37°C to 45°C for 10 min and returned to 37°C. The protein profiles were analyzed by SDS–PAGE in a linear concentration gradient (A) or in a uniform concentration gel (B, C) and fluorography, with each lane containing equal amounts of cell equivalents. The molecular weights of the HSP (HSP28, 71, 73, 80, 90, and 100) are indicated.

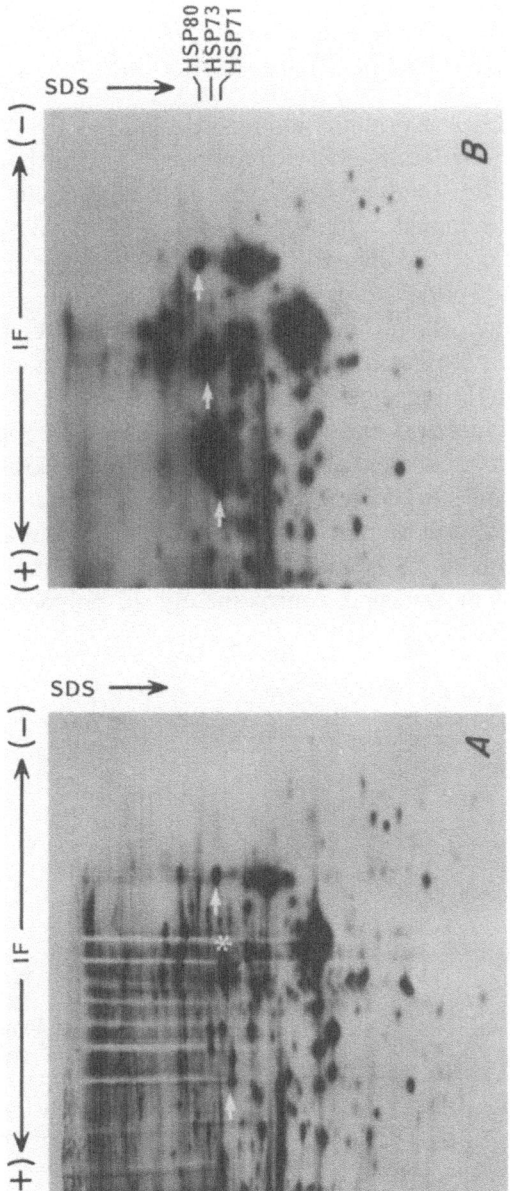

Figure 2. Two-dimensional gel analysis of normal and heat-treated BAEC. BAEC grown at 37°C (A) or transferred from 37°C to 42°C for 2 hr (B) were labeled with [^{35}S]methionine. The cells were solubilized and the proteins analyzed by isoelectric focusing and SDS–PAGE. The HSP are indicated by arrows. The asterisk marks the 73k cognate.

fold (human umbilical vein endothelium) increase in the mRNA, depending on the source of the endothelium.[4,21] Therefore, it appears likely that BAEC and bovine brain capillary endothelial cells grown on plastic dishes may be under an abnormal ''stressed'' or injured state and thus express HSP71.

B. The Effect of Hyperthermia on the Protein Profiles of the Growth Medium of Primary Cultures of Bovine Aortic Endothelium

In an effort to identify the proteins induced during the stress response, the polypeptides in the growth medium of BAEC exposed to hyperthermia were analyzed by SDS–PAGE. Interestingly, BAEC exposed to continuous heat treatment at 41.5°C for up to 4 hr demonstrate an increase of a 180k polypeptide in their growth medium (Fig. 3B) and the appearance of HSP71 in cell monolayers.[38] The amount of the 180k polypeptide increases with increase in the length of hyperthermia. When these same cells are incubated for 4 hr at 41.5°C and returned to 37°C for 24 hr, the protein profiles of the growth medium and monolayers are the same as controls (Fig. 3C,D and data not shown).

Examination by two-dimensional gel electrophoresis of [³H]proline-labeled proteins of the growth medium reveals that a 200k and the 180k polypeptides appear to have identical pI's (pI at 4.5) (data not shown). Trace amounts of the 180k polypeptide are evident in the growth medium of cells incubated at 37°C for 4 hr. Two-dimensional gels of ³⁵S-labeled proteins in the growth medium of cells exposed to 41.5°C for 4 hr revealed approximately 20 polypeptides (data not

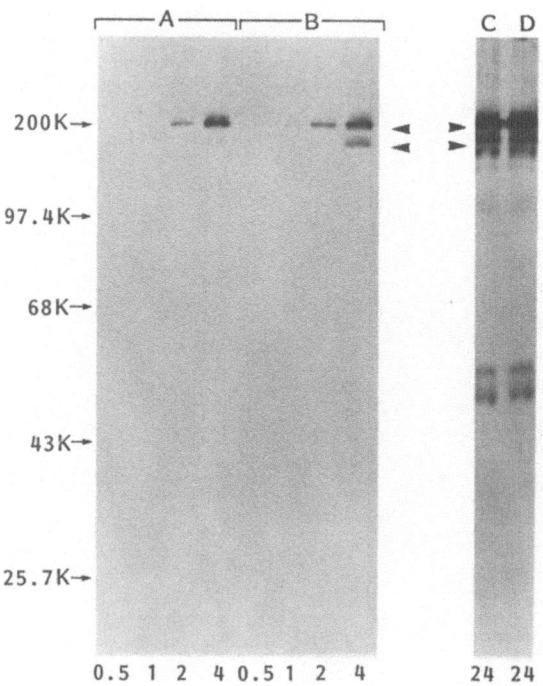

Figure 3. Effect of continuous heat treatment on the protein profiles of the growth medium of BAEC. BAEC were labeled with [³H]proline for the times (hours) indicated after transfer from 37°C to 41.5°C (B) or at 37°C without heat-shock treatment prior to labeling (A, C). (D) The protein profile of the growth medium of cells transferred from 37°C to 41.5°C for 4 hr and returned to 37°C for 20 hr. The protein profiles were analyzed by SDS–PAGE and fluorography. Each lane contains the same amount of total protein.

shown) that were labeled with [^{35}S]methionine whereas two or three were labeled with [^{3}H]proline.

The induced polypeptide can be immunoprecipitated by a monoclonal antibody to human platelet thrombospondin.[38] The 200k polypeptide is fibronectin as determined by immunoprecipitation experiments.

The expression of the 180k polypeptide in the growth medium after hyperthermic treatment of BAEC is blocked by cycloheximide.[38] As such, the expression of this polypeptide appears to require continued translation. However, as the temperature is increased (>41.5°C), the protein profiles of both the growth medium and monolayers are dramatically modified relative to control (data not shown). The 180k comigrates in one-dimensional gels with purified human platelet thrombospondin, whose reported molecular weight on reducing gels is similar.[68] This increase seems not to be caused by an increasing number of dead or dying cells, because the cells are viable as determined by three independent parameters: trypan blue exclusion, LDH release, and ^{51}Cr release (data not shown).

The level of expression of the HSP71 and thrombospondin genes was determined.[38] Assay of mRNA levels coding for thrombospondin after an initial hyperthermic trigger at 45°C for 10 min and the development period at 37°C for 2 hr, revealed a twofold increase in mRNA abundance. Analysis of the HSP71 gene revealed that the abundance of mRNA was greatest during brief hyperthermic treatment followed by a recovery for 2 hr at 37°C. However, the activation level of the HSP mRNA occurred at an earlier time than that of the thrombospondin mRNA.[38] Therefore, it was concluded that expression of thrombospondin is heat-shock stimulated.

By immunofluorescence microscopy, differences in thrombospondin localization and patterns are observed in primary BAEC cultures exposed to hyperthermia. BAEC at control temperature show both intracellular and extracellular fluorescence when probed for thrombospondin (Fig. 4A). Staining is noted in a perinuclear pattern which presumably represents thrombospondin in Golgi vesicles and secretory granules and in a fibrillar extracellular array. An immunoreactive thrombospondin was previously localized by Raugi et al.[73] in BAEC at 37°C. However, after exposure to hyperthermia, there is a notable decrease in the amount of extracellular stained material (Fig. 4B), the pattern of granules of intracellular fluorescence appears more pronounced, and the fluorescence appears to be clustered more about the nucleus than in control cells. The morphological and biochemical data suggest several possibilities for the increase in the 180k polypeptide. During continuous heat stress, it could be derived from a translational alteration, a post-translational modification, or a change in the turnover of the thrombospondin pools. Consistent with the posttranslational regulation of thrombospondin is our observation that the Golgi is rearranged in heat-stressed BAEC as determined by electron microscopy (N. V. Ketis and M. J. Karnovsky, unpublished data). However, cells exposed to brief, hyperthermic challenge followed by a recovery period at 37°C show an increase in the transcription of the thrombospondin gene, and Welch and Suhan have shown that heat-stressed cells regain normal morphology during later times of recovery. Thus, the expression of throm-

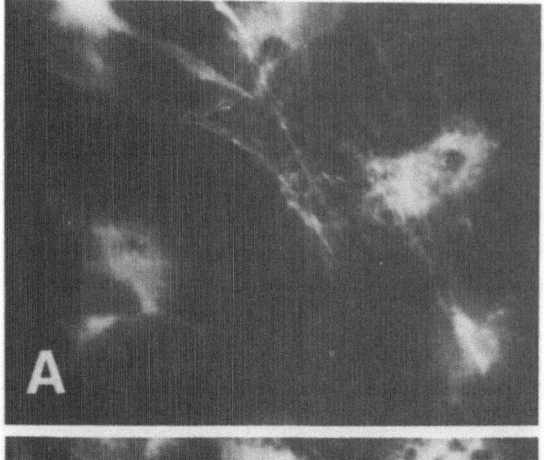

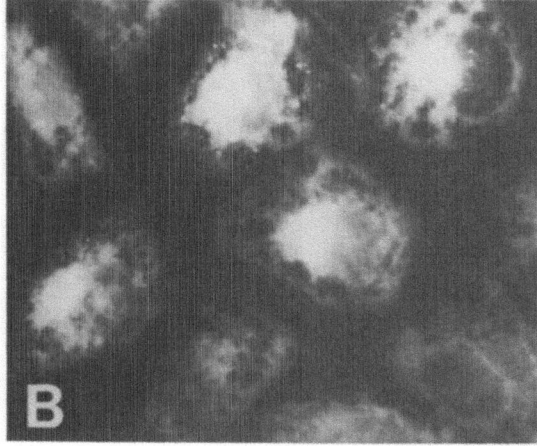

Figure 4. Immunolocalization of thrombospondin in BAEC cultures. Confluent cultures of BAEC were fixed, permeabilized, and stained for thrombospondin. (A) BAEC at 37°C and (B) after exposure to 41.5°C for 4 hr. Note that, after hyperthermic treatment, there is a decrease in the fibrillar thrombospondin staining and that the perinuclear staining appears somewhat more clustered about the nucleus.

bospondin by endothelial cells in response to heat stress is regulated by mechanisms which act at several levels of gene expression and/or protein processing.

Ketis et al.[39] have observed that the growth of endothelial cells in culture is altered after exposure to hyperthermia. Growth is initially retarded followed by exponential growth (data to be presented elsewhere). During the recovery period, the amount of thrombospondin in the growth media increases[38] as does the transcription of the thrombospondin gene.[38] We speculate that thrombospondin may aid in the recovery of cells from heat shock.

Thrombospondin is a glycoprotein consisting of three, apparently identical, disulfide-bonded chains of 180,000 molecular weight. It has been shown to have a binding domain for heparin,[22,49,50] fibrinogen,[52] fibronectin,[47] collagen,[47,68] and possibly growth factors.[60] Thrombospondin appears to interact with fibronectin during platelet adhesion and aggregation.[48,57] Recently, Majack et al.[60] have provided evidence suggesting that thrombospondin may function as an extracellular "integrator" of growth stimulatory and inhibitory signals since the amount of thrombospondin incorporated into the extracellular matrix of smooth muscle cells is regulated by both PDGF and heparin-like glycosaminoglycans. They postulated that the amount of thrombospondin in the smooth muscle cell environment, for any vascular injury, may determine the extent of smooth muscle cell–thrombospondin interaction and the extent of smooth muscle response. We are presently examining the role of thrombospondin in the growth and proliferation of endothelial cells as it may relate to a "stress" or injured state and how endothelial injury or stress may affect the surrounding vascular milieu, specifically smooth muscle cells.

C. In Vitro Translation of Total Cellular mRNA Isolated from Control and Heat-Stressed Primary Cultures of BAEC

HSP synthesis in confluent monolayers of BAEC is blocked by preincubation with 2 μg/ml of actinomycin D but not with 5 μg/ml of cycloheximide (data not shown). This concentration of cycloheximide blocks overall general translation of cellular proteins by 80% in these cells. The results indicate that in this system the expression of HSP requires continued transcription.

A comparison of the polypeptides synthesized by the cells and in a cell-free system from control and heat-shocked cells was made to determine if alterations occur in the pattern of total [^{35}S]protein synthesis of cells exposed to brief hyperthermic challenge. No such alteration is evident in unstressed cells or cells exposed to continuous heat treatment provided the temperature does not exceed 42°C. Figure 1 shows one-dimensional fluorographs of the polypeptides synthesized by BAEC and Fig. 5, the total cellular RNA in a rabbit reticulocyte lysate system. Following heat shock, a shift to the new and/or enhanced synthesis of heat-shock polypeptides is observed in the cellular sample *in vivo* (Fig. 1) and in the *in vitro* translations (Fig. 5). Enhanced synthesis of a 25k polypeptide is observed in *in vitro* translations which is not readily evident in cellular samples *in vivo* when labeled with [^{35}S]methionine (cf. Figs. 1 and 5). It should be noted that the *in vitro* translation conditions used in this study were optimized initially to allow maximum translation of all size classes of RNA. There are, however, at least two or three products generated, one of which is major (45k; asterisk in Fig. 5), from translations carried out in the absence of any exogenous RNA as determined by SDS–PAGE.

Induction of HSP and shutoff of production of normal protein synthesis in BAEC after exposure to temperatures of 43°C or greater is not due to a decrease in transcription of mRNAs. Figure 5 demonstrates that, when total mRNA isolated from stressed and unstressed cells is translated in a reticulocyte lysate system, the protein profiles obtained are the same, suggesting that at 45°C the block is at the level of translation.

At elevated temperatures (>43°C), BAEC curtail the translation of preexisting messages, while HSP mRNAs are translated very efficiently. This dramatic change in the specificity of protein synthesis has been reported in other cells and has been studied extensively in *Drosophila*.[58,81] Thus, controls appear to be exerted on both translation and transcription, and the two are coordinated to produce a HSP response as rapidly as required.

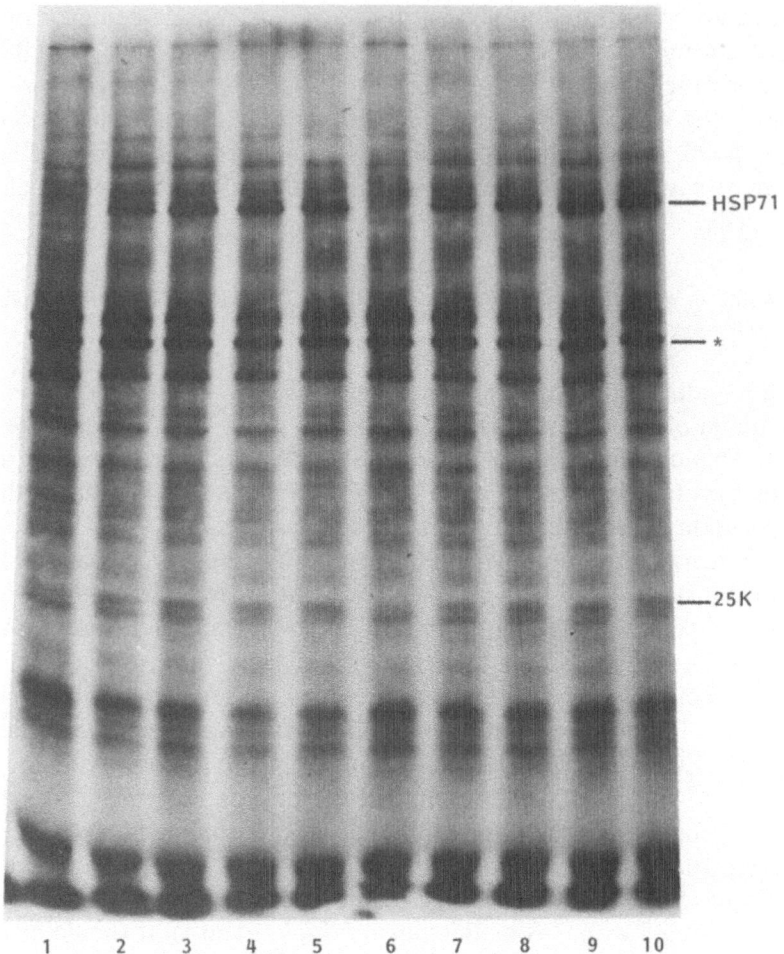

Figure 5. Fluorograms of one-dimensional PAGE separation of products obtained from a cell-free translation of total RNA from BAEC in a rabbit reticulocyte translation system. Lanes 1–5, 2μg RNA. Lanes 6–10, 3 μg RNA. Lanes 1 and 6 contain control RNA from unstressed BAEC; lanes 2 and 7, heat-shock RNA from BAEC exposed to continuous heat treatment (41.5°C) for 2 hr; lanes 3–5, 8–10, heat-shock RNA from BAEC exposed to brief hyperthermia (45°C, 10 min) followed by a return to 37°C for 2 hr. Each lane contains an equal number of TCA-precipitable counts. The HSP are indicated. The asterisk marks the translation product (45k) generated from translations carried out in the absence of any exogenous RNA.

D. Differential Expression of HSP in Endothelial Cells of Different Origins

We have found that HSP are noncoordinately expressed in primary and passage cultures of BAEC, passage cultures of bovine brain capillaries (BBCE), and passage cultures of rat epididymal capillaries (REEC) and that each protein has its own optimal temperature for expression. This is clearly demonstrated in Figs.

6 and 7. Figure 6 displays the electrophoretic distribution of total endothelial cell cultures labeled with [^{35}S]methionine incubated at 37°C (Fig. 6A) or 41°C (Fig. 6B). BAEC primary and passage cells respond to 41°C by induction of HSP with no curtailment of preexisting levels of control cell protein synthesis; however, passage cells synthesize greater amounts of HSP71 than primary cells and demonstrate increased synthesis of HSP73. Passage BBCE synthesize greater amounts of HSP71 than primary BAEC (1°) and show a 56% suppression of preexisting levels of control cell protein synthesis after 2 hr at 41°C. Very late passage REEC (P-36) respond by even greater repression of total cellular protein synthesis, approximately 60% after 2 hr at 41°C with very little induction of HSP71. Thus, endothelial cells of different origins differ in their response (expression of a given stress protein) to a given range of temperatures and duration of treatment (Figs. 6, 7, and data not shown).

An interesting observation deserves mention here: maximum stress protein synthesis is noted 6 to 10 hr after return to control temperatures from 42–43°C in all cell types examined (data not shown). The elevated levels of HSP continue for 10 to 12 hr following a return to 37°C and decrease thereafter. Total cellular protein synthesis increases with increase in time at 37°C. Moreover, it was noted that, following a 20- to 23-hr recovery period from hyperthermia (42–43°C) primary and passage BAEC and passage cultures of BBCE appear very active with respect to total cellular protein synthesis, which appears to exceed control levels (more than 100%) (data not shown). In these experiments, endothelial cells were exposed to [^{35}S]methionine for 1 to 2 hr prior to the time point of interest.

Data in Figs. 2 and 7 confirm that HSP have very distinct induction characteristics in endothelial cells of different origins, and show that normal cellular proteins are also not all affected by temperature in the same way. Figure 7 displays autoradiographs of two-dimensional gels of [^{35}S]methionine-labeled total cellular protein from passage BAEC and passage BBCE at 37°C (Fig. 7A,C) or 42°C (2 hr) (Fig. 7B,D). In passage BAEC cultures, the response at 2 hr at 42°C is the induction of five HSP—HSP71/pI 6.5–7.2, HSP73/pI 5.5–5.6, HSP80/pI 5.0, HSP90/pI 5.2, and HSP100/pI 4.8 (Fig. 7B)—while in primary BAEC only HSP71, 73, and 80 are induced (Fig. 2B). Passage BBCE responded to 42°C (2 hr) by the induction of three HSP: HSP71, 73, and 80.

All HSP are detected at substantially lower levels at 37°C in primary and passage BAEC and passage BBCE. Basal levels of HSP71, the most inducible of the HSP family, is expressed in all control cells (37°C), with the exception of REEC cultures, rather than just the more acidic cognate proteins, which is typical of most other mammalian systems.[39] We believe that endothelial cells grown on plastic dishes may be under an abnormal "stressed" or injured state and therefore express HSP71 (see Section III.A for discussion).

In all endothelial cell types examined, except perhaps REEC,[39] the accumulation of HSP71 is related to the extent of stress. Endothelial cells of different origins differ in their response to a given range of temperatures and duration of treatment.

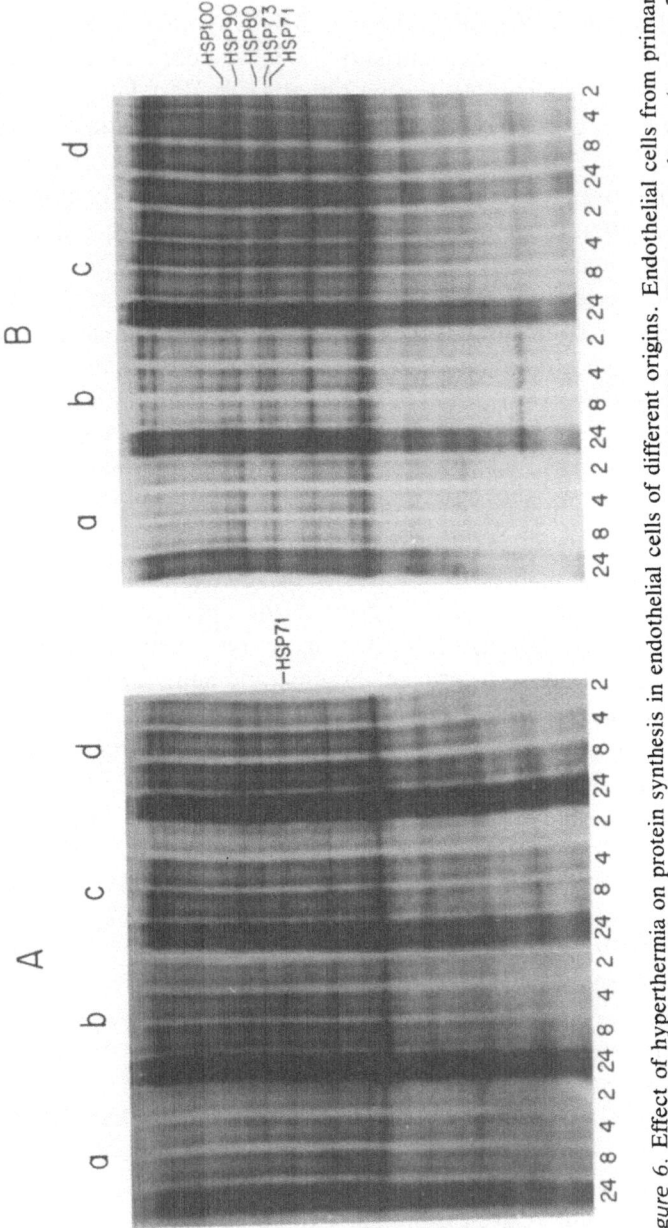

Figure 6. Effect of hyperthermia on protein synthesis in endothelial cells of different origins. Endothelial cells from primary (c) and passage (d) cultures (passage 7) of BAEC, passage cultures (passage 13) of BBCE (b), and passage cultures (passage 36) of REEC (a) at 37°C (A) or 41°C (B) were labeled with [³⁵S]methionine for the times indicated. After 8 hr at 41°C, the cells were transferred to 37°C for 24 hr. Each lane contains equal numbers of cell equivalents. (Reproduced in part from Ketis et al.,[39] with permission.)

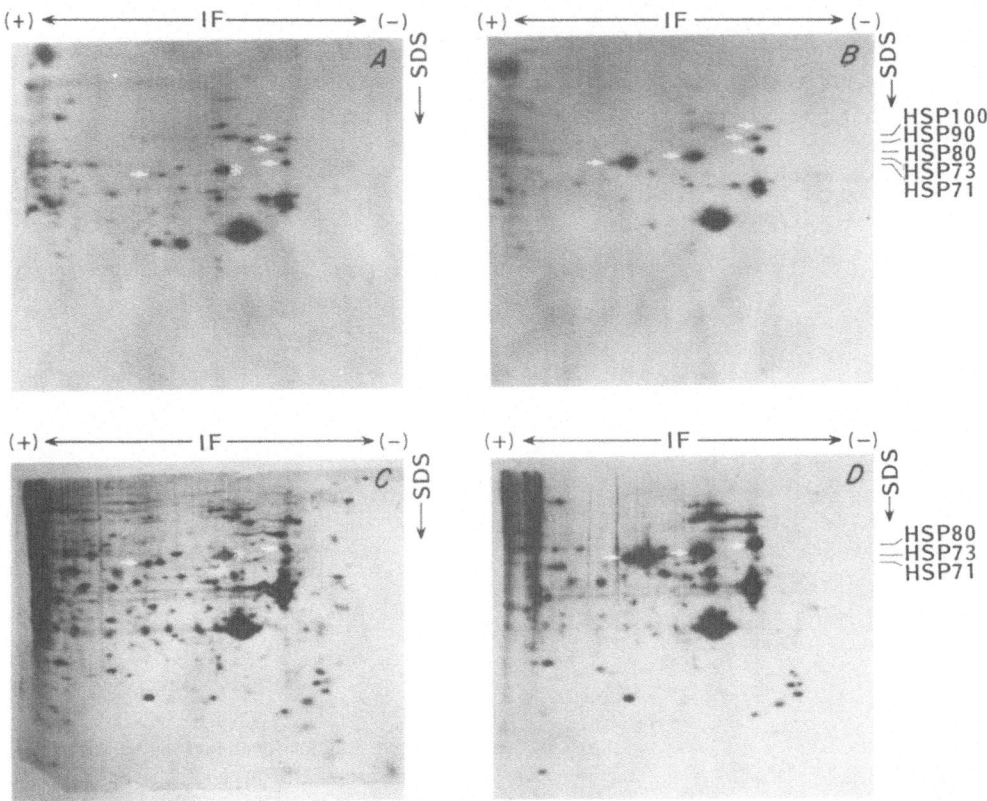

Figure 7. Two-dimensional gel analysis of normal and stressed passage cultures of BAEC and passage cultures of BBCE. Endothelial cells at 37°C (A, C) or 42°C for 2 hr (B, D) were labeled with [35S]methionine. (A, B) Passage 7 of BAEC; (C, D) passage 13 of BBCE. The stress proteins are indicated by arrows. The asterisk marks the 73k cognate. (Reproduced in part from Ketis *et al.*,[39] with permission.)

E. Relationship between Hyperthermia-Induced HSP and Thermotolerance in Endothelium

Thermotolerance, transient resistance to heat induced by heat itself, is generally thought to be linked to the accumulation of HSP71 in eukaryotic cells.[48,57] Nonetheless, there are reports in the literature which show that heat resistance can be acquired without synthesis of measurable amounts of HSP71.[28,94] We report one such case here.

Studies on the kinetics of protein synthesis in endothelial cells from primary and passage cultures of BAEC, passage cultures of BBCE, and passage cultures of REEC after heat treatment were performed in parallel to the cell survival studies (Fig. 8). An aliquot of the above treatment groups used for thermotolerance studies was used for labeling experiments.

Figure 8a demonstrates the kinetics of thermotolerance induced when cells

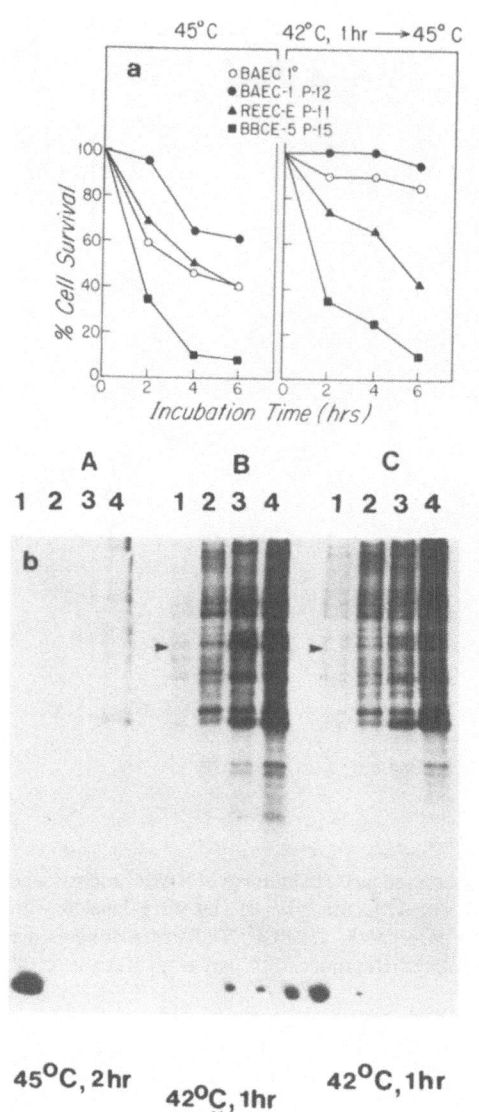

Figure 8. Induction of thermotolerance and synthesis of HSP in endothelial cells of different origins. Endothelial cells were exposed to 45°C for 1 hr followed by a second treatment at 45°C for the times indicated. (a) Cell survival plotted as a function of time after the initial treatment. (b) Autoradiographs of an SDS–polyacrylamide slab gel bearing ^{35}S-labeled proteins from endothelial cells at 45°C (2 hr) (A), 42°C (1 hr) followed by 45°C (2 hr) (B), and 42°C (1 hr) (C). (1) BBCE; (2) REEC; (3) passage BAEC; (4) primary cultures of BAEC. The location of stress HSP71 is indicated by arrowheads. Equal numbers of cell equivalents per lane.

are first treated at 42°C for 1 hr followed by transfer to 45°C for up to 6 hr. All four cell types develop thermotolerance provided they are first exposed to a primary nonlethal heat dose. Thermotolerance reaches its maximum 2 to 4 hr after heat treatment for primary and passage BAEC and REEC and decays thereafter. BBCE show the least degree of thermotolerance. Cell survival is only 10% when BBCE are exposed to a priming heat dose followed by incubation at 45°C for 6 hr, whereas REEC under the same conditions demonstrate a cell survival of 42%.

REEC are very sensitive to heat stress as compared to other endothelial cell lines. These cells develop transient thermotolerance after prolonged heating at 45°C if they are exposed to a priming nonlethal heat dose without notable increase in HSP71 as determined by one-dimensional uniform SDS–PAGE (Fig. 8a and Ketis et al.[39]). Late-passage REEC express low levels of HSP71 as determined by one- and two-dimensional gel electrophoresis with a dramatic decrease in overall protein synthesis (Fig. 8b and Ketis et al.[39]). Late-passage REEC, however, did not appear to be more thermotolerant than their parental cell of earlier passages. In fact, late passages of REEC and BBCE appear somewhat less thermotolerant. REEC, due to their physiological locale, might be expected to be more thermosensitive. In the literature, there are reports showing that heat resistance can be acquired without synthesis of measurable amounts of HSP71. Cycloheximide appears to confer transient thermotolerance on cells exposed to subsequent heat challenge,[94] and it has recently been reported[28] that glucocorticoids (hydrocortisone and dexamethasone) induce transient heat resistance in B15 melanoma cells. Thus, while the expression of HSP may be a good indicator of heat resistance, the reverse is not necessarily true. It appears that thermotolerance can be acquired through mechanisms not related to the synthesis of HSP71 or other classical HSP.

Autoradiographs of SDS–polyacrylamide gels of [^{35}S]methionine-labeled proteins from endothelial cells of different origins exposed to heat stress are shown in Fig. 8b. All endothelial cells after prolonged treatment at 45°C (2 hr) demonstrate dramatic alterations in their total cellular protein synthesis (Fig. 8b, A) relative to control (data not shown). A priming heat dose of 42°C for 1 hr (Fig. 8b, C) shows that primary and passage BAEC show a small increase in HSP71 (indicated by arrowheads) as do BBCE; however, the latter demonstrate a greater decrease in total cellular protein synthesis. Primary BAEC show no curtailment of total cellular protein synthesis relative to control (data not shown). Passage BAEC demonstrate a 40% decrease, REEC 60%, and BBCE 85%. Early passage REEC (P-5 through P-17) do not appear to show an increase in HSP71 as determined by one-dimensional gel electrophoresis (Fig. 8b, C, lane 2, and data not shown). However, after a priming heat dose followed by exposure at 45°C for 2 hr, REEC synthesized HSP73, 90, and 100 (Fig. 8b, B, lane 2). In BBCE, HSP71, 73, and 90 increase (Fig. 8b, B, lane 1); and in passage BAEC, all five HSP (HSP71, 73, 80, 90, and 100) (Fig. 8b, B, lane 3) are increased. In all cell types except BBCE, total cellular protein synthesis is protected if the cells are first exposed to a priming heat dose followed by 2 hr at 45°C (Fig. 8b, B), as compared to when cells are treated directly at 45°C for 2 hr (Fig. 8b, A).

Cells approaching senescence in culture are known to demonstrate many morphological and functional changes.[1,5,29] Several characteristics of endothelial cell aging in culture have been examined.[19,32,33,67] In addition to the aspects described in the literature, we have shown that, as endothelial cells age in culture, they also show a differential heat response, both in terms of thermotolerance and expression of cellular proteins after exposure to heat stress. Although late-passage cells of different origins maintain a classic cobblestone appearance,[34,96] the enzyme activity of angiotensin II-converting enzyme decays with prolonged passage (Table 1). Del Vecchio and Smith[19] obtained similar results with calf aortic en-

Table 1. Angiotensin II-Converting Enzyme
(ACE) Activity in Aging Cultures of Endothelial
Cells of Various Origins[a]

	ACE (U/mg protein)
BAEC 1°	17,864
BAEC P-16[b]	6,540
BBCE P-17	22,509
BBCE P-15	3,883
REEC P-8	16,750
REEC P-9	8,800
REEC P-31	2,560

[a] Note that there is some variation in activity of ACE between clones. The most dramatic effect on the heat-stress response, morphology, and ACE activity was noted with late passages of endothelial cells of different origins.

[b] P, passage number.

dothelial cells in culture. We have clearly demonstrated that at least three parameters have been altered as a function of passage or aging in endothelial cells of different origins: cellular protein synthesis after exposure to hyperthermia, thermotolerance, and expression of angiotensin II-converting enzyme.

F. Possible Role of Oxygen Free Radicals in the Expression of HSP in Endothelium

Organisms which use molecular oxygen must protect themselves against toxic by-products of oxygen metabolism. These include superoxide anion, singlet oxygen, hydrogen peroxide, and hydroxyl radicals. Superoxide anion can be generated by mitochondria[10] and by certain oxidative enzymes, e.g., xanthine oxidase[62] and cytochrome P-450 reductase.[45] Hydrogen peroxide is produced by oxidative enzymes, e.g., xanthine oxidase, and can be generated by the spontaneous or catalyzed dismutation of superoxide. Hydroxyl radicals are most likely to be produced through the interaction of superoxide anion and hydrogen peroxide by the Haber–Weiss reaction.[27] These reactive oxygen species are capable of (1) oxidizing membrane fatty acids initiating lipid peroxidation,[63] (2) oxidizing proteins,[11] and (3) damaging DNA.[14,31,53]

Oxygen radicals have been implicated as mediators in a number of pathological conditions, including inflammation,[61] ischemia,[20] acute hypertension,[42] and traumatic brain injuries.[90]

Because of the compelling evidence in the literature suggesting a relationship between oxidative stress and vascular abnormalities, we began to examine the possible relationship between the two and the expression of stress proteins in bovine aortic endothelium.

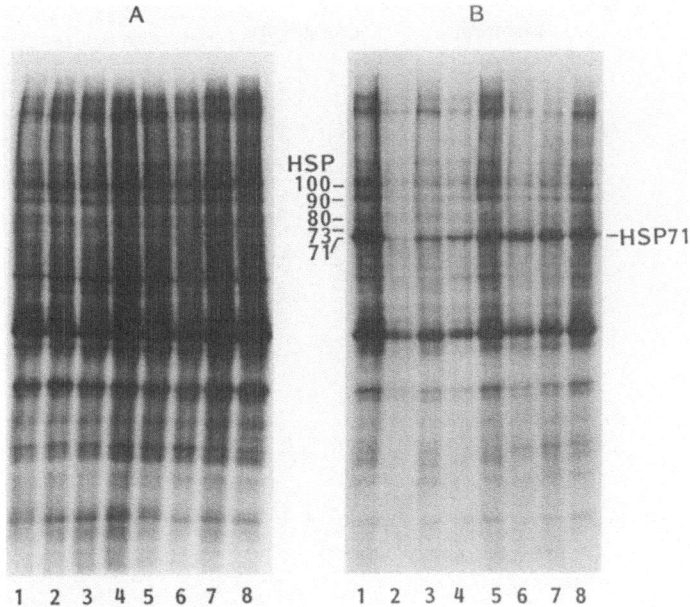

Figure 9. Effect of specific inhibitors of catalase and GSH reductase on protein profiles of control and heat-stressed primary cultures of BAEC. Cells were labeled with [^{35}S]methionine and each lane contained equal amounts of cell equivalents. (A) Cells at 37°C; (B) cells after exposure for 2 hr at 43°C. Lane 1, control cells at 37°C; lane 2, cells in the presence of aminotriazole (AT) at 6.72 mg/ml; lane 3, AT at 3.36 mg/ml; lane 4, AT at 1.68 mg/ml; lane 5, AT at 0.8 mg/ml; lane 6, BCNu at 20 µg/ml; lane 7, BCNu at 10 µg/ml; lane 8, BCNu at 5 µg/ml.

Recently, several lines of evidence suggest that HSP production may be a defense to oxidative stress. In *Drosophila*, upon recovery from anoxic conditions, the cell synthesizes HSP and demonstrates an increase in superoxide dismutase and catalase activity.[75] Lee *et al.*[51] have shown that there is rapid accumulation of AppppA and a series of related adenylylated nucleotides[8] when *S. typhimurium* LT_2 are subjected to a variety of oxidative stresses and hyperthermia. In a recent study, Christman *et al.*[16] showed that some 30 proteins are included in *S. typhimurium* as an adaptation response to hydrogen peroxide, some of which are the classical HSP.

In this section, we report that specific inhibitors of oxygen free radical scavengers affect the expression of HSP and total cellular protein synthesis in endothelial cells exposed to hyperthermia. The implication of these results is discussed.

Total cellular protein and HSP synthesis were determined for BAEC treated for 2 hr at 43°C. HSP71 is clearly evident in these cells (Fig. 9B, lane 1). Upon treatment with aminotriazole (AT), which is reported to specifically inhibit catalase (Fig. 9B, lanes 2–5), and BCNu, which inhibits glutathione (GSH) reductase without affecting catalase and GSH peroxidase (Fig. 9B, lanes 6–8), there are decreased quantities of HSP relative to control (Fig. 9B, lane 1). Densitometric

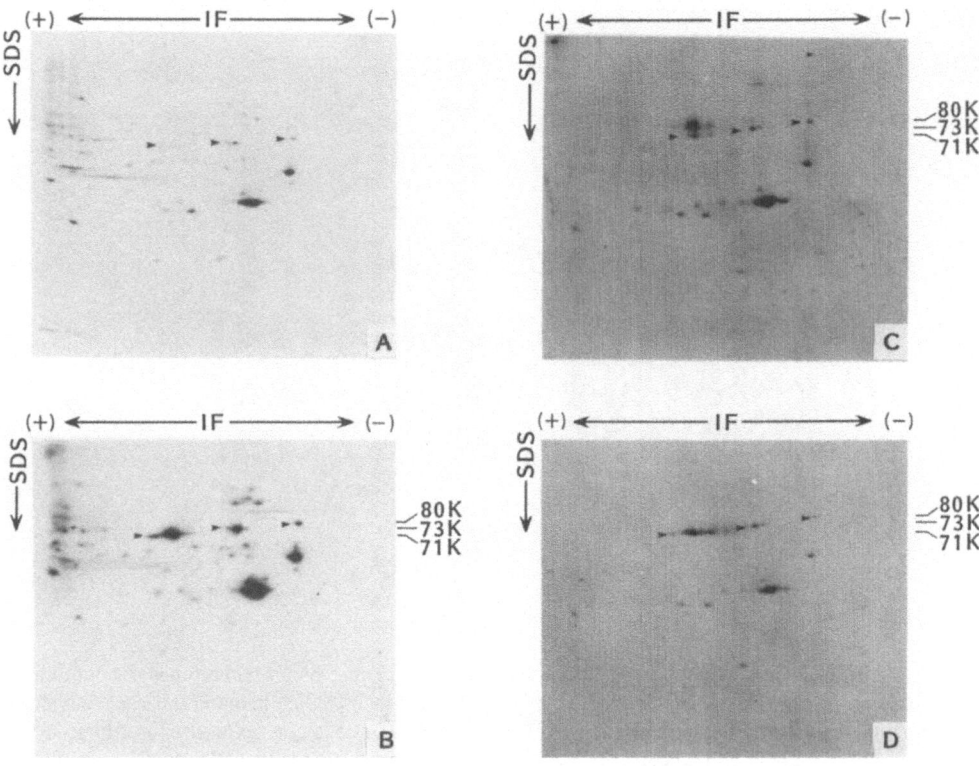

Figure 10. Two-dimensional gel analysis of control and heat-stressed BAEC exposed to aminotriazole or BCNu. Cells were labelled with [^{35}S]methionine for 2 hr at control or heat-stressed temperature. (A) Cells at 37°C; (B) cells exposed for 2 hr at 42°C; (C) cells pretreated with 1.7 mg/ml of AT for 10 min at 37°C followed by hyperthermic treatment for 2 hr at 42°C; (D) cells pretreated with 10 μg/ml of BCNu for 10 min at 37°C followed by heat treatment for 2 hr at 42°C. Each gel contained equal amounts of cell equivalents. HSP are indicated by arrowheads.

scanning studies demonstrated that AT at 0.8 mg/ml decreases HSP by 30% and BCNu at 5 μg/ml by 20% (data not shown). However, BCNu above 20 μg/ml was found to be toxic to BAEC. In addition to a decrease in HSP with AT or BCNu pretreatment, a decrease in total protein synthesis is noted at AT concentrations which exceed 1 mg/ml and BCNu concentrations which exceed 7 μg/ml. This decrease in total protein is not evident in control cultures at the same concentration of inhibitor (Fig. 9A), nor did treatment of BAEC with these inhibitors generate the induction of HSP at 37°C as determined by two-dimensional gel electrophoresis (data not shown).

Two-dimensional gel analysis of control and heat-stressed BAEC exposed to AT or BCNu is shown in Fig. 10. At 42°C, three HSP are induced: HSP71, 73, and 80. Pretreatment with AT (1.7 mg/ml) (Fig. 10C) or BCNu (10 μg/ml) (Fig. 10D) results in a decrease of HSP. All isoforms of each individual HSP are present and are equally reduced. At these concentrations, the specific inhibitors of oxygen free radical scavengers partially inhibit total protein synthesis of heat-stressed cells. These inhibitors are expected to affect the cellular oxidation–reduction state of GSH, a compound that plays an integral role in the maintenance of the cellular redox state[15] and protects cells from oxidative stress brought about by oxygen metabolism. Kosower et al.[43] reported that oxidized GSH at 10^{-4} M inhibits protein translation in a reticulolysate system. Recently, Russo et al.[76] noted that oxidized GSH is raised upon heat stress in Chinese hamster V79 cells and that depletion of GHS by diethyl maleate (sulfhydryl-trapping agent) or inhibition of GSH synthesis results in decreased HSP expression and decreased thermotolerance.

We have examined the expression of superoxide dismutase and catalase activities in BAEC (data not shown) and noted that these cells express four or five superoxide dismutase species of different molecular weight as determined by polyacrylamide gel electrophoresis. Their expression is modulated by heat stress and other agents such as ethanol, a free-radical scavenger; menadione, which causes redox cycling; and hydrogen peroxide. In our hands, ethanol (0.5 to 6%) does not appear to cause a dramatic increase in HSP71 as determined by one-dimensional gel electrophoresis, but short exposure to ethanol does generate many peptides as determined by two-dimensional gel electrophoresis (data not shown). Ethanol decreases the cytotoxic effect of adriamycin, a mediator of anion radical hydroquinone production.[56] This may result from the free-radical scavenger effect of ethanol, or ethanol-induced resistance may be related to changes in cell membrane resulting from ethanol treatment,[56] which may affect adriamycin transport.

Our evidence and that present in the literature suggest that GSH levels within the cells play a role in the expression of total cellular proteins, HSP, and in the modulation of thermotolerance. Ananthan et al.[2] have presented evidence indicating that abnormal proteins trigger the expression of HSP. This implies that modulation of scavengers of oxygen free radicals which results from heat stress in our system and those of other animals[25,59,64,76] may represent a response to the increase in toxic by-products produced during protein degradation. We are currently investigating the induction and regulation of oxygen free radical scavengers induced by various physiological and pathological stresses.

IV. PERSPECTIVES

The functional significance of the heat-shock response is yet to be elucidated. At the simplest level, the heat-shock response functions as a homeostatic mechanism[55] which protects cells against the ravages of the environment and ensures the cells' survival after the crisis has passed. This is best demonstrated by thermotolerance studies, where cells acquire protection against heat and other adverse conditions by first being exposed to a priming heat dose. Although we are almost entirely ignorant of the function of HSP and their target(s) of "protection," vigorous research in the areas of cellular, biochemical, and molecular regulatory mechanisms controlling the HSP response has shed some light on the possible function(s) of some of the stress proteins in cells exposed to hyperthermia or (patho)physiological stress. The generation of mutants in *E. coli* and *S. ty-phimurium* has greatly expedited our understanding of the induction and regulation of the heat-shock response. However, different species appear to differ in the nature of the response and the natural inducers of the response.

It has long been known that tumors are sensitive to heat. Hyperthermia has been used as effective therapy.[18,83,84] However, the response is complex, and not all tumors respond equally well to such treatment.[23] *In vitro* studies, dealing with transformed cells in culture, have shown that these cells are more sensitive to hyperthermia.[88] Although there is a large body of evidence suggesting that the accumulation of HSP71 may be linked to acquired thermotolerance, nonetheless there are reports in the literature showing that transient heat resistance can be acquired without synthesis of measurable amounts of HSP71. The relationship between HSP synthesis, acquired thermotolerance, and hypersensitivity of tumors is clearly of clinical relevance.

Heat shock appears to be an appropriate model for (patho)physiological stress in endothelial cells. For example, endothelial cells approaching senescence in culture show a differential heat response, both in terms of thermotolerance and synthesis of proteins after exposure to hyperthermia. In addition, heat stress modulates scavengers of oxygen free radicals in endothelial cells. Oxygen free radicals are implicated as mediators in a number of pathological conditions which affect endothelium, including inflammation, ischemia, acute hypertension, and traumatic brain injury. It appears that the cellular "stress" response and "heat-shock" response are intimately associated in that they may protect cells (e.g., endothelium) against adverse conditions of a (patho)physiological nature.

REFERENCES

1. Aizava, S., Mitsui, Y., Kurimoto, F., and Nomura, K., 1980, Cell surface changes accompanying aging in human diploid fibroblasts. V. Role of large major cell surface protein and surface negative charge in aging and transformation-associated changes in concanavalin A-mediated red blood cells, *Exp. Cell Res.* **126**:143.
2. Anathan, J., Goldberg, A. L., and Voellmy, R., 1986, Abnormal proteins serve as eukaryotic stress signals and trigger the activation of heat-shock genes, *Science* **232**:522.
3. Ashburner, M., and Bonner, J. J., 1979, The induction of gene activity in *Drosophila* by heat shock, *Cell* **17**:241.
4. Barrett, T. B., Gajdusek, C. M., Schwartz, S. M., McDougall, J. K., and Benditt, E. P., 1984, Expression of the *sis* gene by endothelial cells in culture and *in vivo*, *Proc. Natl. Acad. Sci. USA* **81**:6772.
5. Basler, J. W., David, J. D., and Agris, P. F., 1979, Deteriorating collagen synthesis and cell ultrastructure accompanying senescence of human normal and Werner's Syndrome fibroblast cell strains, *Exp. Cell Res.* **118**:73.
6. Bensaude, O., Badinet, C., Morange, M., and Jacob, F., 1983, Heat-shock proteins, first major product of zygotic gene activity in normal mouse embryos, *Nature* **305**:331.
7. Bensaude, O., and Morange, M., 1983, Spontaneous high expression of heat-shock proteins in mouse embryonal carcinoma cells and ectoderm from day 8 mouse embryos, *EMBO J.* **2**:173.
8. Bochner, B. R., Lee, P. C., Wilson, S. W., Cutler, C. W., and Ames, B. N., 1984, AppppA and related adenylylated nucleotides are synthesized as a consequence of oxidation stress, *Cell* **37**:225.
9. Booyse, F. M., Sedlak, B. J., and Rafelson, M. E., 1975, Culture of arterial endothelial cells: Characterization and growth of bovine aortic cells, *Thromb. Diath. Haemorrh.* **35**:825.

10. Boveris, A., 1977, Mitochondrial production of superoxide radical and hydrogen peroxide, *Adv. Exp. Med. Biol.* **78**:67.

11. Brot, N., Weissbach, L., Werth, J., and Weissbach, H., 1981, Enzymatic reduction of protein-bound methionine sulfoxide, *Proc. Natl. Acad. Sci. USA* **78**:2155.

12. Brugge, J. S., 1985, Interaction of the Rous sarcoma virus protein, pp60^{src} with the cellular proteins pp50 and pp90, *Curr. Top. Microbiol. Immunol.* **122**:1–22.

13. Catelli, M. G., Binart, N., Jung-Testas, I., Renoir, J. M., Baulieu, E. E., Feramisco, J. R., and Welch, W. J., 1985, The common 90 Kd protein component of non-transformed 8S steroid receptors is a heat-shock protein, *EMBO J.* **4**:3131.

14. Cathcart, R., Schwiers, E., Saul, R. L., and Ames, B. N., 1984, Thymine glycol and thymidine glycol in human and rat urine: A possible assay for oxidative damage, *Proc. Natl. Acad. Sci. USA* **81**:5633.

15. Chance, B., Sies, H., and Boveris, A., 1979, Hydroperoxide metabolism in mammalian organisms, *Physiol. Rev.* **59**:527.

16. Christman, R. W., Morgan, F. S., Jacobson, B., and Ames, B. N., 1985, Positive control of a regulon for defense against oxidative stress and some heat-shock proteins in *Salmonella typhimurium, Cell* **41**:753.

17. Cleveland, D. W., Fischer, S. G., Kirschner, M. W., and Laemmeli, U. K., 1977, Peptide mapping by limited proteolysis in sodium dodecyl sulfate and analysis by gel electrophoresis, *J. Biol. Chem.* **252**:1102.

18. Crile, G., 1963, The effects of heat and radiation on cancer implanted on the feet of mice, *Cancer Res.* **23**:372.

19. Del Vecchio, P. J., and Smith, J. R., 1982, Aging of endothelium in culture: Decrease in angiotensin-converting enzyme activity, *Cell Biol. Int. Rep.* **6**:379.

20. Demopoulos, H. B., Flamm, E. S., Pietronigro, D. D., and Seligman, M. L., 1980, The free radical pathology and the microcirculation in the major central nervous system disorders, *Acta Physiol. Scand. Suppl.* **492**:91.

21. Di Corleto, P. E., and Bowen-Pope, D. F., 1983, Cultured endothelial cells produce a platelet-derived growth factor-like protein, *Proc. Natl. Acad. Sci. USA* **80**:1919.

22. Dixit, V. M., Grant, G. A., Santoro, S. A., and Frazier, W. A., 1983, Isolation of a heparin-binding domain from proteolytic digests of platelet thrombospondin, *Fed. Proc.* **42**:1993.

23. Eddy, H. A., 1980, Alterations in tumor microvasculature during hyperthermia, *Radiobiology* **137**:515.

24. Falkner, F. G., Saumweber, H., and Biesswann, H., 1981, Two *Drosophila melanogaster* proteins of vertebrate cells, *J. Cell Biol.* **91**:715.

25. Freeman, M. L., Malcolm, A. W., and Meredith, M. J., 1985, Decreased intracellular glutathione concentration and increased hyperthermic cytotoxicity in an acidic environment, *Cancer Res.* **45**:504.

26. Gimbrone, M. A., Cotran, R. S., and Folkman, J., 1974, Human vascular endothelial cells in culture, *J. Cell Biol.* **60**:673.

27. Haber, F., and Weiss, J., 1934, The catalytic decomposition of hydrogen peroxide by iron salt, *Proc. R. Soc. London Ser. A* **147**:332.

28. Hahn, G. M., Fisher, G., Tao, T. W., Anderson, R., and Caldwood, S. K., 1985, Are heat shock proteins necessary for heat resistance? *Heat Shock Abstr. (CSH)*, p. 119.

29. Hayflick, L., 1976, The cell biology of human aging, *N. Engl. J. Med.* **295**:1302.

30. Hightower, L. E., 1980, Cultured animal cells exposed to amino acid analogue or puromycin rapidly synthesize several polypeptides, *J. Cell. Physiol.* **102**:407.

31. Hollstein, M., Brooks, P., Linn, S., and Ames, B. N., 1984, Hydroxymethyluracil DNA glycosylase in mammalian cells, *Proc. Natl. Acad. Sci. USA* **81**:4003.

32. Hormia, M., 1982, Expression of factor VIII-related antigen and Ulex lectin-binding sites in endothelial cells during long-term cultures, *Cell Biol. Int. Rep.* **6**:1123.

33. Hormia, M., Linder, E., Lehto, V.-P., Vartio, T., Badley, R. A., and Virtamen, I., 1982, Vimentin fibroblasts in cultured endothelial cells from butyrate-sensitive juxtanuclear masses after repeated subculture, *Exp. Cell Res.* **138**:159.

34. Jaffe, E. A., 1983, Culture and identification of large vessel endothelial cells, in: *Biology of Endothelial Cells* (E. A. Jaffe, ed.), Nijhoff, The Hague, pp. 1–13.

35. Jaffe, E. A., Ruggiero, J. T., Leung, L. L. K., Doyle, M. J., McKeown-Longo, P. J., and Mosher, D. F., 1983, Cultured human fibroblasts synthesize and secrete thrombospondin and incorporate it into extracellular matrix, *Proc. Natl. Acad. Sci. USA* **80**:998.

36. Johnston, D., Oppermann, H., Jackson, J., and Levinson, W., 1980, Induction of four proteins in chick embryo cells by sodium arsenite, *J. Biol. Chem.* **255**:6975.

37. Kelley, P. M., and Schlesinger, M. J., 1978, The effect of amino acid analogue and heat shock on gene expression in chicken embryo fibroblasts, *Cell* **15**:1277.

38. Ketis, N. V., Lawler, J., Hoover, R. L., and Karnovsky, M. J., 1988, Effect of heat shock on the expression of thrombospondin by endothelial cells in culture, *J. Cell Biol.* (in press).

39. Ketis, N. V., Hoover, R. L., and Karnovsky, M. J., 1988, Effects of hyperthermia on cell survival and patterns of protein synthesis in endothelial cells from different origins, *Cancer Res.* (in press).

40. Kingston, R. E., Baldwin, A. S., Jr., and Sharp, P. A., 1984, Regulation of heat shock protein 70 gene expression by C-myc, *Nature* **312**:280.

41. Kingston, R. E., Sharp, P. A.. and Kaufman, R. J., 1984, Regulation of gene expression by the adenovirus E1A region and by C-myc, in: *Oncogenes and Viral Genes* (G. Vande Woude, A. Levine, W. Topp, and J. D. Watson, eds.), Cold Spring Harbor Laboratory, Cold Spring Harbor, N.Y., p. 539.

42. Kontos, H. A., Wei, E. P., Dietrich, W. D., Navari, R. M., Poulishock, J. T., Ghatak, N. R., Ellis, E. F., and Patterson, J. L., Jr., 1981, Mechanism of cerebral arteriolar abnormalities after acute hypertension, *Am. J. Physiol.* **240**:4511.

43. Kosower, N. S., Vanderhoff, G. A., and Kosower, E. M., 1972, Glutathione. VIII. The effects of glutathione disulfide on initiation of protein synthesis, *Biochim. Biophys. Acta* **272**:623.

44. Kruger, C., and Beneche, B.-J., 1981, *In vitro* translation of *Drosophila* heat-shock and non-heat-shock mRNA in heterologous and homologous cell-free system, *Cell* **23**:595.

45. Kuthan, H., and Ullrich, V., 1982, Oxidase and oxygenase function of the microsomal cytochrome P450 monooxygenase system, *Eur. J. Biochem.* **126**:583.

46. Laemmeli, U. K., 1970, Cleavage of structural proteins during the assembly of the head of the bacteriophage T4, *Nature* **227**:680.

47. Lahav, J. M., Schwartz, A., and Hynes, R. O., 1982, Analysis of platelet adhesion with a radioactive chemical crosslinking reagent: Interaction of thrombospondin with fibronectin and collagen, *Cell* **31**:253.

48. Landry, J., Bernier, D., Chretien, P., Nicole, L. M., Tanguary, L. M., and Marceau, N., 1982, Synthesis and degradation of thermotolerance, *Cancer Res.* **42**:2457.

49. Lawler, J. W., and Slayter, H. S., 1981, The release of heparin-binding peptides from platelet thrombospondin by proteolytic action of thrombin, plasmin, and trypsin, *Thromb. Res.* **22**:267.

50. Lawler, J. W., Slayter, H. S., and Coligan, J. E., 1978, Isolation and characterization of a high molecular weight glycoprotein from human blood platelets, *J. Biol. Chem.* **253**:8609.

51. Lee, P. C., Bochner, B. R., and Ames, B. N., 1983, AppppA, heat-shock stress and cell oxidation, *Proc. Natl. Acad. Sci. USA* **80**:7496.

52. Leug, L. K. K., and Nachman, R. L., 1982, Complex formation of platelet thrombospondin with fibrinogen, *J. Clin. Invest.* **70**:542.

53. Levin, D. E., Hollstein, M., Christman, M. F., Schwiers, E. A., and Ames, B. N., 1982, A new *Salmonella* tester strain (TA102) with A:T base pairs at the site of mutation detects oxidative mutagens, *Proc. Natl. Acad. Sci. USA* **79**:7445.

54. Levinson, W., Oppermann, H., and Jackson, J., 1980, Transition series metals and sulfhydryl reagents induce the synthesis of four proteins in eukaryotic cells, *Biochim. Biophys. Acta* **606**:170.

55. Lewis, M. J., Helmsing, P., and Ashburner, M., 1975, Parallel changes in puffing activity and patterns of protein synthesis in salivary glands of *Drosophila, Proc. Natl. Acad. Sci. USA* **72**:3604.

56. Li, G. C., and Hakn, G. M., 1978, Ethanol-induced tolerance to heat and adriamycin, *Nature* **274**:699.

57. Li, G. C., and Werb, Z., 1982, Correlation between synthesis of heat-shock protein and development of thermotolerance in Chinese hamster fibroblasts, *Proc. Natl. Acad. Sci. USA* **79**:3219.

58. Lindquist, S., 1981, Regulation of protein synthesis in *Drosophila* during heat shock, *Nature* **293**:311.

59. Loven, D. P., Leeper, D. B., and Oberbey, L. W., 1985, Superoxide dismutase levels in Chinese hamster ovary cells and ovarian carcinoma cells after hyperthermia or exposure to cycloheximide, *Cancer Res.* **45**:3029.

60. Majack, R. A., Cook, S. C., and Bornstein, P., 1985, Platelet-derived growth factor and heparin-like glycosaminoglycan-regulated thrombospondin synthesis and disposition in the matrix by smooth muscle cells, *J. Cell Biol.* **101**:1059.

61. McCord, J. M., 1974, Free radicals and inflammation: Protection of synovial fluid by superoxide dismutase, *Science* **185**:529.

62. McCord, J. M., and Fridovich, I., 1968, The reduction of cytochrome C by mild xanthine oxidase, *J. Biol. Chem.* **243**:5753.

63. Mead, J. F., 1976, Free radical mechnisms of lipid damage and consequences for cellular membranes, in: *Free Radicals in Biology* (W. A. Pryor, ed.), Academic Press, New York, p. 51.

64. Mitchell, J. B., Russo, A., Kinsella, T. J., and Glatstein, E., 1983, Glutathione elevation during thermotolerance induction and thermosensitization by glutathione depletion, *Cancer Res.* **43**:987.

65. Morange, M., Diu, A., Bensaude, O., and Babinet, C., 1984, Altered expression of heat shock proteins in embryonal and mouse early embryonic cells, *Mol. Cell. Biol.* **4**:730.

66. Morimoto, R., and Fodor, E., 1984, Cell-specific expression of heat shock proteins in chicken reticulocytes and lymphocytes, *J. Cell Biol.* **99**:1316.

67. Mueller, S. N., Rosen, E. M., and Levin, E. M., 1980, Cellular senescence in a cloned strain of bovine fetal aortic endothelial cells, *Science* **207**:889.

68. Mumby, S. M., Abbot-Brown, D., Raugi, G., and Bornstein, P., 1984, Regulation of thrombospondin secretion by cells in culture, *J. Cell. Physiol.* **120**:280.

69. Mumby, S. M., Raugi, G., and Bornstein, P., 1984, Interaction of thrombospondin with extracellular matrix proteins: Selective binding to type V collagen, *J. Cell Biol.* **98**:646.

70. Nevins, J., 1981, Induction of the synthesis of a 70,000-dalton mammalian heat-shock protein by the adenovirus E1A product, *Cell* **29**:913.

71. O'Farrell, P. H., 1975, High resolution two-dimensional electrophoresis of proteins, *J. Biol. Chem.* **80**:4007.

72. O'Kimoto, R., Sacks, M. M., Porter, E. K., and Freeling, M., 1980, Patterns of polypeptide synthesis in various maize organs under anaerobiosis, *Planta* **150**:89.

73. Raugi, G. J., Mumby, J. M., Abbot-Brown, D., and Bornstein, P., 1982, Thrombospondin: Synthesis and secretion by cells in culture, *J. Cell Biol.* **95**:351.

74. Riehl, R. M., Sullivan, W. P., Uroman, B. T., Bauer, V. J., Pearson, G. R., and Toft, D. O., 1985, Immunological evidence that the nonhormone binding component of avian steroid receptors exists in a wide range of tissues and species, *Biochemistry* **24**:6586.

75. Ropp, M., Courgeon, A. M., Calvayrac, R., Best-Belpo, M., and Belpomme, M. M. E., 1983, The possible role of the superoxide ion in the induction of heat-shock and specific proteins in aerobic *Drosophila* cells during return to normoxia after a period of anaerobiosis, *Can. J. Biochem. Cell Biol.* **61**:456.

76. Russo, A., Mitchell, J. B., and McPherson, S., 1984, The effects of glutathione depletion on thermotolerance and heat stress synthesis, *Br. J. Cancer* **49**:753.

77. Sanchez, E. R., Toft, D. O., Schlesinger, M. J., and Pratt, W. B., 1985, Evidence that the 90 KDa phosphoprotein associated with the untransformed L-cell glucocorticoid receptor is a murine heat-shock protein, *J. Biol. Chem.* **260**:12398.

78. Schlesinger, M. J., Ashburner, M., and Tissieres, A., (eds.), 1982, *Heat Shock from Bacteria to Man*, Cold Spring Harbor Laboratory, Cold Spring Harbor, N.Y.

79. Schwartz, S. M., and Benditt, E. P., 1977, Aortic endothelial cell replication. I. Effects of age and hypertension in the rat, *Circ. Res.* **41**:248.

80. Scott, M. P., and Pardue, M. L., 1981, Translocational control of lysates of *Drosophila melanogaster* cells, *Proc. Natl. Acad. Sci. USA* **78**:3353.

81. Shiu, R. P. C., Pouyssegur, J., and Pastan, I., 1977, Glucose depletion accounts for the induction of two transformation-sensitive membrane proteins in Rous sarcoma virus-transformed chick embryo fibroblasts, *Proc. Natl. Acad. Sci. USA* **74**:3840.

82. Spatz, M., Bembry, J., Dodson, R. F., Hervonen, H., and Murray, M. R., 1980, Endothelial cell cultures derived from isolated cerebral microvessels, *Brain Res.* **191:**577.

83. Suit, H. D., 1977, Hyperthermic effects on animal tissue, *Radiobiology* **123:**483.

84. Suit, H. D., and Shwayder, M., 1974, Hyperthermia: Potential as an anti-tumor agent, *Cancer* **34:**122.

85. Thomas, G. P., Welch, W. J., Matthews, M., and Feramisco, J. R., 1982, Molecular and cellular effects of heat shock and related treatments of mammalian tissue culture cells, *Cold Spring Harbor Symp. Quant. Biol.* **46:**985.

86. Tsukeda, H., Mackawa, H., Izani, S., and Nitta, K., 1981, Effect of heat shock on protein synthesis by normal and malignant human lung cells in tissue culture, *Cancer Res.* **41:**5188.

87. Ungewickel, E., 1985, The 70-Kd mammalian heat shock proteins are structurally and functionally related to the uncoating protein that releases clathrin triskelions from coated vesicles, *EMBO J.* **4:**3385.

88. Urano, M., 1986, Kinetics of thermotolerance in mammalian and tumor tissues: A review, *Cancer Res.* **46:**474.

89. Wagner, R. C., and Matthews, M. A., 1975, The isolation and culture of capillary endothelium from epididymal fat, *Microvasc. Res.* **10:**286.

90. Wei, E. P., Kontos, H. A., Dietrich, W. D., Povlishock, J. T., and Ellis, E. F., 1981, Inhibition by free radical scavengers and by cyclooxygenase inhibitors of pial arteriolar abnormalities from concussive brain injury in rats, *Circ. Res.* **48:**95.

91. Welch, W. J., 1985, Phorbol ester, calcium ionophore, or serum added to quiescent rat embryo fibroblast cells all result in the elevated phosphorylation of two 28,000-dalton mammalian stress proteins, *J. Biol. Chem.* **260:**3058.

92. Welch, W. J., Garrels, J., Thomas, G. P., Linn, J. J. C., and Feramisco, J. R., 1983, Biochemical characterization of the mammalian stress proteins and identification of two stress proteins as glucose and Ca^{++}-ionophore-regulated proteins, *J. Biol. Chem.* **258:**7102.

93. Welch, W. J., and Suhan, J. P., 1985, Morphological study of the mammalian stress response: Characterization of changes in cytoplasmic organelles, cytoskeleton, and nucleoli, and appearance of intracellular actin filaments in rat fibroblasts after heat-shock treatment, *J. Cell Biol.* **101:**1198.

94. Widelitz, R. B., Magun, B. E., and Gerner, E. W., 1986, Effects of cycloheximide on thermo-tolerance expression, heat-shock protein synthesis and heat-shock protein on mRNA accumulation in rat fibroblasts, *Mol. Cell Biol.* **6:**1098.

95. Wu, B. J., and Morimoto, R. I., 1985, Transciption of the human HSP70 gene is induced by serum stimulation, *Proc. Natl. Acad. Sci. USA* **82:**6070.

96. Zetter, B. R., 1983, Culture of capillary endothelial cells, in: *Biology of Endothelial Cells* (E. A. Jaffe, ed.), Nijhoff, The Hague, p. 14.

97. Zimmerman, L., Petri, W., and Meselson, M., 1983, Accumulation of a specific subset of *D. melanogaster* heat-shock mRNAs in normal development without heat shock, *Cell* **32:**1161.

15

Endothelial Activation

Its Role in Inflammatory and Immune Reactions

Ramzi S. Cotran and Jordan S. Pober

It has long been known that endothelium of postcapillary venules involved in the inflammatory response of a delayed-type hypersensitivity and other cell-mediated immune reactions, undergoes a number of functional and morphological alterations (Table 1).[19,47,57,58] In particular, the vessels become hyperpermeable to macromolecules, and the endothelial cells undergo hypertrophy, acquire a plump cuboidal appearance, protrude into the lumen, and display increased biosynthetic organelles on electron microscopy. These changes have been described collectively as "endothelial activation," and the cells themselves likened to the high endothelial venules of lymph nodes. Although some earlier workers suggested that such activation may be related to the presence of lymphoid cells,[22] the nature of the activation process, and the specific signals which induce it were unknown. However, it has recently become apparent from studies on endothelial cultures that many of these functional and morphological effects can be induced *in vitro* by purified lymphokines, monokines, and by other stimuli. This process of endothelial activation represents either a marked stimulation of constitutive functions of the endothelium or the induction of new activities and molecules. In this chapter we shall briefly review the effects of well-characterized cytokines on endothelial function, morphology, and antigen expression *in vitro*, and examine the role of endothelial activation in several clinical settings *in vivo*.

I. ACTIONS OF SPECIFIC CYTOKINES ON ENDOTHELIAL CELLS IN CULTURE

The recent purification of various lymphokines and monokines and their expression as recombinant gene products, has allowed an increasingly detailed analysis of the effects of these cytokines on a variety of endothelial functions.

Ramzi S. Cotran and Jordan S. Pober • Department of Pathology, Brigham and Women's Hospital and Harvard Medical School, Boston, Massachusetts 02115.

Table 1. Endothelial Alterations in Delayed
Hypersensitivity Reactions

Increased permeability
Increased leukocyte adhesion
Increased replication
Fibrin deposition (mostly extravascular)
Endothelial necrosis (variable)
Enlarged cells   ⎫
Increased organelles ⎬ "Activation"

Table 2 summarizes some of the reported effects of cytokines, and here we shall
only briefly highlight the major points.

A. Interferon-γ (IFN-γ)

IFN-γ is a protein mediator produced by activated T lymphocytes, encoded
by a single gene and secreted as an 18k protein core with one or two glycan
moieties.[54] In cells of the immune system, it has multiple effects, including macrophage activation, induction of B-cell differentiation, and increased expression of
cell surface major histocompatibility complex (MHC) antigens, especially class II
antigens.

IFN-γ has profound effects on endothelial cells in culture. The first effect to
be discovered was the induction of class II MHC antigen synthesis and surface
expression by cultured human umbilical vein and foreskin microvascular endothelial cells.[41,42] Class I (HLA-A,B) antigen expression is also increased, although

Table 2. Endothelial Cell "Activation" by Cytokines[a]

Category	Function	Cytokines
Coagulation	↑ tissue factor	IL-1, TNF, LT*
	↓ activated protein C	IL-1, TNF, LT*
	↓ PA	IL-1, TNF, LT*
	↑ inhibitor of PA	IL-1, TNF, LT*
Inflammation	↑ PAF; ↑ PG	IL-1, TNF, LT
	↑ E-LAM 1 (leukocyte adhesion)	IL-1, TNF, LT
	↑ IL-1 secretion	IL-1, TNF, LT
	↑ CSF secretion	IL-1, TNF, LT
Immunity	"HEV" morphology	IL-1, TNF, LT, IFN-γ
	↑ ICAM-1 (lymphocyte adhesion)	IL-1, TNF, LT, IFN-γ
	↑ class I MHC	TNF, LT, IFN-α, β, γ
	↑ class II MHC (antigen presentation)	IFN-γ
	↑ membrane IL-1	IL-1, TNF, LT

[a] Abbreviations: PA, plasminogen activator; PAF, platelet-activating factor; PG, prostaglandins; CSF,
colony-stimulating factor; HEV, high endothelial venule. The functions are ascribed to each category
arbitrarily and there is obviously a great deal of overlap. The LT effects marked with an asterisk
have not been demonstrated unequivocally, but appear likely.

this is not specific for IFN-γ, being also induced by IFN-α, IFN-β, tumor necrosis factor (TNF),[13] and lymphotoxin (LT).[46] Class II antigen expression is first detectable about 6–8 hr after IFN-γ exposure and reaches plateau levels by 4 to 6 days. It persists as long as IFN-γ remains in the culture, and is associated with the expression of all known class II antigens (i.e., HLA-DR, DP, and DQ).[12]

The role played by induced class II antigens on endothelium in immune reactivity is unclear. Class II expression is necessary for presentation of antigen to T helper cells and IFN-γ-treated endothelial cells are capable of such accessory cell function, but class II antigen expression is insufficient to explain the ability of endothelial cells to activate T lymphocytes, since cultured dermal fibroblasts and vascular smooth muscle cells, which express quantitatively similar amounts and densities of class II antigens in response to IFN-γ, do not activate resting T cells.[23,43]

Besides inducing MHC antigens, IFN-γ also causes increased expression of an antigen on endothelial cells (and fibroblasts) termed ICAM-1 (intercellular adhesion molecule-1).[18,44] This antigen is recognized by monoclonal antibody RR1/1, which inhibits lymphocyte homotypic adhesion stimulated by phorbol esters, and natural adhesion of lymphocytes to fibroblasts. It is an approximately 90k glycoprotein. The time course of enhanced expression of RR1/1 binding in endothelium and its persistence in the presence of mediators resembles the time course of class II MHC modulation.

In addition to its effects on antigen expression, IFN-γ inhibits endothelial cell growth, and in particular causes a profound reorganization in endothelial cell morphology.[52] Forty-eight to seventy-two hours after addition of IFN-γ to confluent endothelial monolayers, the normally polygonal and strictly contact-inhibited endothelial cells become markedly elongated, extensively overlap, and expose focal gaps in the monolayer, revealing the substratum. In addition, there is reorganization of the actin cytoskeletons, and both of these morphological changes persist as long as the IFN-γ remains in the culture. IFN-α and IFN-β have no effect on endothelial cell morphology.

To summarize then, the most important endothelial effects of IFN-γ are the induction of class II antigen expression and the characteristic morphological changes. These effects are relatively slow, peaking at about 4 days, and persist in the presence of mediators. These changes are defined as the "long program" of endothelial activation, to distinguish them from the more rapid transient changes described below for IL-1, TNF, and LT.

B. IL-1, TNF, and LT

IL-1 and TNF are protein products of stimulated macrophages (i.e., monokines). IL-1 was originally defined by its ability to act as comitogen for murine thymocytes stimulated with concanavalin A or phytohemagglutinin. It is now known to be produced by many cells, including endothelial cells, and to have a variety of inflammatory effects.[39] Two separate IL-1 genes have been cloned, sequenced, and expressed: an acidic form (pI 5, also called IL-1α) and a neutral form (pI 7, also called IL-1β).[2,29] Both IL-1 species are processed to 17k proteins which appear to compete for a common cellular receptor.[16,31] TNF, originally defined as a tumoricidal activity in the serum of endotoxin-treated animals, is a noncovalent oligomer of 17k subunits.[5,38] The TNF gene has also been cloned, sequenced, and expressed.[40] Studies on the recombinant protein have revealed the surprising finding that TNF has many IL-1-like inflammatory properties,[5] although human TNF does not have activity in standard IL-1 thymocyte assays. LT is a 24k protein secreted by activated T lymphocytes; it has tumoricidal properties indistinguishable from TNF.[49] Sequences deduced from the cloned gene have revealed structural homologies to TNF,[24] and purified LT will compete with TNF for binding to the same cell surface receptor.[1]

IL-1 and TNF have many similar effects on cultured endothelial cells (Table 2). The two most well-studied effects are the modulation of endothelial cell coagulant properties, and the induction of increased leukocyte adhesivity (reviewed in detail by Gimbrone and Bevilacqua, this volume). With regard to coagulant properties, both IL-1 and TNF markedly enhance surface tissue factor (TF) activity[6,8,34-36]; decrease surface thrombomodulin, thus markedly inhibiting the anticoagulant effects of protein S and protein C[34-36]; and increase secretion of an inhibitor of tissue plasminogen activator by endothelial cells.[9,20,33] All of these effects tend to tip the balance of coagulant–anticoagulant molecules on the surface of endothelium toward fibrin deposition and intravascular coagulation. The second major effect of IL-1 and TNF on endothelium is stimulation of increased surface adhesivity for polymorphonuclear leukocytes, monocytes, lymphocytes, and other leukocyte cell lines.[7,11,21] These functional effects of IL-1 and TNF can be blocked by RNA and protein synthesis inhibitors, but not by cyclooxygenase inhibitors; in the case of leukocyte adhesivity and TF expression, the effects peak at 4 to 6 hr and decline to basal values by 24 hr, whether or not the IL-1 or TNF remains in the medium. After 24 hr in the continuous presence of IL-1 or TNF, there is refractoriness of the endothelium to further restimulation by addition of the same mediator, but not to addition of the other monokine. In addition to these cytokines, it has been noted that bacterial endotoxin is also a weak inducer of both increased procoagulant activity, and leukocyte adherence.

Since IL-1- and TNF-induced functions are dependent on RNA and protein synthesis, monoclonal antibodies were developed against cytokine-stimulated umbilical vein endothelial cells in an effort to identify newly induced antigens. One of these antibodies, H4/18,[45] exhibited reactivity that paralleled the functional changes: it bound to IL-1- or TNF-stimulated endothelial cells, but not to non-stimulated cells; binding was maximal at 4 to 6 hr and declined to basal levels by 24 hr. Induction of maximal binding was also dependent on protein and RNA synthesis, and it was not affected by cyclooxygenase inhibitors. While this antibody only partially blocks IL-1-induced adhesion of HL-60 cells to endothelial cells, it has proven to be an extremely useful marker for the detection of endothelial cell activation *in vivo*, as will be described later. Recently, Bevilacqua *et al.*[10] have described a second antibody, H18/7, which recognizes the same antigen as H4/18 and inhibits neutrophil adhesion to cytokine-stimulated umbilical vein endothelial cells. The antigen recognized by these two antibodies has been called E–LAM 1 (endothelial–leukocyte adhesion molecule 1).

In addition to inducing an antigen recognized by antibody H4/18, IL-1 and TNF also stimulate the surface expression of ICAM-1, recognized by the antibody RR1/1.[44] However, RR1/1 also binds to unstimulated endothelial cells; the increased RR1/1 binding induced by IL-1 is maximal at 24 hr whereas it peaks at 4 to 6 hr for H4/18, it is sustained as long as the cytokines remain in the culture medium, whereas H4/18 binding spontaneously declines. It should also be stressed that RR1/1 can also be induced by IFN-γ and is stimulated on other cell types, including dermal fibroblasts and lymphoid cells, whereas H4/18 binding is endothelial cell-specific and is not affected by IFN-γ.

In addition to these rapidly induced changes, IL-1 and TNF also have a number of more gradual effects (peak 3 days or longer) on cultured endothelium. Specifically, both mediators induce a morphological rearrangement indistinguishable from that caused by IFN-γ.[46,52] Furthermore, combinations of IL-1 or TNF plus IFN-γ appear to synergize in their morphological effects. At optimal concentrations, unique morphological changes are seen by combining mediators.

As noted above, LT was found to share the tumoricidal activities of TNF. The recent availability of recombinant LT has led to the observation that these two mediators are indistinguishable in their actions on endothelial cells in assays of antigenic modulation and morphological rearrangement.[46] Similarly, it appears that the two IL-1 gene products (IL-1α and IL-1β) are indistinguishable from each other in their effects on endothelium,[46] a finding consistent with the observation that both IL-1 species bind to the same cell surface receptor. However, the LT/TNF pair can be distinguished from the IL-1α/IL-1β pair by at least three criteria: (1) only LT or TNF causes a significant increase in expression of class I MHC antigens; (2) effects of LT/TNF plus IL-1α/IL-1β are additive, even at saturating concentrations of individual mediators; and (3) LT/TNF and IL-1α/IL-1β do not induce refractoriness to members of the other pair for restimulation of H4/18 binding[46] or TF expression.[8]

Of great interest is that endothelium itself is capable of IL-1 synthesis and secretion. Such IL-1 production can be induced by endotoxin, TNF, LT, and IL-1 itself,[28,32,37,46,51,55] and may serve to amplify the procoagulative and proinflammatory effects of the original stimulus. However, most IL-1 synthesized by endothelial cells is not secreted, but rather remains associated with the plasma membrane where it is optimally positioned to stimulate marginating leukocytes.[25a] To summarize then, IL-1, TNF, and LT are important modulators of endothelial cell phenotype. They have profound influences on endothelial cell surface thrombogenicity and leukocyte adhesivity and result in the induction of new endothelial cell proteins. Most of these responses are rapid (as compared to those of IFN-γ) and transient. TNF/LT shows additivity with IL-1. These mediators also combine with IFN-γ to produce unique morphological effects.

II. ENDOTHELIAL ACTIVATION IN VIVO

The studies summarized above and in Tables 1 and 2 leave no doubt that cytokines induce endothelial cell activation *in vitro*. The questions to ask are: can the effects also be demonstrated *in vivo*; what regulates the diverse activities in normal vasculature; what is the role of endothelial activation in inflammation and immunity; and does it, in certain instances, result in vascular injury? These questions remain largely unanswered, but have begun to be addressed.

Intradermal injections of crude lymphokine-containing supernatants into the skin of cancer patients have resulted in a pattern of vascular cell change which is histologically similar to that elicited by delayed hypersensitivity reactions.[17] Class II MHC antigens, inducible by IFN-γ *in vitro*, are also induced *in vivo* at the site of an incipient cell-mediated immune response.[50] Infusions of IL-1 and TNF in rabbits[35] and rats[53] cause fibrin deposition on the endothelium, and leukocyte aggregation, and replace the need for a second injection of endotoxin in the localized Shwartzman phenomenon.[4] These studies suggest an effect of these mediators on endothelium *in vivo*.

Recently, we have addressed the question of *in vivo* relevance by examining human tissues for the presence of antigens that will bind monoclonal antibody H4/18, monoclonal antibody RR1/1, and monoclonal antibody Leu 10 (reactive with HLA-DQ). As discussed earlier, H4/18 and Leu 10 bind only to activated endothelium *in vitro*, while RR1/1, although present in normal endothelium, is markedly enhanced by IFN-γ, LT, TNF, and IL-1. Examination of endothelium in normal human tissues by immunoperoxidase techniques showed absence of H4/18 staining, weak to moderate RR1/1 and Leu 10 staining[15]; the presence of HLA-DQ expression on "resting" endothelium *in vivo* has been attributed to low levels of IFN-γ in the normal state.[25] We first showed striking H4/18 labeling in delayed hypersensitivity reaction elicited on the forearm skin of a healthy male volunteer using streptococcal varidase antigen.[14] While control endothelium was routinely negative for H4/18, the well-developed hypersensitivity reaction sites, at 16 and 23 hr, exhibited strong reaction product in numerous small vessels in the hypodermis and dermis (Fig. 1). The staining outlined the lumina of small vessels and on high magnification appeared to be exclusively in the endothelium (Fig. 2). It was most prominent in the deeper venules which exhibited a dense perivascular infiltrate, but was also prominent in the most superficial smaller vessels. Biopsies taken at 6 days showed some persistent perivascular inflammatory infiltrate, but there was no staining with H4/18 in the endothelium. Control antibodies failed to reproduce the endothelial staining obtained with H4/18, while anti-von Willebrand factor (vWF) antibodies stained endothelium in all sections at all time intervals (Table 3). We then examined a large series of inflammatory, neoplastic, and immune reactions in human lymph nodes and skin.[15] As shown in Tables 4 and 5, we have found prominent and frequent endothelial staining in cases of acute granulomatous lymphadenitis, Hodgkin's lymphoma, T-cell lymphoma, and in hypersensitivity conditions of the skin associated with an active inflammatory infiltrate. The most useful marker for activation was H4/18 antibody, since normal endothelium was consistently negative, but in most of these conditions there was in addition accentuation of staining with RR1/1 and anti-HLA-DQ antibodies. This overview of clinical material clearly indicates that endothelial activation, similar in terms of its profile of expression of endothelial antigens,

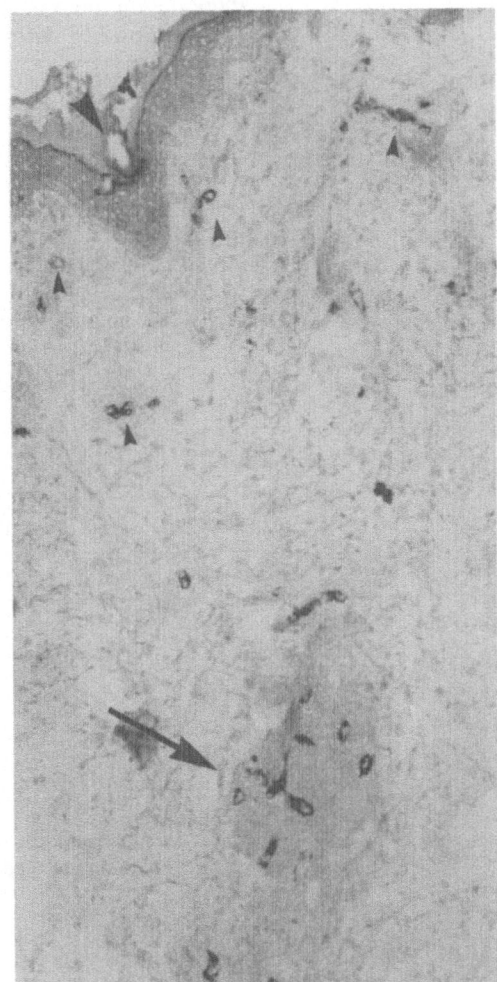

Figure 1. Photomicrograph of frozen section from skin biopsy 23 hr after eliciting a delayed hypersensitivity reaction, reacted with the immunoperoxidase technique with antibody H4/18 and lightly stained with hematoxylin. Note the positively staining venules in the reticular dermis (long arrow) and the smaller vessels in the more superficial papillary dermis (small arrowheads). The large arrowhead points to the epidermis. (A smaller micrograph appears in Cotran *et al., J. Exp. Med.* **164**:661, 1986.)

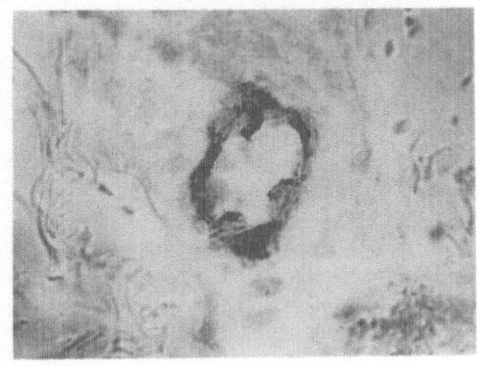

Figure 2. High magnification of positively staining vessels. The staining is confined to the endothelium.

Table 3.   Staining of Endothelium by Three Monoclonal
Antibodies at Intervals after Eliciting a Delayed
Hypersensitivity Reaction in Human Skin[a]

	H4/18	K16/16	Anti-vWF
0 hr	−	−	+
16 hr	+	−	+
23 hr	+	−	+
6 days	−	−	+

[a] H4/18 is described in the text; K16/16 is a control antibody of
the same isotope as H4/18; anti-vWF is a positive control since
it stains endothelium under normal conditions.

Table 4.   H4/18 Staining in Lymph Nodes[a]

	No. positive/No. examined
"Normal" lymph nodes	0/5
Acute granulomatous lymphadenitis	3/3
Hodgkin's lymphoma	12/14
T-cell lymphoma	7/13
B-cell lymphoma	3/23
Chronic sarcoid granulomas	0/5
Dermatopathic lymphadenitis	0/3
Toxoplasma lymphadenitis	0/4

[a] From Cotran et al.[15] and unpublished observations.

Table 5.   H4/18 Staining in Skin Biopsies[a]

	No. positive/No. examined
Induced delayed hypersensitivity	5/5
Erythema multiforme	5/6
Photoallergic eruptions	7/9
Bullous pemphigoid	4/6
Cutaneous vasculitis	4/5
Lichen planus	1/4
Dermatitis herpetiformis	0/3
Nonspecific dermatitis	2/5
Skin burns	0/4
Noninflammatory urticaria	0/2
Others (single cases)	0/6

[a] From Cotran et al.[15] and unpublished observations.

occurs regularly in settings of inflammatory, immune, or neoplastic reactions associated with activated lymphocytes and monocytes.

From these studies, it may be suggested that the changes of activation described by early workers in delayed hypersensitivity reactions and other inflammatory conditions reflect some of the alterations in structure and function induced in endothelium by specific lymphokines and monokines. It might be, for example, that the *in vivo* counterparts of the morphological changes seen in cell monolayers induced *in vitro* with IFN-γ may cause the increased vascular permeability seen in hypersensitivity reactions, as well as the transendothelial migration of lymphocytes into tissues. The fibrin deposition may reflect the effect of cytokines on the coagulant activities (although in delayed hypersensitivity reactions, most fibrin is in fact extravascular) and the cell hypertrophy may represent the combined effects of TNF and IFN-γ shown in culture. The cause of the endothelial replication which has been clearly demonstrated in these reactions[47] is unclear. IFN-γ and TNF are in fact cytostatic for endothelial cells, but other still poorly defined monokines and lymphokines, as well as manophage-derived fibroblast growth factor,[20a] cause endothelial proliferation *in vitro* and angiogenesis *in vivo*.[3,30,48,56]

Another possible relevance of endothelial activation *in vivo* derives from studies on Kawasaki's disease, a disease of childhood associated with fever, mucocutaneous manifestations, and panvasculitis of unknown etiology.[26,27] Children in the acute phase of Kawasaki's disease, but not in the convalescent stage or age-matched controls or febrile controls, have at least two types of circulating antibodies which lyse activated endothelium. The first lyses IFN-γ-treated endothelial cells but not control endothelium, control or IFN-γ-treated fibroblasts, or control or IFN-γ-treated smooth muscle cells.[26] The time course of induction of the target antigen is similar to *in vitro* modulation by IFN-γ, i.e., it is maximal after 3 days of exposure to the lymphokine. The second set of antibodies is also present in acute Kawasaki's syndrome serum, and lyses IL-1- or TNF-treated endothelial cells, but not normal endothelial cells.[27] The time course of induction of the IL-1- or TNF-treated target antigen is similar to IL-1 or TNF modulation of H4/18 binding antigen *in vitro*, i.e., it is maximal at 4 to 6 hr after cytokine treatment, and disappears by 24 hr. These two sets of antibodies recognize completely different antigens as shown by cross adsorption studies. It is tempting to speculate that the immunoregulatory disturbances that are well known to occur in patients with Kawasaki's disease (e.g., increased T4 cells) result in endothelial activation by cytokines and the induction of new surface antigens. The latter in turn elicit an antibody response (enhanced by the polyclonal B-cell activation also characteristic of the syndrome) which results in endothelial cell lysis and vascular injury. In this instance therefore, endothelial activation may play a role in the subsequent development of vasculitis. Studies are in progress to determine whether such a scenario is also present in other forms of vasculitis.

In summary, there is considerable evidence that many of the endothelial effects induced by cytokines and other stimuli *in vitro* may have relevance *in vivo*. But the processes involved are complex and their regulation is still poorly understood. Thus far, the concept of endothelial activation has been a useful one, as it has stimulated a thorough search for inducible endothelial functions in culture and of similar phenomena *in vivo*. However, as in the case of macrophage, lymphocyte, and neutrophil activation, a more precise understanding of the various effector functions of endothelium will necessitate a more exact definition of triggering events, and the molecular and metabolic processes which result in altered endothelial function and structure.

ACKNOWLEDGMENTS. We wish to acknowledge our collaborators Drs. Michael Gimbrone, Michael Bevilacqua, and Donna Mendrick, Mr. George Stavrakis and Ms. Lynne Lapierre for excellent assistance, and Ms. Julie Smith for help in the preparation of the manuscript. Our research was supported by grants from the National Institutes of Health (HL 22602 and HL 36003). J.S.P. is an Established Investigator of the American Heart Association.

REFERENCES

1. Aggarwal, B. B., Fessalu, T. E., and Hass, P. E., 1985, Characterization of receptors for human tumor necrosis factor and their regulation by γ-interferon, *Nature* **318**:665.
2. Auron, P. E., Webb, A. C., Rosenwasser, L. J., Mucci, S. F., Rich, A., Wolff, S. M., and Dinarello, C. A., 1984, Nucleotide sequence of human monocyte interleukin 1 precursor cDNA, *Proc. Natl. Acad. Sci. USA* **81**:7907.
3. Banda, M. J., Knighton, D. R., Hunt, T. K., and Werb, Z., 1982, Isolation of a nonmitogenic angiogenesis factor from wound fluid, *Proc. Natl. Acad. Sci. USA* **79**:7773.
4. Beck, G., Habicht, G. S., Benach, J. L., and Miller, F., 1986, Interleukin 1: A common endogenous mediator of inflammation and the local Shwartzman phenomenon, *J. Immunol.* **136**:3025.
5. Beutler, B., and Cerami, A., 1987, Cachectin: More than a tumor necrosis factor, *N. Engl. J. Med.* **316**:379.
6. Bevilacqua, M. P., Pober, J. S., Majeau, G. R., Cotran, R. S., and Gimbrone, M. A., Jr., 1984, Interleukin 1 (IL-1) induces biosynthesis and cell surface expression of procoagulant activity in human vascular endothelial cells, *J. Exp. Med.* **160**:618.
7. Bevilacqua, M. P., Pober, J. S., Wheeler, M. E., Cotran, R. S., and Gimbrone, M. A., Jr., 1985, Interleukin-1 acts on cultured human vascular endothelial cells to increase the adhesion of polymorphonuclear leukocytes, monocytes, and related leukocyte cell lines, *J. Clin. Invest.* **76**:2003.
8. Bevilacqua, M. P., Pober, J. S., Majeau, G. R., Fiers, W., Cotran, R. S., and Gimbrone, M. A., Jr., 1986a, Recombinant tumor necrosis factor induces procoagulant activity in cultured human vascular endothelium: Characterization and comparison with the actions of interleukin 1, *Proc. Natl. Acad. Sci. USA* **83**:4533.
9. Bevilacqua, M. P., Schleef, R. R., Gimbrone, M. A., Jr., and Loskutoff, D. J., 1986b, Regulation of fibrinolytic system of cultured human vascular endothelium by interleukin 1, *J. Clin. Invest.* **78**:587.
10. Bevilacqua, M. P., Pober, J. S., Mendrick, D. L., Cotran, R. S., and Gimbrone, M. A., Jr., 1987, Identification of an inducible endothelial–leukocyte adhesion molecule, *Proc. Natl. Acad. Sci USA* **84**:9238.
11. Cavender, D. E., Haskard, D. O., Joseph, B., and Ziff, M., 1986, Interleukin 1 increases the binding of human B and T lymphocytes to endothelial cell monolayers, *J. Immunol.* **136**:203.
12. Collins, T., Korman, A. J., Wake, C. T., Boss, J. M., Kappes, D. J., Fiers, W., Ault, K. A., Gimbrone, M. A., Jr., Strominger, J. L., and Pober, J. S., 1984, Immune interferon cultivates multiple class II major histocompatibility complex genes and the associated invariant chain gene in human endothelial cells and dermal fibroblasts, *Proc. Natl. Acad. Sci. USA* **81**:4917.
13. Collins, T., Lapierre, L. A., Fiers, W., Strominger, J. L., and Pober, J. S., 1986, Recombinant human tumor necrosis factor increases on RNA levels and surface expression of HLA-A,B antigens in vascular endothelial cells and dermal fibroblasts *in vitro*, *Proc. Natl. Acad. Sci. USA* **83**:446.

14. Cotran, R. S., Gimbrone, M. A., Jr., Bevilacqua, M. P., Mendrick, D. L., and Pober, J. S., 1986, Induction and detection of a human endothelial activation antigen *in vivo*, *J. Exp. Med.* **164**:661.

15. Cotran, R. S., Stavrakis, G., Pober, J. S., Gimbrone, M. A., Jr., Mendrick, D. L., Mihm, M. C., Jr., and Pinkus, G. S., 1987, Endothelial activation: Identification in lesions of lymph nodes and skin, *Lab. Invest.* **56**:16A.

16. Dower, S. K., Kronheim, S. R., Hopp, T. P., Cantrell, M., Deeley, M., Gillis, S., Henney, C. S., and Urdal, D. L., 1986, The cell surface receptor for interleukin-1α and interleukin-1β are identical, *Nature* **324**:266.

17. Dumonde, D. C., Pulley, M. S., Paradinas, F. J., Soutcott, B. M., O'Connell, D., Robinson, M. R. G., DenHollander, F., and Scuurs, A. H., 1982, Histological features of skin reactions to human lymphoid cell line lympholine in patients with advanced cancer, *J. Pathol.* **138**:289.

18. Dustin, M. L., Rothlein, R., Bhan, A. K., Dinarello, C. A., and Springer, T. A., 1986, Induction by IL-1 and interferon-γ, tissue distribution, biochemistry and function of a natural adherence molecule (ICAM-1), *J. Immunol.* **137**:245.

19. Dvorak, H. F., Galli, S. J., and Dvorak, A. M., 1986, Cellular and vascular manifestations of cell-mediated immunity, *Hum. Pathol.* **17**:124.

20. Emeis, J. J., and Kooistra, T., 1986, Interleukin 1 and lipopolysaccharide induce an inhibitor of tissue-type plasminogen activator *in vivo* and in cultured endothelial cells, *J. Exp. Med.* **163**:1260.

20a. Folkman, J., and Klagsbara, M., 1987, Angiogenic factors, *Science* **235**:442.

21. Gamble, J. R., Harlan, J. M., Klebanoff, S. J., and Vadas, M. A., 1985, Stimulation of the adherence of neutrophils to umbilical vein endothelium by human recombinant tumor necrosis factor, *Proc. Natl. Acad. Sci. USA* **82**:8667.

22. Gell, P. G. H., 1958, Cytologic events in hypersensitivity reactions, in: *Cellular and Vascular Aspects of Hypersensitivity States* (H. S. Lawrence, ed.), Harper & Row, New York.

23. Geppert, T. D., and Lipsky, P. E., 1985, Antigen presentation by interferon-γ-treated endothelial cells and fibroblasts: Differential ability to function as antigen-presenting cells despite comparable Ia expression, *J. Immunol.* **135**:3750.

24. Gray, P. W., Aggarwal, B. B., Benton, C. V., Bringman, T. S., Herzel, W. J., Jarrett, J. A., Leung, D. W., Moffat, B., Ng, P., Dvedersky, L. P., Palladino, M. A., and Nedwin, G. E., 1984, Cloning and expression of cDNA for human lymphotoxin, a lymphokine with tumor necrosis activity, *Nature* **312**:721.

25. Groenewegen, G., Buurman, W. A., and Van der Linden, C. J., 1985, Lymphokine dependence *in vivo* expression of MHC class II antigens by endothelium, *Nature* **316**:361.

25a. Kurt-Jones, E. A., Fiers, W., and Pober, J. S., 1987, Membrane interleukin 1 induction on human endothelial cells and dermal fibroblasts. *J. Immunol.* **139**:2317.

26. Leung, D. Y. M., Collins, T., Lapierre, L. A., Geha, R. S., and Pober, J. S., 1986a, Immunoglobulin M antibodies present in the acute phase of Kawasaki syndrome lyse cultured vascular endothelial cells stimulated by gamma interferon, *J. Clin. Invest.* **77**:1428.

27. Leung, D. Y. M., Geha, R. S., Newburger, J. W., Burns, J. C., Fiers, W., Lapierre, L. A., and Pober, J. S., 1986b, Two monokines, interleukin 1 and tumor necrosis factor, render cultured vascular endothelial cells susceptible to lysis by antibodies circulating during Kawasaki syndrome, *J. Exp. Med.* **164**:1958.

28. Libby, P., Ordovas, J. M., Auger, K. R., Robbins, A. H., Birinyi, L. K., and Dinarello, C. A., 1986, Endotoxin and tumor necrosis factor induce interleukin-1 gene expression in adult human vascular endothelial cells, *Am. J. Pathol.* **124**:179.

29. March, C. J., Mosley, B., Larsen, A., Cerretti, D. P., Braedt, G., Price, V., Gillis, S., Henney, C. S., Kronheim, S. R., Grabstein, K., Conlon, P. J., Hopp, T. P., and Cosman, D., 1985, Cloning, sequence and expression of two distinct human interleukin-1 complementary DNAs, *Nature* **315**:641.

30. Martin, B. M., Gimbrone, M. A., Jr., Unanue, E. R., and Cotran, R. S., 1981, Stimulation of nonlymphoid mesenchymal cell proliferation by a macrophage-derived growth factor, *J. Immunol.* **126**:1510.

31. Matsushima, K., Akahoshi, T., Yamada, M., Furutani, Y., and Oppenheim, J. J., 1986, Properties of a specific interleukin 1 (IL 1) receptor on human Epstein–Barr virus transformed B lymphocytes: Identity of the receptor for IL 1-α and IL 1-β, *J. Immunol.* **136**:4496.

32. Miossec, P., Cavender, D., and Ziff, M., 1986, Production of interleukin 1 by human endothelial cells, *J. Immunol.* **136:**2486.
33. Nachman, R. L., Hajjar, K. A., Silverstein, R. L., and Dinarello, C. A., 1986, Interleukin 1 induces endothelial cell synthesis of plasminogen activator inhibitor, *J. Exp. Med.* **163:**1595.
34. Nawroth, P. P., and Stern, D. M., 1986, Modulation of endothelial cell hemostatic properties by tumor necrosis factor, *J. Exp. Med.* **163:**740.
35. Nawroth, P. P., Handley, D. A., Esmon, C. T., and Stern, D. M., 1986a, Interleukin 1 induces endothelial cell-surface anticoagulant activity, *Proc. Natl. Acad. Sci. USA* **83:**3460.
36. Nawroth, P. P., Bank, I., Handley, D., Cassimeris, J., Chess, L., and Stern, D., 1986b, Tumor necrosis factor/cachectin interacts with endothelial cell receptors to induce release of interleukin 1, *J. Exp. Med.* **163:**1363.
37. Nawroth, P. P., Handley, C. T., Esmon, C., and Stern, D., 1986c, Interleukin 1 induces endothelial cell procoagulant activity while suppressing cell surface anticoagulant activity, *Proc. Natl. Acad. Sci. USA* **83:**3460.
38. Old, L. J., 1985, Tumor necrosis factor (TNF), *Science* **230:**630.
39. Oppenheim, J. J., Kovacs, E. J., Matsushima, K., and Durum, S. K., 1986, There is more than one interleukin 1, *Immunol. Today* **7:**45.
40. Pennica, D., Nedwin, G. E., Hayflick, J. S., Seeburg, P. H., Derynk, R., Palladino, M. A., Kohr, W. J., Aggarwal, B. B., and Goeddel, D. V., 1984, Human tumor necrosis factor: Precursor structure, expression, and homology to lymphotoxin, *Nature* **312:**724.
41. Pober, J. S., Gimbrone, M. A., Jr., Cotran, R. S., Reiss, C. S., Burakoff, S. J., Fiers, W., and Ault, K. A., 1983a, Ia expression by vascular endothelium is inducible by activated T cells and by human γ interferon, *J. Exp. Med.* **157:**1339.
42. Pober, J. S., Collins, T., Gimbrone, M. A., Jr., Cotran, R. S., Gitlin, J. D., Fiers, W., Clayberger, C., Krensky, A. M., Burakoff, S. J., and Reiss, C. S., 1983b, Lymphocytes recognize human vascular endothelial and dermal fibroblast Ia antigens induced by recombinant immune interferon, *Nature* **305:**726.
43. Pober, J. S., Collins, T., Gimbrone, M. A., Jr., Libby, P., and Reiss, C. S., 1986a, Overview: Inducible expression of class II major histocompatibility complex antigens and the immunogenicity of vascular endothelium, *Transplantation* **41:**141.
44. Pober, J. S., Gimbrone, M. A., Jr., Lapierre, L. A., Mendrick, D. L., Fiers, W., Rothlein, R., and Springer, T. A., 1986b, Overlapping patterns of activation of human endothelial cells by interleukin 1, tumor necrosis factor and immune interferon, *J. Immunol.* **137:**1893.
45. Pober, J. S., Bevilacqua, M. P., Mendrick, D. L., Lapierre, L. A., Fiers, W., and Gimbrone, M. A., Jr., 1986c, Two distinct monokines, interleukin 1 and tumor necrosis factor each independently induce biosynthesis and transient expression of the same antigen on the surface of cultured human vascular endothelial cells, *J. Immunol.* **136:**1680.
46. Pober, J. S., Lapierre, L. A., Stolpen, A. H., Brock, T. A., Springer, T. A., Fiers, W., Bevilacqua, M. P., Mendrick, D. S., and Gimbrone, M. A., Jr., 1987, Activation of cultured human endothelial cells by recombinant lymphotoxin: Comparison with tumor necrosis factor and interleukin 1 species, *J. Immunol.* **138:**3319.
47. Polverini, P. J., Cotran, R. S., and Sholley, M. M., 1977a, Endothelial proliferation in the delayed hypersensitivity reaction: An autoradiographic study, *J. Immunol.* **118:**529.
48. Polverini, P. J., Cotran, R. S., Gimbrone, M. A., Jr., and Unanue, E. R., 1977b, Activated macrophages induce vascular proliferation, *Nature* **269:**804.
49. Ruddle, N., 1985, Lymphotoxin redux, *Immunol. Today* **6:**156.
50. Sobel, R. A., Blanchette, B. W., Bhan, A. K., and Colvin, R. B., 1984, The immunopathology of experimental allergic encephalomyelitis. II. Endothelial cell Ia increases prior to inflammatory cell infiltration, *J. Immunol.* **132:**2402.
51. Stern, D. M., Bank, I., Nawroth, P. P., Cassimeris, J., Kisiel, W., Fenton, J. W., Dinarello, C. A., Chess, L., and Jaffe, E. A., 1985, Self regulation of procoagulant events on the endothelial cell surface, *J. Exp. Med.* **162:**1223.
52. Stolpen, A. H., Guinan, E. C., Fiers, W., and Pober, J. S., 1986, Recombinant tumor necrosis factor and immune interferon act singly and in combination to reorganize human endothelial cell monolayers, *Am. J. Pathol.* **123:**16.

53. Tracey, K. J., Beutler, B., Lowry, S. F., Merryweather, J., Wolpe, S., Milsark, I. W., Hariri, R. J., Fahey, T. J., III, Zentella, A., Albert, J. D., Shires, G. T., and Cerami, A., 1986, Shock and tissue injury induced by recombinant cachectin, *Science* **234:**470.

54. Trinchieri, G., and Perussia, B., 1985, Immune interferon: A pleiotropic lymphokine with multiple effects, *Immunol. Today* **6:**131.

55. Wagner, C. R., Vetto, R. M., and Burger, D. R., 1985, Expression of I-region associated antigen (Ia) and interleukin 1 by subcultural human endothelial cells, *Cell. Immunol.* **93:**91.

56. Watt, S. L., and Auerbach, R., 1986, A mitogenic factor for endothelial cells obtained from mouse secondary mixed leukocyte cultures, *J. Immunol.* **136:**197.

57. Wiener, J., Lattes, R. G., and Spiro, D., 1967, An electron microscopic study of leukocyte emigration and vascular permeability in tubercalia sensitivity, *Am. J. Pathol.* **50:**485.

58. Willms-Kretschmer, K., Flax, M. H., and Cotran, R. S., 1967, The fine structure of the vascular response in hapten-specific delayed hypersensitivity and contact dermatitis, *Lab. Invest.* **17:**334.

16

Intimal Responses to Shear Stress, Hypercholesterolemia, and Hypertension
Studies in the Rat Aorta

G. Majno, T. Zand, J. J. Nunnari, M. C. Kowala, and I. Joris

From the perspective of pathology, the three most common challenges to the intima of larger arteries are local disturbances of flow, hypercholesterolemia, and hypertension. We have studied all three in one experimental system, the aorta of the rat. This chapter is a critical summary of these studies. As an introduction, we will examine some relevant features of our experimental model.

I. THE RAT AORTA AS AN EXPERIMENTAL SYSTEM

The distinctive advantage of this vessel, as an experimental preparation, is its thinness (about 200 μm in an adult rat); it can be transilluminated and examined *en face* after suitable surface staining of the intima. This means that cellular events (both above and beneath the endothelium), lipid accumulation, and endothelial loss can be studied simultaneously and evaluated topographically, which cannot be accomplished by scanning electron microscopy. Platelets, in clusters or even singly, can be seen after hematoxylin staining but not quantitated. Flow studies are now also possible on vessels of this size thanks to a miniaturized flow meter.*

The rat aorta also has some special features which must be taken into account.

1. There are *significant differences between the thoracic and the abdominal intima*. Under normal conditions, adherent mononuclear cells, intimal cells, and myoendothelial hernias are more abundant in the thoracic segment (see further). Correspondingly, in hypercholesterolemia the lesions are more severe in the thoracic segment[27,46] whereas in the human aorta the trend is opposite.

2. *Mononuclear cells* (monocytes and a few lymphocytes) *normally* adhere to the intimal surface here and there and migrate into it, becoming *intimal cells*.[23]

---

* Valpey–Fisher flow meter is available from Crystal Biotech, Inc., Holliston, MA.

G. Majno, T. Zand, J. J. Nunnari, M. C. Kowala, and I. Joris • Department of Pathology, University of Massachusetts Medical School, Worcester, Massachusetts 01655.

This mononuclear cell "invasion" occurs in many other species and is accelerated in hypercholesterolemia.

3. The intima and especially the media contain *extracellular debris* (largely vesicles and clusters of vesicles) that have been described also in other arteries and other species. We assume that most of them represent the remains of dead cells that have not been eliminated because arteries do not have an intramural phagocytic system for "internal housekeeping."[20] Since our study of this topic, Haust has proposed that some of the clusters may represent breakdown products of elastin.[16]

4. Last, the intima is peppered with peculiar structures, known as *myoendothelial hernias*[42] which represent club-shaped pseudopodia arising from smooth muscle cells of the media[17] passing through fenestrations of the internal elastic lamina, herniating into the endothelium and sometimes even pushing their way through it and into the lumen. Most of the so-called *stomata* and *stigmata*, as

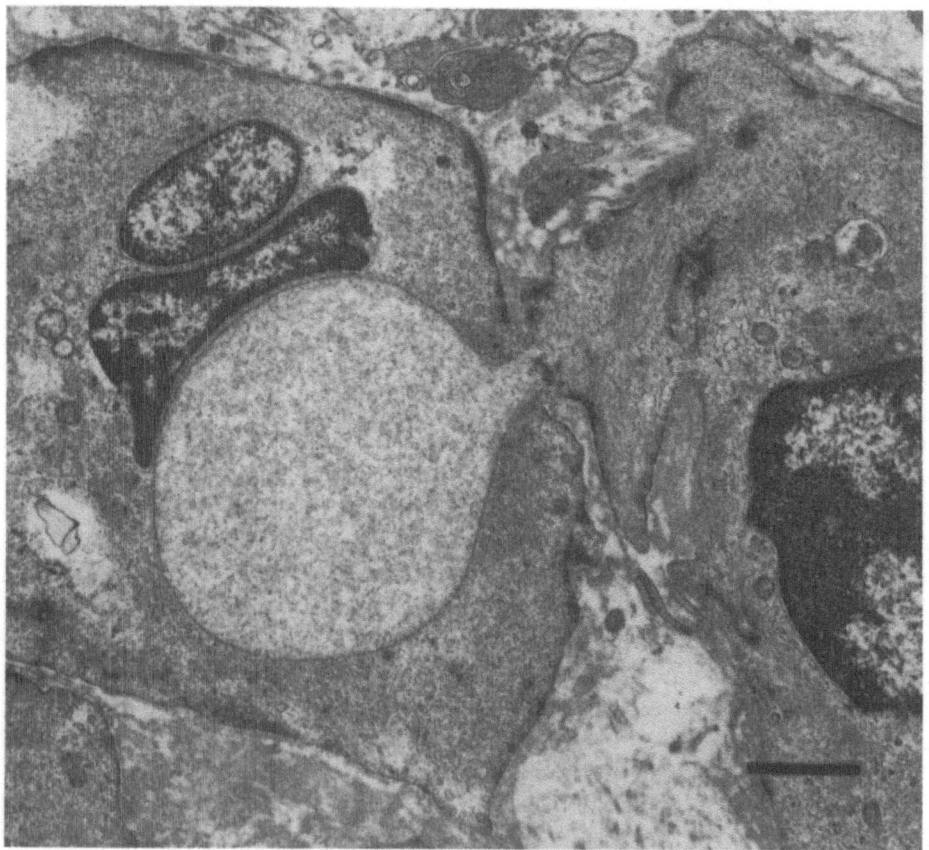

Figure 1. Media of human coronary artery obtained from a heart transplant operation, and kindly supplied by Dr. Margaret E. Billingham. Herniation of one smooth muscle cell into another. The neck of the hernia is clearly visible. Bar = 1 μm.

seen *en face*, are caused by myoendothelial hernias.[36] Whether myoendothelial hernias occur also in human arteries is not known. However, herniations between medial smooth muscle cells exist in human coronaries (Fig. 1). These medial hernias are interesting in that they have been shown to arise through intense muscular contraction.[21,25]

II. INTIMAL CHANGES RELATED TO FLOW DISTURBANCES: EFFECTS OF ACUTE AND CHRONIC STENOSIS

The simplest method for producing a flow disturbance within an artery is to restrict its lumen. The biologic effects of this procedure are especially interesting. In sharp contrast to all the other and more traditional methods of intimal injury (such as pinching, freezing, heating, suturing, ballooning, and the many forms of intimal scratching, just to mention a few), stenosis causes intimal changes by modifying a physiologic force: the shear stress produced by blood flow on the endothelial surface.

Stenosis can be produced by constricting the lumen from within, with a plug[11] or extrinsically, with an external device. For our studies we chose the latter method[26]: the lumen was focally flattened by slipping a U-shaped metal clip astride the abdominal aorta between the two iliolumbar arteries (Fig. 2). The clip is prepared from a $1 \times 7 \times 0.20$-mm flat strip of platinum or titanium, neither of which reacts with the tissues in long-term studies.

This method produces a stenosis with a well-defined throat, limited anteriorly and posteriorly by parallel areas of flattened intima, well suited for morphologic study (Fig. 3). This geometry is also well suited for study by means of hydrodynamic and computer models. By contrast, aortic stenosis obtained with a

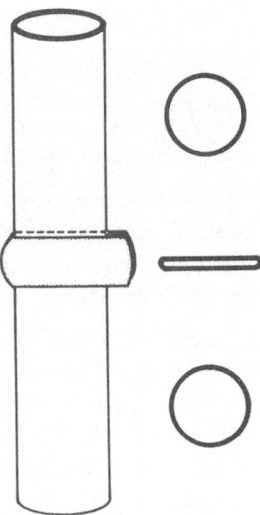

Figure 2. Method for producing stenosis in the rat aorta with a U-shaped flat metal clip. Shapes at right indicate the deformation of the lumen.

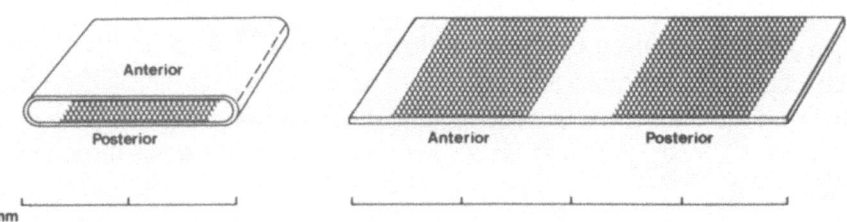

Figure 3. Scheme of the aortic intima in the area of stenosis, as obtained by the method shown in Fig. 2. *Left:* aorta closed; *right:* aorta opened along one of its margins. Stippling indicates the areas of high shear stress (anterior and posterior) which are sampled for morphologic study.

ligature[3,31,33,34] produces a very short throat with a complex accordion-type geometry: the intima is thrown into longitudinal folds which press against each other and can cause endothelial damage independent of shear stress.[24] It is difficult to see how such a stenosis can be reliably sampled for morphologic study.

A. Acute, Severe Stenosis

The 70–80% stenosis produced by the clip method induces (in the throat of the stenosis) a sharply increased shear stress.[18,40,50] The acute effect of this procedure is that within 1 hr most of the endothelial cells within the throat of the stenosis are stripped off.[26] At the margins of the stenotic zone, some cells remain in place but develop microscopic "ulcers" on their surface, presumably because small pieces of the plasma membrane have been torn away. These lesions are apparently not thrombogenic, and the exposed cytoplasm is slightly more electron dense, as if focally coagulated; the rest of the cell appears normal. The subsequent fate of the cells carrying such microulcers is not known. A lesser form of injury of the endothelial cells is suggested by complex myelin figures arising from the luminal surface. We have observed similar figures also in hypercholesterolemic rats.[27]

B. Chronic, Severe Stenosis

If the 70–80% stenosis is allowed to persist, a rather surprising phenomenon occurs: despite the greatly increased shear stress, regenerating endothelium creeps back over the denuded area; after 3 weeks, silver staining shows that the stenotic segment is completely reendothelialized. The endothelial pattern—in the segment of the aorta including the clip site—now shows *five distinct zones*[50]: the throat of the stenosis is covered by more densely packed, elongated endothelial cells; just proximal and distal to the throat there is a transverse band of polygonal cells, larger and wider than normal; last, after the polygonal cells, at the transition toward normal endothelium, there is (both proximally and distally) a band of extremely long, narrow cells. Comparison between these patterns and the results obtained in hydrodynamic and computer models[18,40,50] offered some predictable correlations and some surprises. Predictable was the correlation between the highest shear stress (in the throat of the stenosis) and the morphologic pattern of cells elongated in the direction of flow. Predictable also was the existence of polygonal cells just below the stenosis, where flow is turbulent and reversed; cells of this shape and in the same location were found by Levesque *et al.* distal to an aortic ligature in the dog.[34] Surprising were (1) the presence of polygonal cells *proximal* to the stenosis, along the slope where shear stress is *increasing*; and (2) the presence of a patch of elongated cells above and below the stenosis, in areas where shear stress (according to the models) is not increased. We must conclude that shear stress is not the only determinant of endothelial cell shape. Perhaps stresses within the arterial wall are also involved.

Of the various intimal changes induced by chronic stenosis, the most significant is the appearance of new endothelium in the throat of the stenosis, where all the cells had been wiped away during the acute stage. We conclude that this new endothelium is in some way "adapted." In a recent study we have shown that it differs from normal endothelium in a number of ways: its cells are longer and narrower than normal, and 60% more numerous; they are thicker, and their basal surface carries more prominent attachment plates (*microtendons*)[50] (Fig. 4). Furthermore, the endothelial cytoskeleton has undergone significant changes, best visualized by study *en face* of the aorta treated with rhodamine-conjugated phalloidin, which binds to F-actin.[2,48] Figure 5 shows the aspect of normal endothelium compared with endothelium in the throat of a 3-week stenosis. In the *control areas* the cellular outlines are strongly fluorescent, due to a cortical band of F-actin; they mimic the pattern of the silver lines (indeed, the cellular outline often appears double, as it does in many silver line preparations). The stress fiber system

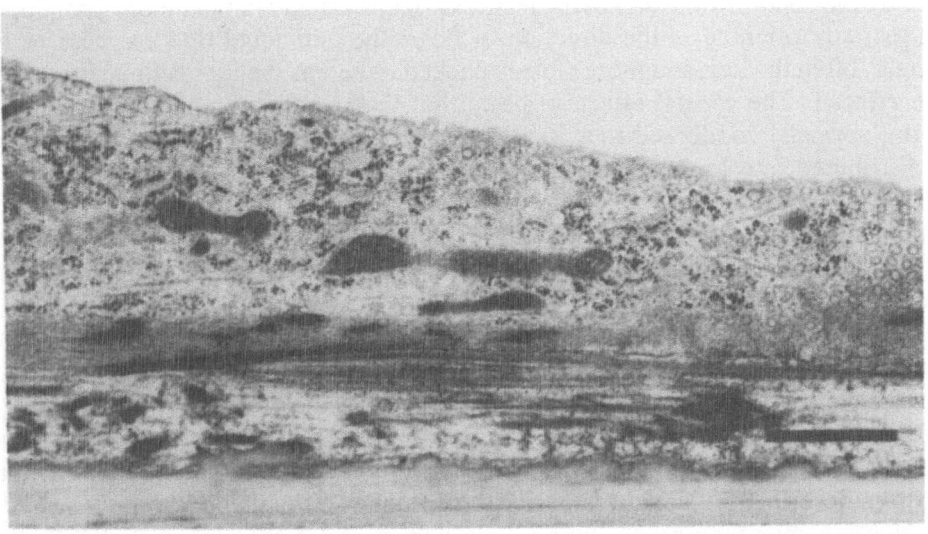

Figure 4. Intima of rat aorta in the zone of 70–80% chronic stenosis (the clip was applied 3 weeks earlier). Longitudinal section. Note tendon-like structure (*microtendon*) emerging from the abluminal surface of the endothelial cell, and continuous with an intracellular stress fiber. Bar = 1 μm.

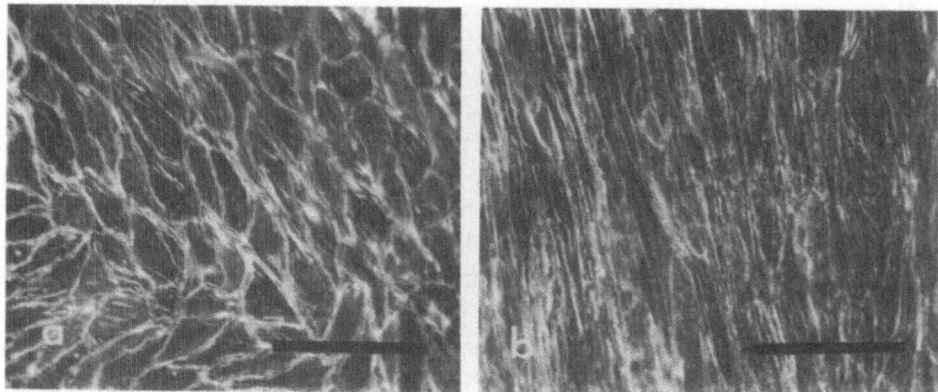

Figure 5. Intima of rat aorta examined *en face* by UV light, after staining with rhodamine-conjugated phalloidin for the demonstration of F-actin. Bar = 50 μm.

(a) *Control zone:* the contours of the endothelial cells are fluorescent, due to a cortical band of F-actin. The irregular cellular pattern is common in the normal aorta. In the cytoplasm, note short segments of stress fibers; not all are oriented in the direction of flow.

(b) *Zone of high shear stress (stenotic zone)* 3 weeks after the application of a clip inducing 70–80% stenosis. The outlines of the cells are no longer clearly visible; most of the cellular F-actin corresponds to bands of stress fibers, more numerous and longer than in normal cells.

is surprisingly variable; some cells have no visible stress fibers at all, others have short or long segments but not always arranged in a direction parallel to the longitudinal axis of the aorta. This variability may reflect local differences in the pattern of flow at the cellular level.

In the throat of the stenosis the stress fiber system is much more prominent and strictly oriented in the direction of flow; the individual fibers appear to be longer, often thicker, and more closely packed, whereas the intercellular junctions are effaced. The overall effect suggests that in the adapted endothelium the F-actin, normally condensed to a great extent at the periphery of the cell, has been redistributed into the stress fiber system. These changes *in vivo* are consistent with data obtained *in vitro*: shear stress applied to cultured endothelium induces an increase in the stress fiber system[6,33,38,45]; a redistribution of F-actin from the cortical zone to a stress fiber system has been observed after single-cell wounds in cultures.[47]

These findings on chronic stenosis in the rat should be relevant to understanding changes in human stenotic coronary arteries. In our experience with human coronaries, even when stenosis is severe, the endothelium is still present.[28] We must assume that here too it has become adapted to the increased shear stress. Its ultrastructure must be interpreted with caution, because human endothelium is often described as being rich in fibrils also under normal conditions. However, in electron micrographs of stenotic coronary arteries the endothelial cell may appear entirely filled with cytoskeletal components (usually loose tangles of filaments of the intermediate 100-Å type).

In summary, our data on the stenotic rat aorta and on human coronary arteries indicate that the endothelium responds to *chronic* stenosis with a number of cellular adaptations.

C. Mild, Acute Stenosis

It might be expected that mild stenosis would induce in a milder form the changes induced by severe stenosis. Such is not the case.

In current work on mild stenosis we are using titanium clips inducing a 35–40% aortic stenosis 3 mm long.[49] When aortas bearing such clips for 24 hr are viewed *en face*, the throat of the stenosis shows little or no denudation; instead, clusters of neutrophils and mononuclear cells adhere to the endothelial surface. Electron microscopy shows them undergoing diapedesis and accumulating in the intima. This behavior of neutrophils is very unusual in arterial pathology. Some neutrophils participate in experimental hypertension[43,44] but not in the massive numbers observed with the loose clip. The aortic endothelium, in particular, attracts preferentially mononuclear cells under a variety of conditions (see further).

There are at least three ways to interpret this unpredicted effect of mild stenosis. (1) The slightly increased shear stress has somehow *stimulated the endothelial cells* (possibly altering their secretory activity) in such a way that neutrophils are more attracted (or less repelled). (2) The mildly increased shear stress has induced a form of *nondenuding endothelial damage* resulting in neutrophil sticking (it is realized, however, that it is difficult to distinguish between "nondenuding damage" and "stimulation"). (3) The clip has somehow *induced the media to produce chemotactic substances*. At present we favor the first interpretation. It stands to reason that a mild increase in shear stress should be able to stimulate the endothelium instead of injuring it; for example, it has been reported that increased shear stress on cultured endothelium stimulates the secretion of PGI_2.[14]

III. INTIMAL CHANGES IN HYPERCHOLESTEROLEMIA: THE "INFLAMMATORY THEORY" OF ATHEROSCLEROSIS

The use of the rat for the study of atherosclerosis was proposed long ago[46] but met with little success; the common objection has been that rats do not spontaneously develop atherosclerosis. The data that have accumulated during the past few years clearly show that hypercholesterolemia induces the same basic changes in all animals tested, including the pigeon.[35] In fact, because the intimal changes in hypercholesterolemia are typically focal and multiple, the aorta of the rat is especially useful since it permits topographic studies *en face*. The ultimate argument in favor of the hypercholesterolemic rat—as a model of atherosclerosis—is in the lesions themselves: the tiny aortic plaques of the hypercholesterolemic rat are surprisingly similar to those of man. To allow the reader the benefit of truly blind choice, we are submitting two electron micrographs (one from a

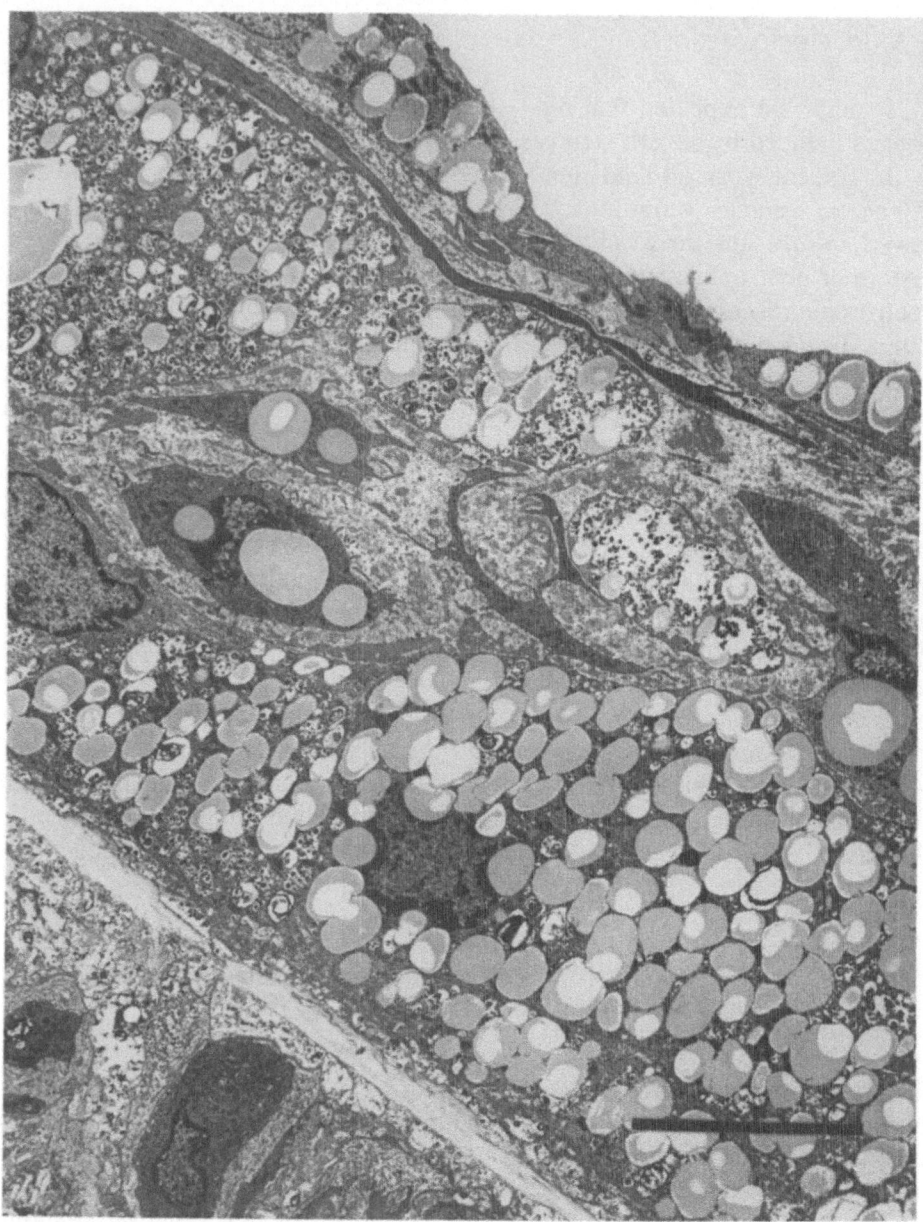

Figures 6 and 7. Two florid atherosclerotic lesions, one from a human coronary, the other from a rat aorta after 12 months on the CCT diet. In Fig. 6, smooth muscle cells contain the larger droplets. These unlabeled illustrations are presented as evidence of the great similarity between human and rat lesions. The key is given at the end of this chapter. Figure 6 bar = 10 μm; Fig. 7 bar = 5 μm.

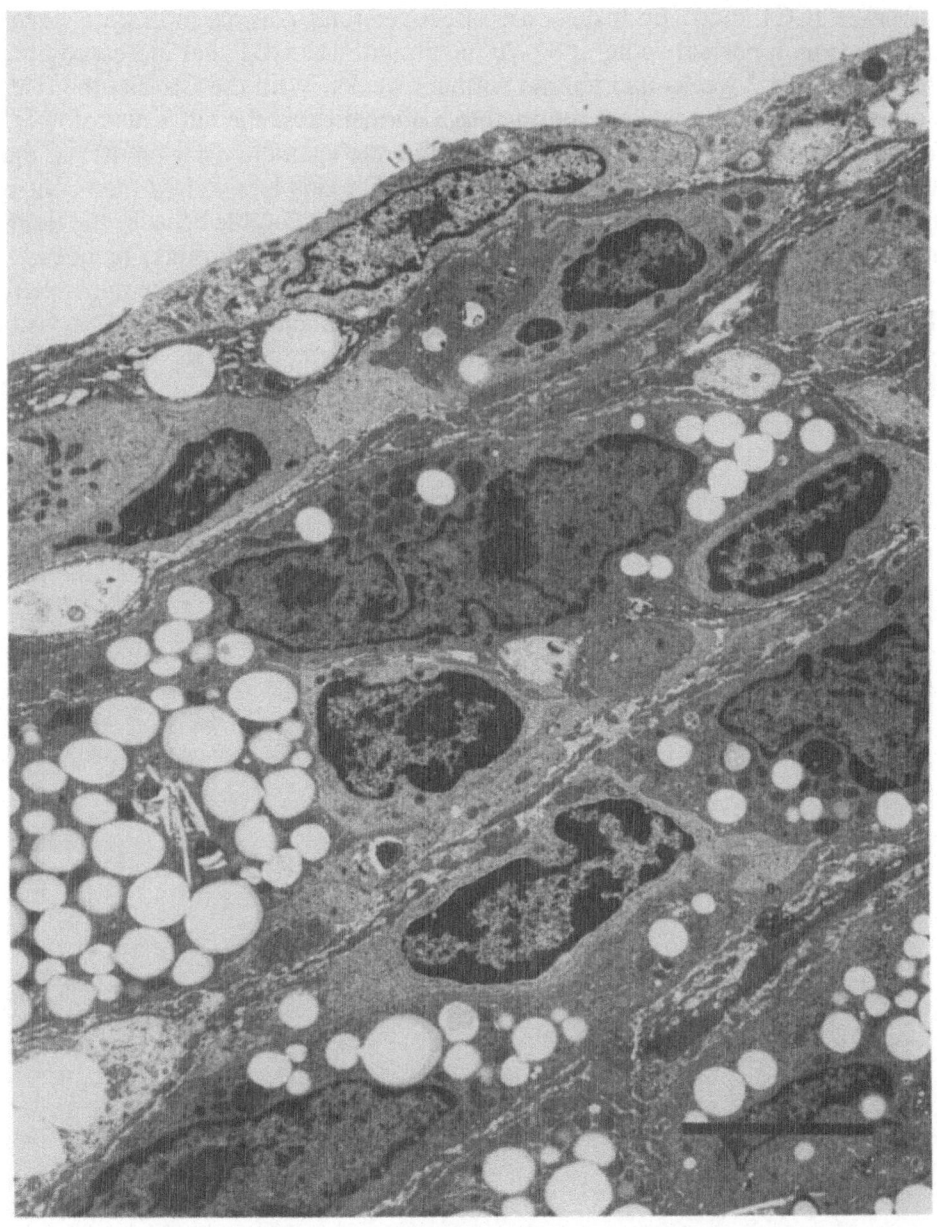

Figures 6 and 7. (*continued*)

human coronary artery, one from a hypercholesterolemic rat) without indicating their source (Figs. 6, 7). The key will be found at the end of this chapter.

To make our male Wistar rats hypercholesterolemic we used two diets[27]: rat chow with 4% cholesterol and 1% cholic acid (CC diet) or the same with 0.5% 2-thiouracil (CCT diet). Both diets are effective: total plasma cholesterol levels, starting from a normal value of 50–70 mg/dl, with the CCT diet increased about fourfold within 2 weeks and tenfold within 3 weeks. With the CC diet the rise is slower, and when the rats are returned to a normal chow the fall is also slower[29]; furthermore, the lesions tend to remain (within the span of 1 year on the CC diet) at the stage of fatty streaks. Rats on the CCT diet gain less weight than controls and CC rats; however, in our experience, this diet is preferable because the lesions appear faster and become histologically similar to human fibrofatty plaques.

Grossly visible lesions do not appear until 3 months. It should be remembered that rat lesions are miniaturized; fatty streaks and plaques that would be visible by the naked eye in a human aorta should be sought in the rat by means of a dissecting microscope. The main observations have already been reported.[27] In

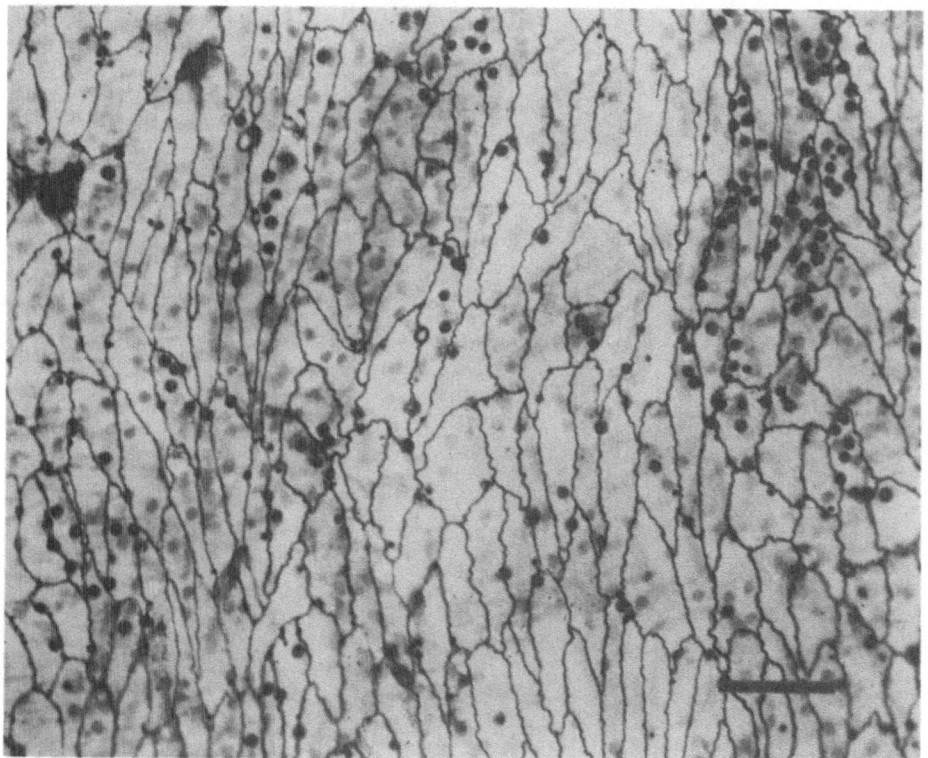

Figure 8. Intima of a rat aorta seen *en face*, after staining *in situ* with silver nitrate and hematoxylin. The rat had been maintained for 1 year on the CCT diet. Note clusters of adherent cells, visible as dark nuclei. The light-stained nuclei belong to cells that have migrated beneath the endothelium. There is no endothelial denudation. Bar = 50 μm.

essence, the first microscopic change seen in aortic preparations studied *en face* was the sticking of mononuclear cells to the endothelial surface, followed by diapedesis into the intima (Figs. 8, 9). About 10% of these cells were lymphocytes. Lipid droplets also appeared in the endothelium. Although we examined large areas of intima stained with silver nitrate (which can demonstrate the loss of a single cell), endothelial denudation was never observed. Platelet participation was minimal; most of the platelets observed were attached to monocytes rather than to the endothelial surface.

These data are consistent with a theory of atherogenesis proposed almost half a century ago whereby the first step is the adhesion and penetration of mononuclear cells into the intima.[32] This mechanism was examined in detail (with studies *en face*) by Duff *et al.*[7] in 1957; it was confirmed histologically by Poole and Florey[39] in 1958, then electron microscopically by Gerrity *et al.*[12] and Lewis *et al.*[35] All these studies and others (for literature see Refs. 27, 37) were fundamental in demonstrating the existence of mononuclear cell attachment and penetration; however, being based on cross sections of the intima, they could not exclude that endothelial denudation did not occur in other areas. Our rat model confirmed once again the mononuclear cell invasion into the intima, and also supplied a topographic overview, showing that denudation did not occur in this model. Current information shows that it fails to occur (as an early event) also in other models, including a nonhuman primate.[9,10]

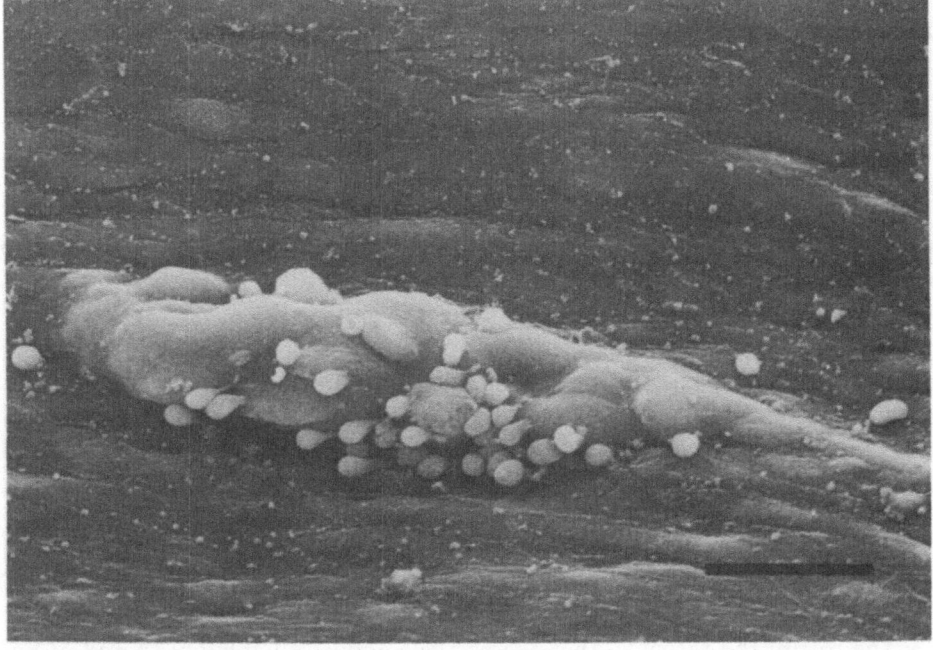

Figure 9. Scanning electron micrograph of the aortic intima from a rat kept for 2 months on the CCT diet. Note adherent cells (tails pointing upstream) over a small lesion. Bar = 50 μm.

Taking into account all these results the best name for a theory of athero-sclerosis would seem to be "the inflammatory theory." This is certainly descrip-tive of the cellular events, and fits with the notion that atherosclerosis—once thought to be a "degenerative" disease—actually displays the principal features of chronic inflammation.[22] The name would need to be qualified in one respect: it describes the earliest cellular feature known to date, but not necessarily the first change at the molecular level. In the rabbit model used by Simionescu et al., transendothelial passage of lipid appeared to be the earliest event.[41]

Assuming that the inflammatory mechanism occurs, it remains to be ex-plained why the mononuclear cells adhere to the endothelium of hypercholester-olemic animals. There are several possibilities[27]: (1) the surface of the endothelial cells has become abnormal; (2) the lipoproteins act as ligands (this theory has received recent support[1]); (3) regenerating cells are sticky for mononuclear cells (for this too there is much evidence; see Ref. 27); (4) the arterial wall produces chemoattractants for monocytes.[13] It is possible that several mechanisms are at play.

Another pressing problem is the *mechanism of lipid accumulation* beneath the endothelium: study *en face* in the rat shows that the two basic events, mono-nuclear cell invasion and subendothelial lipid accumulation, often coincide in time and place, but also that they can occur independently.[37] The passage of lipid presumably occurs by transcytosis; this seems the best way to account for the subendothelial "liposomes" recently described in hypercholesterolemic rabbits.[41]

IV. SHEAR STRESS AND LIPID DEPOSITION: A STUDY BASED ON AORTIC STENOSIS

One of the central dilemmas in the pathogenesis of atherosclerosis is the focal nature of the lesions. Nobody doubts that flow patterns have something to do with their localization, and that shear stress on the endothelial surface must also be involved, but precisely how does shear stress operate? Two opposing theories have evolved: lipid deposition occurs preferentially in areas of high shear stress[11,19] or conversely of low shear stress.[4,5]

To solve this problem we combined aortic stenosis and hypercholesterolemia in the rat; since we knew, from physical studies on hydrodynamic models and computer simulations,[18,40,50] the pattern of high and low shear stress in relation to the stenotic segment in our animal model, the experiment was straightforward.

We used a 70–80% stenosis as described earlier, made the rats hypercho-lesterolemic with both CC and CCT diets, and followed them for periods of 2 weeks to 8 months. Approximately each month a rat was sacrificed, the aorta was fixed by perfusion, stained *in situ* with AgNO₃, hematoxylin, and oil red O, and examined microscopically (see Figs. 10, 11).

The result: the throat of the stenosis was almost completely spared of lipid deposition; only tiny, sparse droplets of lipid could be found. This is the area of the highest shear stress, values being approximately 20 times higher than normal.[50]

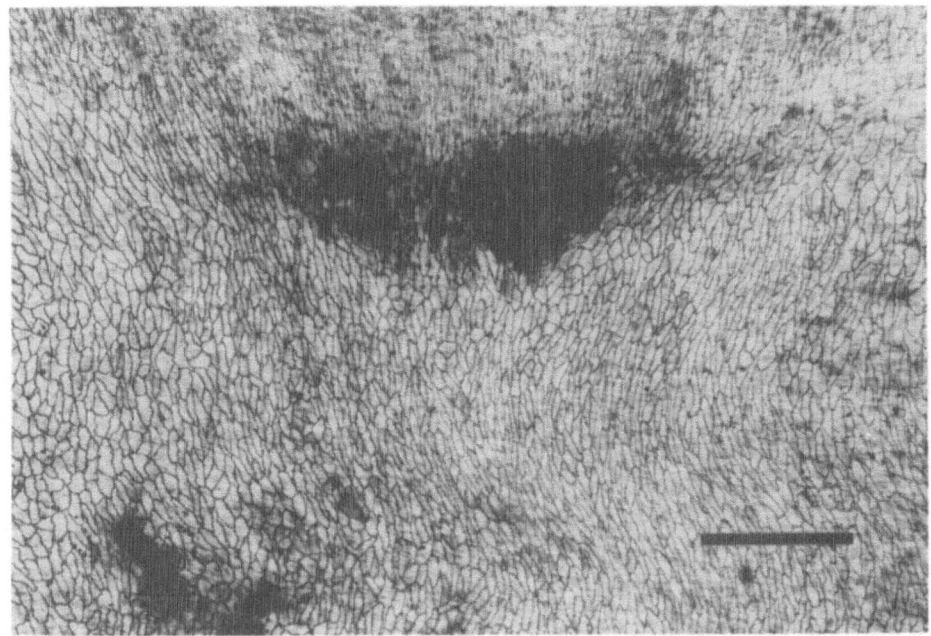

Figure 10. Aortic intima (from a rat with a 70–80% stenosis for 3 months and fed CCT diet) seen *en face* after staining *in situ* with silver nitrate, hematoxylin, and oil red O. The upper quarter of the photograph corresponds to the distal part of a stenotic zone. Note typical shape of the endothelial cells, long and narrow, indicating high shear stress. The lower three-quarters corresponds to the poststenotic zone. Note lipid deposition (black mass) in this zone; little or no lipid is present in the zone of high shear stress. Black spots at lower left are "random" lipid deposits seen in all rats on the CCT diet. Bar = 0.2 mm.

By contrast, two fairly symmetrical transverse patches of lipid deposition developed, one just above and the other just below the clip (the latter was slightly larger). When this pattern of lipid deposition was compared with the pattern of shear stress distribution as observed in the hydrodynamic model, some surprising facts appeared: the distal patch corresponded to an area of low shear, where turbulent and reversed flow prevailed; the proximal patch, instead, occurred in an area of increasing shear along the slope leading into the throat of the stenosis.

We are therefore faced with the fact that lipid deposition can occur in areas of decreased shear as well as in areas of increasing shear; by contrast, *very high shear* prevents lipid deposition. Experiments are under way to confirm and extend these data with the use of stenoses of different degrees.

The use of stenosis for localizing lipid deposits in hypercholesterolemic animals has a further application: it also offers a method for producing an atherosclerotic plaque at a given site at will. This means that it will now be possible to follow the growth and the regression of a fatty streak or plaque of known age and structure. To date, all studies of this kind have depended on the selection of randomly appearing lesions: with such an approach one can never be sure that a

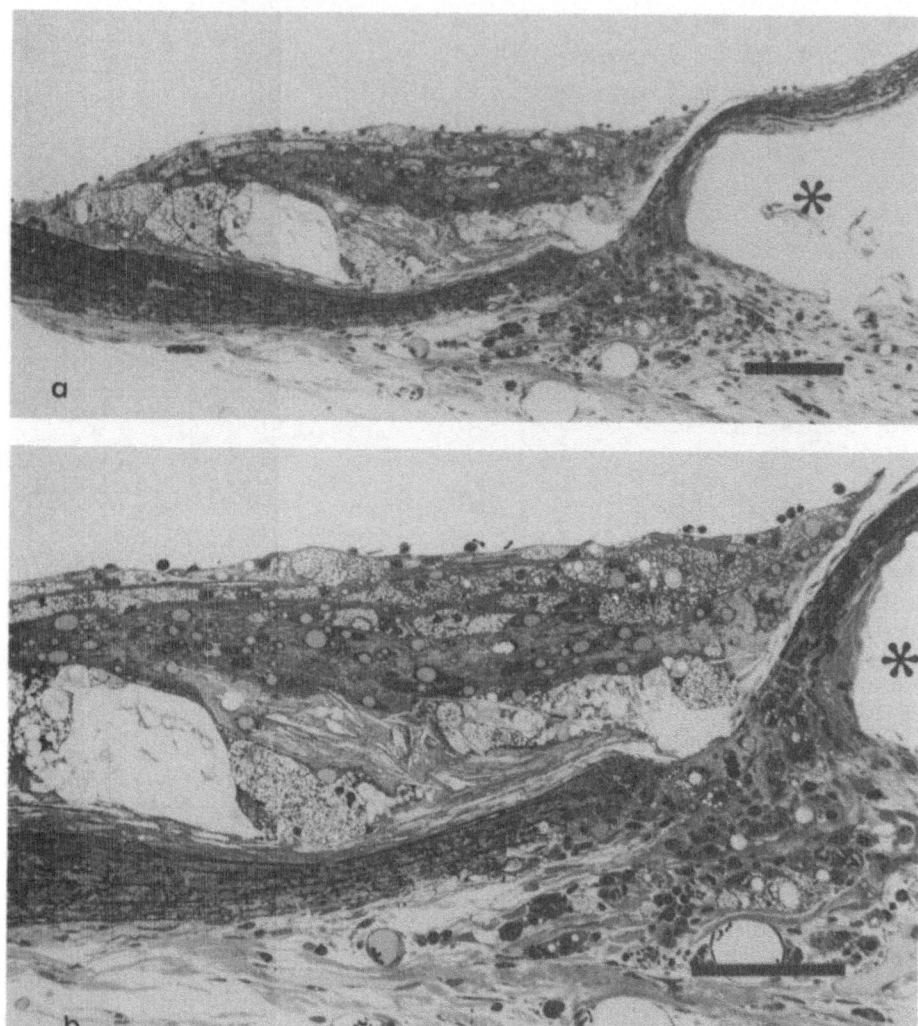

Figure 11. Micrographs demonstrating the localization of atherosclerotic lesions in the poststenotic zone (low shear area); 35% stenosis, 10 months. Longitudinal sections of rat aorta; intima at top; flow from right to left. The asterisks correspond to the clear space left by the (removed) titanium clip. Epon section, 1 μm thick, stained with toluidine blue. (a) *Topographic view*; above the clip, the media (atrophied) and the intima are lifted by the clip; toward the left they form a "step" corresponding to the poststenotic area; here the intima is thickened by a typical plaque. Bar = 0.1 mm. (b) *Detail.* Note adherent cells, typical foam cells, fat-laden smooth cells and deep intimal necrosis with cholesterol crystals. Bar = 0.1 mm.

given lesion, chosen as representative, would have truly progressed or regressed if allowed to follow its own natural history. Studies of lesion progression and regression are planned, using poststenotic fatty streaks and plaques.

V. INTIMAL CHANGES IN HYPERTENSION

Several types of hypertensive rats are available, and it must be anticipated that intimal changes, if any, may not always be the same. As a start, we chose two models: partial ligature of the aorta between the renal arteries; and unilateral nephrectomy coupled with a Goldblatt-type steel clip [with a 0.2-mm gap) on the contralateral renal artery, the so-called one-kidney–one clip method (1K1C)]. We focused on quantitating intimal cell invasion by mononuclear cells, a phenomenon that appears to offer a link between atherosclerosis and hypertension. *En face* preparations were used for the cell counts 2, 4, and 8 weeks after the operation. Mean blood pressure (MBP) was measured in semiconscious rats recovering from light ether anesthesia, through a catheter inserted into the right carotid artery and connected to a Parametron 4 monitor (Roche). Briefly, the following results are pertinent here[30]: (1) MBP in controls was in the range of 90–119 mm Hg; in half of the 1K1C rats it began to rise after 1 and 2 weeks; and by 4 weeks all 1K1C rats were hypertensive (over 120 mm Hg). Rats with aortic ligation were hypertensive at 2, 4, and 8 weeks, except for two animals with infarction of the left kidney. (2) The number of intimal cells (mainly emigrated monocytes) increased up to 15-fold 2 weeks after surgery and remained stationary thereafter. (3) Rather surprisingly, the number of *adherent* mononuclear cells did not rise in parallel with the number of intimal cells. We take this to mean that the adherent cells undergo diapedesis at a faster rate. (4) The thoracic aorta—compared with the abdominal—showed the largest increase of intimal cells. (5) No endothelial denudation occurred. (6) The number of smooth muscle hernias from the media into the intima (*myointimal hernias*) increased 2- to 3-fold, suggesting a smooth muscle dysfunction. (7) The *subintimal space* (as seen by electron microscopy) was widened by two groups of macromolecular materials: some were ostensibly from the blood, such as crystallized hemoglobin and possibly some plasma (explained by an increase in endothelial permeability)[8,15]; others were presumably derived from local cells (clumps of elastin, clumps and sheets of basement membrane-like material, electronlucent masses presumably corresponding to glycosaminoglycans). Where do these materials come from? The likeliest source is the endothelium. We concluded that in the early stages of hypertension (in our models) *the endothelium is somehow activated to synthesize macromolecules*. This also means that the intimal thickening seen in hypertension and other conditions should not be understood solely as a product of smooth muscle cell synthesis (very few smooth muscle cells are present in the intima of normal and hypertensive rats).

Further studies now in progress show that several aortic changes in hypertensive rats (intimal thickening, increase in intimal cells, increase in endothelial mitoses) *precede the onset of measurable hypertension*.

The critical question that must still be pursued is the following: since hypertension is accompanied by intimal changes (cellular and extracellular), how can these changes help in understanding the accelerating effect that hypertension has on the lesions of atherosclerosis? If mononuclear cell invasion were the mechanism, one would expect it to be progressive, but it is not (at least in the two models tested). Stimulation of macromolecular synthesis by the endothelium may be involved. Since both phenomena begin to occur before the onset of hypertension, they may be induced, not by the hypertension *per se*, but by some humoral factor(s) characteristic of the hypertensive state.

VI. PERSPECTIVES

The *adherence of mononuclear cells to arterial endothelium* is emerging as a common theme in many facets of arterial pathophysiology. In contrast to microvascular (venular) endothelium, where polymorphonuclear cells adhere and emigrate at the slightest provocation, the endothelium of larger arteries consistently attracts mononuclear cells in hypercholesterolemia, hypertension, regeneration, and even to some extent under normal conditions. What triggers this phenomenon? Does it depend on one mechanism or on several? Does it occur in smaller arteries? What does it imply for the function of the arterial wall? Is it "good," "bad," or both?

The notion of *endothelial activation* recently introduced (see Gimbrone and Bevilacqua, this volume) promises to be extremely useful in interpreting vascular changes. Activation of the endothelial cell, which is capable of manufacturing so many products, can surely take many different forms. We have offered evidence that in hypertension it may manifest itself by the synthesis of intimal extracellular materials.

The determination of endothelial cell shape (elongated versus polygonal) no longer seems to be prerogative of shear stress. Further work is needed to establish what other factor(s)—such as intramural strains—may be involved.

The relation of shear stress to lipid deposition is an old puzzle that could be approaching a solution; as we indicated, it may be possible to reconcile the two seemingly opposed theories, namely the "high shear" and the "low shear" theory. As so often happens in science, both theories may turn out to be right. However, we are still in the dark as to the molecular mechanism(s) that favors lipid deposition in areas of low as well as high shear.

Last but certainly not least, the role of *transcytosis* in intimal lipid deposition must be worked out. All indications are that it plays a central role; N. Simionescu, who coined the term, has already made a significant step toward proving it.[41]

More generally, the data reported in this chapter offer some perspective on the role of experiments *in vivo*. There is no doubt that many cellular events—such as endothelial secretion—are best analyzed in tissue culture; but it is equally true that no model can replace, after all, the study of the living artery.

ACKNOWLEDGMENTS. The authors' research was supported in part by Grants HL

25973 and HL 33529 from the National Institutes of Health. We are most grateful to Mrs. Jean M. Underwood for technical help, to Mr. Christopher D. Hebert for the photographic prints, and Ms. Jane M. Manzi for preparation of the manuscript.

REFERENCES

1. Alderson, L. M., Endemann, G., Lindsey, S., Pronczuk, A., Hoover, R. L., and Hayes, K. C., 1986, LDL enhances adhesion to endothelial cells in vitro, *Am. J. Pathol.* 123:334–342.
2. Barak, L., Yocum, R. R., Nothnagel, E. A., and Webb, W. W., 1980, Fluorescence staining of the actin cytoskeleton in living cells with 7-nitrobenz-2-oxa-1,3-diazole-phallacidin, *Proc. Natl. Acad. Sci. USA* 77:980–984.
3. Bomberger, R. A., Zarins, C. K., and Glagov, S., 1981, Subcritical arterial stenosis enhances distal atherosclerosis, *J. Surg. Res.* 30:205–212.
4. Caro, C. G., Fitz-Gerald, J. M., and Schroter, R. C., 1969, Arterial wall shear and distribution of early atheroma in man, *Nature* 223:1159–1161.
5. Caro, C. G., Fitz-Gerald, J. M., and Schroter, R. C., 1971, Atheroma and arterial wall shear: Observation, correlation and proposal of a shear dependent mass transfer mechanism for atherogenesis, *Proc. R. Soc. London Ser. B* 177:109–159.
6. Dewey, C. F., Jr., Bussolari, S. R., Gimbrone, M. A., Jr., and Davies, P. F., 1981, The dynamic response of vascular endothelial cells to fluid shear stress, *J. Biomech. Eng.* 103:177–185.
7. Duff, G. L., McMillan, G. C., and Ritchie, A. C., 1957, The morphology of early atherosclerotic lesions of the aorta demonstrated by the surface technique in rabbits fed cholesterol together with a description of the anatomy of the intima of the rabbit's aorta and the "spontaneous" lesions which occur in it, *Am. J. Pathol.* 33:845–873.
8. Esterly, J. A., and Glagov, S., 1963, Altered permeability of the renal artery of the hypertensive rat: An electron microscopic study, *Am. J. Pathol.* 43:619–638.
9. Faggiotto, A., Ross, R., and Harker, L., 1984, Studies of hypercholesterolemia in the nonhuman primate. I. Changes that lead to fatty streak formation, *Arteriosclerosis* 4:323–340.
10. Faggiotto, A., and Ross, R., 1984, Studies of hypercholesterolemia in the nonhuman primate. II. Fatty streak conversion to fibrous plaque, *Arteriosclerosis* 4:341–356.
11. Fry, D. L., 1968, Acute vascular endothelial changes associated with increased blood velocity gradients, *Circ. Res.* 22:165–197.
12. Gerrity, R. G., Naito, H. K., Richardson, M., and Schwartz, C. J., 1979, Dietary induced atherogenesis in swine: Morphology of the intima in prelesion stages, *Am. J. Pathol.* 95:775–792.
13. Gerrity, R. G., Goss, J. A., and Soby, L., 1985, Control of monocyte recruitment by chemotactic factor(s) in lesion-prone areas of swine aorta, *Arteriosclerosis* 5:55–66.
14. Grabowski, E. F., Jaffe, E. A., and Weksler, B. B., 1985, Prostacyclin production by cultured endothelial cell monolayers exposed to step increases in shear stress, *J. Lab. Clin. Med.* 105:36–43.
15. Haudenschild, C. C., Forney Prescott, M., and Chobanian, A. V., 1980, Effects of hypertension and its reversal on aortic intima lesions of the rat, *Hypertension* 2:33–44.
16. Haust, M. D., 1979, Proliferation and degeneration of elastic tissue in aortic explants from normo-, and hypercholesterolemic rats: An ultrastructural study, *Exp. Mol. Pathol.* 31:169–181.
17. Hoff, H. F., and Gottlob, R., 1967, A fine structure study of injury to the endothelial cells of the rabbit abdominal aorta by various stimuli, *Angiology* 18:440–451.
18. Hoffman, A. H., Savilonis, B. J., Adner, D., and Mackay, D. G., 1983, Correlating in vivo experiments with hydrodynamic scale models using a transverse clip stenosis, in: *Advances in Bioengineering*, ASME, New York, pp. 14–15.
19. Houle, S., and Roach, M. R., 1981, Flow studies in a rigid model of an aorta–renal junction: A case for high shear as a cause of the localization of sudanophilic lesions in rabbits, *Atherosclerosis* 40:231–244.
20. Joris, I., and Majno, G., 1974, Cellular breakdown within the arterial wall: An ultrastructural study of the coronary artery in young and aging rats, *Virchows Arch. A* 364:111–127.

21. Joris, I., and Majno, G., 1977, Cell-to-cell herniae in the arterial wall. I. The pathogenesis of vacuoles in the normal media, *Am. J. Pathol.* **87:**375–398.

22. Joris, I., and Majno, G., 1978, Atherosclerosis and inflammation, in: *The Thrombotic Process in Atherogenesis* (A. B. Chandler, K. Eurenius, G. C. McMillan, C. B. Nelson, C. J. Schwartz, and S. Wessler, eds.), Plenum Press, New York, pp. 227–233.

23. Joris, I., Stetz, E., and Majno, G., 1979, Lymphocytes and monocytes in the aortic intima: An electron microscopic study in the rat, *Atherosclerosis* **34:**221–231.

24. Joris, I., and Majno, G., 1981, Endothelial changes induced by arterial spasm, *Am. J. Pathol.* **102:**346–358.

25. Joris, I., and Majno, G., 1981, Medial changes in arterial spasm induced by L-norepinephrine, *Am. J. Pathol.* **105:**212–222.

26. Joris, I., Zand, T., and Majno, G., 1982, Hydrodynamic injury of the endothelium in acute aortic stenosis, *Am. J. Pathol.* **106:**394–408.

27. Joris, I., Zand, T., Nunnari, J. J., Krolikowski, F. J., and Majno, G., 1983, Studies on the pathogenesis of atherosclerosis. I. Adhesion and emigration of mononuclear cells in the aorta of hypercholesterolemic rats, *Am. J. Pathol.* **113:**341–358.

28. Joris, I., Billingham, M. E., and Majno, G., 1984, Human coronary arteries: An ultrastructural search for the early changes of atherosclerosis, *Fed. Proc.* **43:**710.

29. Joris, I., Zand, T., Nunnari, J. J., and Majno, G., 1986, Regression of aortic lesions in hypercholesterolemic rats, *Fed. Proc.* **45:**813.

30. Kowala, M. C., Cuénoud, H. F., Joris, I., and Majno, G., 1986, Cellular changes during hypertension: A quantitative study of the rat aorta, *Exp. Mol. Pathol.* **45:**323–335.

31. Langille, B. W., Reidy, M. A., and Kline, R. L., 1986, Injury and repair of endothelium at sites of flow disturbances near abdominal aortic coarctations in rabbits, *Arteriosclerosis* **6:**146–154.

32. Leary, T,, 1941, The genesis of atherosclerosis, *Arch. Pathol.* **32:**507–555.

33. Levesque, M. J., and Nerem, R. M., 1985, The elongation and orientation of cultured endothelial cells in response to shear stress, *J. Biomech. Eng.* **107:**341–347.

34. Levesque, M. J., Liepsch, D., Moravec, S., and Nerem, R., 1986, Correlation of endothelial cell shape and wall shear stress in a stenosed dog aorta, *Arteriosclerosis* **6:**220–229.

35. Lewis, J. C., Taylor, R. G., Jones, N. D., St. Clair, R. W., and Cornhill, J. F., 1982, Endothelial surface characteristics in pigeon coronary artery atherosclerosis. I. Cellular alterations during the initial stages of dietary cholesterol challenge, *Lab. Invest.* **46:**123–138.

36. Majno, G., Underwood, J. M., Zand, T., and Joris, I., 1985, The significance of endothelial stomata and stigmata in the rat aorta, *Virchows Arch. A* **408:**75–91.

37. Majno, G., Joris, I., and Zand, T., 1985, Atherosclerosis: New horizons, *Hum. Pathol.* **16:**3–5.

38. Nerem, R. M., Sato, M., and Levesque, M. J., 1985, Elongation orientation and effective shear modulus of cultured endothelial cells in response to shear, *Fed. Proc.* **44:**1659.

39. Poole, J. C. F., and Florey, H. W., 1958, Changes in the endothelium of the aorta and the behaviour of macrophages in experimental atheroma of rabbits, *J. Pathol. Bacteriol.* **75:**245–251.

40. Savilonis, B. J., Hoffman, A. H., LaCoy, D. R., Jr., Soebroto, S. P., and Holzman, J. J., Jr., 1984, Using hydrodynamic scale models and computer simulations to study the flow thru a transverse clip stenosis, in: *Advances in Bioengineering*, ASME, New York, pp. 115–116.

41. Simionescu, N., Vasile, E., Lupu, F., Popescu, G., and Simionescu, M., 1986, Prelesional events in atherogenesis: Accumulation of extracellular cholesterol-rich liposomes in the arterial intima and cardiac valves of the hyperlipidemic rabbit, *Am. J. Pathol.* **123:**109–125.

42. Stetz, E. M., Majno, G., and Joris, I., 1979, Cellular pathology of the rat aorta: Pseudovacuoles and myoendothelial herniae, *Virchows Arch. A* **383:**135–148.

43. Still, W. J. S., 1967, The early effect of hypertension on the aortic intima of the rat: An electron microscopic study, *Am. J. Pathol.* **51:**721–734.

44. Still, W. J. S., 1968, The pathogenesis of the intimal thickenings produced by hypertension in larger arteries in the rat, *Lab. Invest.* **19:**84–91.

45. White, G. E., Fujiwara, K., Shefton, E. J., Dewey, C. F., and Gimbrone, M. A., Jr., 1982, Fluid shear stress influences cell shape and cytoskeleton organization in cultured vascular endothelium, *Fed. Proc.* **41:**321.

46. Wissler, R. W., Eilert, M. L., Schroeder, M. A., and Cohen, L., 1954, Production of lipomatous and atheromatous arterial lesions in the albino rat, *Arch. Pathol.* **57:**333–351.
47. Wong, M. K. K., and Gotlieb, I., 1984, *In vitro* reendothelialization of a single-cell wound: Role of microfilament bundles in rapid lamellipodia-mediated wound closure, *Lab. Invest.* **51:**75–81.
48. Wulf, E., Deboben, A., Bautz, F. A., Faulstich, H., and Wieland, T., 1979, Fluorescent phallotoxin, a tool for the visualization of cellular actin, *Proc. Natl. Acad. Sci. USA* **76:**4498–4502.
49. Zand, T., Nunnari, J. J., Joris, I., and Majno, G., 1986, Adhesion of polymorphonuclear neutrophils to arterial endothelium during acute mild stenosis of the rat abdominal aorta, *Fed. Proc.* **45:**694.
50. Zand, T., Nunnari, J. J., Hoffman, A. H., Savilonis, B. J., MacWilliams, B., Majno, G., and Joris, I., Endothelial adaptations in aortic stenosis: Correlation with flow parameters, *Am. J. Pathol.* (submitted).

VIII

Endothelial Cell in Atherogenesis

Endothelial Injury and Atherosclerosis

Russell Ross

I. INTRODUCTION

In 1973, we proposed a hypothesis of atherogenesis[35] that provided a basis for the design of *in vivo* and *in vitro* experiments to determine how the advanced proliferative smooth muscle lesions of atherosclerosis form. Three fundamental biologic phenomena comprise the advanced lesions of atherosclerosis and lead to the development of occlusive lesions. Understanding why and how each of these phenomena occurs is fundamental to the development of this hypothesis. These phenomena are: (1) the accumulation of large numbers of intimal cells, principally proliferated smooth muscle together with numerous macrophages derived from blood monocytes; (2) formation by the proliferated smooth muscle cells of connective tissue matrix macromolecules including collagen, elastic fibers, and proteoglycans; and (3) accumulation of lipid both within the accumulated smooth muscle cells and macrophages, and within the components of the extracellular matrix. Any hypothesis of atherogenesis must take into account these phenomena, and should afford the opportunity to devise experiments to test the various components of the hypothesis.

II. RESPONSE TO INJURY HYPOTHESIS OF ATHEROSCLEROSIS

Since its inception, the response to injury hypothesis has been modified several times,[33,34,36,38] taking into account, each time, new observations that determine whether the several parts of the hypothesis may have been correct or not, as well as newer observations that demonstrate that some components of the hypothesis are proven and hence no longer hypothetical. The original hypothesis was formulated upon observations that endothelial injury led to endothelial desquamation and denudation, followed by platelet interactions that preceded intimal smooth muscle proliferation.[4,18,43] The latest version of this hypothesis[34] provides a more comprehensive definition of endothelial injury, so that it covers a broader

Russell Ross • Department of Pathology. University of Washington, Seattle, Washington 98195.

spectrum of possible changes. At one end of the spectrum the endothelium can be intact and normal in appearance, but altered in functional properties; at the other end, there may be endothelial denudation and possibly platelet interactions. This modified hypothesis also takes into account observations that have resulted from testing this hypothesis and which demonstrate that growth factors can be formed by many different cell types including "injured" endothelium. Like its predecessors, this latest version of the response to injury hypothesis must take into account the fact that some lesions of atherosclerosis are lipid-rich and fibrotic, others are lipid-poor and fibrotic, and others are lipid-rich and contain relatively little fibrosis. The correlation of lesion morphology with the cellular events that precede lesion formation, together with the known risk factors for atherosclerosis, or in some cases specific agents that are known to cause the lesions to form, has improved our understanding of the significance of these observations. For example, it has been observed that chronic cigarette smokers often have extensive fibrotic occlusive lesions of the superficial femoral and coronary arteries which often contain relatively little lipid, particularly if the patients are not hyperlipidemic.[39] In contrast, hyperlipidemic patients may have extensive lipid-filled lesions of the coronary arteries, abdominal and thoracic aortas, sometimes with little involvement of the leg arteries. Hypertensive individuals may have involvement of the carotid and cerebral arteries, and if they are not hyperlipidemic, little involvement of the coronary arteries.[31] Consequently, it is entirely conceivable that different segments of the arterial tree may respond differently to different etiologic factors, and it is also conceivable that under normal circumstances endothelium in different parts of the arterial tree may have different physiologic characteristics. Much of this is dealt with in other chapters in this volume.

The latest version of the response to injury hypothesis of atherosclerosis proposes, as did earlier versions, that the initiating event in the formation of the lesions of atherosclerosis is some form of "injury" to the lining endothelial cells. "Injury" may have different manifestations dependent upon the site, source, nature, and duration of the injury. If the injury results from altered rheologic properties of the blood flow, it may occur principally at branches and bifurcations in an artery. Furthermore, different segments of the arterial tree may have different levels of susceptibility to the same injurious agents.

"Injury" may be manifest as simply an alteration in one or more of the normal functional properties of the endothelial cells. Such changes could include alterations in the nonthrombogenic properties of the endothelium, in the permeability characteristics of the cells, in cell–cell attachment or cell connective tissue attachment, in the formation of vasoactive substance by the endothelium, in connective tissue formation by the endothelial cells, in their capacity to form growth factors, or in any of the other capacities of the endothelial cells discussed in other chapters in this volume.

Dependent upon the alteration in functional capacity of the endothelial cells, the changes in the artery that occur will probably be determined by the anatomic site and by the causative agent. In the case of chronic hyperlipidemia[12,13,15,16] and as demonstrated recently in the case of hypertension,[8] one of the first signs of endothelial injury is an increased attachment of circulating leukocytes, principally monocytes and lymphocytes, to the endothelial cells. In hyperlipidemic animals these monocytes and lymphocytes attach in clusters that are located randomly throughout the arterial tree. The leukocytes migrate over the surface of the endothelium and are chemotactically attracted to enter into the subendothelial space by probing between endothelial cells. When they localize subendothelially, they are activated to become macrophages and they form the first and ubiquitous lesion of atherosclerosis (in hyperlipidemic individuals), the fatty streak. The hypothesis then proposes several possible cellular changes. One of these results from the capacity of the monocyte/macrophage to act not only as a phagocytic cell, but as a cell that can synthesize and secrete a number of different substances,[28] including several different growth factors such as platelet-derived growth factor (PDGF),[26,42] fibroblast growth factor (FGF),[2] transforming growth factor alpha,[24a] interleukin 1 (IL-1),[11] and transforming growth factor-β (TGF-β).[1a] If the macrophages are appropriately stimulated to secrete one or more of these growth factors, they could serve as a source for the chemotactic attraction and migration of smooth muscle cells from the media of the artery into the intima, where the smooth muscle cells could proliferate and form new connective tissue leading to the development of the advanced lesion of atherosclerosis, the fibrous plaque, which can go on to become a complicated lesion.

A further sequence of events is predicted by this hypothesis if injury to the endothelium is more severe. These are represented by endothelial cell–cell dysjunction, which has been observed in cells that overlie fatty streaks in which subendothelial accumulation of fat-filled macrophages in the form of foam cells has occurred. Endothelial dysjunction could lead to retraction of the endothelial cells and exposure of either the macrophages beneath the endothelium, or the subendothelial connective tissue. In either case this would permit platelets to attach to the exposed macrophages, or connective tissue leading to the formation of mural thrombi accompanied by release of growth factors from the platelets similar to those released from activated macrophages. In this case platelet adherence, aggregation, and release could stimulate a more rapid, fulminating form of smooth muscle proliferation and thus lead to the formation of advanced lesions of atherosclerosis. These two pathways are described in the response to injury hypothesis presented in Fig. 1. This diagram also suggests that in some cases etiologic factors may injure the endothelium in such a manner that the endothelial cells themselves would be stimulated to synthesize and secrete growth factors, including a form of PDGF.[3,9,14] This could also lead to migration of smooth muscle cells into the intima from the media, and if the smooth muscle cells are appropriately stimulated, perhaps to secretion of a form of PDGF by the smooth muscle cells themselves.[29,41,45] If this occurred, it would represent a form of autocrine or paracrine stimulation by the smooth muscle cells that could gradually, over a period of time, also lead to progression of the lesions of atherosclerosis. The correlation of different risk factors and/or causative agents that could lead to one or any combination of these events, could, over periods of time ranging from months to years, lead to the development of advanced proliferative, potentially occlusive lesions of atherosclerosis. These possibilities are depicted in Fig. 1.

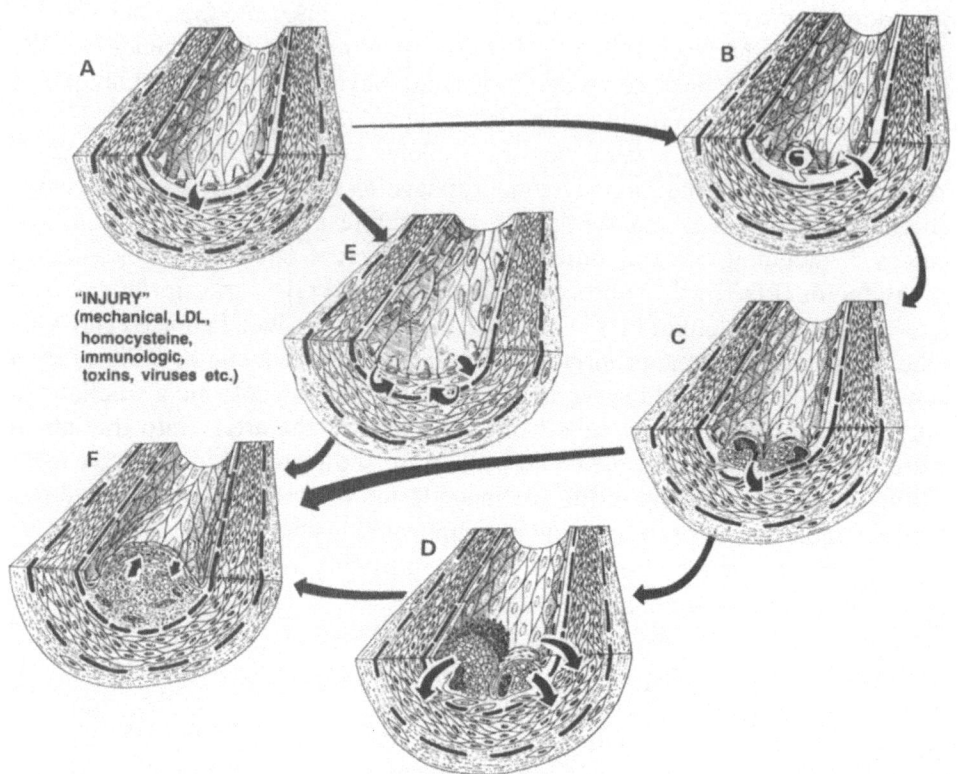

Figure 1. Endothelial injury: the response to injury hypothesis. Advanced intimal proliferative lesions of atherosclerosis may occur by at least two pathways. The pathway demonstrated by the clockwise (long) arrows to the right has been observed in experimentally induced hypercholesterolemia. Injury to endothelium (A) may induce growth factor secretion (short arrow). Monocytes attach to endothelium (B), which may continue to secrete growth factors (short arrow). Subendothelial migration of monocytes (C) may lead to fatty streak formation and release of growth factors such as PDGF (short arrow). Fatty streaks may become directly converted to fibrous plaques (long arrow from C to F) through release of growth factors from macrophages or endothelial cells or both. Macrophages may also stimulate or injure the overlying endothelium. In some cases, macrophages may lose their endothelial cover and platelet attachment may occur (D), providing three possible sources of growth factors—platelets, macrophages, and endothelium (short arrows). Some of the smooth muscle cells in the proliferative lesion itself (F) may form and secrete growth factors such as PDGF (short arrows).

An alternative pathway for development of advanced lesions of atherosclerosis is shown by the arrows from A to E to F. In this case, the endothelium may be injured but remain intact. Increased endothelial turnover may result in growth factor formation by endothelial cells (A). This may stimulate migration of smooth muscle cells from the media into the intima, accompanied by endogenous production of PDGF by smooth muscle as well as growth factor secretion from the "injured" endothelium (E). These interactions could then lead to fibrous plaque formation and further lesion progression. Reproduced by permission of *The New England Journal of Medicine*.[34]

III. CELLULAR INTERACTIONS DURING HYPERCHOLESTEROLEMIA

A number of cellular interactions observed in hypercholesterolemic nonhuman primates,[12,13] swine,[15,16] the endogenously hyperlipidemic Watanabe rabbit (a model of human familial homozygous hypercholesterolemia) versus fat-fed rabbits,[32] the hypercholesterolemic rat[21] (also described in the chapter by Majno *et al.*), and hypercholesterolemic pigeons[20,24] support many of the formulations proposed in the response to injury hypothesis of atherosclerosis.

As discussed in the hypothesis, the first event observed in hypercholesterolemia is increased monocyte adhesion to the endothelium, presumably due to the formation of chemotactic agents either from the smooth muscle cells,[27] from endothelium,[5,10] or possibly due to changes induced by the hypercholesterolemic state (Fig. 2). Monocyte attachment is followed by migration of monocytes and probably lymphocytes between the endothelium into the intima, where they localize and form foam cells due to the accumulation of large amounts of lipid in the macrophages, principally in the form of cholesterol ester (Fig. 3). Such accumulations of intimal foamy macrophages represent the development of the first and ubiquitous lesion of atherosclerosis, the fatty streak. Fatty streaks expand in a dynamic fashion by continued adherence, subendothelial migration, and localization of monocytes and presumably lymphocytes as well, so that the lesions

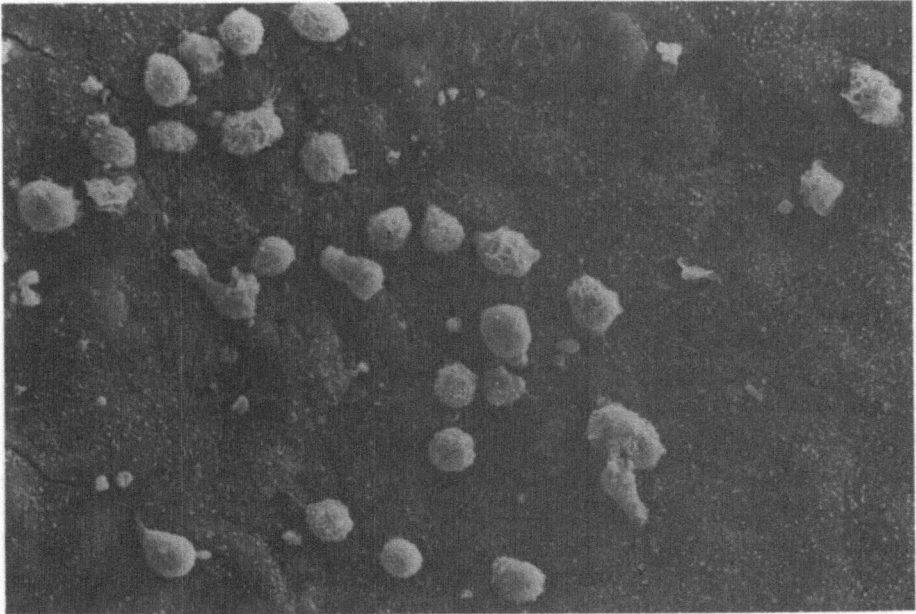

Figure 2. Electron micrograph demonstrating leukocytes adherent to the endothelium after 12 days on an atherogenic diet. Adherent leukocytes (mostly monocytes) were found scattered in patches at all levels of the aortic tree. Some of the cells appear spread on the surface, whereas most are rounded in appearance. Reproduced by permission of the American Heart Association.[13]

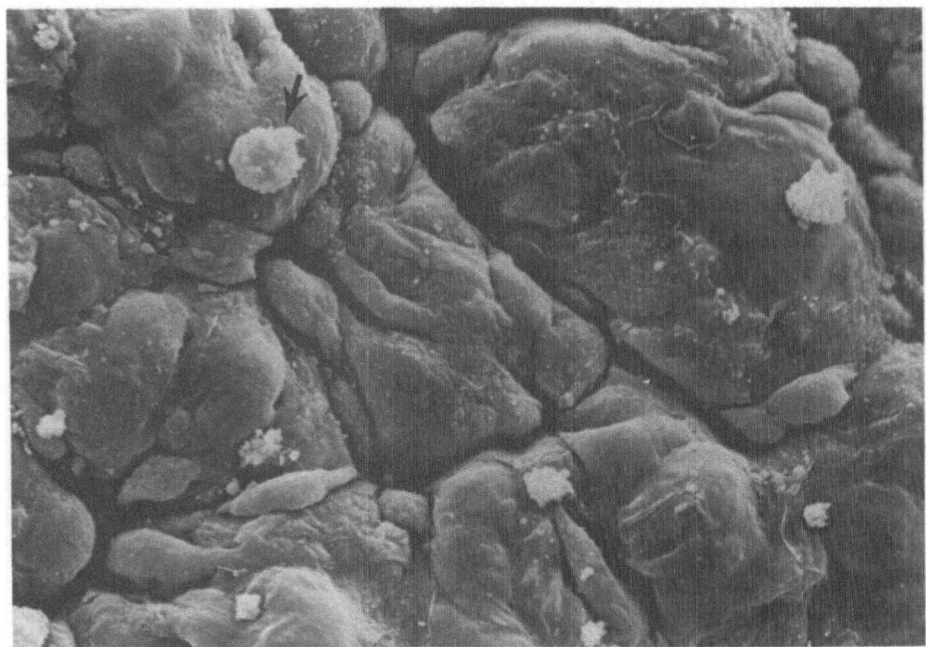

Figure 3. A surface view of a fatty streak showing the endothelium after 3 months of hypercholesterolemia. The surface has become highly irregular and forms a striking nodular pattern with relatively deep crevices between the nodules. Leukocytes (arrow) can be seen adherent to the endothelial cells that are stretched by the accumulated subendothelial macrophages. Continuing leukocyte attachment and subendothelial migration appear to participate in continued fatty streak expansion. Bar = 10 μm. Reproduced by permission of the American Heart Association.[13]

can continue to grow in diameter as well as in thickness ultimately by forming several layers of macrophages (Fig. 4).

In the hypercholesterolemic monkey, a second event has been observed, based upon the anatomic site in the arterial tree, the level of hypercholesterolemia, and the duration of this level of hypercholesterolemia. The higher the level of hypercholesterolemia, the shorter is the time interval required for the change, and conversely the lower the level, the longer the time interval. This change occurs first at branches at bifurcations in the leg arteries of the monkeys and then at other branches and bifurcations, in an ascending fashion with increased time, in the leg arteries; then in the iliac bifurcation, then in the abdominal aorta followed by the thoracic aorta and finally in the coronary arteries.[12,13] The changes consist of endothelial cell–cell detachment followed by retraction of the endothelial cells exposing the subendothelial foam cells to the circulation (Fig. 5). Many of these foam cells are swept into the circulation exposing the connective tissue base upon which they were disposed. Others remain attached to the subendothelium, depending upon the nature of the flow characteristics at the particular site in the artery. In either case, the next event observed was attachment of platelets to the exposed macrophages, or to the exposed subendothelial connective tissue. When

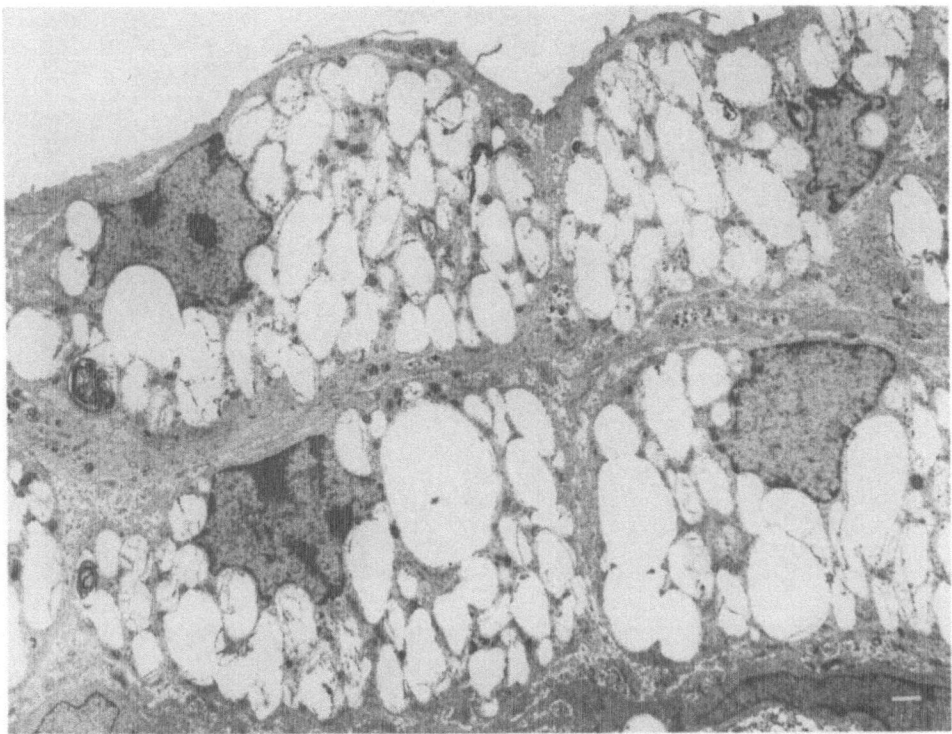

Figure 4. A fatty streak containing two layers of subendothelial foam cells after 2 months of hyper-cholesterolemia. The large lipid-filled macrophages shown in this transmission electron micrograph are distributed focally in multilayers. The cells are four- to sixfold larger than lipid-laden macrophages observed in control animals. There is a small amount of intercellular matrix and some lipid debris. The macrophages maintain a close relationship to the intact endothelium. The endothelium is markedly stretched so that the cells have become very thin. Bar = 1 μm. Reproduced by permission of the American Heart Association.[13]

this occurs, the platelets adhere, undergo shape change, and degranulate. Such platelet interactions are followed within a month or so by migration of smooth muscle cells between fenestrae of the elastic lamina from the media into the intima where they develop an extensive rough endoplasmic reticulum and Golgi complex and where many of them multiply. The smooth muscle cells also form and secrete connective tissue matrix macromolecules into the surrounding environment. This goes on for some time, eventually leading to the formation of an intimal prolif-erative lesion that contains a mixture of smooth muscle cells, monocyte/macro-phages, and lymphocytes surrounded by newly formed elements of connective tissue. Many of the smooth muscle cells at the luminal aspect of the forming lesion become compressed between dense connective tissue fibers and form a fibrous cap.[40]

At other anatomic sites, the endothelium remains intact, where on a some-what slower basis, the smooth muscle cells also migrate and enter into the intima

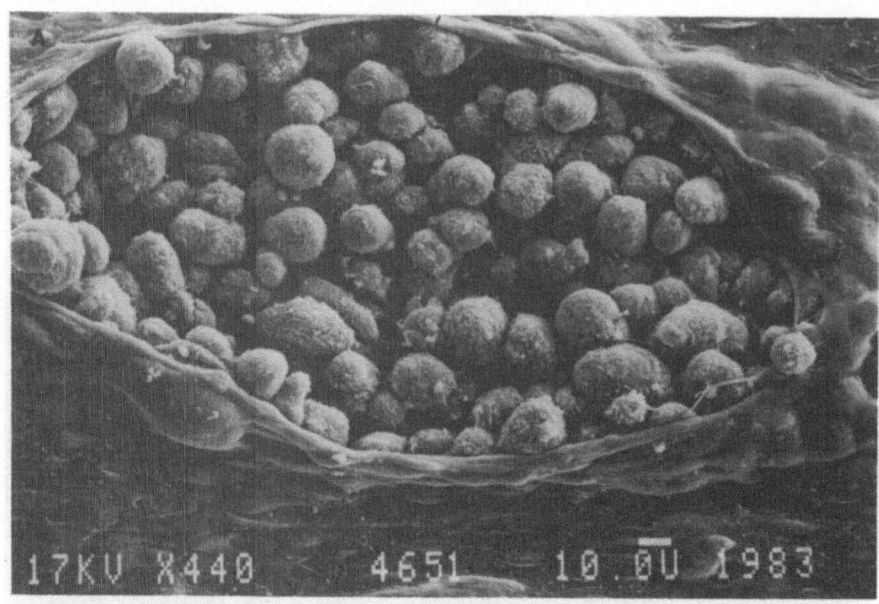

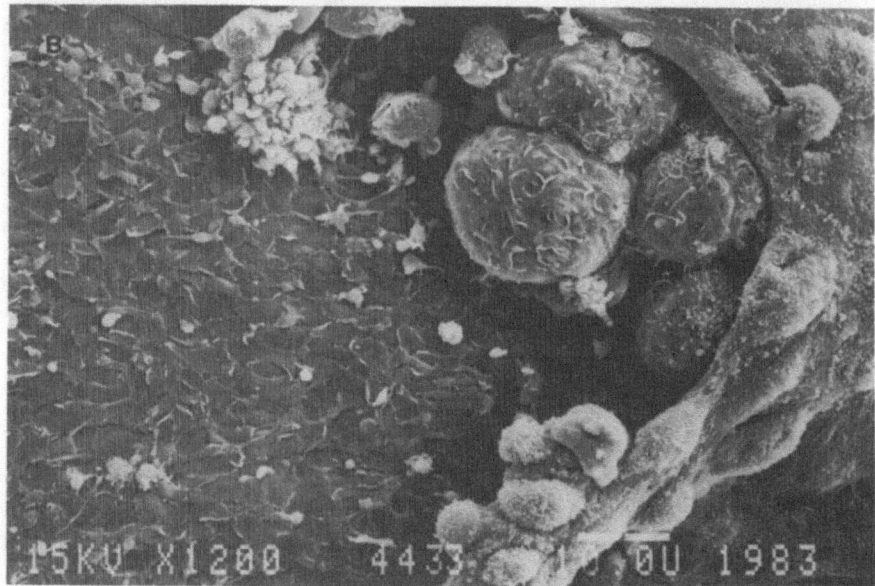

Figure 5. Examples of endothelial dysjunction leading to denudation and exposure of fatty streaks as seen by scanning electron microscopy. (A) Large lipid-laden subendothelial macrophages that constitute the upper part of this fatty streak have become exposed due to the loss of endothelial continuity. Bar = 100 μm. (B) In this fatty streak there has apparently been endothelial dysjunction and contraction to expose the lesion. Many of the exposed macrophages have presumably been shed into the circulation exposing the connective tissue to which platelets have attached, adhered, and spread. They have formed a small thrombus below the upper endothelial edge. A few macrophages remain partially covered by the edge of the endothelium that covers the remainder of the lesion. Bar = 10 μm. Reproduced by permission of the American Heart Association.[12]

in a fashion similar to that described above. In both of these cases, the intimal migration and proliferation of smooth muscle cells suggest that a factor or factors, presumably growth factors, may be important in the migration and proliferation of the smooth muscle cells in the intima.

IV. BLOOD CELLS, ARTERIAL CELLS, AND GROWTH FACTOR FORMATION

As a result of recent observations it is now clear that cells from the artery wall and from the blood, including appropriately activated endothelium, monocyte/macrophages, platelets, and perhaps smooth muscle cells themselves, are capable of forming growth factors.

A. The Platelet

In 1974 it was shown that platelets were the principal source of growth factors responsible for the mitogenic activity of whole blood serum that was absent in cell-free plasma-derived serum.[22,37] Since that time, activated platelets have been shown to be capable of releasing not only PDGF, but also FGF (Baird, unpublished data), a form of EGF or TGF-α,[30] and TGF-β.[1] PDGF is a potent mitogen for mesenchymal connective tissue-forming cells such as fibroblasts, smooth muscle, and glial cells. FGF is a potent angiogenic agent and mitogen for some connective tissue cells, EGF is a potent mitogen for endothelial cells and much less potent for some connective tissue cells, and TGF-β may act as an inhibitor, depending upon the target cells and the tissue involved. Each of these growth factors binds to specific high-affinity cell surface receptors and upon doing so elicits a sequence of intracellular events that ultimately lead to DNA synthesis and cell division or in the case of TGF-β, inhibition of multiplication. Ross et al.,[39] Heldin and Westermark,[19] and Stiles[44] have provided in-depth reviews that discuss most of the molecular characteristics of PDGF as well as its biologic activities.

B. The Monocyte/Macrophage

In 1976 it was demonstrated that activated peritoneal macrophages also secrete growth factors that are capable of stimulating fibroblasts, smooth muscle cells, and endothelial cells to multiply.[23,25] Similar growth factor capacities have since been observed for human pulmonary alveolar macrophages,[6] and for macrophages derived from elutriated human peripheral blood monocytes after appropriate stimulation.[17] A major portion of the growth factor activity secreted by these activated macrophages is a form of PDGF.[42] In addition, these activated macrophages contain a form of FGF[2] and can also secrete a form of TGF,[24a] IL-1,[11] and TGF-β.[1a] Thus, both types of blood cells, the circulating platelet, and the monocyte, after it enters the tissue and becomes a macrophage, are capable of synthesizing, in the case of the macrophage, and releasing in the case of both cell types, growth factors that have as targets each of the particular cells that are capable of entering a wound and dividing during the process of repair. In the case of atherosclerosis, platelets and macrophages can potentially provide growth factors that can stimulate both endothelium and smooth muscle cells in the artery wall. The subendothelial macrophages that constitute the fatty streak, the earliest and ubiquitous lesion of atherosclerosis, may participate in continuing the injury to the endothelial cells, since macrophages can secrete not only growth factors, but also toxic products such as oxygen metabolites and lysosomal enzymes, each of which may be injurious to the overlying endothelium and which may participate in the process of endothelial injury.[7]

Shimokado et al.[42] demonstrated that a major portion of the growth factor secreted by the activated macrophage is a form of PDGF, by separating the PDGF from binding proteins (e.g., α_2-macroglobulin) that macrophages also secrete. They demonstrated that the PDGF-like factor has the appropriate molecular weight ($\sim$ 30,000), competes for binding to the PDGF receptor, is recognized in an ELISA assay using a monospecific anti-PDGF IgG, and that the mitogenesis of smooth muscle cells induced by the macrophage-derived material is neutralized by the monospecific anti-PDGF antibody. In addition, using a cDNA probe for the B chain of human PDGF, activated macrophages contain increased mRNA for both chains by Northern analysis, whereas the circulating monocyte has little to no detectable mRNA for either chain of PDGF.[26,42] These observations point to the potential importance of macrophages in numerous fibrous proliferative responses, since this cell is commonly observed in virtually all of these responses.

C. Endothelium

In 1980 it was demonstrated that cultured arterial endothelial cells also secrete growth factors. It was subsequently shown that approximately 30% of this growth factor activity is due to the presence of a PDGF-like substance. The nature of the remaining growth factor activity has not been identified. Barrett et al.[3] observed that intact endothelial cells from bovine aorta or human umbilical vein have, by Northern analysis using a cDNA probe for the B chain of PDGF, little to no detectable mRNA for this molecule. In contrast, when these same cells are placed in culture and undergo multiple cell divisions, there is a large increase in mRNA for the B chain of PDGF, suggesting that endothelial injury and increased endothelial cell turnover may represent stimulation of arterial endothelium to synthesize and secrete PDGF. Thus, injury to the endothelium may be an important component of growth factor secretion by endothelial cells, putting them in a prime position to participate in responses to injury that lead to the development of lesions of atherosclerosis.

D. Smooth Muscle Cell

Finally, in a surprising turn of events, Seifert et al.[41] observed that cultured newborn rat aortic smooth muscle cells or first-passage adult rat smooth muscle[29] secrete a form of PDGF, in contrast to smooth muscle cells derived from 3-month-old rats of the same strain which secrete no PDGF. This raised the question as to whether this PDGF-like material might be involved in growth and development of the rat aorta and that perhaps the gene may be no longer expressed once the rat has matured. As a sequel to these events, Walker et al.[45] observed that medial arterial smooth muscle cells from the carotid arteries of adult rats secrete no PDGF-like substance, whereas if the carotid artery of such a rat is balloon catheter deendothelialized, this results in the formation of a massive intimal smooth muscle proliferative lesion. The cells derived from such a proliferative lesion secrete a PDGF-like factor into the medium when they are placed in culture. These observations suggested the possibility that once smooth muscle cells are stimulated to proliferate, as occurs in this form of experimentally induced injury, then the gene for PDGF may be turned on and the cells induced to secrete a PDGF-like substance that may stimulate them in an autocrine and/or paracrine fashion.

V. CONCLUDING REMARKS

Thus, the response to injury hypothesis of atherosclerosis has provided numerous opportunities for testing and correlating clinical observations and experimental studies with a number of different animal models as well as tissue culture systems, that have provided us with greater understanding of the cellular processes involved in atherogenesis.

It has become clear that we need to fully understand the nature of the endothelial injury. Although such injury may or may not have morphologic manifestations, alterations in functional characteristics of the endothelium undoubtedly must occur. Appropriate markers for altered endothelial function need to be developed. Injury to the endothelial cells may result in quite subtle alterations ranging from altered permeability or thrombogenicity to growth factor formation, to increased cell turnover, or to the release of vasoactive materials. The factors responsible for monocyte attachment and chemotaxis which have been observed not only in hyperlipidemia but in hypertension as well, may be a result of such injurious phenomena. The activities of the monocyte once it enters the intima and becomes a macrophage, and the release of growth factors as well as possible continued endothelial injury need to be better understood. The nature of the factors and the cells that release them, that are responsible for smooth muscle chemotaxis and intimal proliferation require further clarification.

Understanding the importance of growth factors, and in particular PDGF, and the possible roles of these factors in smooth muscle migration and proliferation within the intima of the artery, should provide opportunities for intervention and possibly for prevention of lesion formation in a number of potentially important circumstances. Thus, the development of specific growth factor antagonists, or modes of prevention of growth factor signal production presents potential approaches that could lead to prevention of the lesions of atherosclerosis.

The response to injury hypothesis has provided an opportunity to ask many of these questions and to test various parts of the hypothesis. Much remains to be done, despite the fact that new experimental approaches have been devised as a result of this hypothesis and much new information has been gained over the past several years.

REFERENCES

1. Assoian, R. K., and Sporn, M. B., 1986, Type β transforming growth factor in human platelets: Release during platelet degranulation and action on vascular smooth muscle cells, *J. Cell. Biol.* **102**:1217–1223.

1a. Assoian, R. K., Fleurdelys, B. E., Stevenson, H. C., Miller, P. J., Madtes, D. K., Raines, E. W., Ross, R., and Sporn, M. D., 1987, Expression and secretion of type β transforming growth factor by activated human macrophages, *Proc. Nat. Acad. Sci. USA* **84**:6020–6024.

2. Baird, A., Morméde, P., and Böhlen, P., 1985, Immunoreactive fibroblast growth factor in cells of peritoneal exudate suggests its identity with macrophage-derived growth factor, *Biochem. Biophys. Res. Commun.* **126**:358–364.

3. Barrett, T. B., Gajdusek, C. M., Schwartz, S. M., McDougall, J. K., and Benditt, E. P., 1984, Expression of the *sis* gene by endothelial cells in culture and in vivo, *Proc. Natl. Acad. Sci. USA* **81**:6772–6774.

4. Baumgartner, H. R., and Studer, A., 1966, Folgen des gefasskatheterismus am normo-und hypercholesterinaemischen kaninchen, *Pathol. Microbiol.* **29**:393–405.

5. Bevilacqua, M. P., Pober, J. S., Cotran, R. S., and Gimbrone, M. A., Jr., 1985, Interleukin 1 (IL-1) acts upon vascular endothelium to stimulate procoagulant activity and leukocyte adhesion, *J. Cell. Biochem. Suppl.* **9A**:148 (abstract).

6. Bitteman, P. B., Rennard, S. I., Hunninghake, G. W., and Crystal, R. G., 1982, Human alveolar macrophage growth factor for fibroblasts: Regulation and partial characterization, *J. Clin. Invest.* **70**:806–822.

7. Cathcart, M. K., Morel, D. W., and Chisolm, G. M., III, 1985, Monocytes and neutrophils oxidize low-density lipoprotein making it cytotoxic, *J. Leukocyte Biol.* **38**:341–350.

8. Chobanian, A. V., Brecher, P. I., and Haudenshild, C. C., 1986, Effects of hypertension and of antihypertension therapy on atherosclerosis: State of the Art Lecture, in: *Inter-American Society Proceedings,* Suppl. I **8**:15–21.

9. DiCorleto, P. E., and Bowen-Pope, D. F., 1983, Cultured endothelial cells produce a platelet-derived growth factor-like protein, *Proc. Natl. Acad. Sci. USA* **80**:1919–1923.

10. DiCorleto, P. E., and de la Motte, C. A., 1985, Characterization of the adhesion of the human monocytic cell line U-937 to cultured endothelial cells, *J. Clin. Invest.* **75**:1153–1161.

11. Dinarello, C. A., 1984, Interleukin-1, *Rev. Infect. Dis.* **6**:51–95.

12. Faggiotto, A., and Ross, R., 1984, Studies of hypercholesterolemia in the nonhuman primate. II. Fatty streak conversion to fibrous plaque, *Arteriosclerosis* **4**:341–356.

13. Faggiotto, A., Ross, R., and Harlan, L., 1984, Studies of hypercholesterolemia in the nonhuman primate. I. Changes that lead to fatty streak formation, *Arteriosclerosis* **4**:323–340.

14. Gajdusek, C., DiCorleto, P., Ross, R., and Schwartz, S. M., 1980, An endothelial cell-derived growth factor, *J. Cell. Biol.* **85**:467–472.

15. Gerrity, R. G., 1981a, The role of the monocyte in atherogenesis. I. Transition of blood borne monocytes into foam cells in fatty lesions, *Am. J. Pathol.* **103**:181–190.

16. Gerrity, R. G., 1981b, The role of the monocyte in atherogenesis. II. Migration of foam cells from atherosclerotic lesions, *Am. J. Pathol.* **103**:191–200.

17. Glenn, K. C., and Ross, R., 1981, Humn monocyte-derived growth factor(s) for mesenchymal cells: Activation of secretion by endotoxin and concanavalin A, *Cell* **25**:603–615.

18. Harker, L. A., Ross, R., Slichter, S. J., and Scott, C. R., 1976, Homocystine-induced atherosclerosis: The role of endothelial cell injury and platelet response in its genesis, *J. Clin. Invest.* **58**:731–741.

19. Heldin, C.-H., and Westermark, B., 1984, Growth factors: mechanism of action and relation to oncogenes, *Cell* **37**:9–20.

20. Jerome, W. G., and Lewis, J. C., 1984, Early atherogenesis in White Carneau pigeons. I. Leukocyte margination and endothelial alterations at the celiac bifurcation, *Am. J. Pathol.* **116**:56–68.

21. Joris, J., Zand, T., Nunnary, J. L., Krolikowski, F. J., and Majno, G., 1983, Studies on the pathogenesis of atherosclerosis. I. Adhesion and emigration of mononuclear cells in the aorta of hypercholesterolemic rats, *Am. J. Pathol.* **113**:341–358.

22. Kohler, N., and Lipton, A., 1974, Platelete as a source of fibroblast growth-promoting activity, *Exp. Cell Res.* **87**:297–301.

23. Leibovich, S. J., and Ross, R., 1976, A macrophage-dependent factor that stimulates the proliferation of fibroblasts in vitro, *Am. J. Pathol.* **84**:501–513.

24. Lewis, J. C., Taylor, R. G., Jones, N. D., St. Clair, R. W., and Cornhill, J. R., 1982, Endothelial surface characteristics in pigeon coronary artery atherosclerosis. I. Cellular alterations during the initial stages of dietary cholesterol challenge, *Lab. Invest.* **46**:123–138.

24a. Madtes, D. K., Raines, E. W., Sakariassen, K. S., Assoian, R. K., Sporn, M. B., Bell, G. I., and Ross, R. 1988, Induction of transforming growth factor alpha in activated human alveolar macrophages, *Cell* (in press).

25. Martin, B. M., Gimbrone, M. A., Unanue, E. R., and Cotran, R. S., 1981, Stimulation of nonlymphoid mesenchymal cell proliferation by a macrophage-derived growth factor, *J. Immunol.* **126**:1510–1515.

26. Martinet, Y., Bitterman, P. B., Mornex, J.-F., Grotendorst, G. R., Martin, G. R., and Crystal, R. G., 1986, Activated human monocytes express the c-sis proto-oncogene and release a mediator showing PDGF-like activity, *Nature* **319**:158–160.

27. Mazzone, T., Jensen, M., and Chait, A., 1983, Human arterial wall cells secrete factors that are chemotactic for monocytes, *Proc. Natl. Acad. Sci. USA* **80**:5094–5097.

28. Nathan, C. F., Murray, H. W., and Cohn, Z. A., 1980, Current concepts: The macrophage as an effector cell, *N. Engl. J. Med.* **303**:622–626.

29. Nilsson, J., Sjolund, M., Palmberg, L., Thyberg, J., and Heldin, C.-H., 1985, Arterial smooth muscle cells in primary culture produce a platelet-derived growth factor-like protein, *Proc. Natl. Acad. Sci. USA* **82**:4418–4422.

30. Oka, Y., and Orth, D. N., 1983, Human plasma epidermal growth factor/β-urogastrone is associated with blood platelets, *J. Clin. Invest.* **72**:249–259.

31. Report of the Working Group on Arteriosclerosis of the National Heart, Lung, and Blood Institute, 1981, Volume 2, Government Printing Office, Washington, D.C. [DHEW Publ. No. (NIH) 82-2035].

32. Rosenfeld, M. E., Faggiotto, A., and Ross, R., 1985, The role of the mononuclear phagocyte in primate and rabbit models of atherosclerosis, in: *Proceedings of the Fourth Leiden Conference on Mononuclear Phagocytes,* Nijhoff, The Hague, pp. 795–802.

33. Ross, R., 1981, Atherosclerosis—A problem of the biology of arterial wall cells and their interaction with blood components, *Arteriosclerosis* **1**:293–311.

34. Ross, R., 1986, The pathogenesis of atherosclerosis—An update, *N. Engl. J. Med.* **314**:488–500.

35. Ross, R., and Glomset, J. A., 1973, Atherosclerosis and the smooth muscle cell, *Science* **180**:1332–1339.

36. Ross, R., and Glomset, J. A., 1976, The pathogenesis of atherosclerosis, *N. Engl. J. Med.* **295**:369–377, 420–425.

37. Ross, R., Glomset, J., Kariya, B., and Harker, L., 1974, A platelet-dependent serum factor that stimulates the proliferation of arterial smooth muscle cells in vitro, *Proc. Natl. Acad. Sci. USA* **71**:1207–1210.

38. Ross, R., and Harker, L., 1976, Hyperlipidemia and atherosclerosis, *Science* **193**:1094–1100.

39. Ross, R., Raines, E. W., and Bowen-Pope, D. F., 1986, The biology of platelet-derived growth factor, *Cell* **46**:155–169.

40. Ross, R., Wight, T. N., Strandness, E., and Theile, B., 1984, Human atherosclerosis. I. Cell constitution and characteristics of advanced lesions of the superficial femoral artery, *Am. J. Pathol.* **114**:79–93.

41. Seifert, R. A., Schwartz, S. M., and Bowen-Pope, D. F., 1984, Developmentally regulated production of platelet-derived growth factor-like molecules, *Nature* **311:**669–671.
42. Shimokado, K., Raines, E. W., Madtes, D. K., Barrett, T. B., Benditt, E. P., and Ross, R., 1985, A significant part of macrophage-derived growth factor consists of at least two forms of PDGF, *Cell* **43:**277–286.
43. Stemerman, M. B., and Ross, R., 1972, Experimental atherosclerosis. I. Fibrous plaque formation in primates, an electron microscope study, *J. Exp. Med.* **136:**769–789.
44. Stiles, C. D., 1983, The molecular biology of platelet-derived growth factor, *Cell* **33:**653–655.
45. Walker, L. N., Bowen-Pope, D. F., Ross, R., and Reidy, M. A., 1986, Production of PDGF-like molecules by cultured arterial smooth muscle cells accompanies proliferation after arterial injury, *Proc. Natl. Acad. Sci. USA* **83:**7311–7315.

18

Prelesional Changes of Arterial Endothelium in Hyperlipoproteinemic Atherogenesis

Nicolae Simionescu

I. INTRODUCTION

A crucial issue in atherosclerosis research has been to identify the earliest lesion and thus to determine the factors and mechanisms which initiate the disease process. The most widely recognized early alteration is the appearance of fatty streaks the hallmark of which is the accumulation of intracellular and extracellular cholesterol. Very little is known about the subtle cellular and molecular changes of the vessel wall that predispose and precede the inception of an atherosclerotic plaque.

Although at present it is still uncertain whether atherosclerosis is produced by a single or multiple factors, one of these, namely hyperlipoproteinemia, is unanimously recognized to play a crucial role both as a risk and an ethiopathogenic factor. The causative linkage between hypercholesterolemia and atherosclerosis has been well documented both in humans and in experimental animals. In the latter, elevated level of plasma β-lipoproteins has proved to be sufficient to induce lipid accumulation in the artery wall; in time this leads to the formation of a fatty streak that can be converted into a fibrous plaque.[22,23,32,33,60,61,78,98,109]

The cellular and molecular mechanisms by which hypercholesterolemia may lead to formation of restricted and preferentially located atherosclerotic lesions are largely unknown. The earliest cellular change so far detected has been the focal adherence of mononuclear leukocytes to arterial endothelium,[32,33,60,61] presumably in response to some chemoattractants.[34,81] Monocytes in particular undergo diapedesis homing in the subendothelium where as activated macrophages they are progressively loaded with cholesteryl ester-rich deposits to become foam cells.[22,32,33,59,60,78] Concomitantly, a focal "plasmatic sero-fibrinous insudation" can be noticed in the intima, usually accompanied by local proliferation of the extracellular matrix. On this general pattern, the evolution of an early lesion can be very polymorphic.

The artery wall appears constitutively poorly equipped with means of pro-

Nicolae Simionescu • Institute of Cellular Biology and Pathology, 79691 Bucharest, Romania.

tection against the massive influx of plasma lipoproteins. The defense is partially accomplished by the continuous recruitment of circulating monocytes and by the conversion of the arterial smooth muscle cells, and to a degree of endothelial cells, to a phagocytic phenotype. To this, the vessel wall adds a protective reaction of its extracellular constituents aimed at localizing the pathologic process. Although, in terms of focal restriction of the lesion, this mechanism appears somewhat successful, the progressive obstruction of the vascular lumen by the established plaque could be seriously detrimental for the tissue blood supply.

The studies reported here were conducted to identify and characterize the biochemical and ultrastructural modifications produced by hypercholesterolemia in the lesion-prone regions of the artery wall before the onset of a detectable bona fide atherosclerotic lesion. Our experiments were carried out on cholesterol-fed rabbits and hamsters in which the investigations were focused on lesion-prone areas of the aortic arch. Focused especially on the stages which precede the adherence and diapedesis of monocytes, the inquiry was particularly addressed to the following issues: (1) changes in endothelial permeability to plasma lipoproteins, (2) early localization in the intima of the accumulated lipoprotein-derived components, (3) modifications of the surface charge and chemistry of arterial endothelium and monocytes, (4) alterations in prostacyclin production. The prelesional modifications detected were correlated with the subsequent atherogenic events such as monocyte emigration and foam cell formation. The investigations were concentrated on the arterial endothelium and the extent of its involvement in the development of the atherosclerotic plaque.

II. MODELING OF EXPERIMENTAL HYPERCHOLESTEROLEMIA

Experimental design. There is an almost general agreement that diet-induced hypercholesterolemia can produce in some animal species arterial lesions which, although not identical, are largely comparable to those in humans. The high concentration of dietary fat usually used is intended to produce an accelerated hyperlipoproteinemic atherosclerosis in a short period of time.[74] Implicitly, the assumption is that at lower levels of cholesterolemia similar changes would occur over longer periods of time, as in humans. For most animals used in our experiments, the cholesterol-rich diet was given up to 12 weeks, and for 32 rabbits and 42 hamsters the diet was extended to 14 months. Tissue specimens were preferentially collected from the inner lesser curvature of the aortic arch. In preliminary experiments we have done a backward screening of the aorta from the advanced lesional stages down to the prelesional or nonlesional stages. We found that under our experimental conditions, in more than 95% of cases the atherosclerotic lesions (fatty dots or streaks) appear preferentially on the intima of the lesser curvature of the aortic arch in both rabbits and hamsters. This region was thus a bona fide "lesion-prone area." Additional preparations were made from thoracic aorta, coronaries, cerebral arteries, heart valves, vena cava, and blood cells. Depending on the purpose of the inquiry, the specimens were examined by radioassay, biochemistry, permeability probes, fluorescence microscopy, electron microscopy, and cytochemistry.

Animals and diet. The experiments were carried out on 286 chinchilla adult male rabbits fed a standard chow supplemented with 0.5% cholesterol and 5% butter, and 775 Golden Syrian adult male hamsters for which the diet contained 3% cholesterol and 5–15% butter. This novel animal model, the hyperlipidemic hamster, proved to be a very useful tool for research on atherogenesis.[28,91] In the hyperlipidemic rabbit the major cholesterol carrier is β-VLDL,[77] whereas in the hyperlipoproteinemic hamster plasma cholesterol is carried mainly by LDL as in humans. Each diet group was paralleled by control animals fed standard chow.

Plasma lipid analysis. From animals fasted overnight, weekly, before sacrifice, blood was collected from the jugular vein. Lipoproteins were isolated as described in Ref. 119 and verified by immunoelectrophoresis, cross-immunoelectrophoresis, and isoelectric focusing. Plasma total lipid distribution was determined by thin-layer chromatography, and serum cholesterol was measured enzymatically with the Sigma reagent kit.

Under relatively large individual variations, rabbits developed rather rapidly hypercholesterolemia characterized by the presence of β-VLDL as the major cholesterol carrier.[77,119] Plasma cholesterol went from 40 mg/ml to 250–500 mg/ml in the 2nd week, 1000 mg/ml in the 4th week, and up to 2000 mg/ml within 8 weeks on the diet. In the hamsters, serum cholesterol doubled in 3 weeks, increased 4-fold in the 4th week, and attained an ~ 17-fold value after 10 months on the diet.

III. LESIONAL STAGES CONSIDERED

Taking into account the focal nature, the difference in the speed of development, and the polymorphism of the early response of the vessel wall to hyperlipoproteinemia, we have arbitrarily classified the ultrastructural aspect of the arterial wall in the lesion-prone areas into six lesional patterns. Since we considered that they represent phases in the progression of an atherosclerotic plaque, we designated them as "stages." Briefly they are characterized as follows[36]:

- *Stage 0: Nonlesional areas without detectable extracellular liposomes and/or matrix proliferation:* any ultrastructurally lesion-free zone in the arterial intima of hypercholesterolemic animals that is indistinguishable from the normal artery.
- *Stage I: Nonlesional areas containing extracellular liposomes and/or proliferated matrix.* In the first 2 weeks of the diet in some apparently normal areas with intact endothelium, the intima displays an accumulation of phospholipid vesicles rich in unesterified cholesterol, tentatively named extracellular liposomes.
- *Stage II: Sites of mononuclear cell adhesion and migration.* Beginning with the 3rd week, monocytes start attaching and migrating underneath endo-

thelium that structurally appears intact. Usually, in these regions the sub-endothelial matrix displays extracellular liposomes.

- *Stage III: Areas of foam cell formation.* Monocyte-derived macrophages, and in later stages proliferated smooth muscle cells are progressively loaded with lipid deposits to become foam cells.
- *Stage IV: Areas with lipid-laden endothelial cells;* can coexist with stage III.
- *Stage V: Sites of foam cell immigration into the blood.*

In all these stages, no endothelial denudation or platelet involvement was noticed.

In a simplified form, for the purpose of this chapter, we shall refer to the *prelesional stage* (including stages 0 and I), the stage of the first waves of *monocyte adhesion and diapedesis* (stage II), and the stage of *foam cell formation* (stages III to V).

Because of the focal nature of the atherogenic process and its relatively slow and uneven progression, nonlesional and lesional areas could coexist in the same region, regardless of the duration of the diet. Monocyte attachment and diapedesis is a continuous process spanning in various degrees the entire period examined.

Our investigations were especially concentrated on the changes occurring during the prelesional stages prior to monocyte involvement.

IV. TRANSPORT OF LIPOPROTEINS BY THE NORMAL ARTERY WALL

Our studies on normal animals addressed the following issues:

1. *The transport of endogenous lipoproteins:* the detection of the interaction between endogenous lipoproteins and the endothelium *in vivo* in various blood vessels
2. *The transport of exogenous LDL* included the following studies

   - Identification of the transendothelial pathways taken by exogenous LDL perfused *in situ*
   - Estimation of the uptake by the artery wall of radiolabeled native LDL as compared to methylated LDL perfused *in situ*
   - Examining whether apoprotein B or methylated apoprotein B (unlabeled or radiolabeled and tagged with gold particles) takes the same transendothelial routes as their LDL counterparts

3. *The transport of β-VLDL*

   - Evaluation of the uptake and accumulation of fluorescent or fluorescent and radiolabeled β-VLDL by the artery wall
   - Detection of the transendothelial pathways followed by β-VLDL tagged with gold particles (Table 1)

Table 1. Probes and Experimental Procedures Used to Study LDL Transport through Normal Vascular Endothelium

Probe	Experimental condition	Method of investigation
Endogenous rat lipoprotein	*In vivo* observations	Electron microscopic visualization of lipoprotein particles in endothelium
Exogenous rat LDL or human LDL	*In situ* perfusion in rat; time course experiments	Immunocytochemical detection of LDL location in arterial endothelium
Double labeling (^{125}I- and cholesteryl[^{14}C]oleate) of rat native LDL and methylated LDL	*In situ* perfusion in rat; time course experiments	Spectrometric measurements of radiolabeled LDL uptake by aorta
^{125}I-native human LDL, and ^{125}I-methylated human LDL	*In situ* perfusion in guinea pig; time course experiments	Spectrometric measurements of radiolabeled LDL uptake by aorta
Unlabeled or radioiodinated apo B–gold, and methylated apo B–gold[a]	*In situ* perfusion of coronary arteries and aorta, in guinea pig	Electron microscopic visualization of tagged apo B pathway through endothelium

[a] Additional experiments aimed to test the biologic activity of apo B–Au particles on bovine aortic endothelial cells in culture, using ^{125}I-apo B–Au followed by electron microscopic localization of tracer interaction with endothelium.

A. Visualization of Lipoprotein–Endothelial Interactions in Vivo

In various experiments in which tissue specimens were exposed to the mordanting effect of tannic acid,[112] we could occasionally see electron-opaque particles which by size and shape were tentatively identified as LDL and HDL (20–25 nm) or VLDL (30–70 nm). A more favorable situation was offered by spontaneously hyperlipidemic Sprague–Dawley rats the vascular lumina of which contained a high number of lipoprotein particles (LP), most being in the dimensional range of LDL and VLDL. Such particles could be detected in structures potentially involved in either endocytosis (morphologically identified as coated pits, coated vesicles, endosomes, multivesicular bodies, lysosomes) or transcytosis (plasmalemmal vesicles or patent transendothelial channels). A certain percentage of such particles occurred in the subendothelial space. In all specimens examined, no LP particles of any size could be observed in the endothelial junctions or intercellular spaces. This pattern of lipoprotein interaction with endothelium was the same in both large vessels (including aorta and coronary arteries), and microvasculature. Morphometric analysis showed that in these hyperlipidemic conditions at a given steady state the LP transcytosis was more pronounced than endocytosis. Of particles in the range of 20–25 nm, ~ 19% were found in endothelial features potentially involved in receptor-mediated endocytosis, ~ 20% within plasmalemmal vesicles (open on either endothelial cell front, or apparently located inside the cytoplasm), and ~ 20% in the subendothelial space. The LP particles were confined exclusively to the endocytotic or transcytotic vesicular carriers and were excluded from the paracellular route. The possibility of a reverse transendothelial pathway—from the subendothelium to the vessel lumen—seemed unlikely because of the steep LP concentration gradient between lumen and perivascular space (about 10:1 as detected by the counting of detectable particles).

When spontaneously hyperlipidemic rats were injected intravenously with cationized ferritin (CF) pI 8.4, the tracer was firmly bound to some LP, forming circulating LP–CF complexes. Examined at 5, 20, and 60 min after CF administration, LP–CF complexes were taken up by endothelium almost exclusively by endocytosis. This seemed to occur by adsorptive uptake via coated pits and coated vesicles, and probably also by fluid-phase uptake through a fraction of plasmalemmal vesicles. Progressively, LP–CF complexes were accumulated within multivesicular bodies and secondary lysosomes. These findings suggested a significant role played by the net surface charge of macromolecular complexes in their intracellular sorting and fate *in vivo*.[114]

B. Transcytosis of Exogenous LDL Perfused in Situ

Native untagged homologous (rat) or heterologous (human) LDL, at concentrations varying from 10 to 80 mg cholesterol/100 ml, was perfused *in situ* in rats for time intervals of 2, 5, and 10 min. After fixation, specimens from aorta and coronary arteries were processed for immunocytochemistry using anti-rat LDL IgG coupled to horseradish peroxidase. The exogenous LDL particles were identified and recorded in various locations within endothelial cells for morphometric analysis. The results showed that the perfused LDL was transported by arterial endothelium via a dual mechanism involving two different parallel compartmented routes.

1. A relatively small amount of particles was endocytosed by (a) a high-affinity receptor-mediated process via coated pits, coated vesicles, and (b) probably by a receptor-independent fluid-phase uptake carried out by a fraction of plasmalemmal vesicles. By both mechanisms, LDL was brought to lysosomes to supply cholesterol for the metabolic needs of the endothelial cell.

2. Most perfused LDL was transported across endothelial cells by transcytosis performed stepwise by plasmalemmal vesicles, thus delivering cholesterol to other cells. Endocytosis was diminished at low temperature and was not augmented by increasing LDL concentration. Transcytosis appeared to occur either mediated by a low-affinity receptor or to be receptor-independent since it was less modified at low temperature, but was markedly enhanced at high LDL concentrations. Endocytosis of homologous LDL was significantly more pronounced than that of heterologous LDL; however, both types of particles were comparably transcytosed. Since in most experiments the LDL concentrations used were above the normal level in rats, it was presumed that at low LDL concentrations, the

saturable high-affinity uptake would occur at higher values.[143-147] The extent of LDL transcytosis was found to vary largely from one region to another of both aortic and coronary endothelium. Other authors have also reported local variation in rabbit aortic wall permeability to LDL,[134] an enhanced LDL uptake occurring near the branch ostia,[104,105] and a significant effect of HDL on this transport.[63] The focal character of LDL accumulation in the aorta was well documented by numerous techniques including the recently introduced immuno-electrotransfer procedure.[54]

Within the inherent limitations of this approach and some uncertainties about the quantitative interpretation of the morphometric data obtained, the conclusion that most of the transendothelial transport of LDL is receptor-independent agrees with the biochemical data reported by others.[120,130,131,151]

C. Role of the LDL Receptor in Transcytosis

It has been shown that unlike native LDL, methylated LDL is not recognized by the LDL receptor.[75,76,154] Based on the assumption that the uptake of native LDL will reflect the sum of receptor-dependent and receptor-independent processes,[131] methylated LDL will give an estimate of the receptor-independent mechanism. We perfused *in situ* in guinea pigs with this probe double-labeled with [^{125}I]- and [^{14}C]oleate cholesterol (10 or 20 mg LDL cholesterol/100 ml), at 37°C for 1 hr. Intima stripped from the aorta and the rest of the artery wall were separately radioassayed. After 1 hr of LDL perfusion, 50–60% of the radioactivity was confined to the intima (although this tunic represents less than 5% of the whole tissue mass of the artery wall). As indicated in Table 2, the mean values recorded for the two probes were not significantly different.[145-147]

Experiments using gold complexes with native apo B or methylated apo B showed that in both cases transcytosis is much more prominent than endocytosis; the difference between these two processes was particularly pronounced in the case of methylated apo B. In similar experiments conducted in Steinberg's laboratory, ^{125}I-labeled reductively methylated LDL was simultaneously injected with ^{131}I-labeled native LDL: since the mean values for the ratios obtained were

Table 2. Uptake of Native LDL and Methylated LDL by Normal Aorta[a]

Probe and animal	LDL concentration (mg LDL cholesterol/100 ml)	ng LDL cholesterol/ mg aortic wet tissue per 1 hr		Ratio methyl-LDL/ native LDL
		Native LDL	Methyl-LDL	
Rat LDL	10	6.02 ± 1.37	5.69 ± 2.76	0.94 ± 0.16
Rat aorta	20	15.74 ± 7.9	14.7 ± 5.2	0.94 ± 0.16
Human LDL– guinea pig aorta	10	4.21 ± 2.5	3.63 ± 1.71	0.87 ± 0.13

[a] Double labeling: ^{125}I- and cholesteryl[^{14}C]oleate-LDL. Results are means ± S.D.

not significantly different from 1.0, it was concluded that the transendothelial transport of LDL could not be to any large extent dependent on a receptor-mediated mechanism.[131,156] Because of the possibility of LDL disassembly within the extracellular matrix,[119] the data obtained with the "trapped ligand" method[12,131] may not exactly reflect either the actual magnitude or the site (extracellular versus intracellular) of LP degradation within the artery wall *in vivo*. Nevertheless, the results from both laboratories showed that aortic endothelium *in vivo* expresses LDL receptors. This is at variance with the observations that highly confluent endothelial cells in culture express little or no LDL receptor activity.[27,43,64,104] Other studies, however, have reported a more significant LDL receptor activity of endothelial cells in culture, but this was generally instrumental only for the endocytotic pathway. Recent studies using more elaborate and better controlled conditions *in vitro*[45,88,89,110] brought information on LDL transcytosis,[111,143,145] such as its saturability[45] and sensitivity to the presence of fatty acids.[46] These discrepant data show that studies based solely on cell culture systems cannot, by their nature, tell us enough about the actual mechanisms of endothelial transport of macromolecules *in vivo*[119,131,135] (for review see Ref. 113). All these data, however, remain only a rough estimate of a very complex process, since they cannot take into account the known LDL structural heterogeneity and several other aspects of its physical chemistry and metabolism[14,30,81] that may be important for its atherogeneity in any experimental condition and any individual animal.[99,100] Another aspect of the overall transport of LDL through the normal artery wall that becomes crucial for understanding atherogenesis is the mechanism(s) of its efflux from the cells and the matrix of the arterial intima.[16,129]

D. Transport of Native or Methylated Apo B

To determine if the interaction of LDL with endothelium is governed by its apoprotein moiety (apo B), we examined the uptake and transcellular routes taken by a gold complex with either native apo B or reductively methylated apo B (Table 2). Due to modification of lysine residues, the latter is assumed not to be recognized by the LDL receptor.[75,76,154] A soluble immunoreactive form of apo B was prepared from human LDL by delipidation in the presence of Triton X-100. Methylated apo B was obtained from previously methylated human or rat LDL, and the degree of methylation was assayed by the trinitrobenzenic sulfonic acid method. Gold complexes with native apo B or methylated apo B were prepared by adsorbing these proteins on 15-nm colloidal gold particles. The interaction of apo B–gold complexes with endothelium was tested by incubating 10–70 μg/ml of [^{125}I]apo B–gold (10 cpm/ng protein) with bovine aortic endothelial cells in culture for 24 hr. Surface binding and cellular uptake of the radiolabeled ligand were measured according to Ref. 38. Although the curves depicting the binding and uptake were indicative of a receptor-mediated process, no detectable degradation could be measured (probably due to the binding of degraded polypeptides to the gold particles which did not enter the cytosol). The binding and uptake of [^{125}I]apo B–gold was efficiently competed with excess cold LDL. In some experiments the endothelial cells were incubated for 2 hr at 37°C with [^{125}I]apo B–

gold (50 µg/ml, protein content), then cooled for 30 min at 4°C, fixed and prepared for electron microscopy.

Apo B–gold or methyl-apo B–gold (at concentrations corresponding to $E_{515 \text{ nm}}^{1.0 \text{ cm}} = 1.4$), at 37°C, was perfused *in situ* in guinea pigs for 5–15 min. After washing and fixation *in situ,* specimens of thoracic aorta and coronary arteries were processed for electron microscopy. In controls, a polyethyleneglycol–gold (PEG–gold) complex ($E_{515 \text{ nm}}^{1.0 \text{ cm}} = 2.8$)[37] was perfused for 15 min, instead of apo B–gold.

Apo B–gold particles perfused *in situ* were found both in features involved in endocytosis (coated pits/vesicles, endosomes, multivesicular bodies, and lysosomes) and in structures performing transcytosis (e.g., plasmalemmal vesicles). At 5 min the ligand predominantly labeled vesicles open on the luminal front, and at 10 and 15 min it also marked vesicles open on the abluminal front and the subendothelial space. In contrast, PEG–gold particles were only occasionally found at the level of endothelial cells.

These findings agree with those from experiments with radiolabeled apo B (native or methylated) in revealing that (1) apo B is important in routing LDL within endothelial cells, (2) for both native LDL (apo B) and methylated LDL (methyl-apo B) transcytosis represents the major pathway followed *in vivo,* being more pronounced for the methylated form.[145–147] Transcytosis appears to be the main endothelial transport process for both native and methylated LDL. Due to the very low endocytotic rate of methylated LDL, the relative amount transcytosed *in situ* appears particularly pronounced though its total accumulation within the aortic wall is comparable to that of native LDL.

E. Transport of β-VLDL

We investigated the uptake, transendothelial pathways and accumulation of β-VLDL by the aortic wall of normal rabbits. The goal of this study was twofold: to gain a better insight into the transport of this LP *in vivo,* and to establish a baseline for the evaluation of changes in β-VLDL transcytosis during experimental hypercholesterolemia.

To this intent we tagged rabbit β-VLDL with DiI (1,1'-dioctadecyl-3,3,3',3'-tetramethylindocarbocyanine perchlorate) to make it fluorescent, with ^{125}I for radioassays, or with colloidal gold for electron microscopic examination. To determine if part of the dietary cholesterol eventually enters the arterial wall, some animals received 1.5 mCi [^{3}H]cholesterol/100 g diet, and after 18 and 40 hr, radioactivity was measured in plasma and in the aortic intima and media, separately (Table 3). The experimental procedures used with these probes are presented in Section V. The results indicated that dietary [^{3}H]cholesterol was incorporated into the aortic wall especially in the intima. *In vivo* and *in situ* uptake of β-VLDL labeled with DiI by the aortic wall was barely detectable. In experiments with [^{125}I]-β-VLDL injected *in vivo,* and allowed to circulate for 24 hr, the radiolabeled protein moiety recovered from the aortic wall was about 3 ng, and the cholesterol (estimated from cholesterol/protein ratio) about 10–15 ng/mg wet tissue. At the cellular level, β-VLDL–gold particles were found in some endothelial structures

Table 3. Probes and Methods Used to Investigate Uptake of Dietary Cholesterol and of Perfused β-VLDL by the Artery Wall in Early Stages of Artherogenesis (Studies on Normal or Hypercholesterolemic Rabbits)

Probe	Mode of administration	Method of investigation	Samples examined
[^{3}H]-Cholesterol	Diet	Radioassay[a]	Plasma, arterial intima and media
[^{125}I]-β-VLDL	i.v. injection in vivo	Radioassay	Plasma, arterial intima and media
[^{125}I]-β-VLDL–DiI[b]	Perfusion in situ	Radioassay	Plasma, arterial intima and media
		Fluorescence microscopy	Artery wall
β-VLDL–DiI	Perfusion in situ	Fluorescence microscopy	Artery wall
β-VLDL–Au	Perfusion in situ	Electron microscopy	Arterial intima

[a] Spectrometry.
[b] DiI, 1,1'-dioctadecyl-3,3,3',3'-tetramethylindocarbocyanine perchlorate.

involved in endocytosis, but especially in those potentially performing transcytosis, i.e., plasmalemmal vesicles in all locations, as well as in subendothelium. No tracer was seen in the intercellular junctions or spaces of the aortic endothelium.[116–119,145–147] There are as yet no conclusive data on the routes taken by VLDL, the fractional contribution of endocytosis versus transcytosis, or the receptor-dependent versus receptor-independent mechanisms.[110] However, the data suggest the existence of a dual process similar to that identified for LDL. It has been shown that cultured aortic endothelium, in addition to LDL receptors and scavenger receptors, also expresses β-VLDL receptors.[3,18] Our results on normal animals were used for comparison with those obtained on cholesterol-fed rabbits.

V. TRANSPORT AND ACCUMULATION OF β-VLDL IN THE ARTERY WALL OF HYPERCHOLESTEROLEMIC RABBITS

The uptake of β-VLDL by the aortic wall in cholesterol-fed rabbits was compared with that in animals fed a standard chow (Table 3). The observations extended over 12 weeks of the fat-rich diet were essentially focused on two early periods: (1) the first 2 weeks when in the lesion-prone regions there were no structurally detectable lesions (the prelesional state designated as stages 0 and I); and (2) the following 2–4 weeks when monocytes started emigrating and foam cell began to form in the intima of the aortic arch and on the ventricular aspect of the atrioventricular and arterial aspect of the sigmoid cardiac valves. As a

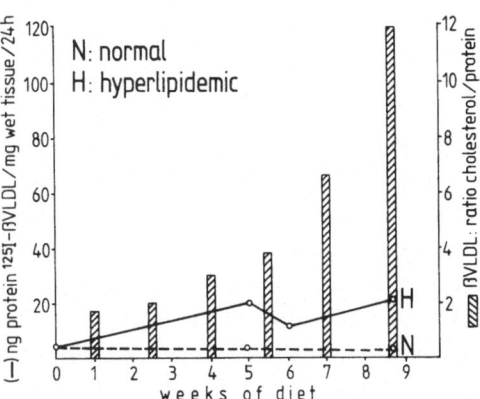

Figure 1. The uptake and accumulation of the protein moiety of [^{125}I]-β-VLDL by the aorta of normal (N) and hyperlipidemic (H) rabbits. The changes are compared with the cholesterol/protein ratio of β-VLDL particles determined during the first 9 weeks of the diet. ([^{125}I]-β-VLDL was injected *in vivo* at different times of the diet and allowed to circulate for 24 hr.)

general pattern, regions with various degrees of foam cell formation coexisted with areas of prelesional (or nonlesional) type.

In the first 2 weeks of the diet, plasma cholesterol was about 7–15 times higher than normal (~ 0.31 mg cholesterol/ml); in the 8th week the values increased to 30–50 times that of controls. In parallel with the progressive decrease in serum content in LDL, β-VLDL became the major cholesterol carrier. Concomitantly, the ratio of mass cholesterol to mass protein content of β-VLDL increased from 1.5–2.0:1 in the 1st week of the diet to ~ 6:1 in weeks 8–9 of the diet (Fig. 1).

The experiments we carried out had two aims: to determine if the cholesterol added in excess to the diet becomes progressively accumulated in the artery wall; and to see if in hypercholesterolemia the aortic wall takes up more β-VLDL, by what process, and where this lipoprotein is deposited.

A. Incorporation of Dietary [^{3}H] Cholesterol

The experiments were conducted in rabbits fed for 10 or 25 days the cholesterol-rich diet: in the last 46 hr, 1–5 mCi [^{3}H]cholesterol was incorporated into 100 g of the diet. At 40 hr, samples were taken from plasma for lipid analysis, and from the aortic arch for electron microscopy, or the intima was prepared for radioassay, as indicated in Ref. 119 (Table 3). After 40 hr of the [^{3}H]cholesterol diet, the serum radioactivity was recovered mainly in the β-VLDL fraction and to a lesser extent in the LDL fraction. Although the intima represented less than 1% of the total weight of the wet aortic wall, 50–60% of the radioactivity was recovered from the intimal tissue. After 10 days of the diet, the level of radioactivity incorporated within the aortic intima was similar to that measured in normal animals. In animals fed the diet for 25 days, the radioactivity accumulated in the aortic intima was double the value recorded both in normal rabbits and in 10-day diet-fed rabbits (~ 1900 cpm versus 950 cpm, respectively). The material extracted from intima, after ultracentrifugation, was recovered as a band in the range densities of LDL and VLDL.

B. Uptake and Accumulation of β-VLDL

1. Experiments with [125I]-β-VLDL Injected in Vivo

Weekly, cholesterol-fed rabbits were injected intravenously with 0.3–0.5 mCi [125I]-β-VLDL (obtained from animals at the same stage of the diet). After 24 hr, blood and segments of aorta were collected and radioassayed. In some specimens, intima and media were separately measured. In the first 2 weeks of the diet, in the aortic wall there was a slight deposition of β-VLDL protein moieties which even at later stages remained at values only slightly higher than those recorded in control animals. Yet, during this period the accumulation of cholesterol in the intima (detected chemically and cytochemically) was very prominent. This apparent discrepancy could be explained by the steep augmentation of the cholesterol content in the VLDL particles: the cholesterol/protein ratio increased from 1.54:1 in the 1st week to 3.8:1 in the 5th week and 12.0:1 in the 8th week. This very rapid and progressive accumulation of the lipid moiety of β-VLDL in lesion-prone areas was demonstrated by experiments in which radioiodinated β-VLDL was concomitantly rendered fluorescent for visualization of its sites of accumulation by fluorescence microscopy.

2. Experiments with [125I]-β-VLDL–DiI

As indicated in Ref. 119, we used either β-VLDL–DiI or [125I]-β-VLDL–DiI and examined the tissue by microscopy alone or microscopy and radioassay, respectively. After removing the blood, 8–10 ml of the probe in PBS at 37°C (4–5 mg cholesterol/ml) was recirculated through the thoracic aorta for 1–2 hr in a closed circuit. After washing with PBS, aorta was removed and examined by fluorescence microscopy; for radioassay, intima and media were separately collected and counted. Some specimens were stained with Nile red. With regard to normal animals, the aortic wall of hypercholesterolemic rabbits displayed an enhanced uptake of β-VLDL as demonstrated by the progressive accumulation of fluorescence in the intima. These findings were confirmed on a large sampling for detection of lipids at the histological and ultrastructural level.

3. Experiments with β-VLDL–Gold Conjugate

Prepared as indicated in Ref. 119, this electron-opaque probe was found, by negative staining, to contain one to four gold particles (5 nm) attached to one particle of β-VLDL. The probe was perfused *in situ* in hypercholesterolemic animals (first 2 weeks of the diet) using the same protocol as for [125I]-β-VLDL (Table 3). The tracer at a concentration corresponding to $E_{525\ nm}^{1.0\ cm} = 1.2$ at 37°C was perfused for periods up to 15 min, followed by washing of unbound tracer, fixation *in situ,* collection of segments of aorta, and processing for electron microscopy. Ligand particles were found predominantly in plasmalemmal vesicles present in various locations and in the subendothelial space (Fig. 2). To a lesser extent, the probe was also detected in endothelial structures involved in endo-

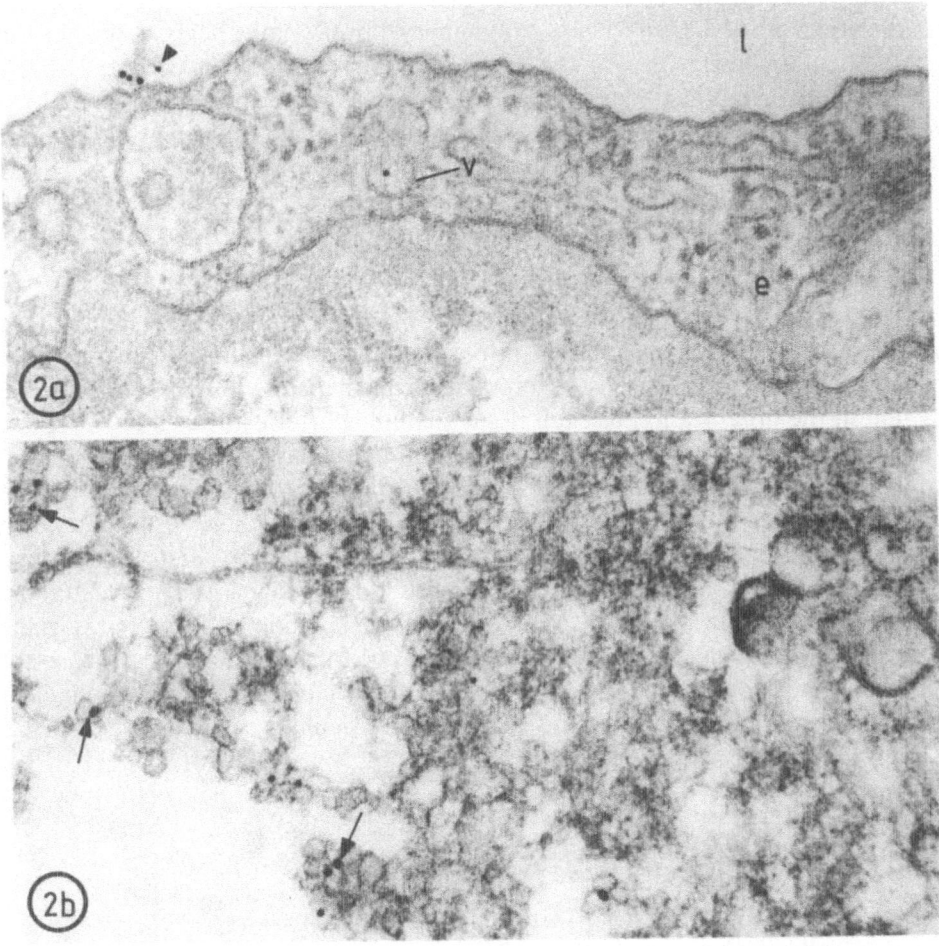

Figure 2. Transcytosis of β-VLDL–Au complex across endothelium in lesion-prone areas of the aortic arch (hypercholesterolemic rabbits at 2 weeks of the diet). (a) After 3 min of perfusion *in situ*, tracer particles appeared attached on plasma membrane (arrowhead) or internalized in plasmalemmal vesicles (v). (b) Starting at 5 min, tracer particles were detected in subendothelium as gold-carrying silhouette [lipoprotein particles or extracellular liposomes (arrows)]. e, endothelial cell; l, lumen. a, × 61,200, b, × 64,800.

cytosis. No tracer particles were observed in the intercellular spaces or junctions of aortic endothelium. At sites of monocyte migration, no β-VLDL–Au particles were detected passing through endothelial junctions. The same finding was reported for LDL in aortic endothelial cells in culture.[139] Our findings support the idea that in the early stages of cholesterol-induced atherogenesis in rabbits, the main transport process of β-VLDL is transcytosis via plasmalemmal vesicles. Endocytosis is generally less pronounced and the paracellular route is not a detectable pathway for β-VLDL particles entering the arterial intima.[119,145–147] It was reported that β-VLDL increases the permeability to LDL and albumin of rabbit aortic endothelial cells in culture[88]; however, albumin does not parallel the accumulation of VLDL *in vivo*.[15] Preliminary observations suggest that in the hyperlipidemic hamster *in vivo* the aortic wall shows an enhanced permeability to [^{125}I]albumin that is markedly augmented by histamine and serotonin.[31] However, the issue of endothelial permeability to albumin and other plasma proteins in the early stages of atherogenesis requires more elaborate study.

4. General Comments

The results reported above indicate that the LDL and β-VLDL interactions with vascular endothelium are only in part comparable to those occurring with fibroblasts, smooth muscle cells, macrophages, and hepatocytes. Due to its special position in the organization of each tissue including arteries, the endothelial cell is endowed with a dual mechanism for handling circulating LP, both in normal and in hyperlipoproteinemic animals (Fig. 3). As do most other cells, endothelium performs receptor-mediated endocytosis of LP that supply cholesterol for its own metabolic needs. Unlike other cells, however, endothelium is also transporting LP-bound cholesterol to other tissues by transcytosis, a low-affinity nonsaturable process involving plasmalemmal vesicles. The transcytotic pathway is the same for LDL, β-VLDL, and apo B. Transcytosis appears as an important means for monitoring the excessive accumulation of LP in the plasma. What is still unclear at the onset of atherogenesis is the contribution of enhanced transcytosis of LP

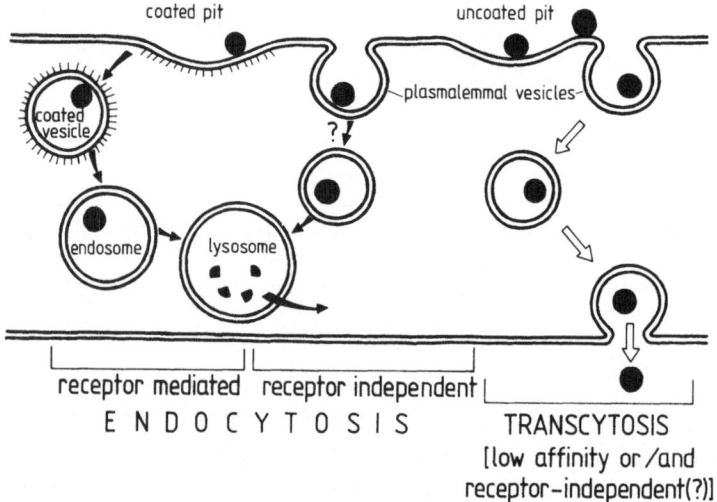

Figure 3. Diagrammatic representation of the dual pathway of lipoprotein (LP) transport by vascular endothelium. LP particles (LDL and β-VLDL) can be endocytosed to lysosomes by a receptor-mediated mechanism involving coated pits, coated vesicles, and endosomes and possibly by a receptor-independent process carried out by some pinocytotic vesicles. LP are also transcytosed by plasmalemmal vesicles via a low-affinity and/or a receptor-independent mechanism.

and what is due to their trapping by the intimal extracellular matrix.[92,119,123,124] During plaque development, lipid deposition depends on the ratio between LP influx, trapping, alteration, and efflux. In advanced stages, opening of endothelial junctions may add a convective pathway, while the intracellular lipid deposits, stromal proliferation, calcification, and necrosis largely impede the efflux of LP-derived material. Although it is firmly established that in most cases elevated levels of plasma LP are sufficient to cause lipid deposits in the artery wall, other separate or concurrent effects of LP on endothelial cells, monocytes, platelets, and smooth muscle cells might initiate or promote atherogenesis.

VI. EXTRACELLULAR ACCUMULATION OF LIPOPROTEIN-DERIVED COMPONENTS

To determine the state of LP material transcytosed in the intima in prelesional stages, we examined lesion-prone areas of the aortic arch and cardiac valves in rabbits and hamsters cholesterol-fed for 1–2 weeks. During this early period, no histological lesion could be detected. However, at the ultrastructural level we observed a characteristic progressive deposition of extracellular lipid material organized as small vesicles formed by phospholipid lamellae. These features, tentatively named "extracellular liposomes" (EL), were rich in unesterified cholesterol,[119] and were closely associated with apo B.[82,84] The deposition of EL continued to increase during monocyte diapedesis and foam cell formation. Over the entire period of these experiments, the endothelium was morphologically intact and no platelet involvement was noticed.

A. Extracellular Liposomes

To localize lipid deposits at the light microscope level, segments from the inner lesser curvature of the aortic arch and cardiac valves were cryosectioned and stained with Nile red or oil red O (to visualize the hydrophobic lipids) or with the fluorescent polyene antibiotic filipin (to detect unesterified cholesterol). For ultrastructural studies, comparable specimens were prepared for thin-section (including mordanting with tannic acid[112]) or freeze-fracture electron microscopy. To visualize proteoglycans, some samples were stained *en bloc* with safranine O.

In the first week of the cholesterol-rich diet, the lipid histochemistry did not reveal any significant modifications in the areas examined. At weeks 2 and 3, fluorescent material became visible in the superficial part of intima that was also Nile red-positive. Oil red O-positive material appeared only in week 4.

The electron microscopy revealed that the earliest modification detectable in the first 2 weeks on the diet was the appearance within subendothelium and then continuous deposition within the extracellular matrix of the intima of numerous, relatively densely packed uni- or multilamellar vesicles, 100–300 nm in diameter, with an amorphous clear content (Fig. 4). The accumulation of these EL was concurrent with the proliferation of layers of basal lamina-like material,

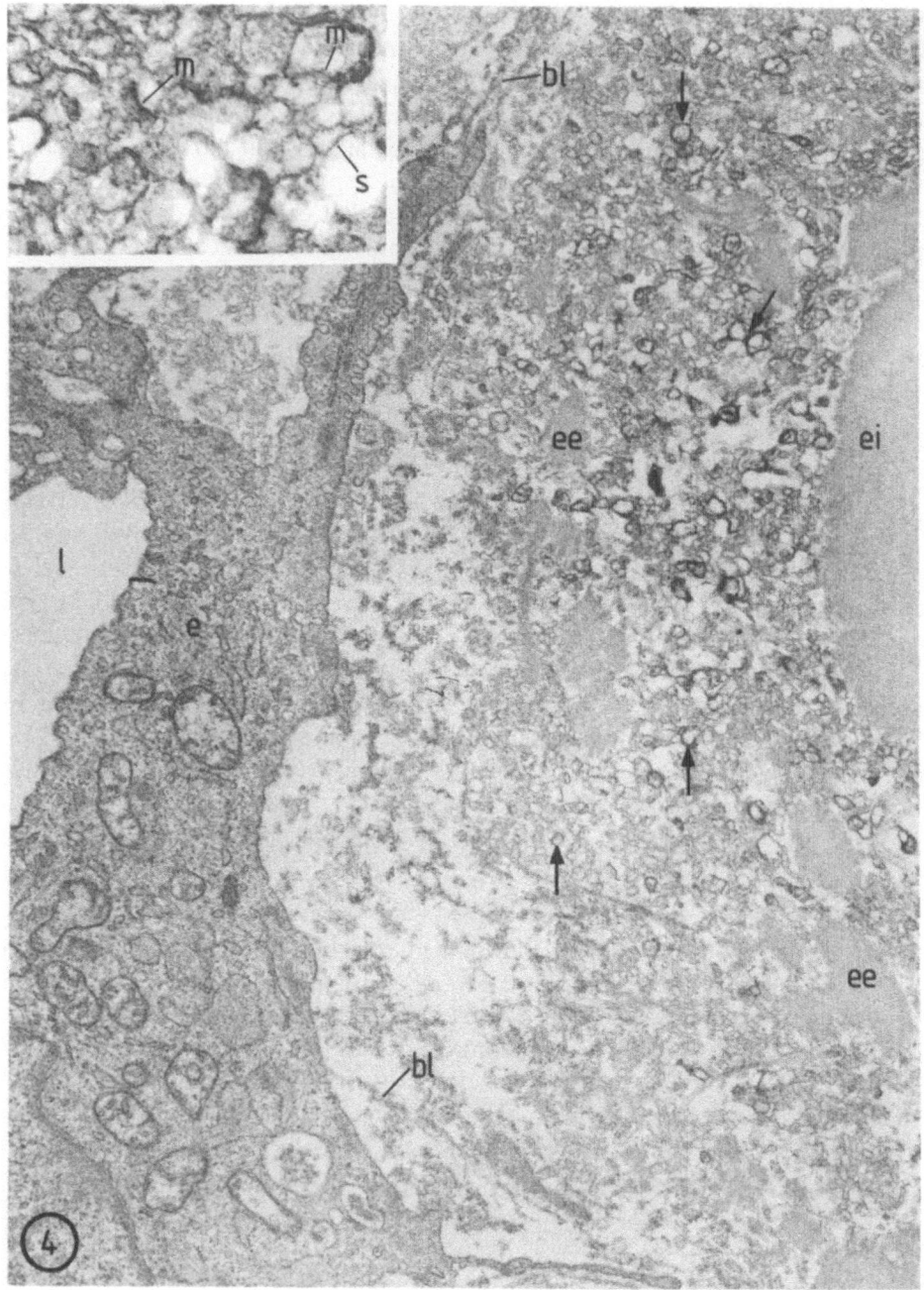

Figure 4. Hypercholesterolemic rabbit at 2 weeks of the diet: lesion-prone area of the aortic arch displaying a subendothelial accumulation of extracellular liposomes (arrows) trapped within a proliferated matrix. (Inset) The vesicle-like structures (extracellular liposomes) are formed by a single (s) or multiple phospholipid lamellae (m). e, endothelium; bl, basal lamina-like material; ee, elastic elements; l, lumen; ei, elastica interna. × 13,500; inset, × 66,600.

microfibrils, and proteoglycans to which EL were associated (Figs. 4 and 6). In freeze fracture, EL appeared as round smooth lamellar vesicles, devoid of particles (Fig. 5), indicating that EL are probably not devoid of translamellar proteins. EL occurred in cell-free subendothelium with no signs of cytolysis. Vesicle-like structures or polymorphic membranous material has been described in the atherosclerotic arteries by other investigators, being commonly ascribed to cell debris or precipitated LP. It was recently demonstrated that during atherogenesis, the rabbit aorta accumulates much sphingomyelin.[158] In areas with EL, apparently intact LDL or β-VLDL (more reliably identified immunocytochemically) were relatively rare. The literature contains conflicting data on the concentration and characteristics of LDL extracted from human atherosclerotic lesions. Some authors found such extracts to contain a relatively low concentration of LDL with unchanged electrophoretic mobility.[122–124] Others have extracted apo B-containing LP associated with glycosaminoglycans and having distinct chemical characteristics, including the capability to stimulate cholesterol esterification by macrophages *in vitro*.[13,39] Such discrepancies might be generated by differences in the extraction procedures used. Data about β-VLDL extracted from early lesions are at this time incomplete.

During foam cell formation, EL continue to accumulate in large amounts; electron microscopic images suggest that they are, at least in part, phagocytosed by macrophages. The latter are able to ingest artificial liposomes *in vitro*[56,103] and are particularly avid for LP complexes with components of the extracellular matrix.[24]

B. Unesterified Cholesterol

In specimens incubated with filipin (known to bind specifically to 3-β-hydroxysterols), characteristic 25-nm deformations revealing filipin–sterol complexes were detected both in thin sections and freeze-cleaved replicas (Figs. 5 and 6). This indicated that EL contain in their phospholipid lamellae a large amount of unesterified cholesterol. During foam cell formation (4–8 weeks) the mass of EL became very prominent, filling the interstitia, and displaying conspicuous filipin–sterol complexes also visible in foam cells. With the increasing number and dimensions of macrophage-derived and at later stages smooth muscle cell-derived foam cells, EL deposits became less pronounced and more polymorphic. In regions with cytolysis they could not be reliably distinguished from cell debris. EL may be the ultrastructural equivalent of the filipin-positive extracellular particles localized by fluorescence microscopy in human or animal atherosclerotic aortas.[64–66] In addition to EL, the extracellular pool of free cholesterol may be incorporated in the plasma membrane of endothelium and foam cells,[57,95] or be associated with proteoglycans,[11,52,53,55,101,127] collagens,[21] elastin[126] or may be transported in the lymph bound to LDL or HDL.[96] The free cholesterol extracted from atherosclerotic lesions in the fractions of nonlipid tissue components[121] may represent EL unspecifically bound to matrix. It is still unclear whether part of the free cholesterol reaches the subendothelium by direct transfer from plasma LP to endothelial cells[26,136,137] or is carried by albumin.[17]

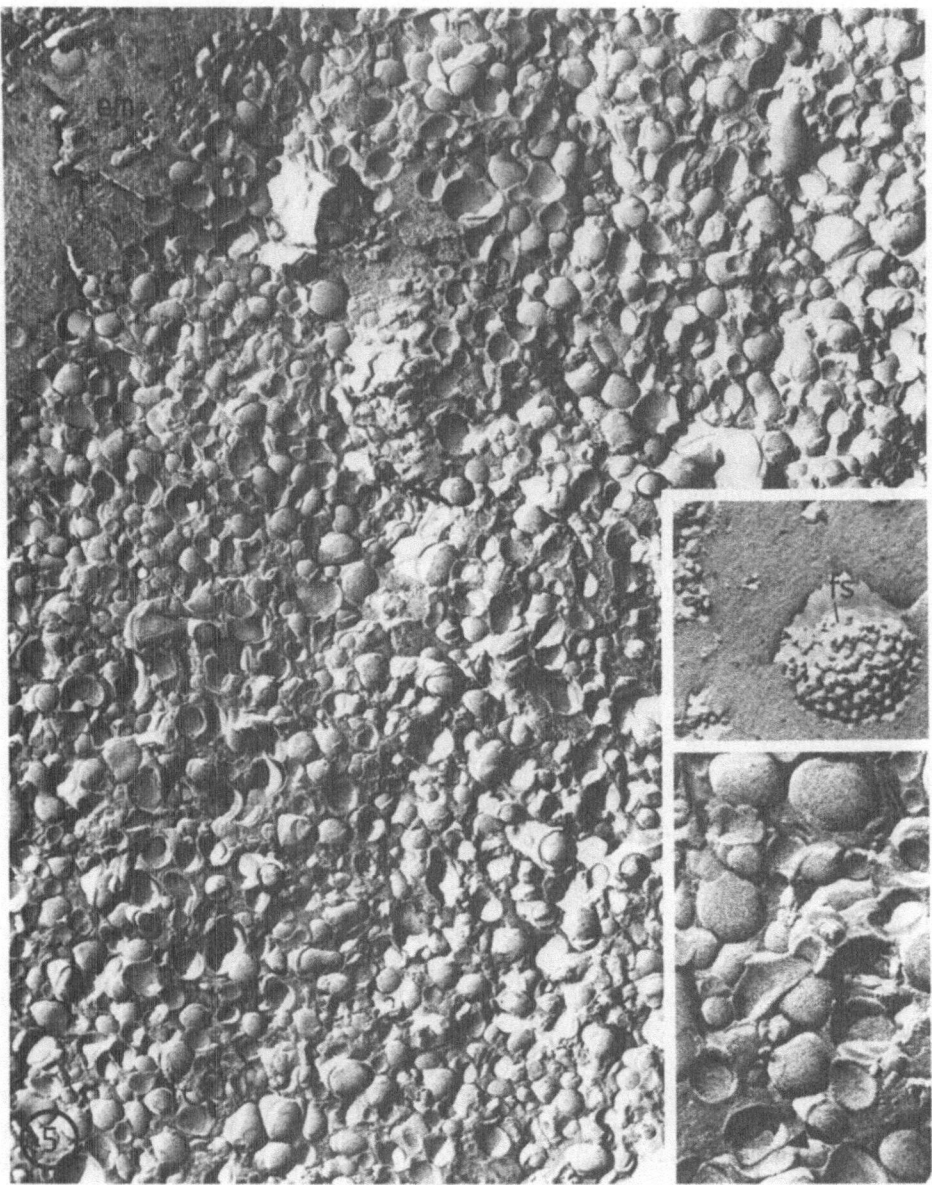

Figure 5. Hypercholesterolemic rabbit at 3 weeks of the diet: lesion-prone area of the aortic arch. Freeze-fracture preparation revealing in the subendothelium densely packed extracellular liposomes (arrows). em, extracellular matrix. (Upper inset) Upon incubation with filipin, the extracellular liposomes exhibit characteristic protrusions of filipin–sterol complexes (fs). (Lower inset) The cleavage planes of vesicles are smooth, devoid of translamellar particles; some vesicles show concentric lamellae (arrowhead). × 63,900; lower inset, × 117,000; upper inset, × 135,000.

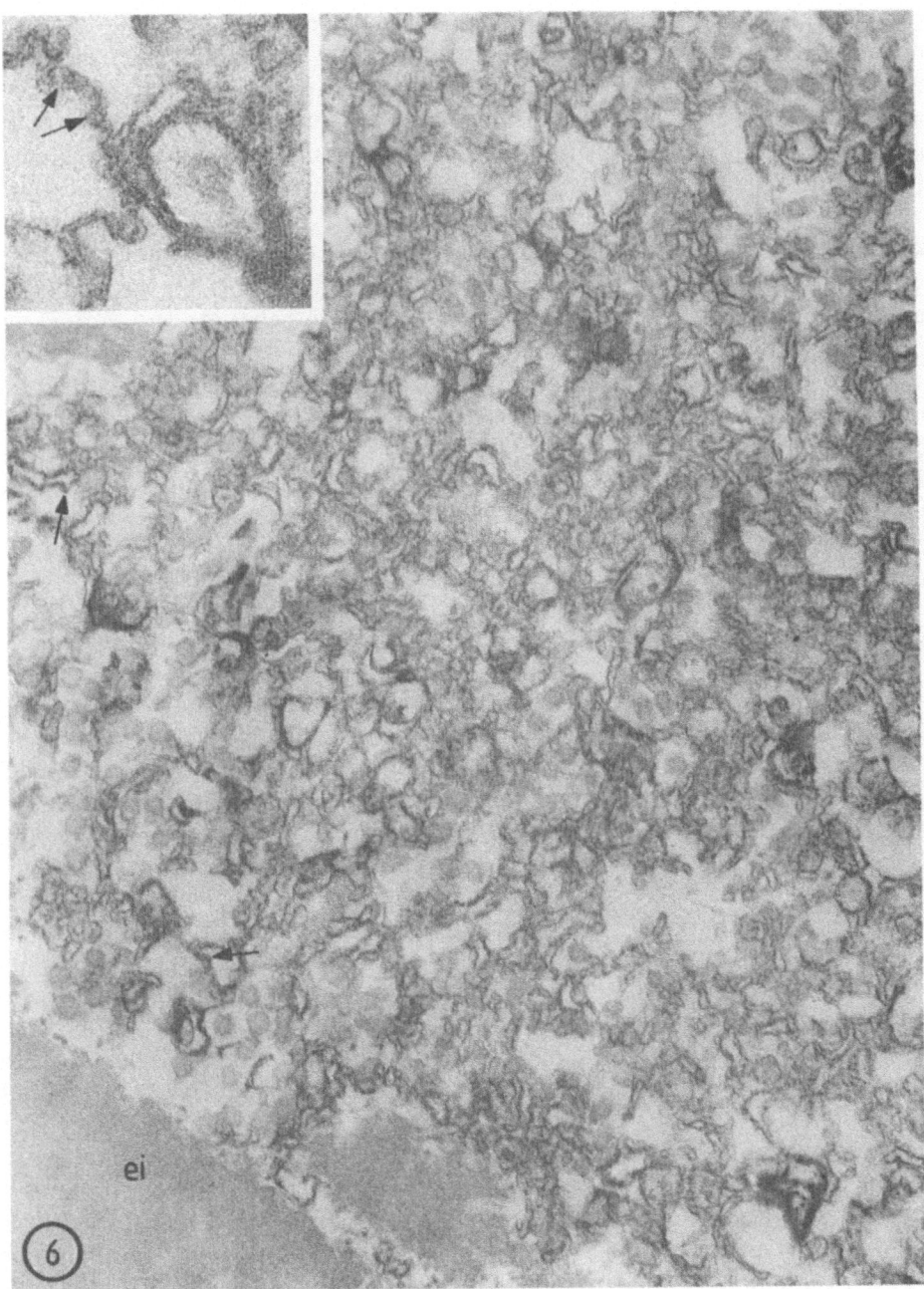

Figure 6. Lesion-prone area of the aortic arch of a hypercholesterolemic rabbit (3 weeks of the diet). Upon incubation with filipin, the extracellular liposomes show characteristic deformations induced by the filipin–sterol complexes, seen at a higher magnification in the inset (arrows). ei, elastica interna. × 75,600; inset, × 169,200.

C. Apo B and Apo E

Apo B was cytochemically detected in advanced atherosclerotic lesions in humans and animals.[8,25,49–51,67] Apo E was generally detected at the light microscope level[86] and found to be present in regions with lipid deposition.[2] In our study, the occurrence of apo B in areas with EL was detected by indirect immunoperoxidase technique using light and electron microscopy. Concomitantly, in the same specimens, unesterified cholesterol (UC) was revealed with filipin and tomatine as specific probes. The latter has the advantage of forming spicule-like complexes with UC that is not associated with membranes. In the prelesional stages, extracellular phospholipid liposomes, UC, and apo B appeared concomitantly and accumulated focally in the same subendothelial regions. Apo B was preferentially located on the contour of EL (Figs. 7 and 8). During lesional stages leading to fatty streak formation, EL, UC, and apo B colocalized, both in the matrix and in some lipid-laden cells derived from macrophages, smooth muscle cells, or endothelial cells.[84–86] A localization similar to apo B was found for apo E. However, since apo E was shown to be synthesized by monocytes, macrophages,[4,155] and smooth muscle cells[19] and appears to be a widely distributed cellular protein,[20,70] the actual source of this protein moiety in those areas is uncertain.

D. Extracellular Matrix

The important role played in the retention and accumulation of LP within the arterial wall was well documented for glycosaminoglycans,[52,53,55,127,148,149] collagens,[21,29,80] and elastin.[29,87,126,151] In our observations, significant modifications of the extracellular components appeared and developed in association with the local occurrence of LP-derived material such as EL. In some locations, restricted areas of a "sero-fibrinous insudate" could be observed in the intima, but always distant from zones containing EL. Such insudate was usually a transient phenomenon. Commonly, the first group of EL appeared interposed between endothelium and its basal lamina that seemed to be repeatedly detached, and new basal lamina-like feltwork material formed. When in sufficient amount, EL had the tendency to accumulate next to bands of proteoglycans or microfibrils, much less to banded collagen. Safranine O-positive fibers occurred frequently in the close vicinity of EL or attached to them and to most elastin bundles. While initially EL accumulated against a compact lamina elastica interna, at later stages small irregular bundles of elastic material were spread throughout the intima. It was unclear whether these features expressed a process of elastolysis or proliferation of preexisting elastic elements. During foam cell formation, EL and their associated UC maintained a close relationship with the stromal fibrillar components, a situation reported to persist also during advanced lesions of the fibrolipid type.[7–9,44,45]

E. General Comments

The EL and their associated material are made up of phospholipids, UC, apo B, and probably apo E which represent the main components of plasma β-VLDL.

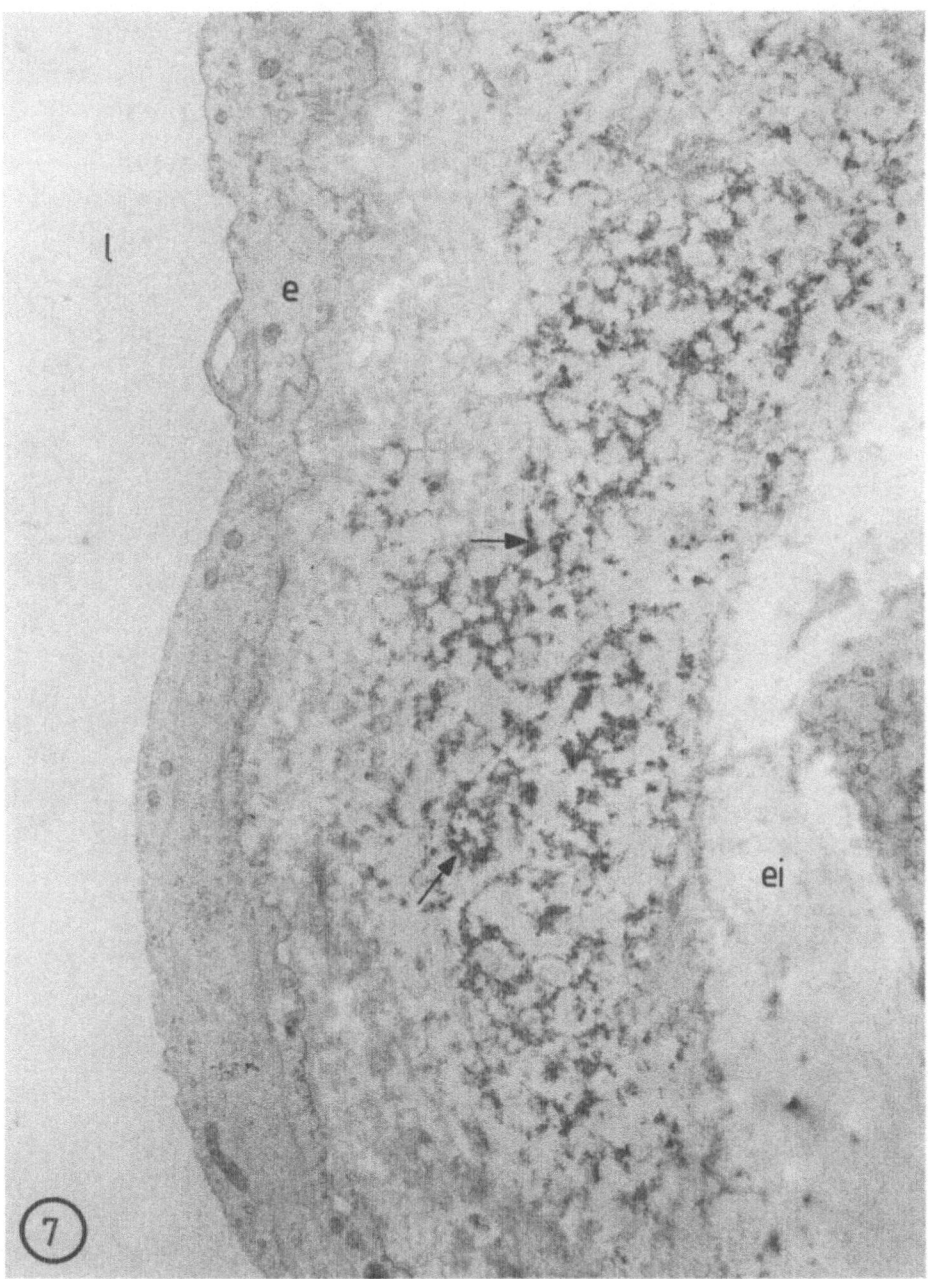

Figure 7. Hypercholesterolemic rabbit at 2 weeks of the diet: lesion-prone area displaying in the sub-endothelium apo B–HRP reaction product (arrows). e, endothelium; ei, elastica interna; l, lumen. × 12,000.

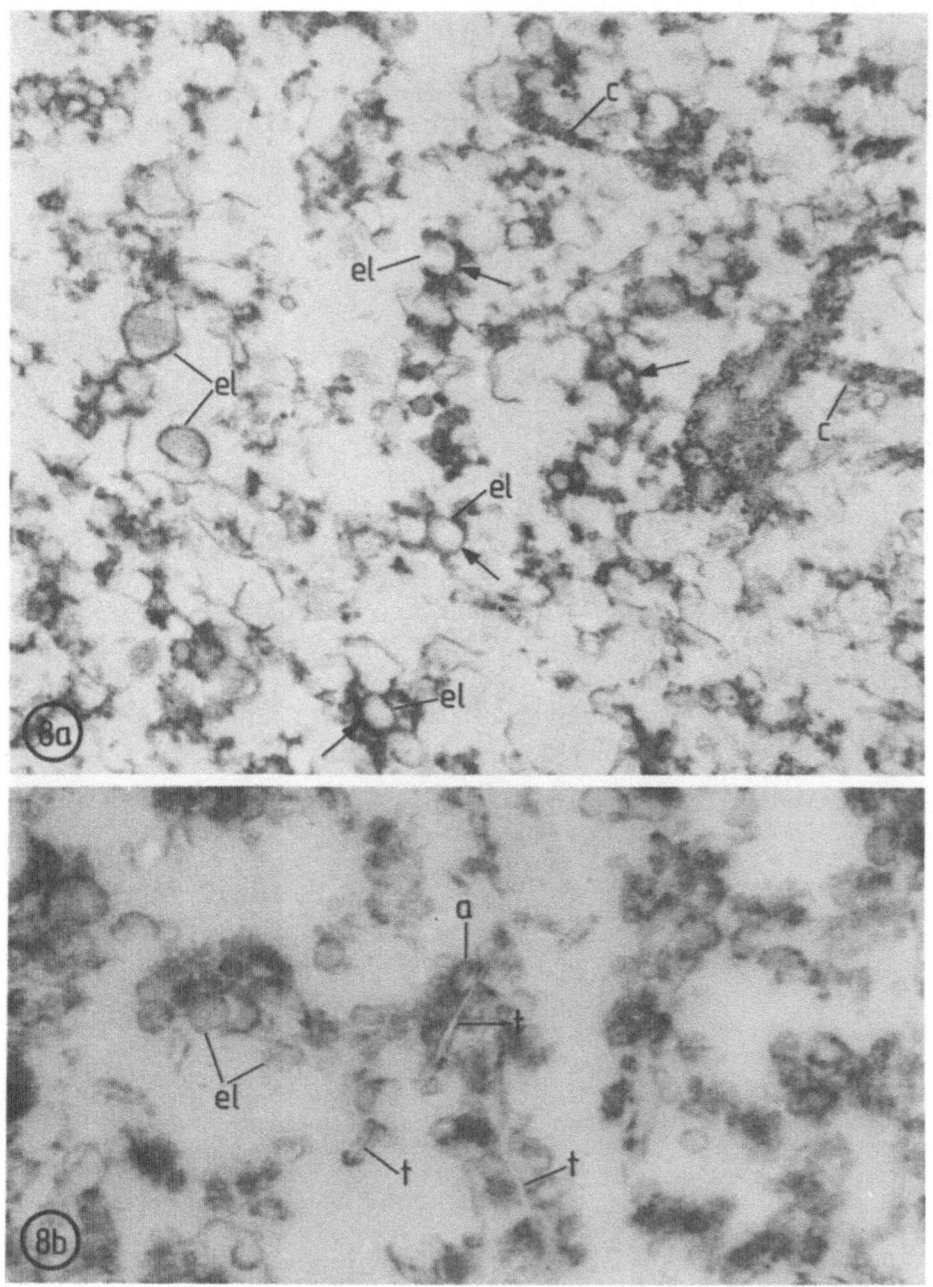

Figure 8. (a) **Apo B–HRP** reaction product (arrows) occurs mostly on the contour of extracellular liposomes (el). c, collagen. × 44,100. (b) Colocalization of phospholipid extracellular liposomes (el), apo B–HRP reaction product on their contour (a) and sterol–tomatine complexes appearing as spicules (t). × 40,500.

The fate of cholesteryl ester in these areas is under investigation. The smooth aspect of EL in freeze-fracture replicas, and the occurrence of immunoperoxidase reaction product with anti-apo B in the EL periphery, suggest that apo B is associated with the contour but not included in the phospholipid lamellae. The extracellular UC is not in direct contact with the hydrophilic components of the matrix but is incorporated in more complex liquid-crystalline phase such as the phospholipid lamellae.[71,121] This may be interpreted as a local defense reaction that renders inoffensive the sclerogenic cholesterol.[70,121] *In vitro* it was shown that apo B can reassociate with phospholipids,[1] and that in atherogenic plasma artificial liposomes can acquire apo E.[157]

The similarity of prelesional changes (such as the appearance of EL) in various locations (e.g., aorta, coronary, cardiac valves) and in different animal species (rabbit and hamster[91]) suggests that a common mechanism is operating at the onset of accelerated hypercholesterolemic atherogenesis. These findings also show that in the early prelesional stages of diet-induced atherogenesis, the transcytosed LDL and/or β-VLDL occurred rarely as intact particles, but rather as partially degraded and probably reassembled LP-derived components. The latter appeared mostly as phospholipid EL rich in UC and associated with apo B and apo E. Reassociation of apo B and phospholipids such as dipalmitoyl phosphatidylcholine was demonstrated *in vitro*.[1] It is still unclear whether LP degradation occurs during their interaction with endothelium (involving lipid peroxidation and phospholipid and apoprotein degradation[85,93,132]), during transcytosis, or in subendothelium. The physiopathologic role of the reassembled material for the plaque development, the fate of cholesteryl esters within the extracellular matrix and their possible deesterification[129] remain to be established.

The prelesional changes revealed by our studies should be considered in the framework of other early atherosclerotic tendencies reported in the literature.

VII. CHANGES IN THE SURFACE CHARGE AND CHEMISTRY OF ARTERIAL ENDOTHELIUM AND MONOCYTES

The evidence that the onset of atherosclerosis requires neither physical damage to endothelium nor platelet involvement[60,61,78,97,119,133] prompted us to search for possible subtle focal endothelial changes that can be linked with atherogenic tendencies. Since the endothelial cell surface participates in all major physiopathologic processes involving the vessel wall, and because the earliest cellular event observed has been the focal adherence of mononuclear cells to arterial endothelium,[32,35,60] we have investigated *in vivo* changes that hypercholesterolemia may induce in the chemistry of the cell coat of aortic endothelium and blood monocytes. The few data so far available were obtained at various separate stages of cholesterol feeding,[59] and were limited to variations in staining with ruthenium red,[33,69] concanavalin A,[69,153] or cationized ferritin.[69] Our studies were focused on the endothelial cell surface of the aortic arch in normal and in hypercholesterolemic rabbits in areas displaying different stages of atherogenesis; concomitantly, we examined the monocytes present either in the circulation or interacting with endothelium.

A. Probes and Experimental Procedures

Weekly (up to 8 weeks), after blood samples were taken for lipid analysis and blood cell preparation, rabbit vasculature was washed free of blood by perfusion with PBS and the endothelial luminal surface exposed to cytochemical probes *in situ*. After fixation *in situ*, specimens collected from lesion-prone regions of aortic arch and coronary arteries (and vena cava as controls) were processed for ultrastructural cytochemistry. Tracer distribution was measured by morphometric analysis. Monocytes removed from circulation were incubated in suspensions with various probes; special attention was given to mononuclear cells found in close association with endothelium. The findings were compared with those recorded in normal rabbits.[68]

The following cell surface moieties were detected cytochemically:

- *Anionic groups* of strongly charged residues of low pK_a values, visualized with cationized ferritin (CF) pI 8.4
- *Cationic groups* of high pK_a values, labeled with hemeundecapeptide (HUP) pI 4.85
- *Sialyl residues* non-σ-acetylated at C-8 or C-9 detected with ferritin hydrazide (FH) on cell surface previously oxidized by Na periodate
- *N-Acetylneuraminic acid and N-acetylglucosamine* visualized with wheat germ agglutinin (WGA) followed by a mucin gold conjugate (M–Au)
- *Mannosyl and glucosyl residues* labeled with concanavalin A (Con A) followed by horseradish peroxidase–gold conjugate (HRP–Au)
- *Terminal galactosyl residues* detected by peanut agglutinin (PA) in sequence with lactosaminated bovine serum albumin–gold (Lac–NBSA–Au)
- *Terminal galactose and N-acetylgalactosamine* marked with the sequence galactose oxidase–ferritin hydrazide (GO–FH)
- *Subterminal galactosyl* reacted with *Ricinus communis* agglutinin (RCA) followed by LacN–BSA–Au

With the exception of HUP, we purposely employed specific ligands conjugated with particulate tracers to allow a quantitative morphometric analysis.

B. Results and Discussion

The data obtained were assembled on various nonlesional and lesional stages found in different regions of the aortic arch. The changes observed in hypercholesterolemic animals were compared (1) with the ligand distribution in normolipidemic rabbits[68] and (2) with the lesion-free zones encountered in the same cholesterol-fed animal. In general, the results revealed a remarkable resistance of the endothelial glycocalyx to very high levels of serum cholesterol. In nonlesional

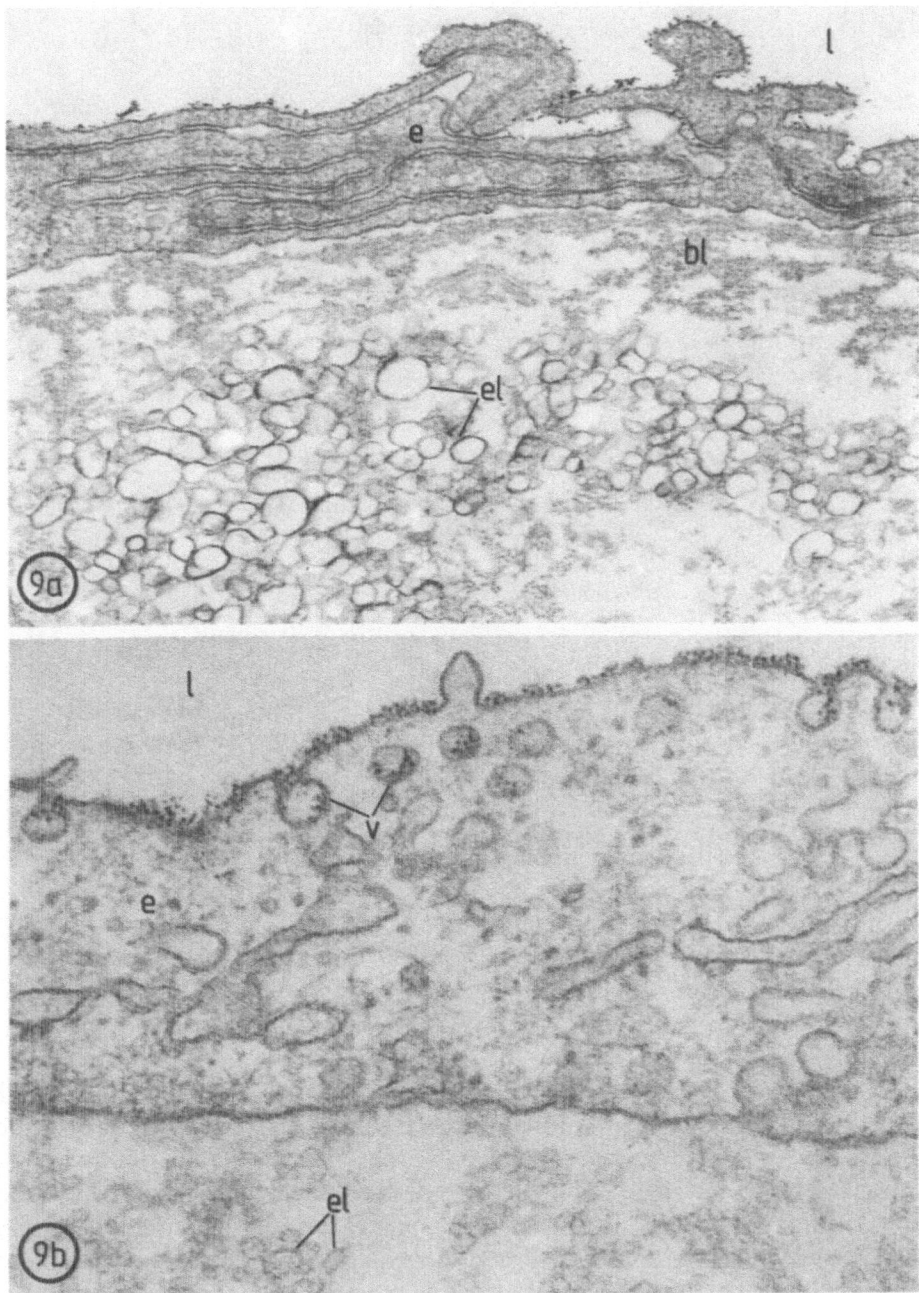

Figure 9. Lesion-prone areas of the aortic arch of hypercholesterolemic rabbits (2 weeks of the diet) showing an intact endothelium (e) and extracellular liposomes (el) in subendothelium. The luminal endothelial surface was labeled for galactosyl residues by RCA/lactosaminated albumin–Au (a), and for anionic groups by cationized ferritin (b). Both decoration patterns are similar to those of normal aorta. l, lumen; v, plasmalemmal vesicles; bl, basal lamina-like material. a, × 16,200; b, × 72,900.

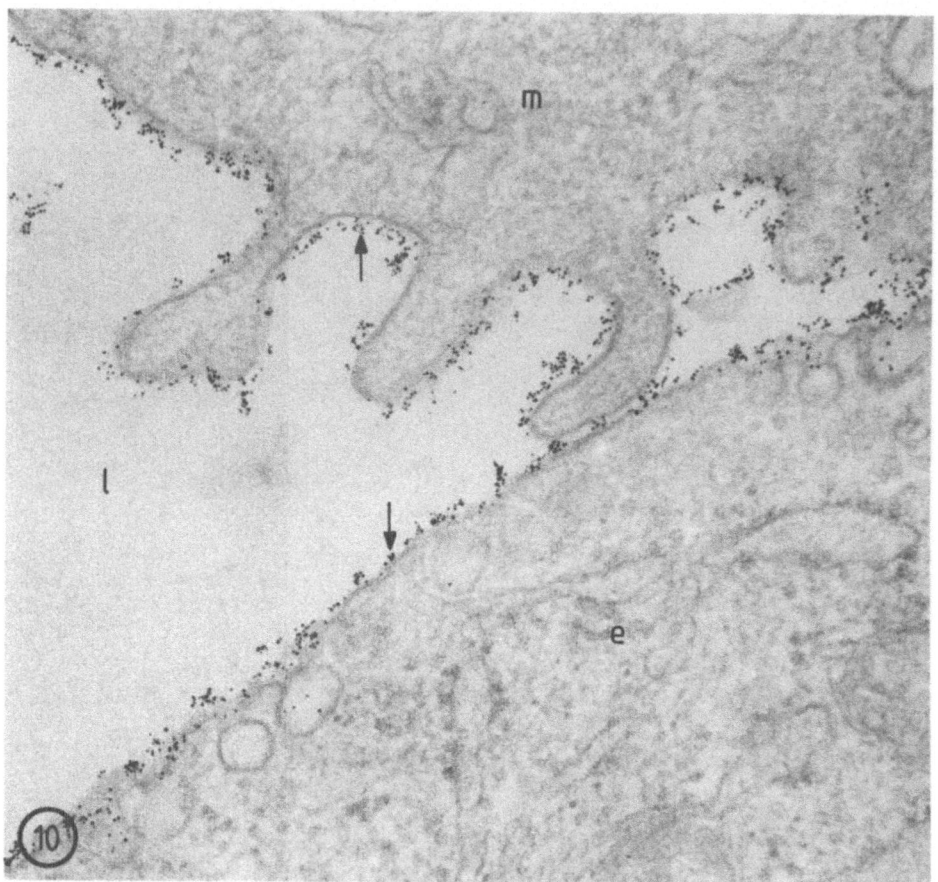

Figure 10. In a region of monocyte (m) attachment to arterial endothelium (e), the apposed cell surfaces show a Con A labeling density (revealed by subsequent reaction with HRP–Au) (arrows) unaffected by hypercholesterolemia. l, lumen. × 113,400.

zones, the endothelial surface charge and glycoconjugates were not significantly affected (Figs. 9 and 10). In lesional areas, including those with forming or advanced fatty streaks, no alterations could be observed in the distribution of cationic groups, anionic sites, sialyl, galactosyl, and N-acetylglucosaminyl residues (Fig. 11). In some specimens, however, mannosyl residues increased in density, a finding for which we do not have a clear explanation. A reduction in anionic groups and sialoconjugates appeared in more advanced stages particularly on endothelial cells heavily loaded with lipid inclusions (Figs. 12–15). Variation in the concentration of surface sialic acid was reported to play a role in LDL interaction with endothelium.[41,42] The changes observed on the endothelial surface were generally paralleled by comparable modulations of the surface moieties of monocytes, especially in those attached to or migrating through endothelium. Nevertheless, irrespective of the extent of surface charge reduction, monocytes

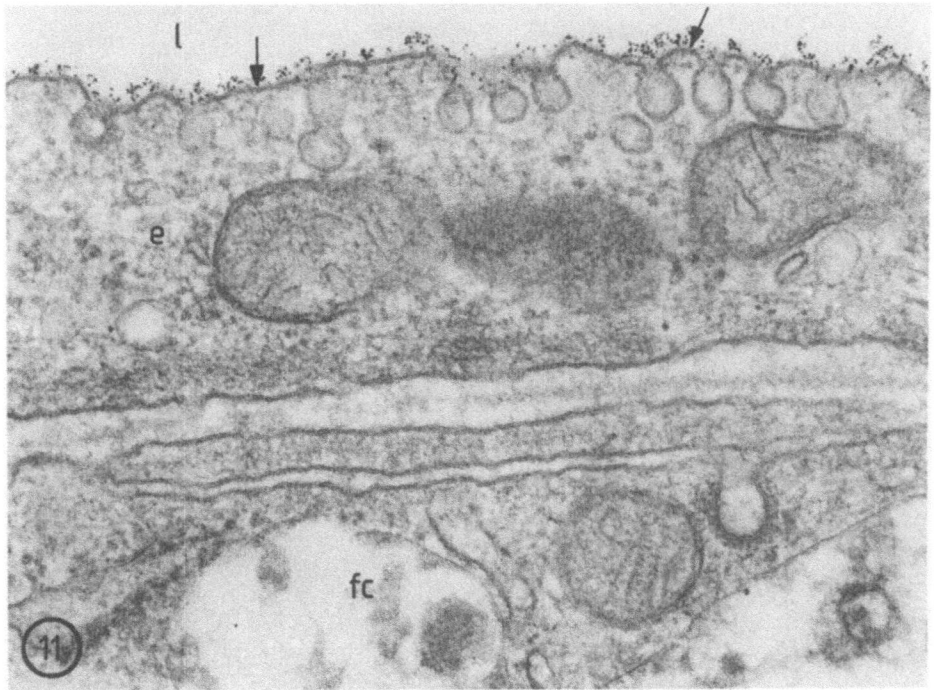

Figure 11. Hypercholesterolemic rabbit at 8 weeks of the diet: endothelial cells (e) overlying foam cells (fc) display on their luminal surface a normal density of mannosyl residues as detected with Con A/HRP–Au (arrows). × 63,000.

continued to migrate and foam cells to egress from the intima, for the entire duration of these experiments.

Commonly, even on the cell surface of endothelium overlying foam cells, some of the moieties detected were expressed at levels very close to normal, irrespective of the severity of hypercholesterolemia. Two examples are given in Fig. 16: FH-detected sialyl residues and mannosyl moieties visualized with Con A/HRP–Au.

These observations revealed that during hypercholesterolemic atherogenesis the inception and progression of early intimal lesions are not preceded by significant alterations of the cell surface charge or glycoconjugates of arterial endothelium and monocytes. It is conceivable that more specific molecules are instrumental in monocyte adhesion and diapedesis.

VIII. ENDOTHELIAL MODIFICATIONS IN THE EARLY STAGES OF ATHEROGENESIS

From recent work including our own, both in humans and in animal models of atherosclerosis, intimal fatty lesions are present long before any recognizable

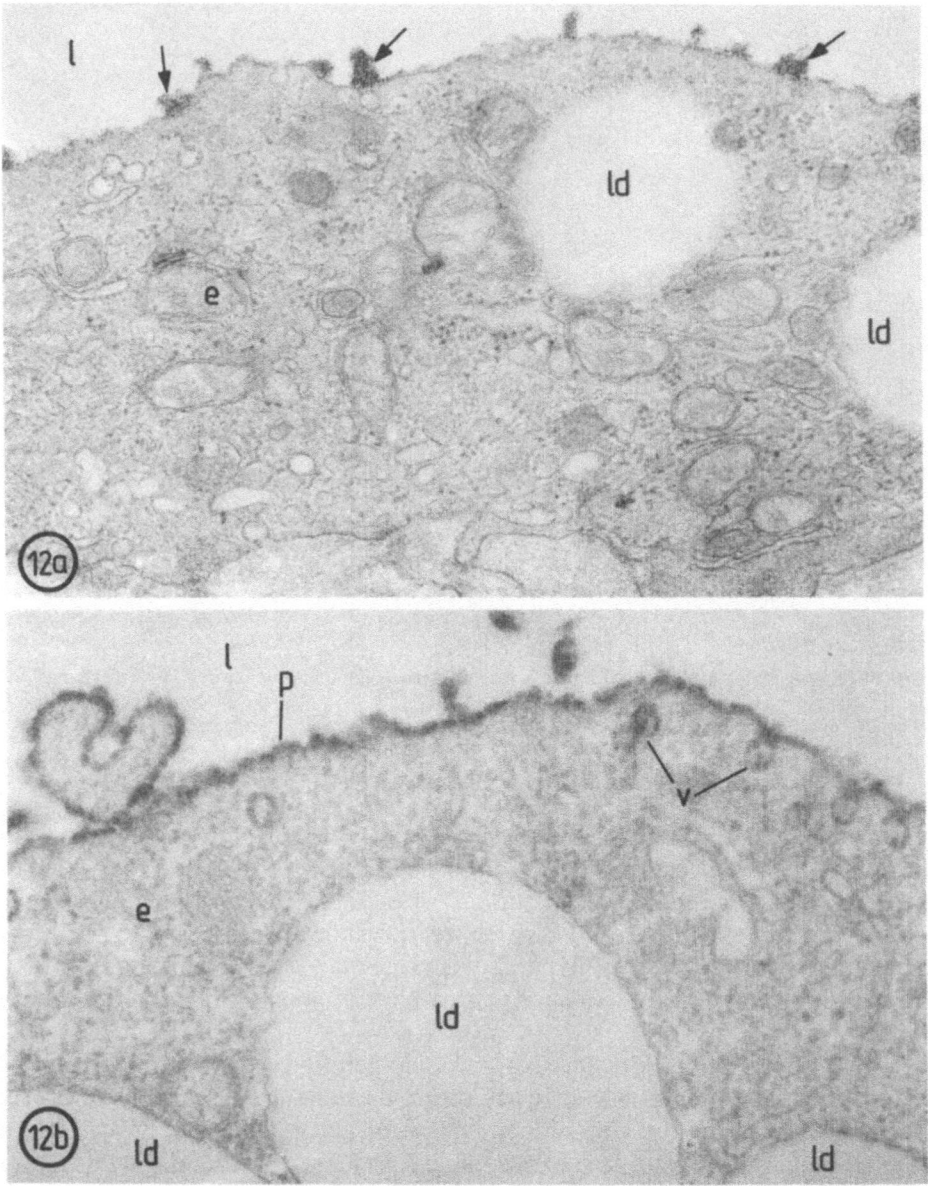

Figure 12. Hypercholesterolemic rabbit at 8 weeks of the diet: the aortic arch with fatty streak in which some endothelial cells (e) are filled with lipid deposits (ld). Such cells usually have a patchy distribution of anionic sites (arrows) (a) but a homogeneous layer of HUP reaction product labeling cationic sites of plasma membrane (p) and some open plasmalemmal vesicles (v) (b). l, lumen. a, × 26,100 b, × 45,900.

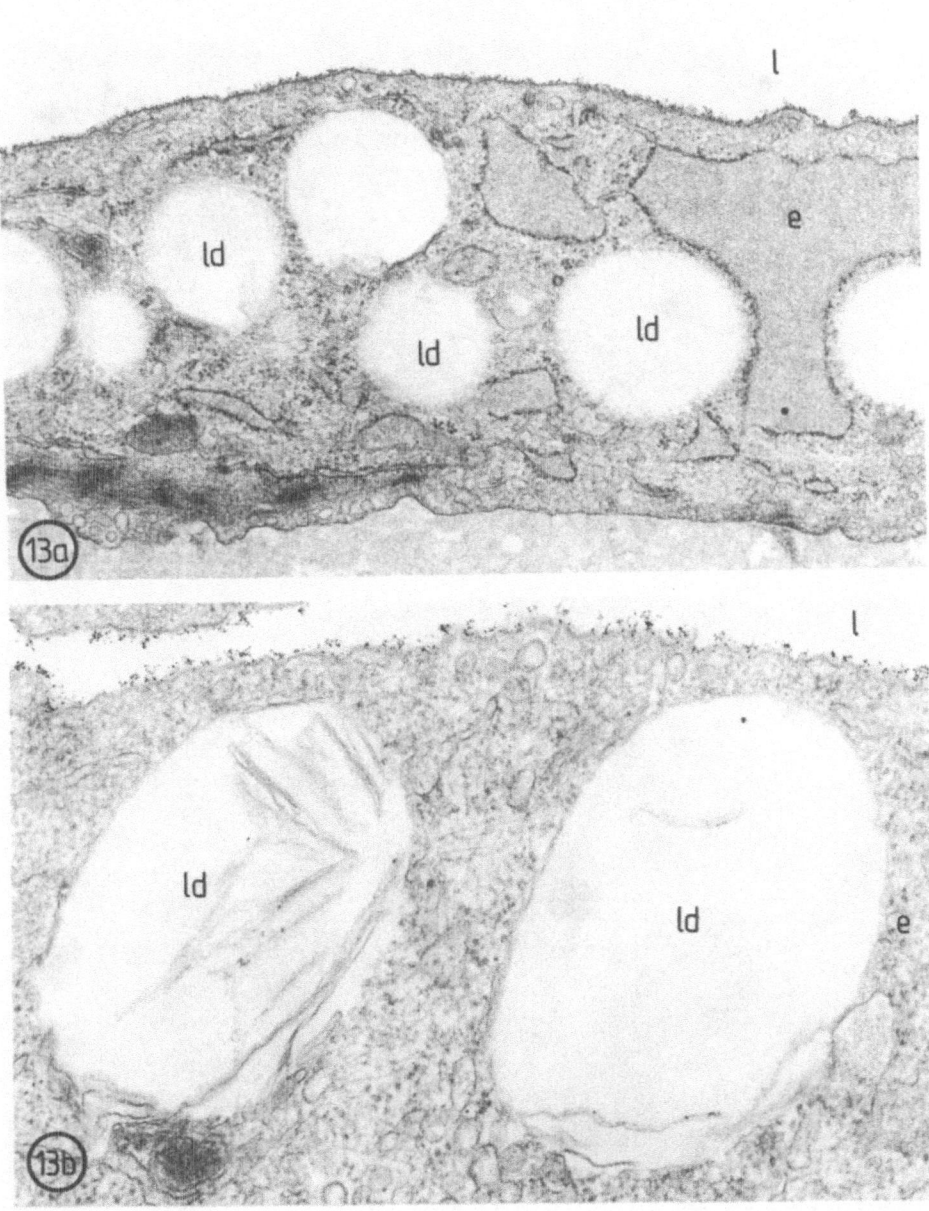

Figure 13. Specimens comparable to those in Fig. 12 but (a) reacted with RCA/lactosaminated albumin for the visualization of terminal galactosyl residues or (b) with Con A/HRP–Au for the detection of mannosyl moieties. Although these endothelial cells (e) contain numerous lipid deposits (ld), the two glycocalyx oligosaccharides appear unaffected by hypercholesterolemia. l, lumen. a, × 17,100; b, × 29,700.

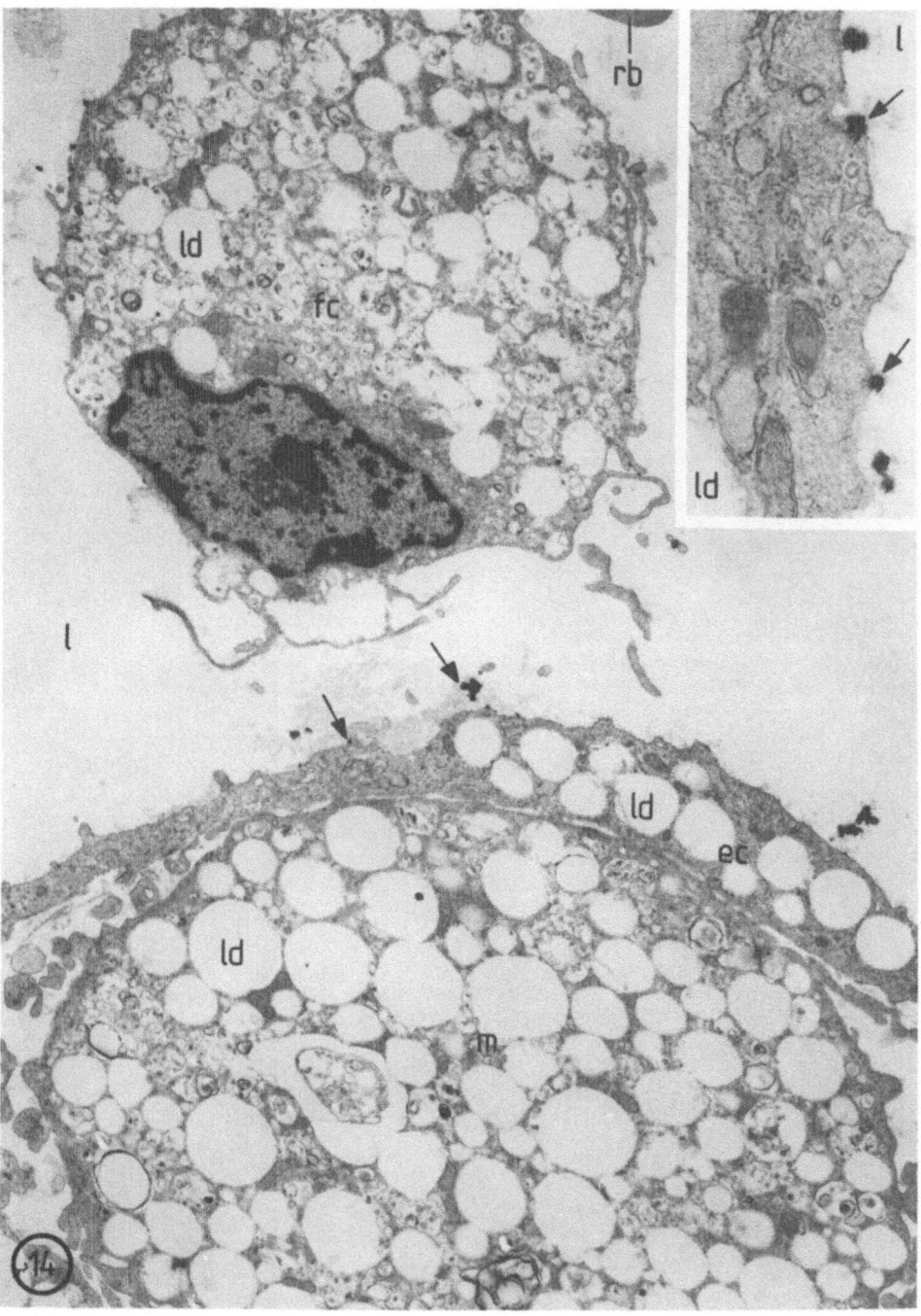

Figure 14. Lesional area of the aortic arch in a rabbit fed 10 weeks with cholesterol-rich diet. Heavy lipid deposition (ld) involves macrophages (m), overlying endothelial cells (e), and foam cells (fc) released in the circulation. On all these cells the CF-detectable anionic sites appear drastically reduced (arrows). l, lumen; rb, red blood cell. × 7125; inset, × 28,500.

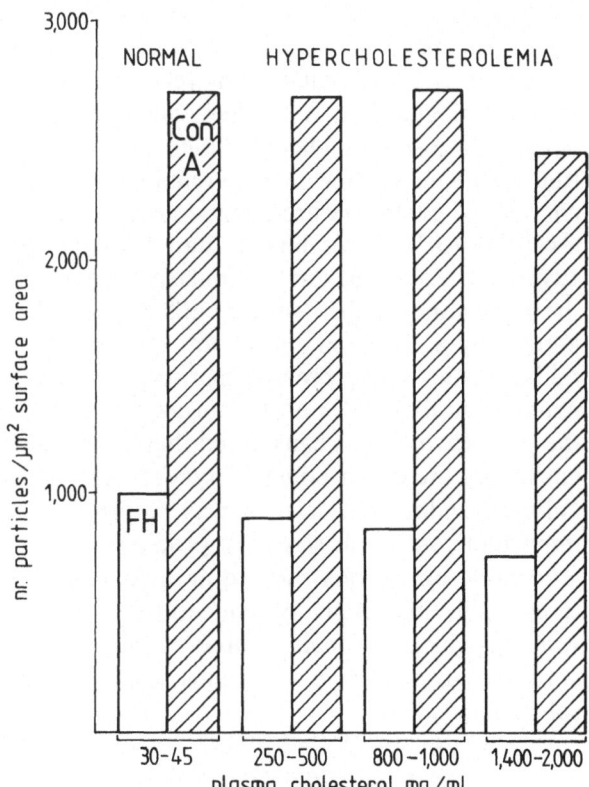

	normal	lesional stages				
		I	II	III	IV	V
Anionic sites: CF	1,500	1,500	1,400	1,470	370	1070
Sialyl residues: FH	880	1,150	n.d.	780	680	720
Man: Con A/HRP-Au	1,980	2,830	2,580	2,700	3,330	n.d.
Gal: RCA/Lac-Au	920	1,100	n.d.	1,010	1,710	1,550
Glc NAc: WGA/M-Au	69	81	n.d.	77	68	n.d.

Figure 15. Mean values of tracer binding densities on endothelial cell surface of normal versus lesional areas of the aortic arch in hypercholesterolemic rabbits. Standard deviation of the mean ($\overline{\text{S.D.}}$) was ± 370 for CF and Con A, ± 180 for FH and RCA, and ± 11 for WGA. The cationic sites detected by HUP, with the exception of a slight decrease in stage IV (in some areas), were not significantly affected.

endothelial denudation or platelet involvement.[60,61,78,97,106,107,115–119] Various factors including hypercholesterolemia can induce alterations in endothelial functions which can be important in atherogenesis. Such endothelial dysfunctions have been detected *in vitro,* but endothelial cells in culture can *a priori* be considered perturbed, and the findings on such systems—although valuable—should very cautiously be accepted as reflecting the situation *in vivo.* Examples of nondenuding modification of some endothelial properties and activities are given in Refs. 40

Figure 16. Histogram illustrating the binding density of ferritin hydrazide (FH) and Con A (revealed by HRP–Au conjugate) to the arterial endothelium overlying foam cells. For both ligands, the binding values are very similar, irrespective of the level of plasma cholesterol.

and 119. Because there is nothing to be gained by either redefining the term "injury" or substituting it for other terms, more substantial progress should be expected from the search aimed at the identification in cellular or molecular terms of those subtle events that predispose or precede the inception of the atherosclerotic lesions *in vivo*.

Although not reproducing exactly the sequence and type of lesions mostly encountered in humans, the accelerated diet-induced atherogenesis in animals, including rabbits, may help to explain how lesions, especially fatty streaks, form and progress. During the hyperbetalipoproteinemic atherogenesis, the early changes reflect to a large extent the outcome of the two-way interaction that successively develops between endothelium and LP, endothelium and monocytes, and endothelium and macrophages or foam cells.

A. Hyperlipoproteinemia–Endothelium Interactions in Prelesional Stages

At the onset of atherogenesis, endothelium is first exposed to elevated levels of plasma LDL or β-VLDL. An approach somewhat simplistic but realistic was to determine (1) the direct effect of hyperlipoproteinemia on endothelial functions, and (2) the chemical alterations suffered by circulating LP upon their interaction with endothelium, being transcytosed or accumulated in a native or modified form in subendothelium. During the first weeks of experimental hypercholesterolemia, while the endothelial lining was found morphologically intact, atherosclerotic tendencies were associated with focal endothelial modifications such as: increased thymidine uptake, enhanced permeability to plasma proteins[145–147] or horseradish peroxidase,[133] accumulation of IgG in endothelial cells, reduction of CF, ruthenium red, and Con A binding, augmentation in the number of plasmalemmal vesicles, transendothelial channels, and Weibel–Palade bodies,[140] and so on (see Ref. 119). Our studies on rabbit and hamster showed that endothelial cell surface charge and chemistry are unexpectedly very resistant even to plasma cholesterol levels 50 to 70 times higher than normal. The modulations observed in more advanced lesions occurred similarly in aorta, coronary, and cardiac valves. In the prelesional stages (serum cholesterol 5–7 times normal), preliminary observations by S. Tasca in our laboratory suggest that prostacyclin production by the rabbit aortic wall decreases markedly in the first 3 weeks of the diet and then plateaus at about 50% of the normal values. Since similar changes were found in the vena cava, it was considered that the metabolic alteration observed was likely caused by the direct effect of hypercholesterolemia on the entire vasculature. In cultured endothelial cells, it was reported that LDL and HDL did not play an important role in prostacyclin production,[125] whereas cells isolated from rabbit atherosclerotic aortas showed an increased level of prostaglandins I and E.[5] Interestingly, this relatively reduced level of prostacyclin production by the vessel wall (in which the main source of PGI_2 is known to be the endothelium) was sufficient to prevent platelet aggregation but not the progression of the atherosclerotic process.

It was demonstrated that incubation of LDL with endothelial cells, smooth muscle cells, or macrophages results in peroxidation of LDL lipid that renders the modified form recognizable by the scavenger receptor.[47,93,132] The oxidized LDL or VLDL is toxic for cultured endothelial cells[48] and reduces their pinocytotic activity.[10] It is not known whether such two-way modification occurs *in vivo*; this avenue needs to be further explored to determine if the modified LDL does activate endothelial cells to express chemotactic properties for monocytes or stimulates their transcytosis.

The observation that in the lesion-prone areas there is a progressive accumulation not of intact LP particles but of a modified form of LP-derived material organized as EL raises the questions of the site and mechanism of this LP degradation. Hypothetically, this may occur at the endothelial cell surface (similar to chylomicron breakdown by lipoprotein lipase), within plasmalemmal vesicles during transcytosis, or in the subendothelial molecular environment. Based on the evidence that at early stages endothelium is structurally intact and there is no platelet involvement, the current view is that the minute lipid lesion starts extracellularly in the matrix of the intima. The LP–endothelial interactions take different aspects in each of the later stages of atherogenesis.

B. Endothelium at Sites of Monocyte Adhesion and Diapedesis

Most investigators consider that in lesion-prone areas the earliest detectable cellular event is the focal attachment of monocytes to arterial endothelium and their diapedesis into the intima.[32,60,106–108] There, they become the prime source of foam cells in the initial phases of atherogenesis.[32,60,61,141,158] Schematically, this process involves four phases presumably governed, at least in part, by different mechanisms: recruitment, attachment, diapedesis, and homing.

1. Recruitment

In lesion-prone areas of arteries it was found that monocyte margination to endothelium is increased in cholesterol-fed animals as compared to controls.[22,23,32,35,58,60,61,106–108,138] However, in pigeons and rabbits no correlation of the severity of this process with β-VLDL or HDL concentrations could be established.[119,128] A possible role in monocyte recruitment was assigned to chemoattractants secreted by the arterial wall including endothelium.[6,32,35,79,92] Recently, it was demonstrated that extracts of blue areas from hypercholesterolemic swine aortas contain two factors that are chemotactic only for hypercholesterolemic swine monocytes.[34,35] It is still poorly understood which functional changes in some population of blood monocytes make them more susceptible for adhering to endothelium. Our observations did not reveal significant differences in the surface charge and major glycoconjugates between circulating and attached monocytes.[36]

2. Attachment

The molecular interactions involved in monocyte adhesion to endothelium are unknown. General alterations in the cell surface of the two partners [charge, sialoproteins, thickness of glycocalyx (for review see Ref. 108)] are unlikely to play a significant role. At sites of monocyte attachment, endothelial glycocalyx is not significantly modified and expresses a normal density of anionic, cationic groups as well as the major oligosaccharide residues. The same applies, at least for the early stages, to monocytes. We observed that in hyperlipidemic rabbits and hamsters, monocyte adhesion to arterial or valvular endothelium occurs in areas containing EL in their intima. The search should now identify the specific molecules involved in monocyte adhesion and diapedesis. One promising avenue is to define such mechanism with monoclonal antibodies.[150]

3. Diapedesis

In our experiments, emigration of monocytes takes places always via intercellular junctions. Questionable images suggesting a transcellular pathway upon examination of serial sections were found to be intercellular. The part of the endothelial intercellular surfaces accessible to our tracers proved to have a normal distribution and density of the residues detected.[36] In early stages, during monocyte diapedesis, the intercellular space was not more permeable to tracers such as LDL–Au, β-VLDL–Au, or albumin–Au. Junctional permeability was increased during foam cell formation especially for Alb–Au complexes. At that time, a sero-fibrinous insudate was focally found in the intima usually in areas with few or no significant accumulation of EL. Monocyte diapedesis was more pronounced in regions rich in foam cells. The state of endothelium was more related to the lesional stage of the underlying intima than to the process of monocyte emigration.

4. Homing

Since there is a net movement of monocytes into the subendothelial space, one may reasonably assume that these cells are capable of locomotion in the presence of a chemotactic gradient across the intima. Although some data suggest that altered constituents of the intima (e.g., elastin polypeptides) may be chemoattractants, this issue is still obscure. Commonly, activated monocytes are present immediately beneath endothelium from which they are separated by an almost inconspicuous basal lamina. Macrophages progressively take up modified LP via their scavenger receptors to become loaded with lipid deposits and to form foam cells. This striking symbiosis between endothelium and successively modified monocyte–macrophage–foam cells is likely to play an important role in the progression of atheroma, which largely depends on the continuous emigration and transformation of monocytes into lesional regions.[22,23,60,98,138] Occasionally, lymphocytes and polymorphonuclear leukocytes interacting with endothelium were seen. At these early stages the pathologic pattern of leukocyte interactions with the vessel wall is largely reminiscent of the protective inflammatory response and so is the presence and scavenger activity of macrophages, the hallmark of chronic inflammation. The inflammatory component represents a facet of the atherogenetic process that will not be discussed here.

C. Endothelium in Areas of Foam Cell Formation

The macrophage-derived foam cells continuously accumulating within the intima establish from the beginning close topographic association with the overlying endothelium. The latter forms a continuous lining sometimes corresponding to the underlying foam cells at a one-to-one ratio. There is no endothelial loss with exposure of subendothelial structures and, except for the sites of monocyte diapedesis or foam cell immigration into the bloodstream, intercellular junctions appear structurally intact. Moreover, these endothelial cells generally maintain their surface charge and distribution of some major oligosaccharide moieties.[36] However, in view of the recognized diversity of macrophage functions and of their close proximity to endothelium, it is conceivable that by some of their secretory products (hydrolytic and proteolytic enzymes, oxidation metabolites)[93,131,142] macrophages may directly affect the functions of the overlying endothelial cells *in vivo*, but such effects have not been clearly substantiated.[48,98] A more detailed examination of the ultrastructural and cytochemical pattern of the endothelial cells overlying foam cells revealed that under a general morphology apparently unchanged (normal set of organelles, normal volume density of plasmalemmal vesicles), the cells display a pronounced polymorphism. Some of them lose to a large extent their cell surface mannosyl residues, while other neighboring cells express such moieties in densities $\sim$ 40% higher than in controls.[36] Within a broad spectrum of appearances, we could distinguish three morphologically characteristic endothelial cell types. (1) *An apparently secretory type* was distinguished by distended cisternae of RER filled with floccular material, and accumulation of large clusters of SER (Fig. 17). Such cells are usually thick ($\sim$ 0.5 μm) and often correspond to regions of proliferated matrix constituents, especially basal lamina material. (2) *The attenuated type* up to $\sim$ 0.1 μm has been observed by others as well.[22] These cells displayed a quasi-normal distribution of cell surface residues; however, the numerical density of plasmalemmal vesicles, coated pits, coated vesicles, and multivesicular bodies was very much diminished. (3) *The foam cell type* was of particular interest not only because of its similarity with the macrophage- or smooth muscle-derived foam cells but also because of its potential physiopathologic significance.[72] The glycocalyx of such cells largely lost its cationic, anionic, and sialyl residues but often express—like a defense reaction—more mannosyl and galactosyl residues.[36] The appearance at a certain stage of atherogenesis of such modified endothelial cells indicates that they are capable of taking up native or modified LDL and/or β-VLDL like intimal macrophages[3,39,77] and foam cells.[94] Endothelium-derived foam cells contained intracellular lipid deposits similar to those described in foam cells of other origins[72,73]; they represent a cytopathologic entity that, with or without exposure of subendothelium, may be removed and released into the circulation. Some regions contained a mixture of these cell types whereas others formed monotypic zones.

D. Endothelium at Sites of Foam Cell Immigration

Heavily loaded foam cells can be seen moving through endothelial junctions or detached free in the adjacent vascular lumen (Fig. 14). Although on the static

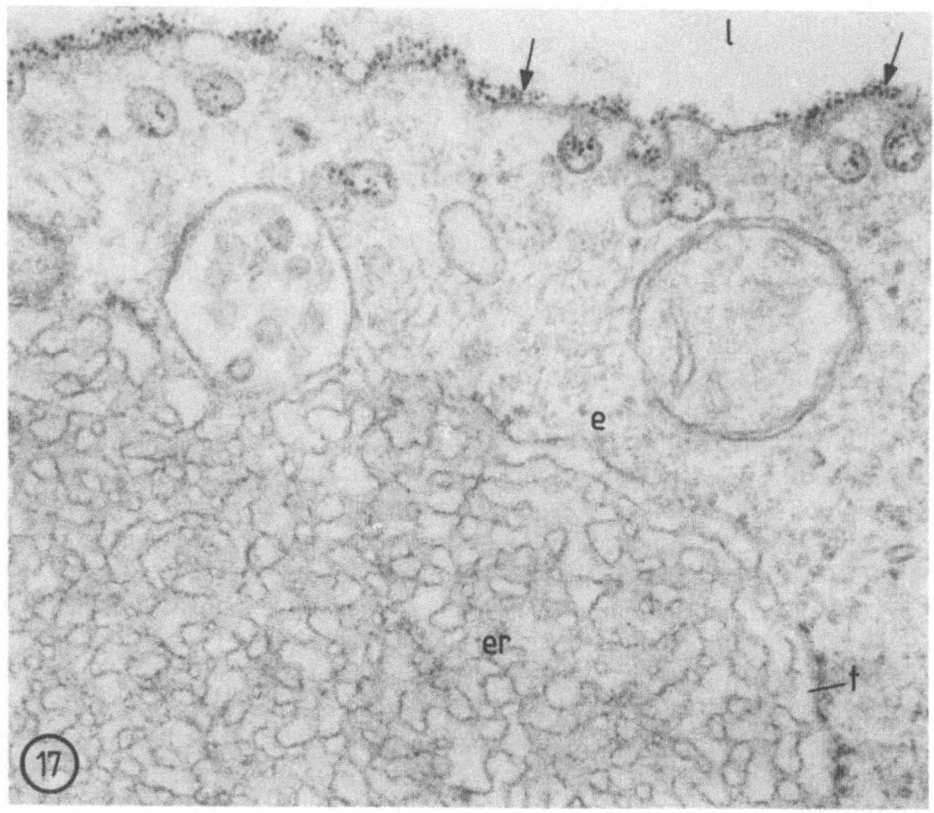

Figure 17. At 8 weeks of the diet, some endothelial cells (e), although displaying a quasi-normal density of CF-detectable anionic sites (arrows), contain abnormally developed convoluted elements of endoplasmic reticulum (er). In t, a presumably transitional element. × 77,400.

images of the electron micrographs one cannot decide whether such a cell is in the process of entering or leaving the intima, there are several reasons to consider them as immigrating into circulation.[22,23,98] The exit occurs through endothelial junctions frequently leakier than those through which monocytes migrate. The surrounding endothelial cells appeared either unaffected structurally and cytochemically, or displayed increased amounts of galactosyl residues. However, these endothelial cells as well as the immigrating foam cells contain fewer anionic sites and FH-detectable sialyl residues[36] (Fig. 15), and intercellular surfaces had very scarce or no CF binding sites. A synopsis of the early molecular and cellular modifications detected by our studies on hyperlipoproteinemic atherogenesis is schematically given in Fig. 18.

IX. CONCLUDING REMARKS

In diet-induced hyperlipoproteinemia in rabbits and hamsters, despite very high levels of plasma cholesterol, arterial endothelium remains morphologically

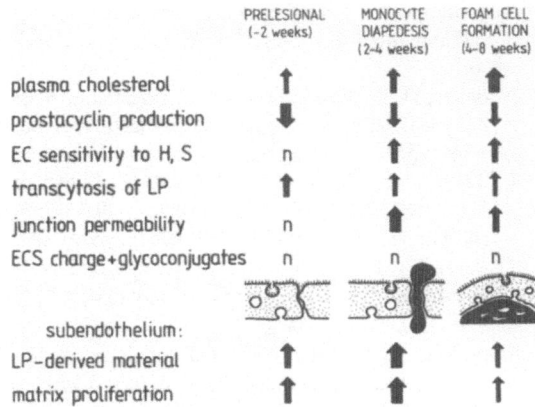

	PRELESIONAL (~2 weeks)	MONOCYTE DIAPEDESIS (2–4 weeks)	FOAM CELL FORMATION (4–8 weeks)
plasma cholesterol	↑	↑	⇑
prostacyclin production	⇓	↓	↓
EC sensitivity to H, S	n	↑	↑
transcytosis of LP	↑	↑	↑
junction permeability	n	⇑	↑
ECS charge+glycoconjugates	n	n	n
subendothelium:			
LP-derived material	↑	↑	↑
matrix proliferation	↑	⇑	↑

Figure 18. The sense of the early molecular and cellular modifications in diet-induced atherogenesis. EC, endothelial cell; H, histamine; S, serotonin; ECS, endothelial cell surface; LP, lipoproteins.

intact and no platelet involvement can be detected until advanced stages of fatty streak formation.

In lesion-prone areas (e.g., the aortic arch), during prelesional stages there are signs of endothelial dysfunctions such as enhanced transcytosis of LDL and β-VLDL associated with a progressive accumulation of LP-derived components reassembled as phospholipid EL. These are rich in UC and are associated with apo B and apo E trapped in bundles of proteoglycans and other proliferated matrix elements.

Endothelial surface maintains practically unaltered its charge, sialic acid and oligosaccharide moieties, even on the thin endothelial cells overlying foam cells. Lipid-laden endothelial cells lose partially their surface anionic groups but gain a higher density of some oligosaccharide moieties. Local adhesion and migration of monocytes marking the inception of lesion formation occur in areas with EL accumulated in the intimal matrix. The monocyte recruitment in such regions and their interactions with endothelium remain to be explained in molecular terms.

It appears that in early stages of atherogenesis the first ultrastructural changes appear extracellularly in the intima. These precede and may eventually contribute to the alteration of endothelium observed in the advanced complicated plaque.

ACKNOWLEDGMENTS. The work on which this review is based was supported by the Ministry of Education (Romania) and National Institutes of Health (USA) Grant HL 26343. The excellent technical assistance of D. Neacsu (word processing), V. G. Ionescu and E. Stefan (photographic work), and C. Neacsu and M. Schean (graphics) is gratefully acknowledged.

REFERENCES

1. Akimova, E. J., and Melgunov, V. I., 1984, Apolipoprotein B: Removal of lipids by sodium cholate and reassociation of a lipid-free apoprotein with dipalmitoyl phosphatidylcholine, *Biochem. Int.* **9**:463–473.

2. Badimon, J. J., Kottke, B. A., Chen, T. C., Chan, L., and Mao, S. J. T., 1986, Quantification and immunolocalization of apoprotein E in experimental atherosclerosis, *Atherosclerosis* **61:**57–66.

3. Baker, D. P., Van Lenten, B. J., Fogelman, A. M., Edwards, P. A., Kean, C., and Berliner, J. A., 1984, LDL, scavenger and β-VLDL receptors on aortic endothelial cells, *Arteriosclerosis* **4:**248–255.

4. Basu, S. K., Ho, Y. K., Brown, M. S., Bilheimer, D. W., Anderson, R. G. W., and Goldstein, J. L., 1982, Biochemical and genetic studies of the apoprotein E secreted by mouse macrophages and human monocytes, *J. Biol. Chem.* **257:**9788–9795.

5. Berberian, P. A., Jenison, M. W., and Roddick, V., 1985, Arterial prostaglandins and lysosomal function during atherogenesis. II. Isolated cells of diet-induced atherosclerotic aortas of rabbits, *Exp. Mol. Pathol.* **43:**36–55.

6. Berliner, J. A., Territo, M., and Fogelman, A. M., 1984, Monocyte chemotactic factor produced by large vessel endothelial cells, *Arteriosclerosis* **4:**524a.

7. Bocan, T. M. A., and Guyton, J. R., 1985, Human aortic fibrolipid lesions: Progenitor lesions for fibrous plaque, exhibiting early formation of the cholesterol-rich core, *Am. J. Pathol.* **120:**193–206.

8. Bocan, T. M. A., Brown, S. A., Krohn, N. J., and Guyton, J. R., 1986, Human aortic fibrolipid lesions: Immunocytochemical localization of apolipoprotein A and B, *Arteriosclerosis* **6:**557a.

9. Bocan, T. M. A., Schifani, T. A., and Guyton, J. R., 1986, Ultrastructure of the human aortic fibrolipid lesion: Formation of the atherosclerotic lipid-rich core, *Am. J. Pathol.* **123:**413–424.

10. Borsum, T., Henriksen, T., and Reisvaag, A., 1985, Oxidized low density lipoprotein can reduce the pinocytic activity in cultured human endothelial cells as measured by cellular uptake of [^{14}C] sucrose, *Atherosclerosis* **58:**81–96.

11. Camejo, G., Hurt, E., and Romano, M., 1985, Properties of lipoprotein complexes isolated by affinity chromatography from human aorta, *Biomed. Biochim. Acta* **44:**389–401.

12. Carew, T. E., Pittman, R. C., Marchaud, E. R., and Steinberg, D., 1984, Measurement in vivo of irreversible degradation of low density lipoprotein in the rabbit aorta, *Arteriosclerosis* **4:**214–224.

13. Clevidence, B. A., Morton, R. E., West, G., Dusek, D. M., and Hoff, H., 1984, Cholesterol esterification in macrophages: Stimulation by lipoproteins containing apo B isolated from human aortas, *Arteriosclerosis* **4:**196–207.

14. Coetzee, G. A., Stein, O., and Stein, Y., 1979, Uptake and degradation of low density lipoproteins (LDL) by confluent, contact-inhibited bovine and human endothelial cells exposed to physiological concentration of LDL, *Atherosclerosis* **33:**425–431.

15. Daugherty, A., Lange, L. G., Sobel, B. E., and Schonfeld, G., 1985, Aortic accumulation and plasma clearance of β-VLDL and HDL: Effects of diet-induced hypercholesterolemia in rabbits, *J. Lipid Res.* **26:**955–963.

16. DeLamatre, Y., Wolfhauer, G., Phillips, M. C., and Rothblat, G. H., 1986, Role of apolipoproteins in cellular cholesterol efflux, *Biochim. Biophys. Acta* **875:**419–428.

17. Deliconstantinos, G., Tsopanakis, C., Karayiannakos, P., and Skalkeas, G., 1986, Evidence for the existence of non-esterified cholesterol carried by albumin in rat serum, *Atherosclerosis* **61:**67–75.

18. Desai, K. S., Gotlieb, A. I., and Steiner, G., 1985, Very low density lipoprotein binding to cultured aortic endothelium, *Can. J. Physiol. Pharmacol.* **63:**809–815.

19. Driscoll, D. M., and Getz, G. S., 1984, Extrahepatic synthesis of apolipoprotein E, *J. Lipid Res.* **25:**1368–1375.

20. Elshourbagy, N. A., Liao, W. S., Mahley, R. W., and Taylor, J. M., 1985, Apolipoprotein E mRNA is abundant in the brain and adrenals as well as in the liver, and is present in other peripheral tissues of rats and marmosets, *Proc. Natl. Acad. Sci. USA* **82:**203–207.

21. Eskenasy, M., Mora, M., and Simionescu, N., 1984, In vitro study of low density lipoprotein-collagen interaction, *Morphol. Embryol.* **30:**147–152.

22. Faggiotto, A., Ross, R., and Harker, L., 1984, Studies on hypercholesterolemia in the nonhuman primate. I. Changes that lead to fatty streak formation, *Arteriosclerosis* **4:**323–340.

23. Faggiotto, A., and Ross, R., 1984, Studies on hypercholesterolemia in the nonhuman primate. II. Fatty streak conversion to fibrous plaque, *Arteriosclerosis* **4**:341–356.

24. Falcone, D. J., Mated, N., Shio, H., Minick, C. R., and Fowler, S. D., 1984, Lipoprotein–heparin–fibronectin–denatured collagen complexes enhance cholesteryl ester accumulation in macrophages, *J. Cell Biol.* **99**:1266–1274.

25. Feldman, D. L., Hoff, H. F., and Gerrity, R. G., 1982, Immunocytochemical localization of LDL in aortas of hyperlipemic swine, *Fed. Proc.* **41**:321.

26. Fielding, C. J., 1984, The origin and properties of free cholesterol potential gradients in plasma, and their relation to atherogenesis, *J. Lipid Res.* **25**:1624–1628.

27. Fielding, P. E., Vlodansky, I., Gospodarowicz, D., and Fielding, C. J., 1979, Effect of contact inhibition on the regulation of cholesterol metabolism in cultured vascular endothelial cells, *J. Biol. Chem.* **254**:749–755.

28. Filip, D. A., Nistor, A., Bulla, A., Radu, A., Lupu, F., and Simionescu, M., 1987, Cellular events in the development of valvular atherosclerotic lesions induced by experimental hypercholesterolemia, *Atherosclerosis* **67**:199–214.

29. Fischer, G. M., Cherian, K., and Swain, M. L., 1981, Increased synthesis of aortic collagen and elastin in experimental atherosclerosis, *Atherosclerosis* **39**:463–467.

30. Gaffney, J., West, D., Arnold, F., Sattar, A., and Kumar, S., 1985, Differences in the uptake of modified low density lipoproteins by tissue cultured endothelial cells, *J. Cell Sci.* **79**:317–325.

31. Georgescu, L., Antohe, F., and Simionescu, N., 1986, The permeability of aortic endothelium to ^{125}I-BSA in hyperlipidemic hamster: Effect of histamine and serotonin, *Rev. Roum. Physiol.* **23**:221–225.

32. Gerrity, R. G., 1981, The role of the monocyte in atherogenesis. I. Transition of blood-borne monocytes into foam cells in fatty lesions, *Am. J. Pathol.* **103**:181–190.

33. Gerrity, R. G., Naito, H. K., Richardson, M., and Schwartz, C. J., 1979, Dietary induced atherogenesis in swine: Morphology of the intima in prelesional stages, *Am. J. Pathol.* **95**:775–792.

34. Gerrity, R. G., and Goss, J. A., 1983, A monocyte chemotactic factor from lesion prone areas of swine aorta, *Circulation* **68**(Suppl. 3):301.

35. Gerrity, R. G., Goss, J. A., and Soby, L., 1985, Control of monocyte recruitment by chemotactic factor(s) in lesion-prone areas of swine aorta, *Arteriosclerosis* **5**:55–66.

36. Ghinea, N., Leabu, M., Hasu, M., Muresan, V., Colceag, J., and Simionescu, N., 1987, Prelesional events in atherogenesis: Changes induced by hypercholesterolemia in the cell surface chemistry of arterial endothelium and blood monocytes in rabbit, *J. Submicrosc. Cytol.* **19**:209–227.

37. Ghitescu, L., Fixman, A., Simionescu, M., and Simionescu, N., 1986, Different mechanisms of serum albumin transcytosis in continuous endothelium of capillaries and large vessels, in: *4th Int. Symp. Biol. Vasc. Endoth. Cell,* Noordwijkerhout, Abstr. Vol., p. 131.

38. Goldstein, J. L., Basu, S. K., Brunschede, V. Y., and Brown, M. S., 1976, Release of low density lipoprotein from its cell surface receptor by sulfated glycosaminoglycans, *Cell* **7**:85–95.

39. Goldstein, J. L., Hoff, H. F., Ho, Y. K., Basu, S. K., and Brown, M. S., 1981, Stimulation of cholesteryl ester synthesis in macrophages by extracts of atherosclerotic human aortas, and complexes of albumin/cholesterol esters, *Arteriosclerosis* **1**:210–226.

40. Gordon, J. L., and Pearson, J. D., 1982, Response of endothelial cells to injury, in: *Pathobiology of the Endothelial Cell* (H. L. Nossel and H. J. Vogel, eds.), Academic Press, New York, pp. 443–454.

41. Görög, P., and Born, G. V. R., 1982, Increased uptake of circulating low density lipoproteins and fibrinogen by arterial walls after removal of sialic acids from their endothelial surface, *Br. J. Exp. Pathol.* **63**:447–451.

42. Görög, P., and Pearson, J. D., 1984, Surface determinants of low density lipoprotein uptake by endothelial cells, *Atherosclerosis* **53**:21–29.

43. Grunwald, J., Hesz, A., Ronenek, H., Brucker, J., and Buddecke, E., 1985, Proliferation, morphology and low density lipoprotein metabolism of arterial endothelial cells cultured from normal and diabetic minipigs, *Exp. Mol. Pathol.* **42**:60–70.

44. Guyton, J. R., Bocan, T. M. A., and Schifani, T. A., 1985, Quantitative ultrastructural analysis

of perifibrous lipid and its association with elastin in non-atherosclerotic human aorta, *Arterio-sclerosis* **5**:644–652.

45. Hashida, R., Anamizu, C., Kimura, J., Ohkuma, S., Yoshida, Y., and Takano, T., 1986, Transcellular transport of lipoprotein through arterial endothelial cells in monolayer culture, *Cell Struct. Funct.* **11**:31–42.

46. Henning, B., Shasby, D. M., and Spector, A. A., 1985, Exposure to fatty acid increases human low density lipoprotein transfer across cultured endothelial monolayers, *Circ. Res.* **57**:776–780.

47. Henriksen, T., Mahoney, E. M., and Steinberg, D., 1982, Interactions of plasma lipoproteins with endothelial cells, *Ann. N.Y. Acad. Sci.* **401**:102–116.

48. Hessler, J. R., Morel, D. W., Lewis, L. J., and Chisolm, G. M., 1983, Lipoprotein oxidation and lipoprotein-induced cytotoxicity, *Arteriosclerosis* **3**:215–222.

49. Hoff, H. F., and Gaubatz, J. W., 1975, Ultrastructural localization of plasma lipoproteins in human intracranial arteries, *Virchows Arch. A* **369**:111–121.

50. Hoff, H. F., Heideman, C. L., Jackson, R. L., Bayardo, R. J., Kim, H.-S., and Gotto, A. M., Jr., 1975, Localization of patterns of plasma apolipoproteins in human atherosclerotic lesions, *Circ. Res.* **37**:72–79.

51. Hoff, H. F., and Gaubatz, J. W., 1977, Ultrastructural localization of apolipoprotein B in human aortic and coronary atherosclerotic plaques, *Exp. Mol. Pathol.* **26**:214.

52. Hoff, H. F., and Gaubatz, J. W., 1982, Isolation, purification, and characterization of a lipoprotein containing apo B from the human aorta, *Atherosclerosis* **42**:273–297.

53. Hoff, H. F., and Morton, R. E., 1985, Lipoproteins containing apo B extracted from human aortas: Structure and function, *Ann. N.Y. Acad. Sci.* **454**:183–194.

54. Hoff, H. F., Dusek, D. M., and Lynn, M. P., 1986, Spatial distribution and accumulation of low density lipoproteins in the abdominal aorta of swine: Determination by a novel electrotransfer procedure, *Lab. Invest.* **55**:377–386.

55. Hoff, H. F., and Wagner, W. D., 1986, Plasma low density lipoprotein accumulation in aortas of hypercholesterolemic swine correlates with modifications in aortic glycosaminoglycan composition, *Atherosclerosis* **61**:231–236.

56. Hsu, M. J., and Juliano, R. L., 1982, Interactions of liposomes with the reticuloendothelial system. II. Nonspecific and receptor-mediated uptake of liposomes by mouse peritoneal macrophages, *Biochim. Biophys. Acta* **720**:411–419.

57. Jackson, R. L., and Gotto, A. M., Jr., 1976, Hypothesis concerning membrane structure, cholesterol and atherosclerosis, *Atheroscl. Rev.* **1**:1–21.

58. Jerome, W. G., and Lewis, J. C., 1984, Early atherogenesis in White Cornean pigeons. I. Leukocyte margination and endothelial alterations at the celiac bifurcation, *Am. J. Pathol.* **116**:56–68.

59. Jerome, W. G., and Lewis, J. C., 1985, Early atherogenesis in White Cornean pigeons. II. Ultrastructural and cytochemical observations, *Am. J. Pathol.* **119**:210–222.

60. Joris, J., Zand, T., Nunnary, J. L., Krolikowski, F. J., and Majno, G., 1983, Studies on the pathogenesis of atherosclerosis. I. Adhesion and emigration of mononuclear cells in the aorta of hypercholesterolemic rats, *Am. J. Pathol.* **113**:341–358.

61. Joris, J., Billingham, M. E., and Majno, G., 1984, Human coronary arteries: An ultrastructural search for the early changes of atherosclerosis, *Fed. Proc.* **43**:710.

62. Kenagy, R., Bierman, E. L., Schwartz, S., and Albers, J. J., 1984, Metabolism of low density lipoprotein by bovine endothelial cells as a function of cell density, *Arteriosclerosis* **4**:365–371.

63. Klimov, A. N., Popov, A. V., Nagornev, V. A., and Pleskov, V. M., 1985, Effect of high density lipoproteins on permeability of rabbit aorta to low density lipoproteins, *Atherosclerosis* **55**:217–223.

64. Kruth, H. S., 1983, Filipin-positive, Oil red O-negative particles in atherosclerotic lesions induced by cholesterol feeding, *Lab. Invest.* **50**:87–93.

65. Kruth, H. S., 1984, Histochemical detection of unesterified cholesterol within human atherosclerotic lesions using the fluorescent probe filipin, *Atherosclerosis* **51**:281–292.

66. Kruth, H. S., and Fry, D. L., 1984, Histochemical detection and differentiation of free and esterified cholesterol in swine atherosclerosis using filipin, *Exp. Mol. Pathol.* **40**:288–294.

67. Kurozumi, T., Imamura, T., Tanaka, K., Yae, Y., and Koga, S., 1984, Permeation and deposition

of fibrinogen and low density lipoprotein in the aorta and cerebral artery of rabbit: Immunoelectron microscopic study, *Br. J. Exp. Pathol.* **65**:355–364.

68. Leabu, M., Ghinea, N., Muresan, V., Colceag, J., Hasu, M., and Simionescu, N., 1987, Cell surface chemistry of arterial endothelium and blood monocytes in the normolipidemic rabbit, *J. Submicrosc. Cytol.* **19**:193–208.

69. Lewis, J. C., Taylor, R. G., Jones, N. D., St. Clair, R. W., and Cornhill, J. F., 1982, Endothelial surface characteristics in pigeon coronary artery atherosclerosis. I. Cellular alterations during the initial stages of dietary cholesterol challenge, *Lab. Invest.* **46**:123–138.

70. Lin, C.-T., Xu, Y., Wu, J.-Y., and Chan, L., 1986, Immunoreaction apolipoprotein E is a widely distributed cellular protein: Immunohistochemical localization of apolipoprotein E in baboon tissues, *J. Clin. Invest.* **78**:947–958.

71. Lundberg, B., 1985, Chemical composition and physical state of lipid deposits in atherosclerosis, *Atherosclerosis* **56**:93–110.

72. Lupu, F., Danaricu, I., and Simionescu, N., 1986, Endothelial cell-derived foam cells in experimental atherosclerosis: A physical, cytochemical, and ultrastructural study, in: *4th Int. Symp. Biol. Vasc. Endoth. Cell,* Noordwijkerhout, Abstr. Vol., p. 124.

73. Lupu, F., Danaricu, I., and Simionescu, N., 1987, The development of intracellular lipid deposits in the lipid-laden cells of the atherosclerotic lesions: A cytochemical and ultrastructural study, *Atherosclerosis* **67**:127–142.

74. Mahley, R. W., 1983, Development of accelerated atherosclerosis: Concepts derived from cell biology and animal model studies, *Arch. Pathol. Lab. Med.* **107**:393–399.

75. Mahley, R. W., Innenarity, T. L., Weisgraber, K. H., and Oh, S. Y., 1979, Altered metabolism (in vivo and in vitro) of plasma lipoproteins after selective chemical modifications of lysine residues of the apoproteins, *J. Clin. Invest.* **64**:743–750.

76. Mahley, R. W., Weisgraber, K. H., Melchior, G. W., Innenarity, T. L., and Hollcombe, K. S., 1980, Inhibition of receptor-mediated clearance of lysine and arginine-modified lipoproteins from the plasma of rats and monkeys, *Proc. Natl. Acad. Sci. USA* **77**:225–229.

77. Mahley, R. W., Innenarity, T. L., Brown, M. S., Ho, Y. K., and Goldstein, J. L., 1980, Cholesteryl ester synthesis in macrophages: Stimulation by beta-very low density lipoproteins from cholesterol-fed animals of several species, *J. Lipid Res.* **21**:970–980.

78. Majno, G., Joris, J., and Zand, T., 1985, Atherosclerosis: New horizons, *Hum. Pathol.* **16**:3–5.

79. Mazzone, T., Jensen, M., and Chait, A., 1983, Human arterial wall cells secrete factors that are chemotactic for monocytes, *Proc. Natl. Acad. Sci. USA* **80**:5094–5097.

80. Modrak, J. B., and Langner, L. O., 1980, Possible relationship of cholesterol accumulation and collagen synthesis in rabbit aortic tissues, *Atherosclerosis* **37**:211–218.

81. Mommaas-Kienhuis, A. M., Krijbolder, L. H., Van Hinsbergh, V. W., Daems, V. T., and Vermeer, B. J., 1985, Visualization of binding and receptor-mediated uptake of low density lipoproteins by human endothelial cells, *Eur. J. Cell Biol.* **36**:201–208.

82. Mora, R., Lupu, F., and Simionescu, N., 1986, Prelesional events in atherogenesis: Colocalization of apoprotein B, unesterified cholesterol and extracellular phospholipid liposomes in lesion-prone areas of aortic intima in hyperlipidemic rabbit, *J. Cell Biol.* **103**(Part 2):197a (abstract).

83. Mora, R., Eskenazy, M., Hillebrand, A., and Simionescu, N., 1986, Immunocytochemical localization of apolipoprotein B in the aorta during prelesional stages of hyperlipidemia, *Acta Biol. Hung.* **37**(Suppl.):253.

84. Mora, R., Lupu, F., and Simionescu, M., 1987, Prelesional events in atherogenesis: Colocalization of apolipoprotein B, unesterified cholesterol and extracellular phospholipid liposomes in the aorta of hyperlipidemic rabbit, *Atherosclerosis* **67**:143–154.

85. Morel, D. W., Di Corleto, P. E., and Chisohu, G. M., 1984, Endothelial and smooth muscle cells alter low density lipoprotein in vitro by free radical oxidation, *Arteriosclerosis* **4**:357–364.

86. Murase, T., Oka, T., Yamada, N., Mori, N., Ishibashi, S., Takaku, F., and Mori, W., 1986, Immunohistochemical localization of apolipoprotein E in atherosclerotic lesions of the aorta and coronary arteries, *Atherosclerosis* **60**:1–6.

87. Noma, A., Takabashi, T., and Wada, T., 1981, Elastin–lipid interaction in the arterial wall, *Atherosclerosis* **38**:373–382.

88. Navab, M., Hough, G. P., Berliner, J. A., Frank, J. A., Fogelman, A. M., Haberland, M. E.,

and Edwards, P. A., 1986, Rabbit beta-migrating very low density lipoprotein increases endothelial macromolecular transport without altering electrical resistance, *J. Clin. Invest.* **78:**389–397.

89. Navab, M., Hough, G. P., Fogelman, A. M., Berliner, J. A., Haberland, M. E., and Edwards, P. A., 1986, Transport of low density lipoprotein across monolayers of human aortic endothelial cells co-cultured with human aortic smooth muscle cells, *Arteriosclerosis* **6:**524a.

90. Nicoll, A., Duffield, R., and Lewis, B., 1981, Flux of plasma lipoproteins into human arterial intima: Comparison between grossly normal and atheromatous intima, *Atherosclerosis* **39:**229–242.

91. Nistor, A., Bulla, A., Filip, D. A., and Radu, A., 1987, The hyperlipidemic hamster as a model of experimental atherosclerosis, *Atherosclerosis* **68:**159–173.

92. Quinn, M. T., Parthasarathy, S., and Steinberg, D., 1985, Endothelial cell-derived chemotactic activity for mouse peritoneal macrophages and the effects of modified forms of low density lipoprotein, *Proc. Natl. Acad. Sci. USA* **82:**5949–5953.

93. Parthasarathy, S., Steinbrecher, V. P., Barnett, J., Witztum, J. L., and Steinberg, D., 1985, Essential role of phospholipase A activity in endothelial cell-induced modification of low density lipoprotein, *Proc. Natl. Acad. Sci. USA* **82:**3000–3004.

94. Pitas, R. E., Innerarity, T. L., and Mahley, R. W., 1983, Foam cells in explants of atherosclerotic rabbit aortas have receptors for beta-very low density lipoproteins and modified low density lipoproteins, *Arteriosclerosis* **3:**2–12.

95. Rapp, J. H., Connor, W. E., Lin, D. S., Inahara, T., and Porter, J. M., 1983, Lipids of human atherosclerotic plaques and xanthomas: Clues to the mechanism of plaque progression, *J. Lipid Res.* **24:**1329–1335.

96. Reichel, D., Myant, N. B., Rudra, D. N., and Pflug, J. J., 1980, Evidence for the presence of tissue free cholesterol in low density and high density lipoprotein of human peripheral lymph, *Atherosclerosis* **37:**489–495.

97. Reidy, M. A., 1985, A reassessment of endothelial injury and arterial lesion formation, *Lab. Invest.* **53:**513–520.

98. Ross, R., 1986, The pathogenesis of atherosclerosis—An update, *N. Engl. J. Med.* **314:**488–500.

99. Rudel, L. L., Bond, M. G., and Bullock, B. C., 1985, LDL heterogeneity and atherosclerosis in nonhuman primates, *Ann. N.Y. Acad. Sci.* **454:**248–253.

100. Rudel, L. L., Parks, J. S., Johnson, F. L., and Babiak, J., 1986, Low density lipoproteins in atherosclerosis, *J. Lipid Res.* **27:**465–474.

101. Salisbury, B. G., Falcone, D. J., and Minick, C. R., 1985, Insoluble low density lipoprotein-proteoglycan complexes enhance cholesteryl ester accumulation in macrophages, *Am. J. Pathol.* **120:**6–11.

102. Sanan, D. A., Strumfer, A. E. M., van der Westhuyzen, D. R., and Coetzee, G. A., 1985, Native and acetylated low density lipoprotein metabolism in proliferating and quiescent bovine endothelial cells in culture, *Eur. J. Cell Biol.* **36:**81–90.

103. Schwendener, R. A., Lagocki, P. A., and Rahman, Y. E., 1984, The effect of charge and size on the interaction of unilamellar liposomes with macrophages, *Biochim. Biophys. Acta* **772:**93–101.

104. Schwenke, D. C., and Carew, T. E., 1986, Enhanced LDL content and degradation near the branch orifices of normal rabbit aorta, *Arteriosclerosis* **6:**527a.

105. Schwenke, D. C., and Carew, T. E., 1986, LDL content and rate of LDL degradation near aortic branch orifices increase with cholesterol feeding, *Arteriosclerosis* **6:**554a.

106. Scott, R. F., Kim, D. N., Schmee, J., and Thomas, W. A., 1986, Atherosclerotic lesions in coronary arteries of hyperlipidemic swine. Part 2. Endothelial cell kinetics and leukocyte adherence associated with early lesions, *Atherosclerosis* **62:**1–10.

107. Scott, R. F., Reidy, M. A., Kim, D. N., Schmee, J., and Thomas, W. A., 1986, Intimal cell mass-derived atherosclerotic lesions in the abdominal aorta of hyperlipidemic swine. Part 2. Investigation of endothelial cell changes and leukocyte adherence associated with early smooth muscle cell proliferative activity, *Atherosclerosis* **62:**27–38.

108. Schwartz, C. J., Sprague, E. A., Kelley, J. L., Valente, A. J., and Suenram, C. A., 1985, Aortic

intimal monocyte recruitment in the normo and hypercholesterolemic baboon (*Papio cynocephalus*), *Virchows Arch. A* **405**:175–191.

109. Shio, H., Haley, N. J., and Fowler, S., 1979, Characterization of lipid-laden aortic cells from cholesterol-fed rabbits. III. Intracellular localization of cholesterol and cholesteryl esters, *Lab. Invest.* **41**:160–167.

110. Simionescu, M., Ghitescu, L., Fixman, A., and Simionescu, N., 1986, Receptor-mediated transcytosis of albumin in vascular endothelium, *Acta Biol. Hung.* **37**(Suppl.):104.

111. Simionescu, M., and Simionescu, N., 1986, Receptor-mediated transcytosis of plasma molecules by vascular endothelium, in: *4th Int. Symp. Biol. Vasc. Endoth. Cell,* Noordwijkerhout, Abstr. Vol. p. 21.

112. Simionescu, N., and Simionescu, M., 1976, Galloyl-glucose of low molecular weight as mordants in electron microscopy, *J. Cell Biol.* **70**:608–621.

113. Simionescu, N., 1983, Cellular aspects of transcapillary exchange, *Physiol. Rev.* **63**:1536–1579.

114. Simionescu, N., and Simionescu, M., 1985, Interactions of endogenous lipoproteins with capillary endothelium in spontaneously hyperlipoproteinemic rats, *Microvasc. Res.* **30**:314–332.

115. Simionescu, N., Vasile, E., Lupu, F., Popescu, G., and Simionescu, M., 1985, Accumulation of extracellular liposomes in the arterial intima as early change in experimental hyperlipidemia, *J. Cell Biol.* **101**:113a.

116. Simionescu, N., and Simionescu, M., 1986, Biopathology of arterial intima in the prelesional stages of atherogenesis, in: *XVIth Int. Congr. Acad. Pathol.,* Vienne, Abstr. Vol., p. 4.

117. Simionescu, N., Lupu, F., Vasile, E., Popescu, G., and Simionescu, M., 1986, Early changes of arterial wall in experimental hypercholesterolemia, in: *IIIeme Congr. Entente Med. Mediterr.,* Palermo, Abstr. Vol. p. 5.

118. Simionescu, N., and Simionescu, M., 1986, Pathophysiological aspects of vascular endothelium in atherogenesis, *Biol. Chem. Hoppe-Seyler* **367**(Suppl.):104.

119. Simionescu, N., Vasile, E., Lupu, F., Popescu, G., and Simionescu, M., 1986, Prelesional events in atherogenesis: Accumulation of extracellular cholesterol-rich liposomes in the arterial intima and cardiac valves of the hyperlipidemic rabbit, *Am. J. Pathol.* **123**:109–125.

120. Slater, H. R., Shepherd, J., and Packard, C. J., 1982, Receptor-mediated catabolism and tissue uptake of human low density lipoprotein in the cholesterol-fed atherosclerotic rabbit, *Biochim. Biophys. Acta* **713**:435–445.

121. Small, D. M., and Shipley, G. G., 1974, Physical chemical basis of lipid deposition in atherosclerosis, *Science* **185**:129–177.

122. Smith, E. B., and Staples, E. M., 1980, Distribution of plasma proteins across the human aortic wall: Barrier functions of endothelium and internal elastic lamina, *Atherosclerosis* **37**:579–590.

123. Smith, E. B., and Staples, E. M., 1982, Plasma protein concentrations in interstitial fluid from human aortas, *Proc. R. Soc. London B Ser.* **217**:59–75.

124. Smith, E. B., and Ashall, C., 1983, Low density lipoprotein concentration in interstitial fluid from human atherosclerotic lesions: Relation to theories of endothelial damage and lipoprotein binding, *Biochim. Biophys. Acta* **754**:249–257.

125. Spector, A. A., Scanu, A. M., Kaduce, T. L., Figard, P. H., Fless, G. M., and Czervionke, R. L., 1985, Effect of human plasma lipoproteins on prostacyclin production by cultured endothelial cells, *J. Lipid Res.* **26**:288–297.

126. Srinivasan, S. R., Jost, C., Radhakrishnamurthy, B., Dalferes, E. R., Jr., and Berenson, G. S., 1981, Lipoprotein–elastin interactions in human aorta fibrous plaque lesions, *Atherosclerosis* **38**:137–147.

127. Srinivasan, S. R., Vijayagopal, P., Dalferes, E. R., Jr., Abbate, B., Radhakrishnamurthy, B., and Berenson, G. S., 1984, Dynamics of lipoprotein–glycosaminoglycan interactions in the atherosclerotic rabbit aorta in vivo, *Biochim. Biophys. Acta* **793**:157–168.

128. St. Clair, R. W., Randolph, R. K., Jokinen, M. P., Clarkson, T. B., and Barakat, H. A., 1986, Relationship of plasma lipoproteins and the monocyte–macrophage system to atherosclerosis severity in cholesterol-fed pigeons, *Arteriosclerosis* **6**:614–626.

129. Stein, O., Halpern, G., and Stein, Y., 1986, Cholesteryl ester efflux from extracellular and cellular elements of the arterial wall: Model systems in culture with cholesteryl linoleyl ether, *Arteriosclerosis* **6**:70–78.

130. Steinberg, D., 1983, Lipoproteins and atherosclerosis: A look back and a look ahead, *Arteriosclerosis* **3**:283–301.
131. Steinberg, D., Pittman, R. C., and Carew, T. E., 1985, Mechanisms involved in the uptake and degradation of low density lipoprotein by the artery wall in vivo, *Ann. N.Y. Acad. Sci.* **454**:195–206.
132. Steinbrecher, U. P., Parthasarathy, S., Leake, D. S., Witztum, J. L., and Steinberg, D., 1984, Modification of low density lipoprotein by endothelial cells involves lipid peroxidation and degradation of low density lipoprotein phospholipids, *Proc. Natl. Acad. Sci. USA* **81**:3883–3887.
133. Stemerman, M. B., 1981, Effect of moderate hypercholesterolemia on rabbit endothelium, *Arteriosclerosis* **1**:25–32.
134. Stemerman, M. B., Morrel, E. M., Burke, K. R., Colton, C. K., Smith, K. A., and Lees, R. S., 1986, Local variation in artery wall permeability to low density lipoprotein in normal rabbit aorta, *Arteriosclerosis* **6**:64–69.
135. Stender, S., and Zilversmit, D. B., 1981, Transfer of plasma lipoprotein components and of plasma proteins into aortas of cholesterol-fed rabbits, *Arteriosclerosis* **2**:115–124.
136. Stender, S., 1982, The in vivo transfer of free and esterified cholesterol from plasma into the arterial wall of hypercholesterolemic rabbits, *Scand. J. Clin. Lab. Invest.* **42**(Suppl. 161):43–52.
137. Stender, S., and Hjelms, E., 1984, In vivo influx of free and esterified plasma cholesterol into human aortic tissue without atherosclerotic lesions, *J. Clin. Invest.* **74**:1871–1881.
138. Taylor, R. G., and Lewis, J. C., 1986, Endothelial cell proliferation and monocyte adhesion to atherosclerotic lesions in White Carnean pigeons, *Am. J. Pathol.* **125**:152–160.
139. Territo, M., Berliner, J. A., and Fogelman, A. M., 1984, Effect of monocyte migration on low density lipoprotein transport across aortic endothelial cell monolayers, *J. Clin. Invest.* **74**:2279–2284.
140. Trillo, A. A., and Prichard, R. W., 1979, Early endothelial changes in experimental primate atherosclerosis, *Lab. Invest.* **41**:294.
141. Tsukada, T., Rosenfeld, M., Ross, R., and Gown, A. M., 1986, Immunocytochemical analysis of cellular components in atherosclerotic lesions: Use of monoclonal antibodies with the Watanabe and fat-fed rabbit, *Arteriosclerosis* **6**:601–613.
142. van Hinsbergh, V. W. M., Scheffer, M., Havekes, L., and Kempen, H. J. M., 1986, Role of endothelial cells and their products in the modification of low density lipoproteins, *Biochim. Biophys. Acta* **878**:49–64.
143. Vasile, E., Nistor, A., Nedelcu, S., Simionescu, M., and Simionescu, N., 1980, Dual pathway of low density lipoprotein transport through aortic endothelium and vasa vasorum, in situ, *Eur. J. Cell Biol.* **22**:181.
144. Vasile, E., Simionescu, M., and Simionescu, N., 1983, Visualization of the binding, endocytosis and transcytosis of low density lipoproteins in the arterial endothelium in situ, *J. Cell Biol.* **96**:1677–1689.
145. Vasile, E., and Simionescu, N., 1985, Transcytosis of low density lipoprotein through vascular endothelium, in: *Glomerular Dysfunction and Biopathology of Vascular Wall* (E. Seno, A. L. Copley, M. A. Ventkatachalam, Y. Hamashida, and T. Tsujii, eds.), Academic Press, New York, pp. 87–102.
146. Vasile, E., Popescu, G., Simionescu, M., and Simionescu, N., 1986, Interaction of low density lipoprotein and beta-very low density lipoprotein with the arterial endothelium in normal and hypercholesterolemic animals, in: *4th Int. Symp. Biol. Vasc. Endoth. Cell,* Noordwijkerhout, Abstr. Vol., p. 123.
147. Vasile, E., Popescu, G., Simionescu, M., and Simionescu, N., 1986, Enhanced transcytosis and accumulation of beta-very low density lipoproteins in the aorta of rabbits with experimental hyperlipidemia, in: *XVIth Int. Congr. Int. Acad. Pathol.,* Vienna, Abstr. Vol., p. 68.
148. Wagner, W. D., 1985, Proteoglycan structure and function as related to atherosclerosis, *Ann. N.Y. Acad. Sci.* **454**:52–68.
149. Wagner, W. D., Salisbury, B. G. J., and Rowe, H. A., 1986, A proposed structure of chondroitin-6-sulfate proteoglycan of human normal and adjacent atherosclerotic plaque, *Arteriosclerosis* **6**:407–414.
150. Wallis, W. J., Beatty, P. G., Ochs, H. D., and Harlan, J. M., 1985, Human monocyte adherence

to cultured vascular endothelium: Monoclonal antibody-defined mechanisms, *J. Immunol.* **135:**2323–2330.

151. Walton, K. W., and Morris, C. J., 1977, Studies on the passage of plasma proteins across arterial endothelium in relation to atherogenesis, *Prog. Biochem. Pharmacol.* **14:**138–152.

152. Watanabe, T., Hirata, M., Yoshikawa, Y., Nagazuchi, Y., Toyoshima, H., and Watanabe, T., 1985, Role of macrophages in atherosclerosis: Sequential observations of cholesterol-induced rabbit aortic lesions by the immunoperoxidase technique using monoclonal antimacrophage antibody, *Lab. Invest.* **53:**80–90.

153. Weber, G., Fabbrini, P., and Resi, L., 1973, On the presence of a concanavalin A-reactive coat over the endothelial aortic surface and its modifications during early experimental cholesterol atherogenesis in rabbits, *Virchows Arch. A* **359:**299–307.

154. Weisgraber, K. H., Innenarity, T. L., and Mahley, R. W., 1978, Role of the lysine residues of plasma lipoproteins in high affinity binding to cell surface receptors on human fibroblasts, *J. Biol. Chem.* **253:**9053–9062.

155. Werb, Z., and Chin, J. R., 1983, Apoprotein E is synthesized and secreted by resident and thioglycolate-elicited macrophages but not by pyran copolymer or bacillus Calmette–Guerin-activated macrophages, *J. Exp. Med.* **158:**1272–1284.

156. Wiklund, O., Carew, T. E., and Steinberg, D., 1985, Role of the low density lipoprotein receptor in penetration of low density lipoprotein into rabbit aortic wall, *Arteriosclerosis* **5:**135–141.

157. Williams, K. J., Tall, A., and Bisgaier, C., 1986, Phospholipid liposomes acquire apo E in atherogenic plasma and inhibit cholesterol loading of macrophages, *Arteriosclerosis* **6:**538a.

158. Williams, R. D., Sgontas, D. S., and Zaatari, G. S., 1986, Enzymology of long-chain base synthesis by aorta: Induction of serine palmitoyltransferase activity in rabbit aorta during atherogenesis, *J. Lipid Res.* **27:**763–770.

19

Response of Blood Vessel Cells to Viral Infection

Nicholas A. Kefalides

I. INTRODUCTION

Injury to blood vessel cells has been implicated in the development of a variety of vascular disorders including atherosclerosis,[2,13] disseminated intravascular coagulation,[7] and immune vasculitis.[44] Endothelial cell (EC) injury can arise from a variety of causes including hemodynamic stress,[50] mechanical trauma,[51] hypercholesterolemia,[45] infectious agents,[14,39,42] oxygen,[28] and other chemical agents such as homocysteine.[56] Renewed interest in the relationship between vascular injury due to viral infection and atherosclerosis stems from recent observations which demonstrated that virus particles, virus antigens, and virus DNA can be detected in vascular lesions of chickens[9,39] and humans[4,21,38] with atherosclerosis. Viruses may produce vascular injury by mechanisms other than direct invasion of endothelium. Antigen–antibody complexes involving hepatitis B surface antigen have been isolated from the sera of some patients with periarteritis nodosa.[17] Tumor formation as a result of viral infection is another possible mechanism by which viruses may induce vascular disease as is the case with cytomegalovirus and Kaposi's sarcoma.[20]

At the time our laboratory began investigating the *in vitro* interaction between virus and EC,[6,15,35] few such studies had been reported.[1,18] In this report I essentially summarize our published work. In our studies we attempted to (1) define further the viruses that are capable of infecting EC *in vitro*, (2) examine the effect of viral infection on (a) adherence of granulocytes to EC,[35] (b) induction of Fc and C3b receptors,[6] and (c) matrix protein synthesis,[27,30,34,58] and (3) determine the effect of altered matrix substratum on the phenotypic expression of EC.

II. EXPERIMENTAL PROCEDURES

A. Endothelial and Smooth Muscle Cell Cultures

Human EC were isolated from umbilical cord vein according to the method of Gimbrone *et al.*[19] Human EC were grown in tissue culture flasks that were

Nicholas A. Kefalides • Connective Tissue Research Institute and Department of Medicine, University of Pennsylvania and University City Science Center, Philadelphia, Pennsylvania 19104.

coated with 1% gelatin in modified medium 199 supplemented with 90 μg/ml heparin[53] and 3 μg/ml EC growth factor.[36]

Bovine arterial and venous EC were isolated and grown as described by Macarak et al.[33] and Friedman et al.[16] Bovine smooth muscle cells (SMC) were obtained from the Institute for Medical Research, Camden, NJ, and grown as recommended by the provider institute.

B. Viral Infection of Blood Vessel Cells

Virus pools for EC or SMC inoculation were prepared on human embryonic lung fibroblasts (MRC-5), rhesus monkey kidney or human epidermoid carcinoma cells (HEP-2) as described by Friedman et al.[15] Twenty-four hours before infection the growth medium was removed and replaced with heparin-free medium and the concentration of the EC growth factor increased to 15 μg/ml. Confluent monolayers were infected with virus for 1 hr at 37°C at a multiplicity of infection (MOI) ranging from 0.003 to 50 depending on the nature of the experiment. After allowing the virus to adsorb to the EC or SMC, the medium and any unadsorbed virus were removed and the monolayer washed and refed with growth medium. Control cultures were either noninfected or mock-infected.

C. Indicators of Viral Infection

Viral infection of blood vessel cells was determined by immunofluorescence using specific antisera to detect virus antigens within EC or SMC cells and by observing monolayers for virus-induced cytopathology. To determine whether virus replication occurred, infected cultures were harvested for viral titrations 2 hr to 10 days after inoculation. Viral titers of infectious particles were expressed either as the highest dilution of fluid which produces cytopathology in 50% of the indicator cell cultures inoculated—50% tissue culture infectious dose endpoint $(TCID_{50})$[15]—or as the number of infectious particles determined by a plaque assay.[47]

D. Assessment of Cell Response to Viral Infection

1. Granulocyte Adherence Assay

Granulocyte adherence to the endothelial monolayers grown in 35-mm-diameter petri dishes was measured as described by MacGregor et al.[35] Medium 199 was decanted from the cultures and 1 ml of heparinized whole blood (5 U/ml) was added to triplicate plates, just covering the surface. Human blood was used in all experiments. The endothelial monolayer–blood overlay was incubated at 37°C and 100% humidity for 15 min without agitation and then the blood was aspirated. Comparison of the granulocyte counts before and after incubation permitted determination of the percentage of granulocytes to the endothelium. In some experiments, a pure suspension of granulocytes in Hanks's balanced salt solution (HBSS) was used rather than whole blood. Granulocytes were separated by Hypaque–Ficoll density gradient sedimentation, washed three times in modified Hanks's solution, and suspended in HBSS at a concentration of $5–10 \times 10^6$ cells/ml. Human umbilical vein EC or bovine aorta EC monolayers were infected with either adenovirus 7 (MOI = 0.3), polio 1 (MOI = 0.03), measles (MOI = 0.03), or herpes simplex 1 (MOI = 0.03) viruses.[35] EC cultures were harvested for viral titrations 2 hr after inoculation of cultures and then at daily intervals for up to 10 days.

2. Assay for Fc and C3 Receptors

To assay for Fc and C3 receptors, cells were grown to confluency in 24-well plates.[6] One-half of the wells were infected with HSV-1 at a MOI of 0.5–1.0 (10^5 $TCID_{50}$/well). The remaining half of the wells were mock-infected to serve as controls. After a 1-hr adsorption at 37°C, the monolayers were washed and refed with 1 ml of medium. The cells were then incubated for 24 hr at 37°C, examined for cytopathic effect by phase-contrast microscopy, and used for the studies described below. Monolayers were also prepared from human mononuclear cells separated by Hypaque–Ficoll density centrifugation. These cells were utilized as positive controls for the binding of antibody or complement-coated erythrocytes. The ability of uninfected or virus-infected EC to bind immunoglobulin or complement-coated particles was studied using fresh sheep erythrocytes radiolabeled with $^{51}Cr(Na_2CrO_4)$.[6]

3. Assessment of Matrix Protein Synthesis

Matrix protein synthesis was measured by metabolic labeling of the EC or SMC cultures as described by Kefalides and Ziaie[27] and Ziaie et al.[58] For metabolic studies, the cells were washed with PBS once and labeling medium was added at different time points postinfection. The labeling medium was Eagle's minimum essential medium supplemented with 0.1% glucose, 1 mM glutamine, 50 μg/ml ascorbic acid, 50 μg/ml β-aminopropionitrile fumarate, and 15 mM N-2-hydroxyethylpiperazine-N-2-ethanesulfonic acid. For ^{14}C labeling, 6 μCi/ml [^{14}C]proline (250 mCi/mmole; Amersham Corp., Arlington Heights, IL) was added. For ^{35}S labeling, the minimum essential medium was replaced by Eagle's methionine-free medium supplemented with 50 μCi/ml [^{35}S]methionine (1400 Ci/mmole; Amersham Corp.). In the latter case, all cultures were starved with methionine-free medium for 1 hr prior to labeling. Following the labeling, at the specified times, the cell layer, medium, and matrix fractions were collected separately and analyzed.

For analysis, the medium was collected, protease inhibitors [N-ethylmaleimide (NEM) 10 mM and phenylmethanesulfonyl fluoride (PMSF) 1 mM] added, and the medium centrifuged at 500g for 20 min at 4°C to remove cell debris. Any virus that was released in the medium was either inactivated by the addition of Triton X-100 in 50 mM tris(hydroxymethyl)aminomethane (Tris), pH 7.2, to a final concentration of 1% or removed by centrifugation at 115,000g for 1 hr at 4°C. The cell layer was extracted by either of the following two methods: (1) 1%

Triton X-100 in 50 mM Tris, pH 7.2, in the presence of inhibitors (NEM and PMSF) was added and the lysate cleared by centrifugation at 18,000g for 40 min at 4°C or (2) extraction buffer containing 2 M urea, 50 mM NaCl, 5 mM NEM, 1 mM PMSF, 15 mM dithiothreitol, and 15 mM N-2-hydroxyethylpiperazine-N-2-ethanesulfonic acid was added to the cell layer, incubated at room temperature for 3 hr, and the layer removed by scraping. Complete virus particles and cell debris were removed by centrifugation at 115,000g for 1 hr at 4°C. The matrix was dissolved in 1% SDS in 50 mM Tris, pH 7.2, containing protease inhibitors (NEM and PMSF) and clarified by centrifugation at 18,000g for 40 min at 4°C. For analytical purposes, the samples were dialyzed against appropriate buffers.

Protein analysis and identification of specific matrix components was carried out by polyacrylamide gel electrophoresis (SDS–PAGE), electroimmunoblot, and enzyme-linked immunosorbent assay (ELISA). Proteins were analyzed by SDS–PAGE essentially according to Laemmli[29] as described by Kefalides and Ziaie.[27] After electrophoresis, the gels were stained with Coomassie brilliant blue. Fluorography was carried out according to Bonner and Laskey[5] using En³Hance (New England Nuclear, Boston). Dried gels were exposed to sensitized Kodak X-O-Mat XAR-5 film.[31]

For electroimmunoblot, proteins were separated by SDS–PAGE and transferred electrophoretically onto a nitrocellulose sheet (Bio-Rad Laboratories, Richmond, Calif.) according to Towbin et al.[54]

ELISA was performed essentially according to Engvall[8] as modified by Rennard et al.[59] To collect medium samples of control and infected cultures for competitive ELISA, growth medium was replaced by serum-free medium at 4 hr postinfection and the cultures were incubated for 13 hr (corresponding to 17 hr postinfection). Following this, the medium was collected as described previously and dialyzed against PBS–0.05% Tween and analyzed.

III. RESULTS

A. Replication of Virus

Results of viral infection of human and bovine EC are shown in Table 1 and Fig. 1. The data show that several human viruses can replicate in vascular cells. Although herpes simplex 1 and 2, adenovirus 7, measles, mumps, parainfluenza 3, polio 1, and echo 9 replicate in human venous EC, others, such as cytomegalovirus, influenza A1 Victoria, respiratory syncytial virus, and coxsackie B4, do not. The same viruses which grew in the human venous EC replicated in bovine arterial and venous EC, except that polio 1 did not grow whereas coxsackie B4 did grow in the latter. The failure of cytomegalovirus to grow in either the human or bovine EC may be unique for these cells since replication of this virus has been reported in human arterial SMC.[55] These findings suggest that EC may provide an active barrier to some viruses, thus preventing those viruses from infecting specific tissues supplied by a given vessel. Moreover, the development of significant cytopathology by some viruses implies that viral infection can po-

Table 1. Infection of Human Venous and Bovine Arterial Endothelium by Common Human Viruses[a]

Virus	Human endothelium				Bovine endothelium			
	Inoculum titer (MOI)	Viral replication	CPE	Positive IFA	Inoculum titer (MOI)	Viral replication	CPE	Positive IFA
Herpesviruses								
Herpes simplex virus	10^6 (5)	Yes	Yes	Yes	10^4 (0.03)	Yes	Yes	Yes
Cytomegalovirus	$10^{4.7}$ (0.3)	No	No	No	10^3 (0.003)	No	No	No
Respiratory viruses								
Adenovirus type 7	$10^{2.7}$ (0.003)	Yes	Yes	Yes	$10^{2.7}$ (0.001)	Yes	Yes	Yes
Measles	10^4 (0.05)	Yes	Yes	Yes	10^4 (0.03)	Yes	Minimal	Yes
Mumps	10^4 (0.05)	Yes	Yes	Yes	10^4 (0.03)	Yes	Minimal	ND
Parainfluenza type 3	10^7 (50)	Yes	Yes	Yes	10^7 (30)	Yes	Yes	Yes
Influenza A1/Victoria	$10^{4.7}$ (0.3)	No	No	No	$10^{4.7}$ (0.1)	No	No	No
Respiratory syncytial	10^3 (0.005)	No	No	No	10^3 (0.003)	No	No	No
Enteroviruses								
Poliovirus type 1	$10^{2.7}$ (0.003)	Yes	Yes	ND	$10^{2.7}$ (0.001)	No	No	ND
Echovirus type 9	10^6 (5)	Yes	Yes	ND	10^6 (3)	Yes	No	ND
Coxsackie virus B4	10^5 (0.5)	No	No	ND	10^5 (0.3)	Yes	No	ND

[a] MOI (multiplicity of infection) = the ratio of the titer of infectious virus in the inoculum to the number of cells in the monolayer; IFA, immunofluorescence; ND, not done. (Modified from Friedman et al.[15])

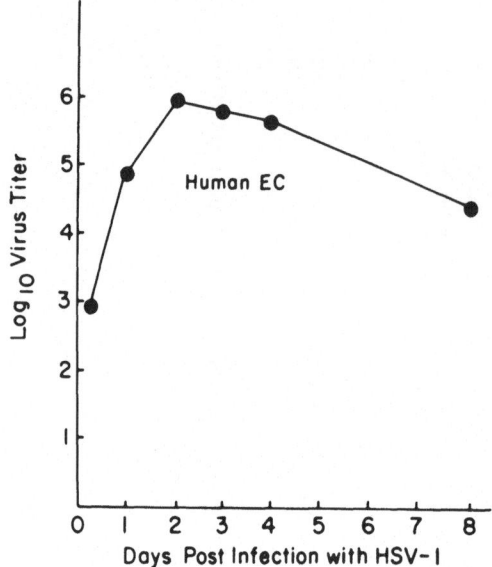

Figure 1. Growth curve of HSV-1 inoculated onto human EC cultures.

tentially influence the functions of the host cell, as will become evident in later sections.

To evaluate the possibility that EC from different vessels vary in their susceptibility to viral infection, cultures of EC from different major vessels of a single species, i.e., bovine, were examined for their susceptibility to viral infection and replication.[16] Table 2 shows the results with five common viral pathogens of humans (herpes simplex 1, measles, mumps, echo 9, and coxsackie B4 viruses). These

Table 2. Comparative Growth of Viruses in Bovine Pulmonary Artery, Thoracic Aorta, and Inferior Vena Cava Endothelial Cell Cultures

	Peak viral titer ($\log_{10}$ TCID$_{50}$/ml)[a]			
Virus	Pulmonary artery	Thoracic aorta	Vena cava	Comment
Measles	5.2 (3.7)	2.7	2.7 (3.7)	Only minimal cytopathology was detectable
Mumps	(6.0)		1.0 (5.7)	Only minimal cytopathology was detectable
Herpes simplex 1	5.7	5.7	5.0	Cytopathology characterized by rounding of cells and retraction of monolayer
Coxsackie B4	4.2	3.7	4.2	No cytopathology was detectable
Echo 9	3.7	4.4	4.2	No cytopathology was detectable

[a] Viral titers are calculated by determining the highest dilution of material harvested from infected endothelial cell cultures which produces viral cytopathic changes when inoculated onto indicator tissue culture cells which are known to be susceptible to the virus. (Data obtained from Friedman *et al.*[16])

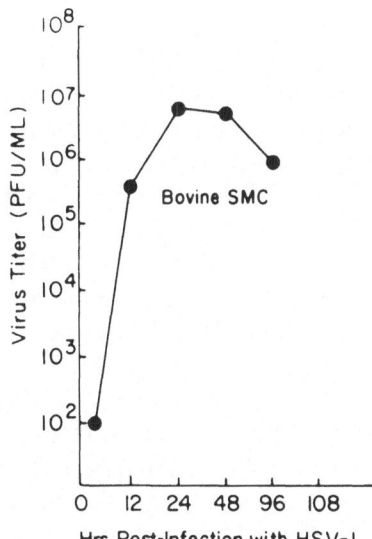

Figure 2. Growth curve of HSV-1 inoculated onto bovine SMC cultures.

viruses were evaluated for growth in EC derived from bovine fetal pulmonary artery, thoracic aorta, and vena cava. All five species replicated in each type of EC. There were apparent differences in the quantities of measles and mumps viruses produced in pulmonary artery endothelium compared with thoracic aorta and vena cava when EC were obtained from different animals. However, when pulmonary artery EC cells were compared with vena cava cells from the same animal, growth of each virus was similar in the two cell types. Four of the viruses replicated in the various endothelial cells without producing appreciable changes in cell morphology. These results indicate that EC from different blood vessels are equally susceptible to the human viruses evaluated and that replication of virus can occur without major alterations in cell morphology. In the latter case, EC could serve as permissive cells permitting viruses to leave the circulation, establish a persistent infection, and initiate infection in adjacent tissues, including the subendothelial SMC.

Studies in our laboratory have demonstrated that bovine aorta SMC will support growth of herpes simplex virus type 1.[30] Figure 2 shows a growth curve of HSV-1 in bovine SMC infected at a MOI of 5. Maximum titer was achieved at 24 hr postinfection after which it began to decrease slowly. Tumilowicz *et al.*[55] have shown that SMC from human umbilical cord artery will support replication of cytomegalovirus. These authors found that the cytopathic effect of CMV in SMC was delayed as compared to infected IMR-90 fibroblasts, resulting in prolonged survival of a fraction of the starting cell population.

B. Effect of Viral Infection of Granulocyte Adherence

In collaboration with Dr. R. R. MacGregor we examined the adherence of granulocytes on monolayers of human and bovine EC before and after infection with different viruses.[35] Adherence to uninfected human EC using 1.0 ml of whole blood averaged 17.9 ± 3.7% and to bovine EC, 20.3 ± 3.7%. The impact of infection of the endothelium by different viruses on subsequent granulocyte adherence was notable. Poliovirus produced an acute lytic infection of human EC, with associated increased adherence to 185.4% of control 24 hr postinoculation. Significantly increased adherence was noted at 6 hr, before detectable cytopathic effect. HSV-1 caused a similar rapidly lytic infection of bovine endothelium associated with increased adherence to 213.7% of control 6 hr postinoculation. This augmented adherence could be demonstrated when granulocytes were suspended in physiologic saline solution, showing that antibody and complement need not be present. Trypsin treatment of infected monolayers did not prevent the augmentation, and supernate from infected monolayers increased the adherence of polymorphonuclear leukocytes to normal, uninfected monolayers. Chronic, slowly lytic infections, lasting 7 days or more, were induced with adenovirus in human endothelium and with measles virus in bovine cells. Adherence increased as virus was noted in the cell cultures on day 4, several days before cytotoxicity was seen. Thus, chronic viral infection of the endothelium appears possible and results in increased granulocyte adherence. The authors suggested that naturally occurring disease such as infection may act synergistically with adherent granulocytes to damage the endothelium and may represent an *in vitro* model of vasculitis.[35]

C. Induction of Fc and C3 Receptors

Studies by Westmorland and Watkins[57] demonstrated the induction of receptors for the Fc portion of IgG on fibroblasts infected with HSV. Fc and C3 receptors are sites for attachment of antigen–antibody complexes and can be demonstrated on the surfaces of monocytes, granulocytes, and B lymphocytes. Although most reports indicate that Fc receptors are not readily detectable in EC from various tissues,[25,48] when a highly sensitive radioimmunoassay was used, Lyss et al.[32] showed binding of IgG to cultured human umbilical vein EC.

When human EC were infected with HSV-1, there was induction of receptors for the Fc portion of IgG and for C3.[6] In these studies, it was noted that when human EC were incubated with ^{51}Cr-labeled sheep erythrocytes sensitized with IgG, IgM, or IgM plus complement, no preferential binding of IgG or complement-coated erythrocytes to EC monolayers was observed. In contrast, significant binding of erythrocytes coated with IgG or IgM plus complement was observed after viral infection. Phase-contrast and scanning electron microscopy demonstrated erythrocyte adherence around the infected EC in a rosette pattern. Binding of IgG-coated erythrocytes was fully inhibited by Fc (0.31 mg/ml) but not Fab' fragments of nonimmune IgG. Binding of complement-coated cells was unaffected by the presence of IgG (1 mg/ml). With purified individual components, binding of complement-coated erythrocytes was proportional to the concentration of C3. Binding of IgG or C3-coated cells was detected beginning 4 hr postinfection.

These studies indicate that HSV-1 infection can induce IgG and C3 receptors on human EC thus promoting the deposition of immune complexes in vascular tissue. It is conceivable that other forms of injury or infection with other viruses may also induce similar receptors which may play a role in immunologically mediated vessel disease.

D. Suppression of Host-Cell Matrix Protein Synthesis

A number of viruses suppress and eventually shut off the synthesis of cellular proteins. The mechanism(s) is not clearly understood.[10,49] Competition for ribosomes by the large excess of viral mRNA has been offered as one possibility. Even when viral mRNA is not in excess, it may have a selective advantage in binding to ribosomes and initiating translation.[26] In cells infected with adenovirus, late shutoff of cellular protein may in part be due to inhibition of transport of cellular mRNA out of the nucleus.[3] Early studies in which permissive cells were infected with HSV-1 demonstrated a disruption in polysomes[52] which was followed by a rapid decline in host-cell protein synthesis.[12,43] HSV-1 also mediates the inhibition of cellular RNA[60] and DNA synthesis.[43] Studies by Fenwick and Clark[11] suggest that in Vero cells infected with a temperature-sensitive mutant of HSV-1 (tsB7) there is an early suppression of host-cell protein synthesis which is virion protein dependent and a delayed shutoff which requires virus protein synthesis. It has also been shown that several strains of HSV-2 cause a more rapid decrease in host-cell protein synthesis than certain HSV-1 strains [22,41] and that the shut off by HSV-1 and HSV-2 is accomplished by different mechanisms.[22]

We chose to study the effect of viral infection on the synthesis of matrix proteins by vascular cells because EC and SMC rest on or are surrounded by a basement membrane matrix which contains several proteins having important functions in cell adhesion, cell proliferation, cell differentiation, and gene expression. We propose that *in vivo* cell injury, whether in the form of viral infection, radiation, or some immune process, can initiate a series of specific cellular responses that can result in the synthesis of altered extracellular matrix components and thus lead to long-term sequelae of chronic vascular disease.

Studies on the effects of HSV-1 infection (MOI = 1) on matrix protein synthesis by bovine EC were recently reported by our laboratory.[34] Protein synthesis was measured in infected and uninfected cultures by following the incorporation of [^{14}C]proline into nondialyzable protein at different times postinfection. Medium proteins were analyzed by PAGE, gel filtration, and electroimmunoblotting. No change was detected in total ^{14}C incorporation into nondialyzable protein in infected as compared to uninfected EC; however, there was a significant decrease in the synthesis of collagen and fibronectin in infected cultures. The hydroxy[^{14}C]proline content of fractionated medium proteins showed no significant differences between infected and uninfected cultures. It is suggested, therefore, that the decrease in collagen synthesis cannot be explained by increased collagen degradation.

The above studies examined the effects of prolonged viral infection of bovine EC on matrix protein synthesis. We thought it would be important to determine the early changes in human EC matrix protein synthesis following infection with HSV-1 or HSV-2.[27,58] Our results demonstrate that during infection, synthesis of host-specific proteins decreases with time, while an increase in the synthesis of

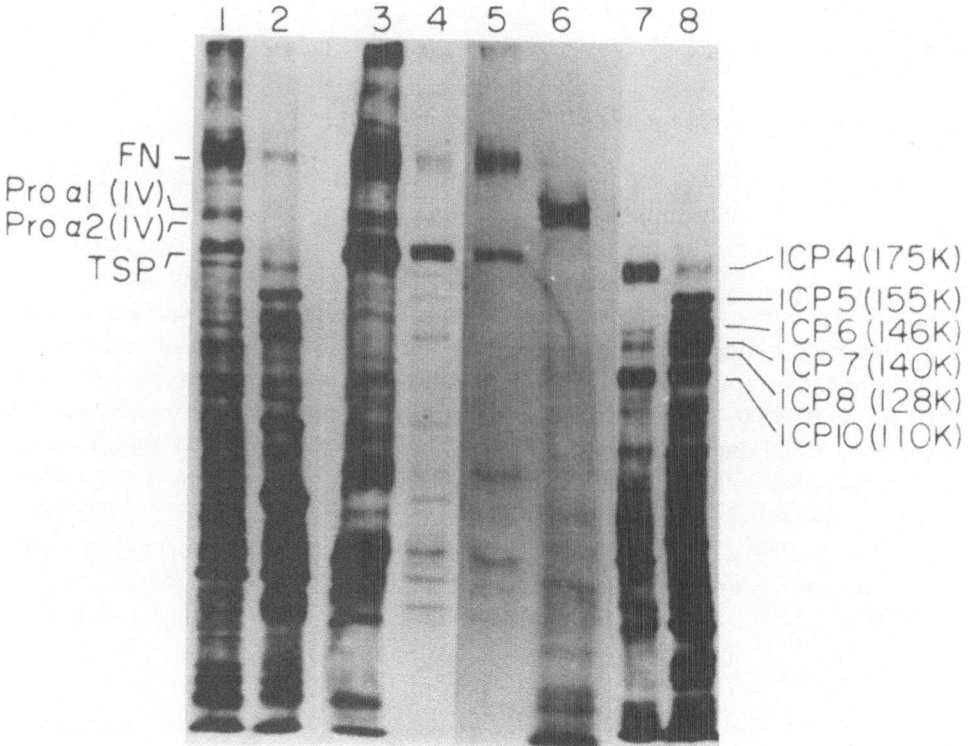

Figure 3. Pattern of protein synthesis by EC cultures infected with HSV-1. EC cultures were infected with HSV-1 at a MOI of 5 for 1 hr and then labeled with [^{14}C]proline at 5 hr postinfection for a period of 10 hr. The cell layer and medium fractions were collected and analyzed by SDS–PAGE. Cell layer: mock-infected (lane 1), infected (lane 2); medium: mock-infected (lane 3), infected (lane 4); FN and TSP standards (lane 5); procollagen type IV standard (lane 6); ICP (infected cell proteins), α- and β-polypeptides of HSV-1 (lanes 7 and 8, respectively). Note the substantial reduction in the level of FN, type IV collagen, and TSP in infected cultures (lanes 2 and 4) as compared to controls (lanes 1 and 3). (Data from Ziaie *et al.*[58])

virus proteins is observed. Figure 3 shows the electrophoretic analysis of samples from infected and mock-infected human EC cultures. In infected cultures, the synthesis of fibronectin (FN), type IV collagen, and thrombospondin (TSP) has decreased (Fig. 3, lanes 2 and 4 of the cell layer and medium fractions, respectively). However, the decrease in the level of TSP in the medium fraction is not as pronounced as compared to FN and type IV collagen (Fig. 3, lanes 3 and 4). These data suggest that there is a suppression of host-cell protein synthesis and, specifically, of matrix proteins. The degree of this suppression appears to vary with the protein. Although the level of cellular proteins decreases during infection, new species of proteins appear in the infected cultures. These new proteins correspond to newly synthesized HSV-1 polypeptides (compare lane 2 with lanes 7 and 8 in Fig. 3).[23,24]

In order to determine the time of onset of inhibition of cellular protein syn-

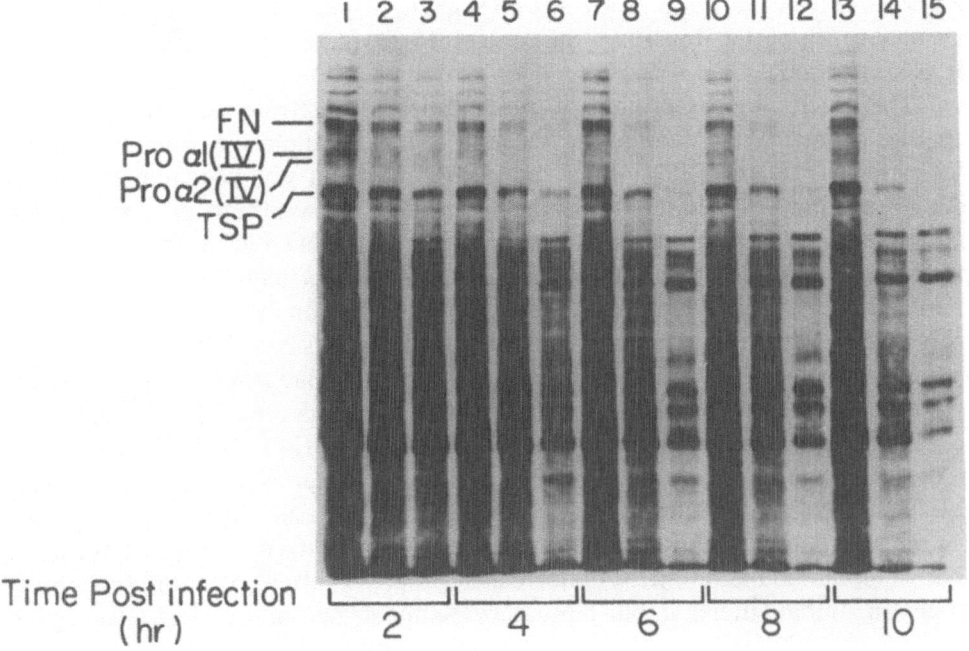

Figure 4. Effect of virus dose on the time and degree of inhibition of host protein synthesis. EC cultures were infected with HSV-1 at a MOI of 5 or 20 for 1 hr and then pulse-labeled with 50 μCi/ml [^{35}S]methionine for 2 hr at different times postinfection (2, 4, 6, 8, and 10 hr). An SDS–PAGE fluorogram of the cell fraction is shown. Control cultures, lanes 1, 4, 7, 10, and 13; cultures infected with a MOI of 5, lanes 2, 5, 8, 11, and 14; cultures infected with a MOI of 20, lanes 3, 6, 9, 12, and 15. Positions of FN, type IV collagen, and TSP are indicated. Decline in the level of these proteins is observed at 2 hr postinfection with a MOI of 20 (lane 3) and continues throughout. The inhibition is much more effective with a MOI of 20 at any time point postinfection, but varies with the protein inoculum size. (Data from Ziaie *et al.*[58])

thesis and the effect of the size of virus inoculum, human EC cultures were pulsed for 2 hr with [^{35}S]methionine at 2, 4, 6, 8, and 10 hr postinfection. One set of experimental cultures was infected with HSV-1 at a MOI = 5 and another at a MOI = 20. For the cell layer fractions, the results are shown in Fig. 4. In cultures infected at a MOI of 20, inhibition of host-cell protein synthesis begins at 2 hr and becomes almost complete at 10 hr postinfection. However, at a MOI of 5, there appears to be some synthesis of matrix proteins, particularly of TSP even at 10 hr postinfection. Regardless of the size of inoculum, it appears that the degree of inhibition of synthesis after HSV-1 infection of human EC varies with each protein. At 10 hr postinfection, with a MOI of 5, TSP synthesis continued at a low level in the infected cells while FN and collagen type IV synthesis were not detectable under the same conditions. Densitometric analysis of SDS–PAGE fluorograms showed that at 6 and 10 hr postinfection, TSP decreases by 33.3 and 68.0% at a MOI of 5 and by 91.5 and 96.6% at a MOI of 20, respectively. The values obtained for FN for the same time points are decreased by 83.1 and 86.2%

at a MOI of 5 and 96.6 and 99.4% at a MOI of 20, respectively. The value obtained for collagen type IV was 12.5% of control (87.5% decrease) at a MOI of 5 at 6 hr postinfection.

The specific matrix proteins were identified by electroimmunoblot using antibodies against type IV collagen, FN, and TSP.[27] Quantitative estimation of the decrease in FN and TSP synthesized during infection was accomplished in an ELISA system.[27,58] Although we have focused on the synthesis of three protein components of EC during viral infection, it is apparent that a variety of host proteins are affected by viral infection. The components migrating slightly slower than FN (Fig. 4) have tentatively been identified as von Willebrand factor, and as is evident, its synthesis is inhibited by the virus (lanes 2, 4, 6, 8, and 10). These results are consistent with the findings of Schek and Bachenheimer[46] who demonstrated that HSV-1 and HSV-2 infection of Vero cells decreased the cytoplasmic levels of β- and γ-actin, β-tubulin, and histone H3 and H4 mRNA, though not all at the same rate. Similarly, Mayman and Nishioka[37] observed that in Friend erythroleukemia cells infected with HSV-1, the ratio of the steady-state level of histone H3 mRNA to total RNA remained unchanged for the first 4 hr postinfection, while that of globin and actin mRNAs decreased progressively at early intervals postinfection. Initial studies in our laboratory demonstrate that the decrease in type IV collagen, FN, and von Willebrand synthesis correlates with decreases in the steady-state levels of mRNAs for these proteins.[62]

Studies by Nishioka and Silverstein,[40] who used murine erythroid cells transformed by Friend leukemia virus, and Fenwick and Clark,[11] who infected Vero cells with HSV-1, suggest that early suppression of cellular protein following infection with these viruses is not dependent on protein synthesized after infection. Similar results were obtained by us[27] when human EC were infected with UV-inactivated HSV-1 and HSV-2. At 4 hr postinfection, control and infected cultures were pulsed for 1 hr with [^{35}S]methionine. Figure 5 shows the electrophoretic pattern of the cell-matrix fraction of cultures infected with either intact or UV-irradiated HSV-1 and HSV-2. It will be noted that although in cultures infected with intact virus at MOIs of 5 or 20 (Fig. 5, HS1 5, HS2 5, HS1 20, HS2 20) there was suppression of matrix protein synthesis, viral protein synthesis was quite evident. Infection with UV-irradiated HSV-1 or HSV-2 at a MOI of 20 caused suppression of matrix as well as viral protein synthesis (Fig. 5, HS1 irr 20, HS2 irr 20). In order to show that suppression of host-cell matrix protein synthesis with UV-irradiated virus was not occurring at low levels of viral DNA transcription, infection was carried out with intact virus in the presence and absence of actinomycin D, an inhibitor of mRNA synthesis.[27] Table 3 shows the total ^{35}S-labeled protein content of human EC cultures infected with HSV-1 or HSV-2 in the presence and absence of actinomycin D. It will be noted that protein-bound radioactivity decreased by only 60% in the presence of actinomycin D in control uninfected cultures whereas in infected cultures actinomycin D resulted in further decrease, down to 2.8% with HSV-1 and 0.7% with HSV-2, respectively. These findings suggest that additional inhibition of host-protein synthesis occurred as a result of the viral infection but in the absence of new protein synthesis.

The electrophoretic pattern of the cell-matrix fraction of human EC cultures

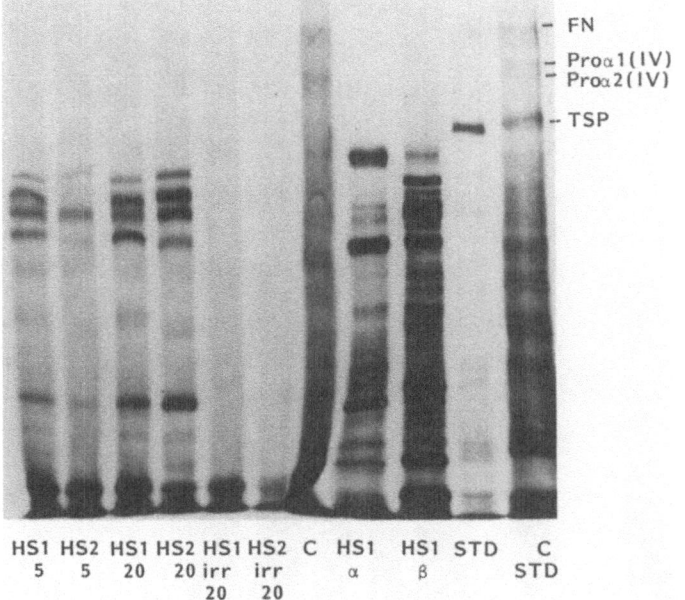

HS1 HS2 HS1 HS2 HS1 HS2 C HS1 HS1 STD C
5   5 20 20 irr irr   α   β   STD
           20 20

Figure 5. Effect of host-cell protein synthesis following infection of EC with intact HSV-1 (HS1) and HSV-2 (HS2) and UV-irradiated virus (HS1 irr and HS2 irr). Infection with HSV-1 or HSV-2 was carried out at a MOI of 5 or 20. Uninfected (C) and infected cultures were pulsed with [^{35}S]methionine for 1 hr at 4 hr postinfection and the cell-matrix fraction analyzed by SDS–PAGE. Standards: HS1 α, HS1 β = immediate early and early polypeptides, respectively, of HSV-1; STD = labeled standard of thrombospondin; C STD = matrix proteins in a cell fraction from a control uninfected EC culture that was labeled for 2 hr. Note the suppression of host-cell and virus proteins with the irradiated virus. (Data from Kefalides and Ziaie.[27])

Table 3. Effect of Actinomycin D on Protein Synthesis of HSV-1- and HSV-2-Infected Endothelial Cells

EC cultures	^{35}S cpm/10^6 cells[a]
Uninfected	2,488,740 (100%)
Uninfected plus actinomycin D	995,490 (40%)
Infected	
HSV-1	1,098,000 (44%)
HSV-2	962,000 (38%)
HSV-1 plus actinomycin D	69,390 (2.8)
HSV-2 plus actinomycin D	17,880 (0.7%)

[a] Average of two experiments. Duplicate EC cultures were infected with HSV-1 or HSV-2 in the presence (2 μg/ml) or absence of actinomycin D. At 4 hr postinfection the cultures were pulsed with [^{35}S]methionine for 1 hr and the cell-matrix fraction was processed for SDS–PAGE. Actinomycin D was present in the culture throughout the duration of the experiment.

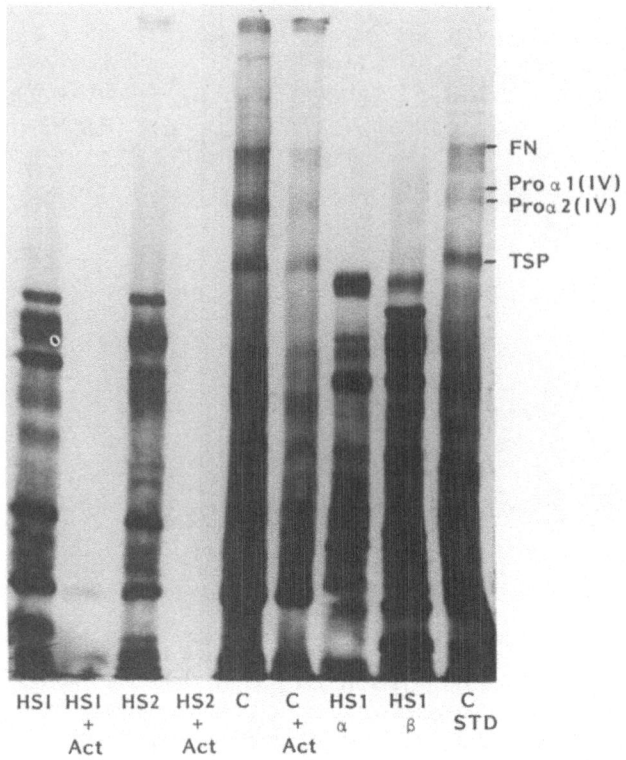

Figure 6. Effect of inhibition of mRNA synthesis on EC cultures infected with HSV-1 (HS1) and HSV-2 (HS2). Actinomycin D was added at the beginning of infection (HS1 + Act, HS2 + Act) and maintained until the end of the experiment. Cultures were pulsed with [^{35}S]methionine for 1 hr at 4 hr postinfection and the cell-matrix fraction analyzed by SDS–PAGE. C = control culture; C + Act = control + actinomycin D. Standards: HS1 α, HS1 β = immediate early and early polypeptides, respectively, of HSV-1; C STD = matrix proteins in a cell fraction from a control uninfected EC culture that was labeled for 2 hr. Note the suppression of host-cell and virus proteins in the cultures infected in the presence of actinomycin D (HS1 + Act, HS2 + Act). (Data obtained from Kefalides and Ziaie.[27])

infected in the presence and absence of actinomycin D is shown in Fig. 6. In the presence of actinomycin D there was only partial shutoff of protein synthesis in control cultures (Fig. 6, C + Act) due to continued translation of preexisting mRNA; however, in the presence of virus plus actinomycin D, there was complete suppression of both host-cell and viral protein synthesis. Human EC cultures infected with HSV-1 or HSV-2 in the absence of actinomycin D (Fig. 6, HS1, HS2) showed suppression of host-cell matrix protein synthesis but active viral protein synthesis. These results indicate that under conditions of inhibition of synthesis of new mRNA, i.e., inhibition of viral protein synthesis, synthesis of host-cell proteins from preexisting host-cell mRNA is blocked. The data further suggest that in human EC cultures infected with HSV-1 or HSV-2, inhibition of host-cell protein synthesis most likely occurs at the translational level and does not require prior synthesis of virus protein. The presence of virus proteins (virions) appears to be sufficient to shut off host-protein synthesis.

The effect of HSV-1 infection on matrix protein synthesis was also investigated in our laboratory using bovine aorta SMC.[30] Monolayers of SMC were infected at a MOI ranging from 0.1 to 20. Indicators of infection included cytopathology, viral titer, and immunofluorescence. Cytopathology varied with the time postinfection and was directly proportional to the MOI. Virus replication occurred at 0.1, 10, and 20 MOI and maximum titers were achieved by 24 hr postinfection. Infected and uninfected cultures of SMC were pulse-labeled with either [14C]proline or [35S]methionine at different hours postinfection. Incorporation of radioactivity into nondialyzable protein was determined in fluorograms following SDS–PAGE of the cell-matrix or medium fractions. The synthesis of fibronectin and collagen types I and III was suppressed and the degree of suppression was dependent on the duration of infection and on the virus dose. These data suggest that SMC can support HSV-1 replication *in vitro* and that such infection can lead to altered matrix protein synthesis. Unpublished results from our laboratory also demonstrate that HSV-1 infection of SMC *in vitro* leads to the qualitative alteration in the protein composition of the extracellular substratum. SDS–PAGE analysis of the extracellular substratum matrix after labeling infected SMC with [35S]methionine for 2 hr and chasing for 4, 24, and 48 hr postlabeling revealed a decrease in the deposition of cell matrix proteins but considerable deposition of polypeptides corresponding to HSV-1 proteins. This is analogous to what we observed with HSV-1-infected human EC (see below).

E. Qualitative Alteration in the Composition of the Extracellular Matrix

Initial studies from our laboratory (Kefalides and Ziaie[61]) demonstrate that HSV-1 infection of human EC at MOI of 0.05, 0.1, and 1.0 followed by a 2-hr pulse with [35S]methionine at 19 hr postinfection and a chase for 4, 24, and 48 hr resulted in a qualitatively altered extracellular matrix substratum. This was demonstrated by the presence of HSV-1 polypeptides in the matrix fraction upon SDS–PAGE. The matrix fraction of the EC cultures infected with either 0.05 or 0.1 MOI contained reduced amounts of cell-matrix proteins as well as proteins corresponding to HSV-1 polypeptides. The matrix fraction from the human EC cultures infected at a MOI of 1.0 contained labeled proteins corresponding to HSV-1 polypeptides only; host-cell proteins were almost completely absent.

When the substratum from infected human EC cultures, freed of their cell layer and any virus inactivated, was subsequently used for the attachment of fresh, uninfected human EC, we noted that attachment was not significantly different between "uninfected" and "infected" EC substratum. However, the amount of cell-matrix proteins in the substratum deposited by EC growing on the infected substratum was significantly diminished. These are exciting findings and suggest that viral infection can qualitatively alter the extracellular substratum which in turn could influence cellular gene expression.

IV. DISCUSSION

The way one individual's tissue responds to injury is a very important factor in determining whether chronic sequelae shall develop and tissue function shall be compromised. There are several types of vascular pathology, including vasculitis, thrombosis, atherosclerosis, aneurysmal dilation, capillary basement membrane thickening, vascular tumors, and others whose pathogenesis is incompletely understood. Little is known about the early, initiating events which lead to discernible pathologic and functional changes in blood vessels. We chose to study the *in vitro* infection of vascular cells with human viruses in order to learn more about the way EC and SMC respond to this particular form of injury. The parameters we have studied are affected in the early stages following viral infection *in vitro* and may reflect analogous responses of the same vessel cells in *in vivo* situations.

The studies reviewed here suggest that EC and SMC are susceptible to a number of common human viruses. Infection of EC leads to virus replication but not all viruses cause cytopathic changes. EC form our first line of defense in preventing the spread of a virus from the vascular lumen to the abluminal side of the EC and the subsequent spreading of the infection to tissues supplied by a given vessel.

Viral infection of EC leads to increased adherence of granulocytes which in turn may increase the chances of further injury to EC through immune mechanisms. Additional mechanisms of EC injury may be brought into play after viral infection as a result of induction of Fc and C3 receptors by HSV-1. Binding of immune complexes to such receptors may be critical in bringing about EC injury. Other early changes associated with viral infection of EC and SMC include the suppression of host-cell matrix protein synthesis and the incorporation of virus polypeptides into the extracellular matrix substratum. Using cDNA probes for human type IV procollagen, human fibronectin, and human von Willebrand factor, we have determined that the steady-state levels of mRNA specific for these proteins are significantly decreased in the early hours postinfection. The mechanism(s) by which this comes about is not clearly understood. It is important to note, however, that new viral protein synthesis is not required for this effect. It has been suggested that disruption of polysomes and inactivation of mRNA are definite possibilities.

The demonstration of virus polypeptides in the matrix fraction of the EC cultures was an interesting and potentially significant finding. The altered substratum may cause detachment of newly deposited cells, may influence gene expression and result in further alteration of the extracellular matrix composition, or it may attract other cells which could release substances that stimulate or inhibit cell proliferation. Finally, the virus itself may persist in a latent state without destroying its host but it could perturb cell functions by disordering the synthesis of specific enzymes or hormones.

It appears, therefore, that a clear understanding of the mechanisms that initiate the response of vascular cells to injury is an important requisite to understanding the pathogenesis of vascular disease.

ACKNOWLEDGMENTS. The author wishes to thank Ms. Sherí Johnson for typing the manuscript. This work was supported by NIH Grants HL 29492 and AM 20553 from the United States Public Health Service.

REFERENCES

1. Andrews, B. S., Theophilopoulos, A. N., Peters, C. J., Loskutoff, D. J., Brandt, W. E., and Dixon, F. J., 1978, Replication of dengue and junin viruses in cultured rabbit and human endothelial cells, *Infect. Immun.* **20**:776.
2. Ashford, T. P., and Freiman, D. G., 1968, Platelet aggregation at sites of minimal endothelial injury: An electron microscopic study, *Am. J. Pathol.* **53**:559.
3. Beltz, G. A., and Flint, S. J., 1979, Inhibition of HeLa cell protein synthesis during adenovirus infection: Restriction of cellular messenger RNA sequences to the nucleus, *J. Med. Biol.* **131**:353.
4. Benditt, E. P., Barret, T., and McDougall, J. K., 1983, Viruses in the etiology of atherosclerosis, *Proc. Natl. Acad. Sci. USA* **80**:6386.
5. Bonner, W. M., and Laskey, R. A., 1974, A film detection method for tritium-labeled proteins and nucleic acids in polyacrylamide gels, *Eur. J. Biochem.* **46**:83.
6. Cines, D. B., Lyss, A. P., Bina, M., Corkey, R., Kefalides, N. A., and Friedman, H. M., 1982, Fc and C3 receptors induced by herpes simplex virus on cultured human endothelial cells, *J. Clin. Invest.* **69**:123.
7. Colman, R. W., Robboy, S. J., and Minna, J. D., 1972, Disseminated intra-vascular coagulation (DIC): An approach, *Am. J. Med.* **52**:679.
8. Engvall, E., 1980, Enzyme immunoassay ELISA and EMIT, *Methods Enzymol.* **70**:419.
9. Fabricant, C. G., Fabricant, J., Litrenta, M. M., and Minick, C. R., 1978, Virus-induced atherosclerosis, *J. Exp. Med.* **148**:335.
10. Fenwick, M. L., 1984, The effects of herpes viruses on cellular macromolecular synthesis, in: *Comprehensive Virology* (H. Fraenkel-Conrat and R. R. Wagner, eds.), Vol. 19, Plenum Press, New York, pp. 359–390.
11. Fenwick, M. L., and Clark, J., 1982, Early and delayed shut-off of host protein synthesis in cells infected with herpes simplex virus, *J. Gen. Virol.* **61**:121.
12. Fenwick, M. L., and Walker, M. J., 1978, Suppression of the synthesis of cellular macromolecules by herpes simplex virus, *J. Gen. Virol.* **41**:37.
13. French, J. E., 1966, Atherosclerosis in relation to the structure and function of the arterial intima, with special reference to the endothelium, *Int. Rev. Exp. Pathol.* **5**:253.
14. Friedman, H. M., 1984, Viral infection of endothelium and the induction of Fc and C3 receptors, in: *Biology of Endothelial Cells* (E. A. Jaffe, ed.), Nijhoff, The Hague, pp. 268–276.
15. Friedman, H. M., Macarak, E. J., MacGregor, R. R., Wolfe, J., and Kefalides, N. A., 1981, Virus infection of endothelial cells, *J. Infect. Dis.* **143**:266.
16. Friedman, H. M., Wolfe, J., Kefalides, N. A., and Macarak, E. J., 1986, Susceptibility of endothelial cells derived from different blood vessels to common viruses, *In Vitro* **22**:397.
17. Fye, K. H., Becker, M. J., Theophilopoulos, A. N., Moutsopoulos, H., Feldman, J., and Talal, N., 1977, Immune complexes in hepatitis: β-antigen associated periarteritis nodosa, *Am. J. Med.* **62**:783.
18. Gimbrone, M. A., and Fareed, G. C., 1976, Transformation of cultured human vascular endothelium by SV40 DNA, *Cell* **9**:685.
19. Gimbrone, M. A., Cotran, R. S., and Folkman, J., 1974, Human vascular endothelial cells: Growth and DNA synthesis, *J. Cell Biol.* **60**:673.
20. Giraldo, G., Beth, E., and Huang, E. S., 1980, Kaposi's sarcoma and its relationship to cytomegalovirus (CMV), *Int. J. Cancer* **26**:23.
21. Gyorkey, F., Melnick, J. L., Guinn, G. A., Gyorkey, P., and DeBakey, M. E., 1984, Herpesviridae in the endothelial and smooth muscle cells of the proximal aorta in arteriosclerotic patients, *Exp. Mol. Pathol.* **40**:328.

22. Hill, T. M., Sinden, R. R., and Sadler, J. R., 1983, Herpes simplex virus types 1 and 2 induce shut-off of host protein synthesis by different mechanisms in Friend erythroleukemia cells, *J. Virol.* **45**:241.
23. Honess, R. W., and Roizman, B., 1974, Regulation of herpesvirus macromolecular synthesis. I. Cascade regulation of the synthesis of three groups of viral proteins, *J. Virol.* **14**:8.
24. Honess, R. W., and Roizman, B., 1975, Regulation of herpesvirus macromolecular synthesis. II. Sequential transition of polypeptide synthesis requires functional viral polypeptide, *Proc. Natl. Acad. Sci. USA* **72**:1276.
25. Johnson, P. M., Faulk, W. P., and Wang, A. C., 1976, Immunological studies of human placenta: Subclass fragment specificity of binding of aggregated IgG by placental endothelial cells, *Immunology* **31**:659.
26. Joklik, W. K., 1981, Structure and function of the reovirus genome, *Microbiol. Rev.* **45**:483.
27. Kefalides, N. A., and Ziaie, Z., 1986, Herpes simplex virus suppression of human endothelial matrix protein synthesis is independent of viral protein synthesis, *Lab. Invest.* **55**:328.
28. Kistler, G. S., Caldwell, P. R. G., and Weibel, E. R., 1967, Development of fine structural damage to alveolar and capillary lining cells in oxygen poisoned rat lungs, *J. Cell Biol.* **33**:605.
29. Laemmli, U. K., 1970, Cleavage of structural proteins during the assembly of the head of bacteriophage T4, *Nature* **227**:680.
30. Lashgari, M., Friedman, H. M., and Kefalides, N. A., 1987, Suppression of matrix protein synthesis by herpes simplex virus in bovine aorta smooth muscle cells, *Biochem. Biophys. Res. Commun.* **143**:145.
31. Laskey, R. A., and Mills, A. D., 1976, Quantitative film detection of ^{3}H and ^{14}C polyacrylamide gels by fluorography, *Eur. J. Biochem.* **56**:335.
32. Lyss, A. P., Finko, R., Knight, K., Bina, M., Reeber, M., and Cines, D. B., 1982, Interaction of IgG with human endothelial cells, *Clin. Res.* **30**:323A.
33. Macarak, E. J., Howard, B. V., and Kefalides, N. A., 1977, Properties of calf endothelial cells in culture, *Lab. Invest.* **36**:62.
34. Macarak, E. J., Friedman, H. M., and Kefalides, N. A., 1985, Herpes simplex virus type 1 infection of endothelium reduces collagen and fibronectin synthesis, *Lab. Invest.* **53**:280.
35. MacGregor, R. R., Friedman, H. M., Macarak, E. J., and Kefalides, N. A., 1980, Virus infection of endothelial cells increases granulocyte adherence, *J. Clin. Invest.* **65**:1469.
36. Maciag, T., Cerundolo, J., Ilsley, S., Kelley, P. R., and Forand, R., 1979, An endothelial cell growth factor from bovine hypothalamus: Identification and partial characterization, *Proc. Natl. Acad. Sci. USA* **76**:5674.
37. Mayman, B. A., and Nishioka, Y., 1985, Differential stability of host mRNAs in Friend erythroleukemia cells infected with herpes simplex virus type 1, *J. Virol.* **54**:1.
38. Melnick, J. L., Petrie, B. L., Dreesman, G. R., Burek, J., McCullum, G. H., and DeBakey, M. E., 1983, Cytomegalovirus antigen within human arterial smooth muscle cells, *Lancet* **2**:644.
39. Minick, C. R., Fabricant, C. G., Fabricant, J., and Litrenta, M. M., 1979, Atheroarteriosclerosis induced by infection with a herpesvirus, *Am. J. Pathol.* **96**:673.
40. Nishioka, Y., and Silverstein, S., 1977, Degradation of cellular mRNA during infection by herpes simplex virus, *Proc. Natl. Acad. Sci. USA* **74**:2370.
41. Pereira, L., Wolff, M. H., Fenwick, M. L., and Roizman, B., 1977, Regulation of herpesvirus macromolecular synthesis. V. Properties of polypeptides made in HSV-1 and HSV-2 infected cells, *Virology* **77**:733.
42. Robbins, S. L., 1967, *Pathology*, Saunders, Philadelphia, pp. 293–294.
43. Roizman, B., Borgman, G. S., and Kamali-Rousta, M., 1967, Macromolecular synthesis in cells infected with herpes simplex virus, *Nature* **206**:1374.
44. Rosenberg, J. C., Hawkins, E., and Rector, F., 1971, Mechanisms of immunological injury during antibody-mediated hyperacute rejection of renal heterografts, *Transplantation* **11**:151.
45. Ross, R., and Harker, L., 1976, Hyperlipidemia and atherosclerosis: Chronic hyperlipidemia initiates and maintains lesions by endothelial cell desquamation and lipid accumulation, *Science* **193**:1094.
46. Schek, N., and Bachenheimer, S. L., 1985, Degradation of cellular mRNAs induced by a virion-associated factor during herpes simplex virus infection of Vero cells, *J. Virol.* **55**:601.

47. Schmidt, N. J., 1979, Laboratory diagnosis of viral infection, in: *Antiviral Agents and Viral Diseases* (G. J. Galasso, T. C. Merigan, and R. A. Buchanan, eds.), Raven Press, New York, pp. 209–252.
48. Shingu, M., Hashimoto, Y., Johnson, A. R., and Hurd, E. R., 1981, The search for Fc receptors on human tissues and human endothelial cells in culture, *Proc. Soc. Exp. Biol. Med.* **167**:147.
49. Shatkin, A. J., 1983, Molecular mechanisms of virus mediated cytopathology, *Philos. Trans. R. Soc. London Ser. B* **303**:167.
50. Stehbens, W. E., 1976, Hemodynamic production of lipid deposition, intimal tears, mural dissection and thrombosis of blood vessel walls, *Proc. R. Soc. London Ser. B* **185**:357.
51. Stemerman, M., Spaet, T. H., Pitlick, F., Cintron, J., Lejnieks, I., and Tiell, M. L., 1977, Intimal healing: The pattern of re-endothelialization and intimal thickening, *Am. J. Pathol.* **87**:125.
52. Sydiskis, R. J., and Roizman, B., 1967, The disaggregation of host polysomes in productive and abortive infection with herpes simplex virus, *Virology* **32**:678.
53. Thornton, S. C., Mueller, S. N., and Levine, E. M., 1983, Human endothelial cells: Use of heparin in cloning and long-term serial cultivation, *Science* **222**:623.
54. Towbin, H., Staehelin, T. H., and Gordon, J., 1979, Electrophoretic transfer of proteins from polyacrylamide gels to nitrocellulose sheets: Procedure and some applications, *Proc. Natl. Acad. Sci. USA* **76**:4350.
55. Tumilowicz, J. J., Gawlik, M. E., Powell, B. B., and Trentin, J. J., 1985, Replication of cytomegalovirus in human arterial smooth muscle cells, *J. Virol.* **56**:839.
56. Wall, R. T., Harlan, J. M., Harker, L. A., and Striker, G. E., 1980, Homocysteine-induced endothelial cell injury *in vitro*: A model for the study of vascular injury, *Thromb. Res.* **18**:113.
57. Westmorland, D., and Watkins, J. F., 1974, The IgG receptor induced by herpes simplex virus: Studies using radiolabeled IgG, *J. Gen. Virol.* **24**:167.
58. Ziaie, Z., Friedman, H. M., and Kefalides, N. A., 1986, Suppression of matrix protein synthesis by herpes simplex virus type 1 in human endothelial cells, *Collagen Relat. Res.* **6**:33.
59. Rennard, S. I., Berg, R., Martin, G. R., Foidart, J. M., and Gehron-Robey, P., 1980, Enzyme-linked immunoassay (ELISA) for connective tissue components, *Anal. Biochem.* **104**:205.
60. Stenberg, R. M., and Pizer, L. I., 1982, Herpes simplex virus-induced changes in cellular and adenovirus RNA metabolism in an adenovirus type 5-transformed human cell line, *J. Virol.* **42**:474.
61. Ziaie, Z., and Kefalides, N. A., 1987, Herpes simplex virus (HSV-1) incorporated in the extracellular matrix of HSV-1 infected endothelial cells (EC) influences behavior of freshly seeded EC, *Fed. Proc.* **46**:975.
62. Brinker, J. M., Ziaie, Z., and Kefalides, N. A., 1987, Differential stability of host poly A^+ mRNAs in endothelial cells infected with herpes simplex type 1, *Fed. Proc.* **46**:1993.

Index